WAGON RO
WEST

W. TURRENTINE JACKSON

WITH A FOREWORD BY WILLIAM H. GOETZMANN

WAGON ROADS WEST

A STUDY OF FEDERAL ROAD SURVEYS AND CONSTRUCTION IN THE TRANS-MISSISSIPPI WEST 1846-1869

University of Nebraska Press   Lincoln & London

First Bison Book printing: 1979
Most recent printing indicated by first digit below:
1 2 3 4 5 6 7 8 9 10

Library of Congress Cataloging in Publication Data

Jackson, William Turrentine, 1915–
Wagon roads west.

Reprint of the ed. published by Yale University Press, New Haven, which was issued as 9 of Yale Western Americana series.
Bibliography: p. 379
Includes index.
1. Roads—The West—History. I. Title. II. Series: Yale Western Americana series ; 9.
[HE356.A17J13 1979] 388.1'0976 79–13959
ISBN 0–8032–4405–3
ISBN 0–8032–9402–6 pbk.

Reprinted by arrangement with Yale University Press

Manufactured in the United States of America

To

THE STAFF OF THE NATIONAL ARCHIVES

WHOSE SERVICE TO ME

AND THE MEMBERS OF MY PROFESSION

IS FAR GREATER THAN THE

CALL OF DUTY

PART 1st

Memorial

to the

CONGRESS of the UNITED STATES

for the Construction of a

WAGON ROAD across the PLAINS

And for the Establishment

of

MILITARY POSTS.

CALIFORNIA,

1856.

California Memorial to Congress, 1856

Foreword

IT IS A PARTICULAR PLEASURE to introduce the Yale paperbound edition of William Turrentine Jackson's *Wagon Roads West* because it is one of the important books produced by what might loosely be considered a new "school" of Western historians. Bearing a colorful title that conjures up that perennial romantic symbol of the Old West, and often classified by librarians under "transportation," Professor Jackson's book is actually neither a work in the "whoop and holler" tradition nor a representative slice of life from one of Frederick Jackson Turner's economic frontiers. Rather it belongs to the "Imperial School" of Western historiography, which seeks to explain the westward movement in terms not only of its relationship to the rest of the nation but specifically of its relationship to the national government. Just as some decades ago Charles M. Andrews and others began to write the history of the American colonies from the steps of the British Public Record Office, so too have certain Western historians sought to see the West as America's first colonial empire—an empire that owed its development and its peculiar tensions in some sense to policies and plans formulated in the national capital. This "school"—and it is presumptuous to term it as such because some members are hardly aware of affiliation at all—has produced such works as Earl Pomeroy's *The Territories and the United States,* Averam B. Bender's *The March of Empire,* Henry P. Beers' *The Western Military Frontier,* F. P. Prucha's *Broadaxe and Bayonet,* Wallace Stegner's *Beyond the Hundredth Meridian,* W. H. Goetzmann's *Army Exploration in the American West,* Howard R. Lamar's *Dakota Territory,* Robert Athearn's *General William Tecumseh Sherman and the Far West,* and Richard Bartlett's *Great Surveys of the American West,* to name only a few of the more recent books. Individually, the significance of these works may be hard to assess, but collectively they represent a departure from the fixed positions and models of previous Western historiography.

Serious Western history undoubtedly began when Frederick Jackson Turner became disenchanted with the germ theories of Herbert Baxter Adams and embarked on his own private campaign of historiographical nationalism, which emphasized the influence of the American frontier rather than European precedents or "germs." He rejected the characteristic German evolutionism of Adams, which took embryology as its model and saw whole cultures growing from social and intellectual seeds; instead he championed environmentalism, which was fast becoming fashionable in the late nineteenth century. In adopting an environmental approach, Turner did not, however, reject evolutionism. Rather his work paralleled the Spencerian evolutionism of Lewis Henry Morgan, whose *Ancient Society* (1877) portrayed human culture as evolving through successive stages from savagery to barbarism to civilization. Turner's stages of evolutionary growth were his successive occupational frontiers: those of the hunter, the fur trader, the miner, the cattleman, the homesteader etc., each one proceeding toward the goal of an industrial civilization and presumed cultural maturity. The difference was that for Turner the survival of the fittest on each frontier produced the boldest, most democratic citizens and those who best shaped the destiny of the nation. So appealing was this idea that most serious Western history has been concerned ever since with amplifying this model or modifying it in certain minor instances. Western writers have been largely dedicated to describing in infinite detail each of the successive Western frontiers—frontiers that were largely, in Turner's terms, economic. A glance at the organization of the major textbooks in Western history will serve to confirm this. Thus it is no surprise that Professor Jackson's book has usually been classified as "transportation frontier" history—as if there were actually such a thing as a "transportation frontier" in the sense that transportation as an activity was confined to any one particular cultural or temporal horizon.

The chief significance of *Wagon Roads West,* however, and Professor Jackson so states this as his intention in his preface, is to indicate the degree of dependence of the West upon government aid in some form. Thus his book, as are many others of this genre, is directed toward revising the twin myths of environmentalism, with its concomitant local pride, and the rugged individualism of an age of laissez-faire that are functions of the Turner evolutionary model. The materials of *Wagon Roads West* are economic, but its over-all conception places particular stress upon political decisions—decisions made in Washington to aid the Western emigrant by building all-important primary roads.

In a curious way this new school of Western historiography seems roughly analogous to the rise of diffusionism in anthropology, which altered forever the simple classical evolutionism of an earlier day by tracing the transference of ideas, concepts, and modes of behavior from culture to culture and from region to region. In this instance the impact of national goals and national politics in the form of economic subsidies has demonstrably changed Turner's universal sequence of development. To place it in this context is perhaps to claim too much for Professor Jackson's book, for it is but a step in a new direction that hopefully will break new trails in serious Western historiography.

Of possible further interest is the genesis of this new approach. Some years ago in a paper entitled "Who Put the Litter in Western Literature," John Alexander Carroll shrewdly observed the degree to which the so-called "color" writing in Western history stemmed from the many colorful and often inaccurate sources available to careless writers who had only to thumb the pages of Wagner and Camp's bibliography to find grist for their gaudy tales. Other more sober historians have based their work on census data, newspapers, overland diaries, business records, and state and local archives. By the same token, the Imperial School possibly owes its genesis, as Professor Jackson's dedication indicates, to the vast resources of the federal archives—a relatively untapped source even today. The records of the Interior Department, the War Department, the State Department, the Territorial Papers of the United States, the files of the Smithsonian Institution, and numerous other archives afford and have afforded some historians a new vantage point from which to view Western experience. Perhaps this is a reflection of the increasing centralization of American life. At the very least it suggests how one new idea has grown out of new sources in a field long thought to have been exhausted by a generation of giants and Hollywood script writers.

Science has been described as a many-brained, cooperative, and cumulative enterprise. To the degree that history is a social science and shares many of the characteristics of science itself, Professor Jackson's *Wagon Roads West* remains significant as a living cell in the many-brained enterprise of American historiography.

W. H. Goetzmann

Austin, Texas
February 1965

Preface

THIS STUDY *is an attempt to describe and assess the role of the federal government in the location, survey, and improvement of routes for wagons in the trans-Mississippi West before the railroad era.*

The covered wagon has become the foremost symbol of the westward movement. Although pioneers describing their wagon travels were an infinitesimal percentage of those who migrated, enough recorded their impressions in diaries to occupy the endeavors of a long line of editors. These accounts have justifiably served to glorify the American frontiersman. Much attention has also been given to the western businessmen who organized the overland mail and freighting companies, to the intrepid riders of the Pony Express, and to the hardened bullwhackers driving ox teams with a lash. In the process historians have conveyed the impression that private citizens, endowed with a pioneering or entrepreneurial spirit, were primarily responsible for building the transportation pattern of western America before the construction of the Pacific Railroads.

This study is written in an effort to redress the balance. The national government's contribution to western transportation was continuous and dominant throughout the nineteenth century. The constitutional controversy over internal improvements early highlighted the importance of this phase of federal activity. With the opening of the Great West after the Mexican War, the cry for government assistance in establishing communication lines over vast distances grew ever louder. Through contracts for forwarding supplies to the Army posts, and subsidies for mail delivery and the transmission of telegraphic dispatches, the government attempted to guarantee the financial success of private enterprise. Although the grants-in-aid were temporary and the effects transitory, without them communication and transportation in the West would have been not only inadequate but almost impossible.

More important than these indirect government aids was the program of reconnaissance, exploration, survey, and improvement of routes usable by emigrants, traders, mail carriers, and soldiers. Throughout the period 1846–1869, there was a constant search by government agents for passages through the western terrain whereby wagons could travel from the Mississippi to the Pacific. Where movement was not blocked by swift flowing rivers, where mountain grades were not too steep for mules or oxen to pull a heavily loaded wagon, where the terrain did not undulate sufficiently to overturn the load, where the soil was not so marshy as to bog down the wheels, and where the timber and underbrush were not too dense to hinder transit, the wagons were free to roll. More emphasis was placed on the discovery of a natural passage than on construction. For this reason great difficulty is encountered in determining when the work of exploration and reconnaissance ceased and that of road survey and construction began. A chain relationship was established between scientific exploration, wagon road survey, and railroad construction.

Three government agencies were chiefly responsible for the wagon road program: the Corps of Topographical Engineers, the Office of Explorations and Surveys of the War Department, and the Pacific Wagon Road Office of the Interior Department. The reports of these bureaus published in the documents of the congressional series have naturally provided the basic source of evidence. Unpublished correspondence, reports, petitions, and cartographic representations have been sought out in The National Archives. To the best of my belief no investigator has previously used the manuscript records of the Pacific Wagon Road Office.

In surveying the great amount of literature on the subject, emphasis has been placed on those wagon roads launched as national projects at the national level; that is, by the United States Congress or by order of a cabinet officer such as the Secretaries of War or Interior. Each endeavor has been considered as a segment of the evolving pattern of transportation in the trans-Mississippi West. To mention all the roads, or even all those built by the federal government, would prove an impossible task. For example, a discussion of dozens of temporary military roads constructed by detachments of soldiers under officers of the western garrisons has been studiously avoided unless essential to present a picture of the total transportation network or to represent an otherwise disregarded geographic area of size, such as West Texas. The disadvantages encountered in a purely geographic or chronological approach to the study were sufficient, in my opinion, to justify the combined geographic-chronological presentation that has been used.

Wagon Roads West is, in reality, the work of many minds. Few fields of *historical investigation have interested students of western America as much as transportation. The publications of Ralph P. Bieber, Averam B. Bender, and LeRoy R. Hafen on the Greater Southwest are well known. Oscar O. Winther has an established reputation as a scholar on Pacific Northwest transportation. Many writers of state histories have emphasized the problems of transportation as essential to an understanding and appreciation of their region: Leland H. Creer, on the Great Basin of Utah; Merrill G. Burlingame, on the Montana frontier; Arthur J. Larsen, on the Minnesota Territory. To present a complete coverage of the trans-Mississippi West, the conclusions of these men have necessarily been assimilated and integrated into this study. My indebtedness to their scholarship is apparent on many pages.*

In the initial stages of the investigation, a tentative outline of the proposed project and bibliography was submitted to several professional historians for critical review. Without exception, valuable suggestions on organization, relevant subject matter, and additional sources were forthcoming. For this aid I am indebted to Carl Coke Rister of the University of Oklahoma; Ralph P. Bieber, Washington University, St. Louis; Robert Riegel, Dartmouth College; Averam B. Bender, Harris Teachers College; and Henry P. Beers, of Washington, D.C. The first drafts of several chapters have been improved by the critical appraisal of competent scholars: James C. Malin on Kansas and Nebraska, Charles M. Gates on Washington, Winther on Oregon, and Creer on Utah. I am indebted to Professor George P. Hammond, director of the Bancroft Library of the University of California, for a critical reading of the manuscript.

Herman Kahn, at the time director of the Interior Department Records in The National Archives, called my attention to the unused records of the Pacific Wagon Road Office. My research in the War Department Records was greatly facilitated by Elizabeth Drewry. Each member of the Archives staff, from administrators to messengers, eagerly and cheerfully assisted my search through the sources.

Officials of the Library of Congress, the Yale University Library, the Bancroft Library of the University of California, the Newberry Library of Chicago, and the Minnesota Historical Society Library have coöperated to the fullest extent in making available pertinent materials in their manuscript collections.

Everett Graff of Chicago pointed out several rare books on the Sawyers expedition along the Niobrara River and permitted me to use personal copies, otherwise unavailable.

Jack Lee Cross and W. Sheridan Warrick were introduced to the subject in the preparation of their theses for the Master of Arts degree at the University of Chicago. Their interest has been continuous, and I am particularly indebted to Mr. Warrick, my research assistant, not only for his constant care in checking quotations and citations, but also for his tenacity and enthusiasm in pushing the work to completion. Charles W. Strong, graduate student of Chicago University, has contributed his cartographic skill in the preparation of the maps and has painstakingly strived for accuracy in recording the results of my research.

The Science Research Council of the Iowa State College purchased the microfilm and photostatic reproductions of unpublished road maps in The National Archives. The Social Science Research Committee of the University of Chicago provided funds for the cartographic work, research and clerical assistance.

My wife, Barbara, with characteristic understanding of my preoccupation, has been a constant source of encouragement. She has also served a long tour of duty as prime typist and proofreader.

W. TURRENTINE JACKSON

Davis, California
September 17, 1951

Contents

Foreword by William Goetzmann vii

Author's Preface xi

CHAPTER I

An Introduction: The Army Gains Experience in Road Survey and Construction 1

Part One

WAGON ROAD SURVEYS AND CONSTRUCTION IN THE WESTERN STATES AND TERRITORIES BY THE UNITED STATES ARMY, 1846–1861

CHAPTER II

Exploring Routes for Roads in the New Western Domain, 1846–1850 17

CHAPTER III

Wagon Roads on the West Texas Frontier 36

CHAPTER IV

Minnesota Territory, an Experimental Laboratory, 1850–1861 48

CHAPTER V

Federal Military Roads in Oregon Territory, 1853–1859 71

CHAPTER VI

Federal Aid to Transportation in Washington Territory, 1853–1860 89

CHAPTER VII

Federal Road Projects in New Mexico Territory, 1853–1860 107

CHAPTER VIII

Shortening and Improving Routes Across the Plains of Kansas and Nebraska 121

CHAPTER IX

Road Locations in Utah Territory by the United States Army, 1854–1859 140

Part Two

WAGON ROAD CONSTRUCTION BY THE DEPARTMENT OF THE INTERIOR, 1856–1861

CHAPTER X

The Thirty-fourth Congress and the Creation of the Pacific Wagon Road Office 161

CHAPTER XI

From the Minnesota Frontier Toward the South Pass 179

CHAPTER XII

Improvement of the Central Overland Route: Fort Kearny to Honey Lake via the South Pass 191

CHAPTER XIII

Interior Department Improvements in New Mexico: El Paso and Fort Yuma Road 218

CHAPTER XIV

Interior Department Construction to Meet a Local Need: Road Along the Missouri River, Nebraska Territory 233

Part Three

THE ARMY CONTINUES TO BUILD ROADS WEST

CHAPTER XV

Beale's Wagon Road Along the Thirty-fifth Parallel: Fort Smith to the Colorado River, 1857–1859 241

CHAPTER XVI

Across the Northern Plains: The Mullan Road, 1853–1866 257

Part Four

AN ATTEMPT AT COÖPERATION IN WAGON ROAD BUILDING BY THE INTERIOR AND WAR DEPARTMENTS

CHAPTER XVII

The Sawyers Wagon Road from the Mouth of the Niobrara River to Virginia City 281

CHAPTER XVIII

The Big Cheyenne River Road: The Dakota Territory Route to the Northwest 297

CHAPTER XIX

Route from Lewiston, Idaho, Across the Bitterroots, to Virginia City, Montana 312

CHAPTER XX

Wagon Roads West 319

Notes 331

Bibliography 379

Index 401

Maps

Wagon Roads West, 1846–1869 between 5-6

Federal Roads in Iowa Territory 10

Wagon Road Surveys and Improvements East of the Great Salt Lake and Through the Rocky Mountains, 1849–1859 30

Wagon Roads on the West Texas Frontier 38

Federal Roads in Minnesota, 1850–1861, Built by the Army Engineers 50

Military Roads in Southern Oregon 74

Astoria-Salem Military Road 78

Military Roads of Washington Territory, 1854–1860 94

Military Wagon Roads in New Mexico Territory Constructed by the Army Engineers, 1854–1861 114

Routes Surveyed and Improved by the Army Engineers in Kansas and Nebraska, 1854–1858 122

Road Locations in Utah Territory by the United States Army, 1849–1859 142

Fort Kearny, South Pass, Honey Lake Wagon Road: Eastern Division–Fort Ridgely Toward the South Pass Wagon Road, 1857 180

Fort Kearny, South Pass, Honey Lake Wagon Road: Central Division 194

Fort Kearny, South Pass, Honey Lake Wagon Road: Western Division 202

Routes from the Arkansas Border to New Mexico 224

The Wagon Road from Platte River Mouth via Omaha Reserve and Dakota City to the Mouth of the Running Water River (Niobrara) 236

Wagon Roads Surveyed and Improved Across New Mexico Territory, 1846–1860 248

Route of the Mullan Road from Fort Walla Walla on the Columbia to Fort Benton on the Missouri 262

Route of Wagon Road from the Mouth of the Niobrara River to Virginia City, Montana, 1865 and 1866 288

Wagon Road from Sioux City, Iowa, via Yankton to Fort Randall, Dakota Territory, 1865–1867—Route of Wagon Road Survey from Western Boundary of Minnesota to Mouth of the Big Cheyenne River, 1865 300

Route from Lewiston, Idaho, to Virginia City, Montana, 1866, Located by Wellington Bird, Superintendent 314

CHAPTER I : *An Introduction: The Army Gains Experience in Road Survey and Construction*

THE PIONEER, no matter of what date or locality, was always a traveller before he was a producer or shipper of goods," once wrote Seymour Dunbar, the transportation historian.[1] Individualism and adaptability characterized all those who participated in America's westward movement. Frontiersmen evinced this as they sought out new routes toward the West and more convenient means of transport.

Despite this individualism, the Westerner always sought the aid of the federal government in solving his transportation problem. Such a vast undertaking as the construction of wagon roads from the Mississippi west to the Pacific required more than a half century for completion. The task was too great for individual planning. Federal sponsorship was essential, since there must be exploring expeditions, reconnaissance of trails, and the survey, building, and improving of roads. The federal government met the challenge. The common experience of the pioneers, gained on their journeys, provided but one basis for the planning of future routes and methods of travel. In road building, much depended upon the judgment of the national government's agents, both military and civil.

The principal instrumentality for the earliest road construction by the federal government was the United States Army. Existing Indian trails were usually followed through the wilderness to Army outposts, and only necessary improvements made for the movement of artillery or supply trains. First roads on the frontier were often known locally as *military* roads. More important for western development, these routes became the migratory wagon roads for early settlers, and when a community was occupied they were quickly used for commercial purposes.[2] Many roads built by the War Department in the western territories, politically justified on the basis of national defense, were of much greater significance in facilitating access to the public lands.

For many years the engineers of the Army were the only men technically qualified to carry out a planned program of road building in the United States. In March, 1802, Congress created the Corps of Engineers within the United States Army. Eleven years later, several European-trained officers were attached to the General Staff as topographical engineers, and by 1830, there had developed two separate bureaus in the War Department responsible for engineering activities. No attempt was made to specify the duties of the Corps of Topographical Engineers as distinct from those of the Corps of Engineers, though its chief, Colonel John J. Abert, received assurances from John H. Eaton, Secretary of War, that the Topographical Corps was to take charge of all civil works, including harbor and river improvements together with road construction. Lewis Cass, the new Secretary of War appointed in 1831, favored a specification of duties for each engineering group but doubted that the Topographical Engineers, entirely comprised of officers, had sufficient personnel to assume responsibility for all the nonmilitary engineering activities of the Army.[3]

Abert, always aggressive, reported in congressional committees that his corps, which was called upon to make the surveys and submit plans and estimates for civil works, should also be assigned the duty of construction. The already overworked Corps of Engineers registered no protest. Soon after assuming office as Secretary of War in 1837, Joel R. Poinsett clarified the hitherto confused organization by assigning fortifications for defense to the Corps of Engineers and transferring works of civil improvement to the Topographical Bureau. By 1838 this future road-building agency had reached maturity as an organization.

The work on "the national road" was the Corps of Engineers' noteworthy achievement east of the Mississippi River. As early as 1806 Congress had recognized the necessity for connecting the waters of the Atlantic with the western rivers by means of a great road. In launching this first national internal improvement, the United States was prompted primarily by political considerations; the individual states by commerical reasons. The decision to construct at federal expense signalized public realization that the magnitude of interstate and territorial road building was beyond the scope of local endeavor.[4]

A federal appropriation of $30,000 was expended on an initial survey from Cumberland on the Potomac to Wheeling on the Ohio, and additional money was made available by Congress for construction, as needed. Work ceased on this Cumberland Road during the War of 1812. Resumed with vigor at the conflict's close, the highway reached the Wheeling terminus in

1818. In May, 1820, Congress appropriated $10,000 for a new survey from Wheeling to the Mississippi River opposite St. Louis. For the next eighteen years liberal annual grants were made for extending the construction of this National Road through the Old Northwest. Bitter debates over the constitutionality of federal allotments for internal improvements and perorations on the significance of government "compacts" and of implied powers did not check congressional financial support. Before the accounts were closed, $6,824,919 in federal funds had been approved.[5] The road was thought of, in time, as a great commercial, postal, and national thoroughfare.

Private contractors were largely responsible for the building of the eastern segment of "Uncle Sam's Pike." When survey and construction moved west of Wheeling, Congress gradually gave the War Department greater responsibility for the administration of expenditures. Direct supervision went to the Army Engineers. Trained military men made the surveys, or assigned them to civilians and, in turn, inspected and approved their work. To facilitate actual building, the road was divided into the Ohio, Indiana, and Illinois sectors, each under a superintendent. The jobs of clearing the forest, grubbing out the stumps and underbrush in the right of way, the grading of the route and its drainage, including the building of ditches and culverts where necessary, were tasks performed by civilians under contract. Technical decisions involving the construction of bridges or culverts of masonry and the laying of segments of "McAdamized metal," or stone surfacing, were handled by the Army Engineers.[6] They investigated grievances, real or imagined, that caused local groups to complain to the War Department. Congress placed upon the shoulders of the Secretary of War ultimate responsibility for the reconciliation of sectional rivalries along the route, the wise expenditure of government funds, and the general progress of construction. Yet, as Jeremiah S. Young has written, the National Road "... never became a great military thoroughfare, as its greatest use was for the transportation of the mails. With its radiating or connecting lines, it formed a vast system of commercial arteries, the importance of which can hardly be weighed or overestimated. In the early and palmy days of the road there came a class of hardy pioneers that blazed the way for an expanding civilization ..."[7]

A vast amount of building was completed on the National Road in Ohio and Indiana. Congress appropriated $1,136,600 for the route in the latter state between 1829 and 1838, but gradually interest in the project diminished in Washington.[8] Differences of opinion, in and out of Congress, on the constitutionality of federal road building within state boundaries and squabbles over the actual route to be followed eventually led to the curtailment of appropriations. In 1843 the Topographical Engineers' chief complained that no allotment for construction had been made since 1838. The federal government had, in the meantime, used the combined engineering skill of both Army corps on this major project.

The United States road construction program elsewhere east of the Mississippi and concentrated, at the time, in Michigan and the Wisconsin Territory, had become the exclusive responsibility of the Topographical Bureau in 1838. Eight Michigan roads inaugurated during the territorial period converged on the population center at Detroit. Only one, the Detroit–Fort Gratiot Road, could be justified on the basis of military necessity, because it provided communication between the town and the St. Clair River fort during periods when ice suspended water transportation. The improvement of these routes to guarantee the direct and speedy delivery of the mails often justified the Engineers' plea for additional congressional support. The road to Saginaw and another to Grand River were to provide access to the unclaimed agricultural lands, facilitate immigration to these regions, and thereby induce additional sales of public lands. Various sectors along the great thoroughfare from Detroit to Chicago and its tributaries, all constructed as federal projects, provided central communication across the state for commercial and agricultural purposes. The Army considered the road's military advantages negligible, but again advocated federal expenditure on the basis of mail deliveries.[9]

More military justification is discernible in the federal road-building program in Wisconsin. The highway connecting Fort Crawford at Prairie du Chien with Fort Howard at Green Bay was declared to be of unquestionable military importance in joining the forts upon which the safety of the people of Wisconsin and northern Illinois depended. A second road from Fort Howard to the northern boundary of Illinois by way of Milwaukee and Racine was a means of concentrating quickly the militia of Wisconsin and Illinois in an Indian attack. Again, the majority of road projects were to open up the forests to facilitate their penetration by hardy settlers with wagonloads of household effects. One thoroughfare from Racine to the Mississippi River opened up copper ore lands and provided the farmers of the southern counties with a means of transporting their surplus to the river- and lake-shipping points. Milwaukee was connected with Madison, the permanent seat of government, to ease the travel of those who had business in the young capital, and an extension of this road was surveyed southwestward to meet the Mississippi at Dubuque.[10]

Although roads were built for migration and mail deliveries, commercial and agricultural advantages, more often than for military purposes east of the Mississippi, west of the river great emphasis was placed on the location and maintenance of essential lines of communication along the military frontier. Neither the Corps of Topographical Engineers nor the Engineers Corps directly concerned itself with the defense of this international boundary in the West. As a result, the road building there was usually the work of line officers and soldiers. Road connections between the outlying posts were of prime importance. An early survey of 1819–1820 was made between Fort Missouri at Council Bluffs and the Minnesota River mouth, later the site of Fort Snelling, and the western segment was improved for more than 300 miles. In 1823 the new fort at the junction of the Minnesota and Mississippi was connected with Fort Crawford by a road along the western bank of the Mississippi. Far to the south, Army officers considered building a road from the Mississippi River to the frontier town of Natchitoches. All agreed that the entrance of American settlers over this highway would be of greater value in holding the military and international frontier than the construction of Army posts.[11]

Great publicity was given to the operations of the Army in protecting emigrant Indians and in facilitating their removal farther west beyond the permanent Indian frontier. To assist the Choctaw migration from Arkansas to the Indian Territory and to improve means of transport for supplies to Fort Towson in the southeastern corner of the territory, a detachment of forty soldiers from Fort Gibson, near the Neosho River mouth, built a road from Fort Smith southward to the Red River in 1832.[12]

Dragoons were concentrated in the Indian Territory during 1832–1833 for patrol purposes. They were to impress upon the civilized tribes and the plains Indians farther to the west the necessity for living together harmoniously. These cavalrymen campaigned west to the Rocky Mountains to assure the Comanches and Kiowas that those guilty of depredations would answer to the United States Army. To prepare for the movement of one major dragoon expedition, General Mathew Arbuckle ordered detachments from the First and Seventh Infantry to build roads from Forts Gibson and Towson to the Washita River and from Fort Gibson to the north fork of the Canadian River. The roads were to provide communication with military forts soon to be constructed beyond the settlements of the five civilized tribes. When the road survey was completed, soldiers erected blockhouses at Camp Washita, at Camp Holmes 60 miles from the mouth of the Canadian, and at Camp Arbuckle at the mouth of the Cimarron.[13]

In the Indian Territory the Army also cleared the way to the sites of

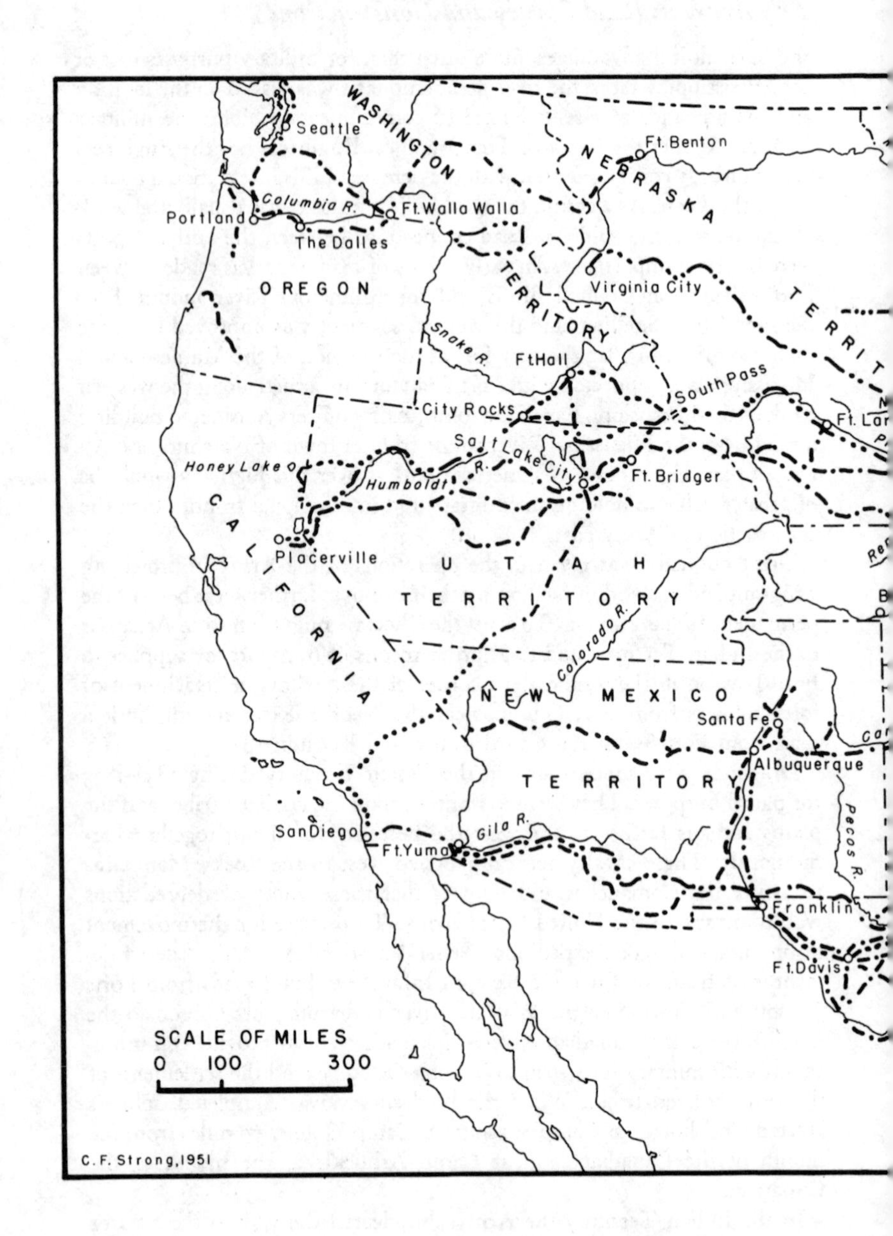
WASHINGTON
Seattle
Ft. Benton
NEBRASKA
Columbia R.
Ft. Walla Walla
Portland
The Dalles
TERRITORY
OREGON
Virginia City
TERRIT
Snake R.
Ft Hall
South Pass
City Rocks
Ft. Lar
Salt Lake City
Honey Lake
Ft. Bridger
Humboldt R.
CALIFORNIA
Placerville
UTAH
TERRITORY
Colorado R.
NEW MEXICO
Santa Fe
Albuquerque
TERRITORY
Pecos R.
Gila R.
San Diego
Ft. Yuma
Franklin
Ft. Davis
SCALE OF MILES
0 100 300
C. F. Strong, 1951

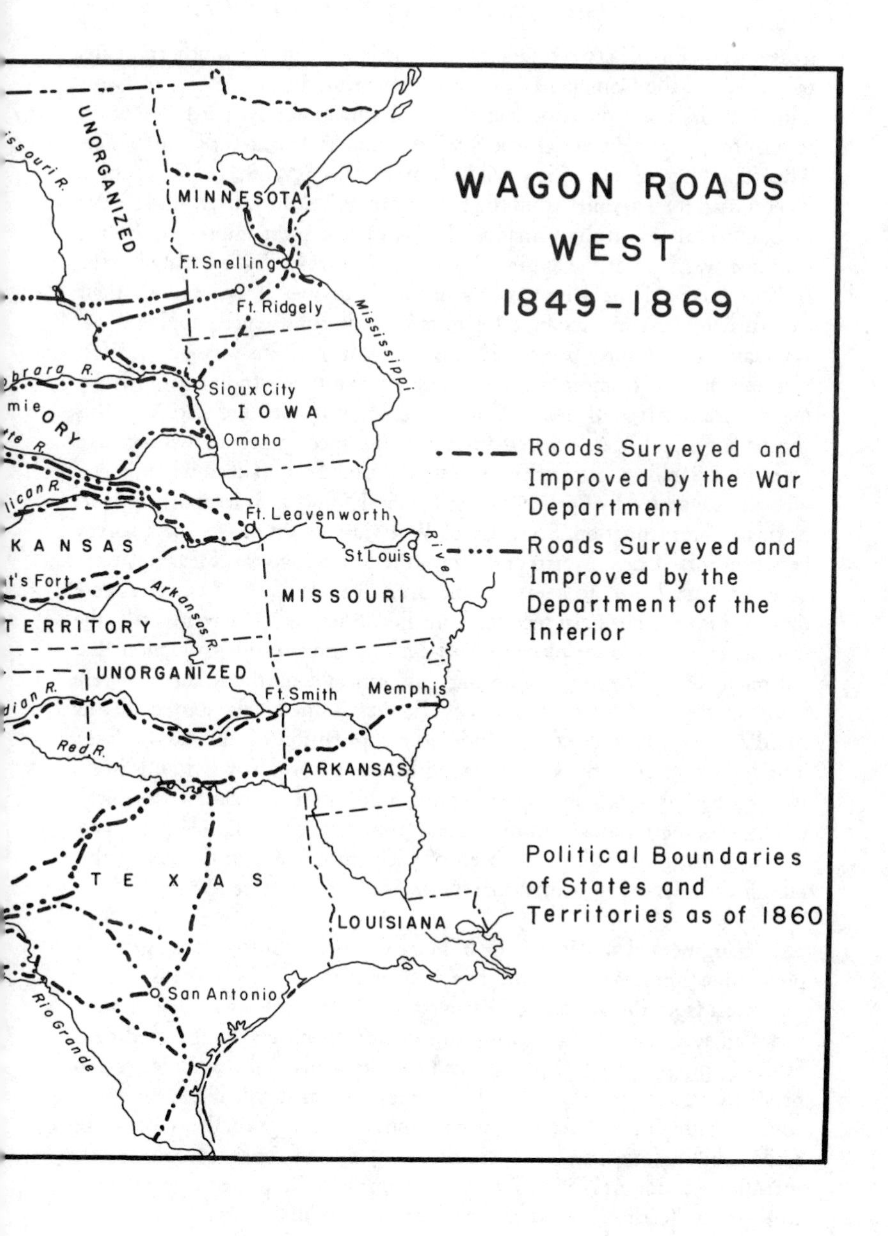
WAGON ROADS WEST 1849-1869
Roads Surveyed and Improved by the War Department
Roads Surveyed and Improved by the Department of the Interior
Political Boundaries of States and Territories as of 1860
UNORGANIZED
MINNESOTA
Ft. Snelling
Ft. Ridgely
Mississippi
Sioux City
IOWA
Omaha
Ft. Leavenworth
St. Louis
River
KANSAS
MISSOURI
TERRITORY
UNORGANIZED
Ft. Smith
Memphis
Red R.
ARKANSAS
TEXAS
LOUISIANA
San Antonio
Rio Grande
Arkansas R.

treaty negotiations between federal commissioners and the southern plains tribes. When the Comanches and Wichitas refused to come east of Cross Timbers for treaty discussion at Fort Gibson, Major Richard B. Mason sent a group of dragoons to Little River to establish a camp for a conference. The following month thirty soldiers from the Seventh Infantry cut a wagon road for 150 miles from the fort to Mason's headquarters and thence to Camp Holmes on the Canadian River. Here a most impressive Indian conclave was held at the summer's close, and a treaty signed whereby the Indians accepted the white man's demands to live at peace with their eastern neighbors in exchange for gifts—chiefly subsistence supplies.[14]

Greatest of the early projects of the Army in the trans-Mississippi West was the plan to connect the posts along the military-Indian frontier by means of a north-south road. Although the advisability and nature of this construction had been debated for years, the specific proposal for a road from Fort Snelling to Fort Towson was not made until 1836 when it was officially suggested by Quartermaster General Thomas S. Jesup. In support of this recommendation, Secretary of War Cass issued orders for a survey between Forts Leavenworth and Smith the following year. Measures were taken in April, 1838, to complete the over-all survey. The 850 miles were divided into the northern section, from Fort Snelling to Leavenworth; the middle, from the latter place to Fort Smith; and a southern section that led on to Fort Towson. While the southern and northern surveys were being made in 1838, construction work was immediately started on the middle section by contractors under the supervision of an Army officer. The following year, the 140-mile road between Forts Towson and Smith was finished. Progress on improving the middle part was somewhat slower, but by 1844 there was a continuous road from Fort Towson to Fort Leavenworth. This undertaking had been of such importance to the Army that all officers referred to it in their reports as *the* military road.[15]

The augmented activity on the military frontier by 1830 and the growing population immediately to the east increased the need for roads leading westward from the Mississippi River. In Arkansas and Iowa, the two territories adjacent to the river during this decade, Congress inaugurated road-building programs which recognized the demands and necessities of pioneer settlers just as they had acknowledged them in Michigan and Wisconsin. Although the Army, on occasions, was assigned the supervising responsibility for survey and construction, military men recognized the intention of Congress in launching certain of these projects by referring to them as "territorial" rather than "military" roads.

A road from Little Rock to the St. Francis River in the direction of Memphis was the major Arkansas construction. Territorial settlers in memorializing Congress for this internal improvement summarized all the earlier arguments justifying federal road construction on the frontier. In the first place, elevating a highway above the swamp that extended for thirty-odd miles west of the Mississippi obviously was a task too great for citizens of a new area. The requested route passed through public domain controlled by the government, and the value of adjacent lands would be enhanced by the construction. With obstacles to the movement of wagons removed, settlers would rapidly populate the territory. Roads would eliminate the necessity of migration by water routes, an expensive process not suitable for the numbers who wished to make homes. In many cases, household goods, agricultural implements, and livestock had to be sacrificed before the migration by water, no matter how badly they were needed at the journey's destination. The road was a commercial necessity for the transportation of produce to the Mississippi at Memphis and on to Natchez and New Orleans. With an increased and commercially prosperous population, the territory would soon be ready for statehood, and the United States relieved from the annual appropriations for territorial government. To clinch their argument, the petitioners presented a statement of military necessity emphasizing the exposed conditions of scattered settlements, and the great mass of Indians clustered on their western border. The government could economize in the movement of mail, troops, and Indians migrating to the West and avoid the expense of suppressing a possible conflict between the races.[16]

As early as 1824 Congress had recognized the validity of the pleas for assistance from Arkansas residents and appropriated $15,000 for the exploration and survey of a route from Memphis to Little Rock.[17] Periodic allotments made possible the construction of the road east of Little Rock as far as the St. Francis River by local contractors.[18] The eastern terminus of this route, in the short sector between the St. Francis and the Mississippi, was in a low-lying swampy area, impassable most of the year. The task of locating and raising an embankment above the periodic flood level along which the road could run was an expensive undertaking, requiring the technical skills of a trained engineer. Local residents had proposed to raise funds by a lottery, authorized by Congress, and to regain the cost of construction by charging tolls for the use of the road. Congress resolved the problem in March, 1833, by authorizing a generous appropriation of $100,000 for the work, and instructing the President to appoint a military engineer to survey the best line for improvement from the Mississippi River at Memphis to the St. Francis.[19]

Lieutenant Alexander H. Bowman of the Army Engineers was placed in charge of the work. A survey of the route was completed in the spring and summer of 1833, and construction contracts issued for the following year. At the close of 1834 a track 160 feet wide had been cleared of timber and underbrush and a central line of 34 feet prepared for the embankment. The dampness of the swamp made it difficult to maintain laborers, who became ill or deserted.[20] The inadequate working crew and wet weather ultimately combined to force the contractors, after financial loss, to forfeit their contracts. Realizing that they could not complete the embankment, Lieutenant Bowman abandoned the contract system and proceeded with the construction by hired labor under direct employment of the United States. Rains and disease continuously harassed the workers. Their numbers dwindled in 1836 to such an extent that the engineer estimated the working party would take two and a half years to finish what was originally thought to be a year's task. The Negroes, who appeared to be less vulnerable to the diseases of the swamp during the summer season, remained healthy and on the job. Eventually slaves provided the nucleus of the working crew.

The Army Engineers pressed forward with the assignment because all realized that this road, when finished, was destined to be the principal channel through which the tide of emigration to Arkansas would pour its thousands. Once through travel was available to Little Rock, the southern and western sections of the state would be opened, and here the land was more desirable for farmers than that adjacent to the route. During September 1836, 1,127 emigrants passed the working crews on the road.[21]

Travelers north and south through Arkansas used a federal government road first improved in 1831 between the settlement of Jackson in the northeast part of the territory and Washington in the southwest corner.[22] The original appropriation of $15,000 proved inadequate, and the following year $2,000 additional was made available to finish the work.[23] In February, 1835, Congress extended this route so that an improved road should run from the southern boundary of Missouri to Jackson, thence to Washington by way of Little Rock, and on to the town of Fulton on the Red River.[24] This became the major avenue for transportation of munitions and supplies by the Quartermaster Corps to the southwestern frontier posts.

Several constructions in the territory grew out of settlers' demands for improved mail facilities. The committee on post office and post roads in the House of Representatives noted that emigrants could use mail roads to gain access to the unsettled lands and that pioneer farmers could transport surpluses to market over them.[25] Three such multipurpose roads were authorized by Congress in June, 1834: the village of Columbia, in Chicot

County, was connected with Little Rock; a road was built from Batesville to a favorite crossing of the St. Francis; Helena, on the Mississippi, was provided with communication inland to the mouth of the Cache River.[26] The general responsibility for building and maintenance of these roads was accepted by the Quartermaster Corps of the Army.[27]

Settlers crossing the Mississippi River farther north to secure the rich farm lands of the Iowa prairie discovered that their greatest need was for passable roads leading westward from the river towns. Iowa pioneers, through their territorial delegate, appealed to the federal government for aid. In 1839 Congress approved $20,000 for construction of a "military road" from Dubuque southward to the Missouri border at the point best suited for extensions to Jefferson City and St. Louis. Congress also appropriated $5,000 for the "opening and construction of a road from Burlington through the counties of Des Moines, Henry, and Van Buren, towards the seat of the Indian agency on the river Des Moines."[28]

The task of constructing these roads was assigned to the Topographical Engineers. As the Bureau Chief, Abert had no officer available to superintend the Iowa improvements, a civilian agent and engineer, R. C. Tilghman, was employed. Upon completing the survey of the "military road," Tilghman employed Lyman Dillon of Cascade, Iowa, to plow a furrow along the surveyed route from Iowa City to Dubuque as a guide to the contractors. Starting at Iowa City, Dillon used a large breaking plow drawn by five yoke of oxen, and under the guidance of the engineer made a furrow some 86 miles long connecting the territorial capital with the Mississippi River town. Construction was then concentrated on the northern section of the road, and various contracts were given residents to improve segments. By the close of the season, patches of timber along the entire route had been cut out for 40 feet and the center 20 feet had been cleared of stumps and underbrush. The road was immediately the most important in the territory.

The "agency road," running from Burlington west to the vicinity of present-day Ottumwa, was surveyed in the same season and construction work concentrated on the eastern section. This improvement not only provided communication with the Indian Agency, but aided emigrants moving westward to settle in the Des Moines River valley. Abert noted that these roads were built for the convenience of the public and such parts as were completed were to be immediately opened for travel.

In 1839 Burlington citizens raised $2,500 to improve a three-mile stretch just east of the Mississippi River opposite their town, and Congress appropriated an equal sum. The majority of emigrants coming into Iowa from

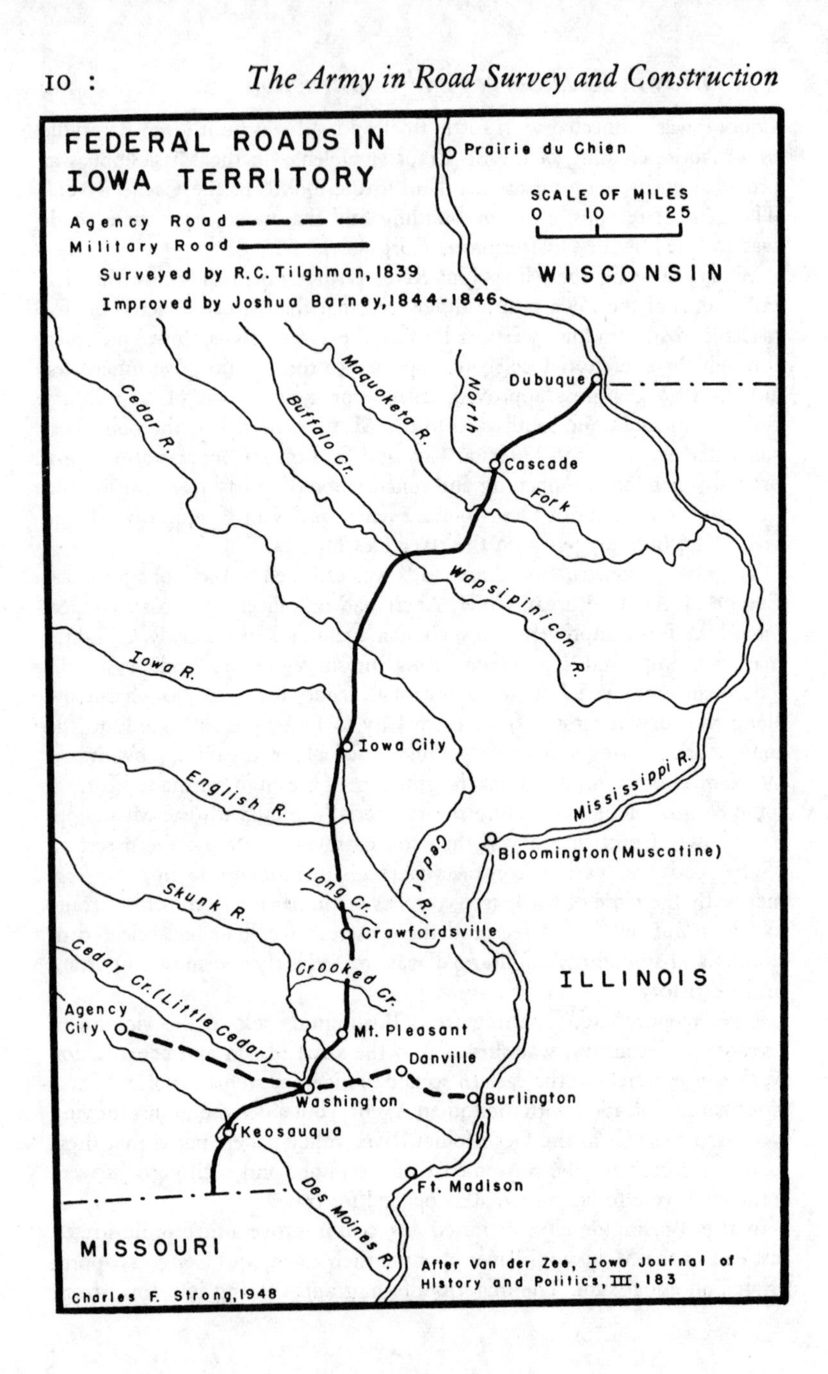
FEDERAL ROADS IN
IOWA TERRITORY
Agency Road
Military Road
Surveyed by R.C. Tilghman, 1839
Improved by Joshua Barney, 1844-1846
Prairie du Chien
SCALE OF MILES
0 10 25
WISCONSIN
Dubuque
Cascade
North Fork
Maquoketa R.
Buffalo Cr.
Cedar R.
Wapsipinicon R.
Iowa R.
Iowa City
English R.
Mississippi R.
Bloomington (Muscatine)
Cedar R.
Long Cr.
Skunk R.
Crawfordsville
Crooked Cr.
ILLINOIS
Cedar Cr. (Little Cedar)
Agency City
Mt. Pleasant
Danville
Washington
Burlington
Keosauqua
Ft. Madison
Des Moines R.
MISSOURI
After Van der Zee, Iowa Journal of History and Politics, III, 183
Charles F. Strong, 1948

Illinois were compelled to use this road to Burlington. For their benefit Tilghman spent the combined funds raising an embankment leading eastward through the marshy ground adjacent to the bluffs. During the season, the engineer considered ways and means of improving the mail route from the Missouri boundary along the west bank of the Mississippi to a point between Dubuque and Prairie du Chien. After a preliminary survey, he recommended a shortening of the old main road by connecting only the major towns and then running feeder lines to the smaller communities. In a single year, the federal government had launched four road projects serving Iowa Territory. In each case, however, the funds allotted were inadequate to complete the construction.

In June, 1844, Congress approved $5,000 for construction and repair on bridges on the Agency Road and $10,000 for the same purpose on the Military Road. Joshua Barney, civilian agent of the Topographical Bureau, reported that the roads surveyed by Tilghman five years before were no longer used in many sections, and most of the bridges had washed away. Land along the original survey had been taken up, farms located and fenced in, and inhabitants had substituted roads which better answered their needs. The War Department ordered him to make detailed cost estimates of the needed repairs on bridges along each route. His reports, with War Department sponsorship, induced Congress to allot an additional $5,000 for the Agency Road and $8,000 for the Military Road during 1845. Five thousand was also appropriated for the construction of a road from the Mississippi bluffs opposite Bloomington (Muscatine after 1849) to Iowa City in compliance with a request from the territorial legislature made in 1841. Barney was convinced that this small appropriation could be spent most advantageously at the bluffs on the east bank of the Mississippi, and the War Department approved his plan. The United States government accepted its obligation in Iowa, as elsewhere on the frontier, to build roads for defense, for the mails, and incidentally for the settlers. Between 1839 and 1845, $60,000 had been allocated by Congress for Iowa roads which were undoubtedly the best constructed and most widely used in the territory.

Between 1820 and 1845 the United States Army had worked out, through experimentation, a most successful procedure for federal road building in the trans-Mississippi West. In earlier years, Congress merely appropriated funds and made the President responsible for proper expenditure. The chief executive invariably turned to the War Department for the administration of military road constructions. When possible, these routes of the 1820's were surveyed by a trained Army Engineer, and when his task was completed construction was supervised by a line officer attached to the nearest military establishment. The actual road building was performed by soldiers, and if the improvement was a vital artery .of communication, detachments of men were periodically sent out to rebuild and improve the route. Once completed, the over-all supervision of military roads usually became the responsibility of the Quartermaster Corps. Concern over the transportation of supplies to the frontier caused the assistant quartermaster to instigate many Army requests for additional federal funds. If more appropriations were forthcoming, the Army Engineers returned to the scene, determining where and how the federal moneys should be used in improving the roads.

In the 1820's when Congress appropriated funds for territorial roads, nonmilitary in character, a distinctly different procedure was followed in construction. Congress authorized the President to name civilian commissioners, usually three in number, to make surveys. Plats, diagrams, and field notes were submitted to Washington, and the President, on the advice of the trained engineers in the War Department, accepted or rejected proposed routes. Only then did construction begin. Funds permitting, civilian contractors built the roads with hired labor. Soldiers could be used, but Congress assumed the responsibility of specifically authorizing the President to utilize their services on nonmilitary routes. Somewhat later, Congress tried to simplify administration by placing the responsibility for the expenditures of nonmilitary federal road appropriations in the hands of the territorial governor. In spite of the emphasis on civilian control, the Army Engineers were assigned the unusually difficult engineering tasks associated with all federal road building, whether military or territorial.

After 1838 the Topographical Engineering Bureau gave meticulous attention to its road-building assignments. If possible, the surveys and constructions were supervised by an officer of the Corps. With limited personnel and the concentration of the federal road-building program in the territories east of the Mississippi River, officers were seldom available for the new western assignments during the early 1840's. In the absence of Army officers, a civilian engineering agent was appointed, as in the Iowa Territory. His first responsibility was a preliminary survey to determine how the appropriation could be spent most advantageously and the preparation of cost estimates for the most urgent and feasible construction. Upon review and approval by the Bureau, building contracts, which invariably went to local residents, might be let. The officer or agent continued the role of supervisor and inspector until released from his duties by the War Department.

Henry P. Beers has shown that by 1840 one could travel from the Louisiana state border to the Great Lakes along the network of military and territorial roads built by the federal government. After traveling from the Sabine or the Red River across Arkansas, military men, mail contractors, and emigrants met Missouri state roads at the southern boundary. From their termini at the northern Missouri boundary, the Iowa Military Road led to Dubuque. From the Mississippi opposite this town, or from Prairie du Chien connected with Dubuque by a mail route, Wisconsin roads, constructed at federal expense, ran in many directions to the lakes. Federal funds had also been used to open main thoroughfares for pioneer families moving into Arkansas and Iowa territories with their wagons. From Memphis to Little Rock, from Dubuque to Iowa City, from Burlington to the Des Moines River valley—settlers were on the move. The federal road construction program eased and speeded their trips and became, in time, a major contribution to the development of the prairie states just west of the Mississippi River.

Part I

WAGON ROAD SURVEYS AND CONSTRUCTION IN THE WESTERN STATES AND TERRITORIES BY THE UNITED STATES ARMY, 1846–1861

CHAPTER II : *Exploring Routes for Roads in the New Western Domain, 1846-1850*

With the outbreak of the Mexican War, the attention of the United States Army was quickly turned toward the West. Foremost among the forces leading the nation into this conflict had been a desire to win the northern Mexican provinces for an expanding democracy. To make certain the attainment of this objective, New Mexico and southern California were to be speedily occupied. The task was assigned to "The Army of the West" under command of Colonel Stephen W. Kearny, who assembled his forces at Fort Leavenworth on the Missouri frontier in the spring of 1846. Secretary of War William L. Marcy had instructed the colonel to seize Santa Fe, establish civil government, and move on to California as soon as possible. Marcy suggested that Kearny should make use of volunteers from a large group of Mormon emigrants who were encamped at Council Grove. The War Department had recently received information that many of these Mormon men, eager to continue their move to the west but temporarily halted at the frontier, would welcome the opportunity to go with Kearny. If nothing more, they would at least learn the best routes over which they could lead their wagons to the far west at the war's end. Kearny, however, was cautioned not to accept Mormon volunteers in numbers exceeding one-third of his total force.[1]

The commander dispatched Captain James Allen of the First Dragoons to the Mormon camps in Iowa Territory with instructions to enlist four or five companies of volunteers. These men were to be sworn into the service of the United States for twelve months and organized into a battalion under the command of Allen, as lieutenant colonel. Infantry volunteers were easily obtained and marched overland to Fort Leavenworth where arms and equipment were assigned with the understanding that these could be retained at the journey's close. The battalion commander, having fallen ill at Leavenworth, ordered his men to proceed to Council

Grove and there await his arrival. Only when they reached the Grove did the Mormons learn that their commander was dead, and an acting lieutenant colonel, Andrew J. Smith, named to lead them to Santa Fe. By this time Kearny, with three hundred regular dragoons, had occupied the New Mexican capital.[2]

Lieutenant William H. Emory of the Topographical Engineers was detailed to accompany Kearny, collect data for the government on the unexplored regions that were to be crossed, and map the route traversed. Lieutenants James W. Abert and William G. Peck assisted Emory in his mapping work on the first leg of the trip from Leavenworth to Santa Fe, but when Abert fell sick at Bent's Fort on the Arkansas River, and Peck became ill at Santa Fe, it was necessary to assign their duties to Lieutenant William H. Warner and Norman Bestor, a civilian engineer. These last two men charted the route from Santa Fe to San Diego while Peck and Abert remained in Santa Fe to explore and map the surrounding country.[3]

After initiating the processes to establish law and order, to protect the inhabitants from Indian depredations, and to organize a territorial government, Kearny organized his forces for their march to California. Leaving Santa Fe on September 25, 1846, he authorized Emory to direct the expedition down the Rio Grande as far as Tomé over ground that the cartographer had surveyed when Kearny was occupied with local affairs. On October 2, at the encampment on the Rio Grande near La Joya, the colonel learned of Allen's death and instructed Captain Philip St. George Cooke to return to Santa Fe and await the arrival of the Mormon Battalion en route from Council Grove.[4]

Cooke was to lead these volunteers to California over the route being used by the First Dragoons under Kearny, but their late arrival prevented Cooke from following immediately. At first Kearny had intended to carry his supplies across the country in wagons. These plans were modified after intercepting Lieutenant Kit Carson who had just come from California over the route he intended to use. Carson, ordered to return to the coast as guide, quickly noted the slow progress of the command and predicted that a four-month journey would be required to make the trip to Los Angeles. Moreover, Carson informed Kearny of the impossibility of taking wagons over his recently explored trail. The general was easily convinced of the wisdom of sending his wagons back, to be brought forward by Cooke and his battalion.[5] Kearny reported to the Adjutant General in Washington that hunting for a road suitable for wagon travel in the broken and sandy country would require much time and, therefore, he had resolved to expedite his march to the Pacific by transferring supplies to pack mules. He planned, however, to send forth his guides to examine the country for the location of the best wagon road. He would then order their return to direct the march of Cooke's Mormon Battalion along such a route.[6]

Kearny's column of men continued on horseback with pack mules, unencumbered by vehicles except for two small mounted mountain howitzers. They marched south along the Rio Grande's banks until October 14, then turned southwest for two days before heading west to intercept the headwaters of the Gila River. Following that river as it cut a course a few miles north and south of the thirty-third parallel, they attempted only one short cut before reaching the Colorado River. The Big Bend in the Gila was avoided by veering west by southwest beyond the Pima villages. The dragoons were at the confluence of the Gila and Colorado rivers by November 22. After an arduous crossing of the Colorado, they moved westward again while gradually edging northward to pass over an area that Emory labeled on his map the "Sandy Desert." Halting at a small but welcome stream, Carisǒn Creek, on the twenty-eighth after an exhausting and dry march, the column spent the next week working through the small coastal mountain range. The worn and bedraggled group reached San Diego December 12, 1846, having averaged about twelve and a half miles a day on the 1,043-mile trip from Santa Fe to San Diego.[7]

Military demands justified Kearny's rapid march, and the location of a route of travel was not a specific purpose. Yet the possibility of making such a wagon trail had not escaped the notice of the expedition's official scribe, Emory. On one occasion he recorded in his journal, "A few pounds of powder would blast the projections of rock from the cañon, and make it passable for packs, and possibly for wagons also. The route which the wagons are to follow is, however, to the south of this."[8] The wagon route to which he referred was the one then being located by Cooke.

While Kearny's mule train had been moving down the Rio Grande River, Cooke impatiently awaited the Mormons in Santa Fe. All had reported by October 12, but their wretched condition was immediately apparent to their commander. The mules were broken down and footsore due to the long haul from Council Grove, and sixty of the four hundred and eighty-six volunteers were so ill that they had been carried most of the way in wagons. To add to Cooke's difficulties, twenty-five women and many children had accompanied the Mormon group. Under escort, the disabled men, together with the women and children, were sent for the winter to Pueblo, a Mormon encampment on the Arkansas River above Bent's Fort. Five of the women, wives of members of the expedition who owned their own wagons, were allowed to remain with the column.[9]

Cooke spent one week in readying the battalion for the journey. He took along a sixty day ration of flour, sugar, coffee, and salt, thirty of salt pork, and twenty of soap. Each of the five companies had three mule-team wagons and six ox wagons; the quartermaster, paymaster, hospital department, and the staff each used a single mule wagon. Four or five private wagons also accompanied the party. Learning of the commander's decision to entrust all the other wagons to his care, Cooke wrote in his journal: "... I am informed that the wagons have been left rather as a matter of convenience. I have brought road tools, and am determined to take through my wagons. But the experiment is not a fair one, as the mules are nearly broken down at the outset."[10] In retrospect, when writing a report to Kearny after his arrival in San Diego, Cooke remarked: "The constant tenor of your letters of instruction made it almost a point of honor to bring wagons through to the Pacific; and so I was retarded in making and finding a road for them."[11]

On the trip south along the Rio Grande after his Santa Fe departure, Cooke became convinced that unless his load was lightened and his rations increased, the possibility of success would be small. He therefore sent back fifty-eight ill men, reduced his load by an estimated 20 per cent, and increased his food supply through purchases of 308 sheep and several beeves. Continuous efforts were made at the river settlements and ranches to exchange worn and jaded animals for fresh ones, but even at the two-for-one ratio practiced on the frontier, the natives were not readily agreeable to exchange. New Mexico had virtually been drained of supplies by the Army.

Cooke, upon his arrival at the Ranches of Albuquerque on October 24, met the guide Charbonneau, left behind by Kearny, who claimed that he had examined a route farther down the river that joined an established trail from El Paso to the Arizonac copper mines. Cooke, however, was convinced that no thorough examination had been made and so seriously doubted the practicability of the route west of El Paso that he refused to follow it. The other guides returned by Kearny came into camp on November 2. Except for one man, they were drunk. The sober guide's reports were not encouraging. He estimated the distance to San Diego to be 1,200 miles, necessitating ninety days' travel and with animals "not half so well fitted out to carry wagons as the General was."[12]

When Cooke left the Rio Grande near present-day Rincón, New Mexico, about 30 miles below the point where Kearny had headed west, the command included 339 men, 5 women, 15 wagons, and a herd of sheep and cattle. He proceeded southwest by a newly discovered spring which he named for himself and on to the well-known Ojo de Vaca, or Cow Spring, where camp was made on November 19. After long consultation with his guides, he resolved to swing southward in a wide semicircular loop before returning to the Gila River by way of its tributary, the San Pedro. Kearny had urged a direct westerly crossing from the Rio Grande to the Gila's headwaters, but the guides were unwilling to attempt it with wagons. "What difference if this distance is doubled," recorded Cooke, "if it is a better route?"[13] Marching southwest from Ojo de Vaca, the Mormon Battalion crossed the mountains near the present New Mexico–Arizona border at Guadalupe Pass and thence turned westward to strike the San Pedro River. After turning north along this stream, the wagon train traveled for fifty-five miles along the river banks before changing the course due west to the presidio at Tucson. From this settlement the volunteers continued northwest to the Gila just above the Pima villages. There the trail of Kearny was picked up and followed along the south side of the Gila to its mouth, across southern California, and to San Diego where they arrived on January 29, 1847.[14]

The Cooke caravan did little wagon-road building on its journey west. The battalion was primarily concerned with the location of a trail along ground hard and smooth enough for the wagons to pass over and without inclines or declines that would cause the wagons to overturn. Physical obstructions that might block the passage of this train, or those of future travelers, were carefully avoided. Cooke wrote in his journal on one occasion: "The road, or rather country, was smoother than usual today; the same gravel and clay, well covered with grass. It has been mostly a gentle descent. After all have passed, we leave a very good road."[15] Wherever there were wagon tracks there was a road. The cursory nature of road improvements is manifest in Cooke's entry of November 20: ". . . it should be noticed that making a wagon road for thirty or forty miles without water, is equal to going fifty or sixty with a road."[16]

When the going was smooth, it appears that they marched along, moving from water hole to water hole, convinced that the marks left by their turning wheels had established a road. Along the difficult stretches "the pioneers," as Cooke called his advanced detachment, were sent ahead to clear the way for the procession of straggling wagons. On occasions, the commander noted, "The road was very bad . . . The pioneers did much work and straightened the trail much."[17] Cooke assigned ten to twenty men to each wagon. These men, carrying full knapsack and musket, pushed and pulled the wagons through the sand.[18] The colonel was by no means satisfied with his route in its entirety. After leaving Tucson for the Gila River, he noted that the road went through the most extensive desert he had ever seen, and yet "far away to the west, as far as the eye could follow, it was the same." He had little hope in the later selection of a more acceptable way.[19]

Cooke's men made some road improvements. Just before entering the San Diego coastal plains, the battalion found itself halted in the mountains near San Felipe by a narrow canyon of solid rock, a defile too narrow for the smallest wagon to squeeze through. Grabbing an ax, the leader inspired his men to widen the opening by hewing away the projecting rocks. The volunteers set to work chipping out the passage, and after dismantling two wagons, the rest were pulled through without great difficulty. Cooke warned that the section from the Colorado River over the "Sandy Desert" to this improved canyon was the most difficult of his entire route.[20]

To celebrate this epic accomplishment of the Mormon Battalion in bringing eight of the wagons through to California, Cooke issued an Army order of congratulations:

> History may be searched in vain for an equal march of infantry. . . . With almost hopeless labor, we have dug deep wells which the future traveller will enjoy. Without a guide who had traversed them, we have ventured into trackless prairies where water was not found for several marches. With crowbar and pick and ax in hand we have worked our way over mountains which seemed to defy aught save the wild goat, and hewed a passage through a chasm of living rock more narrow than our wagons. To bring these first wagons to the Pacific, we have preserved the strength of our mules by herding them over large tracts, which you have laboriously guarded without loss. . . . Thus, marching half naked and half fed, . . . we have discovered and made a road of great value to our country.[21]

"Indeed," states Professor Ralph P. Bieber, "the saints and their commander had reason to be proud of their achievement, for they had opened the first wagon road through the southwest to California."[22]

The road was used by thousands of emigrants in the following years, and, as a practical railroad route, was one of the principal reasons for the Gadsden purchase of 1853 and 1854. Today it is followed, in a general way, by the Atchison, Topeka, and Santa Fe Railway from Rincón to Deming in New Mexico, and by the Southern Pacific Company's lines across Arizona by way of Douglas, Tucson, and Yuma to San Diego, California.

As the Mexican War drew to a close, settlers in the frontier states of Missouri, Arkansas, and Texas began to speculate as to what the treaty terms might indirectly accomplish for their regions. Nowhere was there more optimism than on the western Arkansas frontier. During December, 1847, a writer for the *Fort Smith Herald,* describing the trading activities of the town, suggested: "Should we acquire Northern Mexico by treaty, of which event there is very little doubt, Fort Smith will be the great rendezvous of emigrants and traders. It possesses advantages over every other point, and public attention must be directed to it."[23] News that the formal negotiation of the Treaty of Guadalupe Hidalgo, by which Mexico had confirmed the American title to Texas as far as the Rio Grande and had ceded California, New Mexico, and most of present-day Arizona to the United States, was conveyed to the people of Fort Smith on March 1, 1848.[24] By the end of the month the *Fort Smith Herald* had begun a systematic campaign to interest the community in a direct road to Santa Fe.[25] The editor proudly recommended the following August that a public meeting be held "... for the purpose of getting up a petition to the next legislature, to have that body bring before the President and Congress, the subject of opening a road up the Arkansas and Canadian River to Santa Fe, Oregon, and California."[26]

The proposed meeting was held in September, 1848. In adopting resolutions to the state legislature, the citizens pointed out the commercial advantages of the projected road and emphasized a new justification for the survey: "The subject of a public road ... has become a matter of increased importance, as well as to the United States Government, for the purpose of transporting troops and Government stores into the newly acquired territory ..."[27] Not long after this meeting, news of the discovery of gold in California reached Fort Smith, and its people had additional incentive for promoting the survey and establishment of a national wagon road to the Pacific.[28] Gold findings on the west coast meant that thousands in the East and in the Mississippi Valley would be planning to follow the long overland trek to the new El Dorado. As the gold mania gripped the nation, some citizens in the Arkansas frontier realized that if a number of these emigrant throngs could be routed by way of Fort Smith, the small town would become a thriving business and commercial center.[29]

Local military men promoted the endeavor. The *Herald,* during November, printed letters from Brigadier General Mathew Arbuckle, commander of the Seventh Military Headquarters located at Fort Smith, and from Major Benjamin L. E. Bonneville, well-known western military and trading figure, revealing their interest in locating a road west from Fort Smith. Arbuckle suggested that it follow along the south side of the Canadian River. The Washington correspondent of the *Fort Smith Herald* reported in January, 1849, that extracts from these letters had been reprinted by newspapers in the national capital with favorable comment.[30]

On January 3, 1849, a memorial from the Arkansas state legislature, containing in substance the resolutions adopted at the public meeting of September in Fort Smith, was presented to the United States Senate and referred to the Committee on Military Affairs.[31] Although the committee reported favorably on the memorial, no action was taken by the Senate nor was there much hope that the two houses of Congress would agree upon a program of action. Meanwhile, Senator Solon G. Borland of Arkansas communicated directly with Secretary of War Marcy requesting a military escort for the western emigrants planning to rendezvous at Fort Smith in April.[32] The senator had in mind the safety of the emigrants when he applied for the military escort, but he was more interested in drawing national attention to the Arkansas route by giving it an "official character." He also made reference to the Arkansas memorial with the hope of securing the accompaniment of an officer of the Corps of Topographical Engineers *"to make, and report, a reconnaissance* of the route, in direct reference to the future location of a national road . . ."[33]

The War Department evinced immediate interest in the proposals. Definitive information was desired about the way to Santa Fe along the Canadian River as a possible route of travel for supply trains headed for the New Mexico forts. The protection of the California emigrants moving west through the Indian country was an additional concern.[34] The gold enthusiasm by now was approaching its zenith, and the department had good reason to place great importance upon the multiple objectives. On January 22 of that year orders were sent to General Arbuckle for the organization of a military escort to accompany the California emigrants as far as Santa Fe. The general was also informed that an officer of the Topographical Engineers would accompany the detachment. The *Fort Smith Herald,* jubilant with this news, either in excitement or misinformation, reported that a "corps of *Engineers* with a detachment of U. S. dragoons, and one of Infantry, have been ordered by the Department at Washington, to *survey, mark,* and CUT OUT a road from FORT SMITH TO SANTA FE."[35] While General Arbuckle made preparation for the military escort, the Bureau of Topographical Engineers in Washington ordered Lieutenant James H. Simpson "to repair as soon as practicable to Fort Smith, Arkansas, and report for duty to General Arbuckle, as engineer officer of the expedition to be fitted out at that place."[36] Congress had appropriated $50,000 to defray the expenses of the Topographical Engineers in surveying routes from the valley of the Mississippi to the Pacific Ocean.[37]

Local residents expected Colonel Bonneville to command the military escort and their disappointment was made public when General Arbuckle named Captain Randolph B. Marcy, a United States Military Academy graduate with Mexican War experience, to the post.[38] Marcy moved from Fort Smith before Simpson's arrival. The escort comprised twenty-six dragoons, fifty infantrymen, three junior officers, a physician, and a quartermaster. A train of eighteen six-mule-team wagons, one six pounder, and one blacksmith's traveling forge accompanied the command west.[39] Simpson's arrival at Fort Smith immediately after Marcy's departure was the signal for speculation about his proposed reconnaissance. One citizen suggested to the readers of the *Herald* that he was to determine "... whether a great National *Railroad* is to run through this state or not."[40]

The engineer at once hastened to join Marcy some 26 miles west of Fort Smith. En route he received instructions from the Bureau to continue his exploration and survey all the way to California with the emigrant party, after receiving a relief escort at Santa Fe.[41] Marcy's orders were to go only to Santa Fe and then return by the same route, or by a new trace that he might locate after starting east from the Rio Grande 200 miles south of Santa Fe. The swelling number of gold seekers gathering on the Arkansas frontier had made the Marcy-Simpson military escort of less importance. The group was large enough to minimize the danger of Indian attack, yet with increased numbers came problems of organization and delay in departure. Marcy left Arkansas with the understanding that the Fort Smith Company of California emigrants would join his escort at Choteau's Trading Post.[42]

The Marcy-Simpson route to Santa Fe proceeded along the south side of the Canadian River through the Choctaw and Chickasaw Indian districts, on through the present panhandle of Texas, where ranged the Kiowa and Comanche Indians, past present-day Tucumcari, New Mexico, across the Pecos River at Anton Chico, thence to Galisteo, and northwest to Santa Fe.[43] This was by no means the first excursion westward from the Arkansas frontier along the Canadian River. Indian traders during the 'twenties and 'thirties, and United States troops from Fort Gibson in the 'thirties and 'forties had traveled the route in whole or in part. Josiah Gregg had conducted a trading expedition from Van Buren, Arkansas, to Chihuahua City in 1839 and in 1840 along the Canadian River. In 1845 Lieutenants William G. Peck and James W. Abert of the Topographical Engineers had moved eastward over the route returning to St. Louis from Bent's Fort, and the following year an emigrant party had passed over it headed toward Santa Fe.[44] The route followed by Marcy's command, and called the "new" road by Simpson, represented the first official attempt of the United States government to survey a road for wagons directly from Fort Smith to Santa Fe, and differed from the other routes in that it lay entirely on the south side of the Canadian River.

Lieutenant Simpson's duties consisted of "... surveying the route, furnishing the commanding officer with geographical information when necessary, taking notes of the country, and, in general, in getting up all the data necessary for a full and complete map and report of the route."[45] He had the assistance of a civilian engineer. Marcy pointed out the location of the road and directed its construction.[46] A Delaware Indian, Black Beaver, was engaged as guide and interpreter. Although Lieutenant Simpson and Captain Marcy always referred to their work as the "construction" of a wagon road, they did no more than locate and survey a route passable for wagons. They selected prominent landmarks by which the emigrant and the military could be guided in the future, and marked out day-to-day camp sites approximately close to the three essentials of wood, water, and grass. The route was divided into three sectors on the basis of terrain. The first 120 miles from Fort Smith to Shawneetown was very unsatisfactory because of hills and quicksand mires, particularly during rainy spells like that experienced by this expedition. From Shawneetown through the Cross Timbers for 153 miles, the road was good, and beyond that point the terrain so excellent that Simpson pronounced it *"very fine* . . . a better road I never saw anywhere."[47] Dissatisfied with the route to Shawneetown, the engineer recommended that some future command might grade the road over some of the hills and make it smoother with crushed sandstone. At all stream crossings, with the exception of the Poteau River where a ferry was used, Simpson directed the troops in grading down embankments to facilitate the crossing of wagons. Occasionally, causeways had to be laid. More substantial improvements were proposed for the future which, on account of the necessity of opening an immediate way for the crowd of emigrants pressing them, the escort troops did not have the time to perfect.[48] Though progress was slow, Captain Marcy was convinced that the road would be good ". . . when it has been travelled sufficiently to beat down the earth and pack it."[49]

As the command approached the Cross Timbers, the advance guard was slowed to wait for the arrival of the California bound emigrants. These pioneers had been expected to catch up at an earlier point. Instead of following the Marcy-Simpson road, they had crossed the Canadian River to the north side to follow for a distance an older and more extensively traveled route. Because the road on that side was in much worse condition than the new survey, they had caused themselves discomfort and delay.[50] While the command awaited the emigrant party, Captain Marcy, with the aid of Black Beaver, located a feasible route through the Cross Timbers and the troops were employed in bridging two streams. The combined expedition quickly traversed the timbered stretch and once more passed through open prairie country along the divide between the Canadian and Washita rivers. The farther west they traveled, the easier the wagons rolled. The capital of New Mexico was reached on June 28 after a trip of eighty-five days. The *Fort Smith Herald* announced to its readers: "the road is now open and marked all the way, and it is as plain as any wagon road in the country . . ."[51] Captain Marcy recorded in his official report that ". . . for so long a distance, I have never passed over a country where wagons could move along with so much ease and facility, without the expenditure of any labor in making a road, as upon this route."[52] Simpson, fully aware of the agitation for a railroad to the Pacific, felt compelled to pass judgment on the possibility of using the route for the location of a national railroad. Although the valley of the Canadian was pronounced practical, he was convinced that "the time had not yet come when this or any other railroad can be built over this continent." His solution of the transportation problem to the Pacific was the construction of great highways across the central and southern plains as military roads and the promotion of them by legislation of the national government. Railroads, he insisted, belonged to the future and could be more effectively constructed along wagon routes proved desirable by proper survey and extensive use.[53]

Immediately after his arrival in Santa Fe, Simpson made preparations for the second phase of his assignment—the exploration of a route to California with the emigrants and a relief escort. New orders from the Bureau of Topographical Engineers arrived on July 2, 1849, canceling the proposed expedition, demanding the immediate preparation and submission of a report on the Fort Smith–Santa Fe road, and advising an exploration across the northern part of the Ninth Military Department which embraced the Territory of New Mexico.[54] At this time the Bureau had information on only two routes of importance between Santa Fe and California that lay entirely within the limits of the United States. One of these routes was that located by Kearny in 1846. The other was in the northern part of the territory by way of Cañada, Abiquiu, St. Joseph's Spring, and thence to Los Angeles. The part of this latter route between St. Joseph's Spring and Abiquiu, which was located on the "caravan route" or the Old Spanish Trail, was comparatively unknown to the Bureau, and it was this that Simpson was advised to explore.[55] This New Mexico exploration did not materialize because the Ninth Military Department witnessed such an increase in Indian depredations during the spring and summer that Lieutenant Colonel John M. Washington, departmental commandant and governor of New Mexico, finally resolved to undertake a campaign against the particularly offensive Navajos. Lieutenant Simpson was ordered to accompany the expedition and make such a survey of the country as the movement of troops would permit.[56] The road-building activities of the Topographical Engineer in New Mexico Territory were thus brought to a close.

After his return from the Indian country, Simpson began to speculate about the possible location of a new wagon route from Santa Fe to California. He concluded that such a route could go by way of the Pueblo of Zuñi, then along the Rio de Zuñi to its junction with the Colorado, and, after crossing the Colorado, continue westward to Los Angeles or San Diego. From his recent reconnaissance, he had learned that the country between Santa Fe and Zuñi could be made practicable for wagons with a little labor. The lieutenant became more convinced about the potentialities of this route when Richard Campbell of Santa Fe related his account of traveling to San Diego along the proposed trace in 1827.[57] Campbell had made the journey without difficulty with a party of thirty-five men and believed that sufficient quantities of water, wood, and grass could be found, though he was not certain the Colorado River would always be fordable at his crossing near the mouth of the Rio de Zuñi. Campbell also knew of a route that led from Zuñi to the Ford of the Fathers on the Colorado. This went via the pueblos of the Moquis or Hopi Indians, but was believed to be practicable only for pack trains.[58] Simpson reported his conclusions to the Bureau of Topographical Engineers and recommended that an exploration west of Zuñi be authorized. He was convinced that such an exploration would locate a wagon route shortening the distance between Santa Fe and Los Angeles by more than 300 miles. This was to be a middle course between Cooke's wagon route to the south and the Old Spanish Trail to the north. Simpson's proposals coincided with the views of the Bureau of Topographical Engineers, since in 1851 Captain Lorenzo Sitgreaves was ordered to explore the northern part of the present state of Arizona along the Zuñi River to its junction with the Colorado.[59] He failed to discover a good route for a wagon road.

Captain Marcy, in compliance with his orders, had in the meantime refitted his command for the return march to Fort Smith. Leaving Santa Fe on August 25, his column turned southward down the Rio Grande Valley to Doña Ana, a river settlement 60 miles north of El Paso. From this point he moved eastward, leading the detachment through the San Augustine Pass in the Organ Mountains onto the famous Llano Estacado. Doña Ana was within 15 miles of the point on the Rio Grande where Cooke's wagon train left the valley toward the west. Thus, Marcy's return route became a direct link with Cooke's route to California. The command struck the Pecos River near the thirty-second parallel. In moving northeast across Texas along the headwaters of the Colorado and Brazos rivers, the group traveled more rapidly than on the outward trip along the south bank of the

Canadian. They averaged eleven and a third miles a day on the 895 measured miles of the journey. Their greatest difficulty over the new route was in finding sufficient water on the Llano Estacado. Emigrants, however, en route to California could cross the Red River at Preston, follow Marcy's trail across Texas and through San Augustine Pass to the junction with Cooke's route, and then proceed over the latter to the Pacific. This was an alternate choice to the longer Canadian River route to Santa Fe, and thence southward to Cooke's route west.[60]

Although the attention of the Army, the Congress, and the nation was primarily focused on the Southwest during the Mexican War and the years immediately after, Congress did not overlook the continued importance of the central overland route to California and Oregon. During the same summer months of 1849 that Marcy and Simpson were in New Mexico, another member of the Topographical Engineers, Captain Howard Stansbury, was ordered to escort a party of Oregon emigrants from Fort Leavenworth as far as Fort Hall, Oregon Territory. There Stansbury was to leave the Oregon Trail, turning southward into the new Mexican Cession. He was to complete his assignment by making an accurate survey for a wagon road from Fort Hall to the Great Salt Lake.[61]

Leaving Fort Leavenworth on the Missouri, May 10, 1849, Stansbury and his escort of fifteen traveled northwest across the Big Blue River to the Little Blue and along the headwaters of that stream to Fort Kearny on the Platte River. To the west of Fort Kearny the expedition followed the regular emigrant route to a ford across the south fork of the Platte and thence overland to, and along the north fork to Fort Laramie. From this fort they moved up the North Platte and its Sweetwater branch through the South Pass and descended the western slope of the Continental Divide by way of the Sandy, across the Green River and its numerous forks, to Fort Bridger.[62]

Stansbury, having been instructed to locate new routes for western travel, both for emigrant wagons and a future railroad, decided to leave the emigrant party at this fort rather than to proceed to Fort Hall. The long-range program of the government to connect the Missouri River settlements and California by the shortest possible route justified a new survey between Fort Bridger and Great Salt Lake. Both previously used routes to California appeared unsatisfactory. The northern route along the Oregon Trail turned abruptly northwest after leaving Fort Bridger and followed the Bear River valley by way of Soda Springs and Fort Hall. Here those bound for California could turn again to the southwest into the headwaters of the Humboldt. The Mormon Trail, on the other hand, turned southwest from Fort

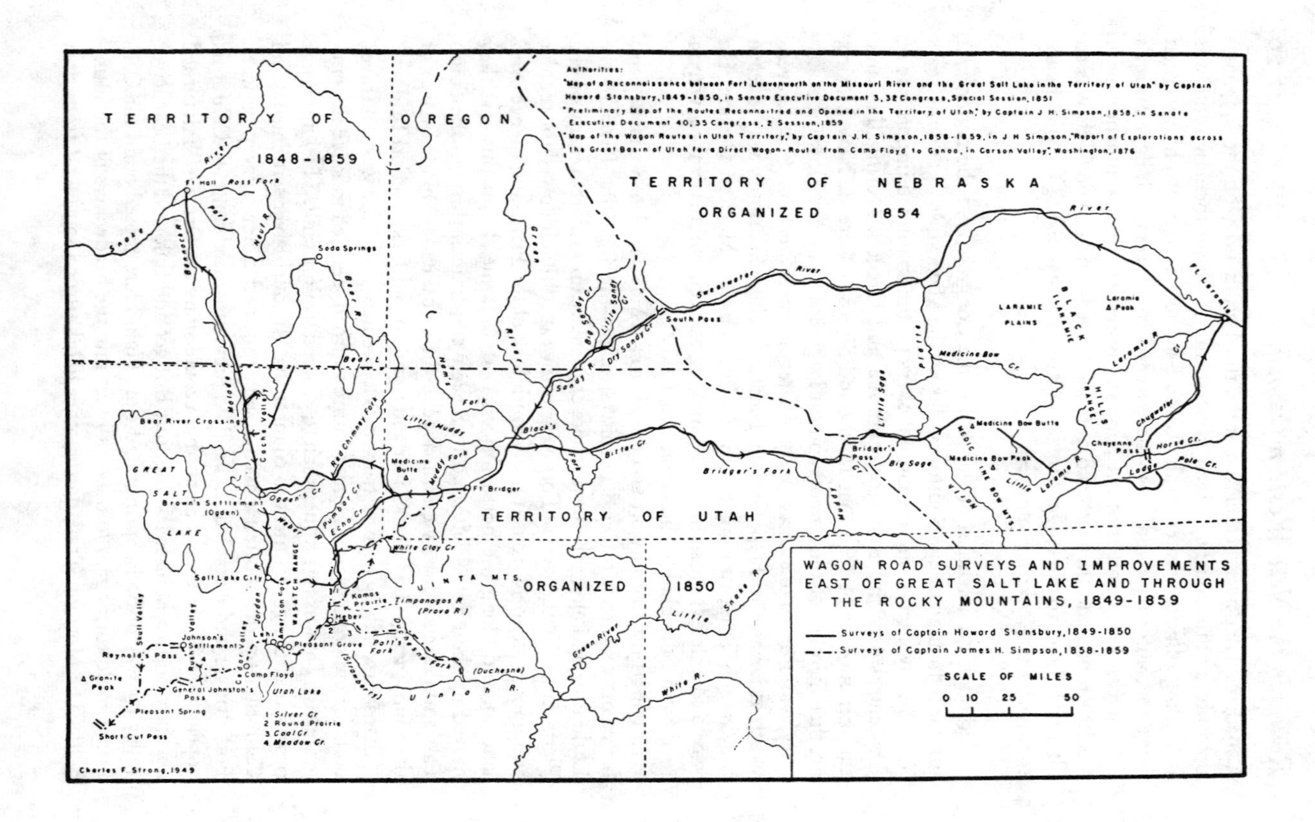

WAGON ROAD SURVEYS AND IMPROVEMENTS
EAST OF GREAT SALT LAKE AND THROUGH
THE ROCKY MOUNTAINS, 1849-1859
Surveys of Captain Howard Stansbury, 1849-1850
Surveys of Captain James H. Simpson, 1858-1859
SCALE OF MILES
0 10 25 50
Authorities:
"Map of a Reconnaissance between Fort Leavenworth on the Missouri River and the Great Salt Lake in the Territory of Utah" by Captain Howard Stansbury, 1849-1850, in Senate Executive Document 3, 32 Congress, Special Session, 1851
"Preliminary Map of the Routes Reconnoitred and Opened in the Territory of Utah," by Captain J. H. Simpson, 1858, in Senate Executive Document 40, 35 Congress, 2 Session, 1859
"Map of the Wagon Routes in Utah Territory," by Captain J. H. Simpson, 1858-1859, in J. H. Simpson, "Report of Explorations across the Great Basin of Utah for a Direct Wagon-Route from Camp Floyd to Genoa, in Carson Valley," Washington, 1876
TERRITORY OF OREGON
1848-1859
TERRITORY OF NEBRASKA
ORGANIZED 1854
TERRITORY OF UTAH
ORGANIZED 1850
Ft Hall
Ross Fork
Port Neuf R.
Snake
Bannock R.
Soda Springs
Bear R.
Bear L.
Malade
Cache Valley
Bear River Crossing
GREAT SALT LAKE
Brown's Settlement (Ogden)
Ogden's Cr.
Weber R.
Red Chimney Fork
Pumbar Cr.
Echo Cr.
Medicine Butte
Muddy Fork
Little Muddy
Ham's Fork
Ft Bridger
White Clay Cr.
Salt Lake City
Jordan R.
American Fork
WASATCH RANGE
Kamas Prairie
Timpanogos R. (Provo R.)
Heber
Lehi
Pleasant Grove
Pott's Fork
Duchesne Fork
(Duchesne)
(Strawberry)
Uintah R.
UINTA MTS.
Utah Lake
Camp Floyd
Cedar Valley
Rush Valley
Johnson's Settlement
Skull Valley
Reynold's Pass
Granite Peak
General Johnston's Pass
Pleasant Spring
Short Cut Pass
1 Silver Cr.
2 Round Prairie
3 Coal Cr.
4 Meadow Cr.
Green River
Black's Fork
Bitter Cr.
Sandy R.
Big Sandy Cr.
Little Sandy Cr.
Dry Sandy Cr.
South Pass
Sweetwater River
Bridger's Fork
Bridger's Pass
Big Sage
Little Sage
Muddy
Platte
Medicine Bow Cr.
Medicine Bow Butte
Medicine Bow Peak
MEDICINE BOW MTS.
North
LARAMIE PLAINS
BLACK (LARAMIE HILLS RANGE)
Laramie Peak
Laramie R.
Little Laramie R.
Ft. Laramie
Chugwater Cr.
Cheyenne Pass
Horse Cr.
Lodge Pole Cr.
Little Snake R.
White R.
River
Charles F. Strong, 1949

Bridger, passing through Echo Creek Canyon of the Weber and on to Salt Lake City at the southern part of the lake basin. California travelers following this route were forced to travel northward from the city along the entire eastern shore of the Great Salt Lake, across its northern end, and turn northwest to meet the California Trail southwest of Fort Hall. The first of these routes ran approximately 120 miles too far to the north, the second an unnecessary 60 miles toward the south. Stansbury desired to locate a route, suitable for wagons, directly west of Fort Bridger to the lake, primarily for those bound for California.[63]

Under the guidance of Lieutenant John W. Gunnison, the bulk of Stansbury's command was sent forward along the Mormon Trail to Salt Lake City, but the captain, with a selected few, left the trail by crossing the ridge between the Muddy fork of the Green and the Bear rivers to the banks of the latter stream. With Jim Bridger as guide, he turned northward down the Bear River within sight of Medicine Butte. West of the Bear, the party came upon Pumbar (Plumber) Creek, a tributary of the Weber. The canyon of this creek proved impassable for wagons. Stansbury discouragingly noted that "... the road would have to be *made* all the way up, and considerable quantity of small cottonwood timber cut out."[64] A satisfactory route to the west was finally located by way of the valley of the Red Chimney Fork of the Pumbar, across the divide to Ogden Creek, flowing into the Weber, thence across Ogden Hole through North Ogden Canyon to the much traveled north-south road between Salt Lake City and the head of the lake. No grades too steep for wagons were encountered along this way and in the valleys of the Red Chimney Fork and Ogden Creek there was sufficient room for a wagon road. At Brown's settlement, the present site of Ogden, Stansbury turned south along the Salt Lake City road, skirting the eastern shore of the lake for 40 miles to the Mormon capital, arriving on August 28. The result of the reconnaissance convinced the topographical engineer that a good road could be built from Fort Bridger to the head of Salt Lake, but he thought a further investigation might find a more useful route farther north through the south end of Cache Valley.[65]

After explaining the nature of his survey to the Mormon leaders and securing the support of Brigham Young and the Governing Council of Twelve, Stansbury departed for Fort Hall September 12 to locate a road from the head of Salt Lake to that fort. He followed the California emigrant route along the eastern shore to Bear River, crossing at the lake's head. The ford at the Bear River was believed unsatisfactory because of the high steep banks on both sides of the stream and because the rapid flow of water in the spring and early summer made fording impossible and ferries necessary.

Leaving the emigrant road, he next drove his pack mules that were to haul supplies back from Fort Hall, up the Malade River to its head, across a dividing ridge to the Pannock (Bannock), down that stream, and across the Portneuf to Fort Hall. The captain observed that the valley of the Malade was "... extremely level, free from underbrush, with very little artemisia, and affords ground for an excellent wagon road."[66] "The result of this exploration," he wrote, *"has been to demonstrate the entire practicability of obtaining an excellent wagon-road from Fort Hall to the Mormon settlement upon the Great Salt Lake."*[67] He was convinced that this route was the best natural road he had ever seen. The transport wagons, loaded with 3,500 pounds each, had traveled the distance with little difficulty. Sufficient timber was available at the Bear and Portneuf crossings for the construction of bridges should the volume of traffic ever justify the expense.

Upon arrival at the Bear River crossing on the return trip, Stansbury turned eastward to make a reconnaissance of Cache Valley. The engineer knew that this valley had served as a rendezvous of the American Fur Company trappers for years, and mountain men assured him that the climate was ideal for wintering stock. With irrigation the rich soil could also produce worth-while crops. His primary purpose was to determine the advisability of establishing a military post. A secondary purpose was to locate a route through the mountains on the valley's eastern rim that would afford passage for wagons. He was disappointed on both counts. The possible severity of winter storms precluded the recommendation of a fort in this valley. However, should a road between Fort Bridger and the head of the lake be built along the Blacksmith's Fork of the Bear, he believed that enough Oregon and California emigrants would be attracted to the route to make a military post necessary for their protection. He retained the view that this more northerly route west of Fort Bridger might prove even more useful than the one which he had already located.[68]

On October 19 Captain Stansbury, accompanied by four men, began a reconnaissance of the deserts around the western shores of Great Salt Lake. Starting from the Mormon capital, their route followed a counterclockwise course around the lake. From the north shore, the party traveled southwestward to Pilot Peak on the present Utah-Nevada boundary, and then returned across the desert wastes west and south of the lake to Salt Lake City. This party was thus the first to complete a circuit of Salt Lake by land.[69]

Upon his return to Salt Lake City, Stansbury decided that further field work was impossible with the arrival of winter. He used these months among the Mormons to write official reports and to make preparations for a survey of the Salt Lake in the spring. When April arrived Stansbury and his assistants, Gunnison and Albert Carrington, a local Mormon, began an examination of Salt and Utah lakes. This exploration engaged the attention of the men for five months.[70]

Planning his return journey to Fort Leavenworth at the close of August, 1850, Captain Stansbury turned once more to road location. Determined to seek another route between Salt Lake City and Fort Bridger, he headed east across the Wasatch Range to the headwaters of the Weber at Kamas Prairie. After traveling down the right bank of the Weber in a northerly direction to the mouth of Echo Creek, Stansbury's party turned northeastward along that stream and then crossed the dividing ridge to Bear River valley. From here the Mormon Trail was followed to Fort Bridger.

On his outward journey Stansbury had been primarily interested in routes around the *north* shore of Salt Lake, and he had located an acceptable wagon road and recommended an alternate route. From his observations on the return trip he concluded that the most acceptable route leading from Fort Bridger to Salt Lake City *south* of the lake, either for a mail route or a railroad, should cross to Bear River, ascend that stream to a tributary of the Weber which, in turn, should be followed westerly to the main stream at Kamas Prairie. From here two branches were deemed practicable: one following his return journey across the Wasatch Range; the other a route down the Timpanogos through Provo Canyon to Lake Utah.[71]

On the eastward journey from Fort Bridger, Stansbury sought a shorter route to the Missouri. Before leaving the fort on September 10, he announced his intention of attempting to locate a road across the North Platte, near Medicine Bow Butte, which would skirt the south end of the Laramie Plains, cross the Black Hills (Laramie Range) near the head of Lodgepole Creek, and ascend that stream to its junction with the south fork of the Platte. Such a road would avoid the northern detour via the Sweetwater and the South Pass and likewise the most rugged section of the Laramie Range.[72] The Stansbury party, with Jim Bridger again serving as guide, departed from the fort along the emigrant road or Oregon Trail which the engineer had used on his outward journey. After about 18 miles they left this road and traveled directly east to cross the Black's Fork of the Green River and the divide between that tributary and the main stream. Stansbury recorded in his journal that "an excellent wagon-road can be traced from Green River at this point to Fort Bridger, and by a very direct route."[73]

From the Green River valley the explorers continued due east by ascending the tributary, Bitter Creek. Their route of travel was between Green River and Rock Springs, Wyoming. As Stansbury ascended this stream to its head he noted that there was "... present no serious obstacle to the easy construction of a good road." From Bitter Creek, the party made a gentle ascent of the dividing ground between that stream and Muddy Creek, a tributary of the Little Snake, and continued across its headwaters to the Continental Divide at Bridger's Pass.[74] The ground was very rough and occasional ascents and descents were judged too steep for a good wagon road. At these places Stansbury always sought out an acceptable way before traveling farther. From the Continental Divide the surveying party descended Little Sage Creek, an affluent of the north fork of the Platte. After crossing the North Platte, they turned northeast skirting the Medicine Bow Mountains to the north. The party entered the Laramie Plains and turned slightly to the southeast to cross the Little Laramie and Laramie rivers at their headwaters. The plains were described as "a beautiful rolling country covered with grass," and although the trace was "rather undulating" the engineer saw no difficulty in making an excellent road.[75]

On September 29 Stansbury turned abruptly north by way of Cheyenne Pass to the headwaters of the Chugwater. In this valley he sustained a severe injury from a fall and was unable to mount his horse. The party, therefore, ceased exploration and traveled directly from the Chugwater camp to Fort Laramie. The captain was confined to an ambulance bed until his arrival at Fort Leavenworth.

The explorers had located the headwaters of Lodgepole Creek in the vicinity of Cheyenne Pass. Stansbury was bitterly disappointed in being unable to follow this stream to its confluence with the South Platte. The results of his work were clear and gratifying, however, and he recorded in his report:

> It has been ascertained that a practicable route exists through the chain of the Rocky mountains, at a point sixty miles south of that now generally pursued.... The entire route through that long distance [from Salt Lake City to the head of Lodgepole creek] *varies but a trifle from a straight line....* The distance from Fort Bridger to Fort Laramie, by the present route, is four hundred and eight miles; while, by the new route from Fort Bridger to the eastern base of the Black Hills, it is but three hundred and forty-seven miles: so that a saving is effected in the total distance, of just sixty-one miles.[76]

Stansbury's expedition had indeed been a remarkable success from the standpoint of road location. On his return, he worked out the trace of a road east of Fort Bridger through southern Wyoming and Nebraska destined to be used by the Overland Stage, the Pony Express, and at the present time, by the Union Pacific Railroad.[77]

These three major exploring and surveying parties sent out by the United

States government to locate wagon roads west had proved that the national domain was too vast, and the distances of travel too great for the engineering officers of the Topographical Bureau to think or act in terms of their road-building experiences east of the Mississippi River or in the prairie states just west of that river. True, the interest of the emigrant remained paramount, but those bound for the Pacific Coast after the news of gold discovery in California were often more intent on speed than on comfort. They demanded only the minimum essential of a trace over which their wagons could pass on solid terrain without mires of mud or sand, free from rocks and other natural obstacles precluding rapid movement. To seek the shortest route possessing these characteristics and the three basic requirements of grass, fuel, and water was a task of the United States Army. Few hours were lost in making improvements.

CHAPTER III : *Wagon Roads on the West Texas Frontier*

THE GREAT WAGON-ROAD SURVEYS of Philip St. George Cooke, Randolph B. Marcy, James H. Simpson, and Howard Stansbury, during and immediately after the Mexican War, signalized the formation of a general policy by the federal government to aid the emigrant, settler, and soldier by road constructions into the new domain obtained by the war effort. By the terms of the Treaty of Guadalupe Hidalgo the nation's hold upon Texas to the Rio Grande River had likewise been secured, but the vast expanse of central and western Texas, where the Apaches and Comanches roamed, remained an unknown area. Here the United States Army launched an extensive program for general exploration, the location of sites for new military forts, and the maintenance of another Indian-military frontier.

Before the Army's decision to open a highway into west Texas, commercially minded citizens of San Antonio raised funds by subscription for an exploring party to establish wagon-road communication with El Paso. For two decades the Anglo-Americans in east Texas had considered ways to divert the Chihuahua–Santa Fe trade in their direction and to make San Antonio the eastern terminus of the New Mexican trade route rather than St. Louis. With $800 collected during the summer of 1848, the San Antonio merchants were able to defray the expenses of a small detachment assigned to seek wagon routes between their city and Chihuahua by way of El Paso. John C. Hays, a Texas Ranger with experience as a surveyor, was placed in charge of the expedition, and he secured an escort of thirty-five Rangers under the command of Captain Samuel Highsmith.

Hays' small party of civilians, accompanied by Indian guides, marched northward from San Antonio on August 27, 1848, and went into camp on the Llano River. Here they were joined by Highsmith's men. The Ranger escort and the road party then headed westward to the headwaters of the Llano, on to the head of the Nueces, thence abruptly southward and across the Pecos River near its Rio Grande outlet. With incompetent Indian guides, the route followed was a devious one. The men suffered from lack of food and water in the inhospitable trans-Pecos country, and to prevent starvation they crossed the Rio Grande into Mexico when the Big Bend of that river was reached. Traveling up the Mexican side of the river they arrived at the village of San Carlos and, after receiving sustenance, proceeded to Presidio del Norte where apologies were made to the Mexicans for trespass. The explorers recrossed the Rio Grande into the United States and went into camp near Fort Leaton, a trading post and ranch five miles southeast of Presidio. Hays by now had been on the march for over fifty days. When his party was still too fatigued to continue to El Paso after ten days' rest, he decided to return to San Antonio.

The ineffectual wanderings on the outward journey made it desirable to seek a new and more direct return. Headed home, the Hays-Highsmith expedition moved northeast along the base of the Sierra Madre Range for 150 miles to the Pecos River. After descending this stream for 70 miles, they turned again to the northeast hoping to reach the headwaters of the Concho and San Saba rivers. This plan was quickly abandoned, however, and the group struck out toward the southeast for San Antonio.

The entire trip lasted one hundred and seven days. Both Hays and Highsmith optimistically reported to Colonel Peter H. Bell of the Texas Rangers that a road, usable at all seasons and with the necessary wood, water, and grass, could be established along their return trace.[1] This exploit was an obvious failure, and it is difficult to understand why events had not dampened the enthusiasm and optimism of its leaders.

In the winter of 1848–1849, citizens of Austin, Texas, also became interested in establishing a direct wagon road to El Paso for commercial purposes. A prominent doctor, John S. Ford, assumed leadership in organizing an exploring party of citizens. Before the final preparations were made, the United States government inaugurated its program of exploration on a far larger scale than private enterprise could hope for, and Major Robert S. Neighbors, the United States Indian agent for Texas, was sent west with the Ford party. Major General William J. Worth, responsible for military activity in Texas, instructed Neighbors to locate a practicable wagon route from Austin to El Paso for the movement of troops and supplies as well as for trading purposes. Several friendly Indians and three Austin citizens comprised his party. Their route was north and westward to the headwaters of the Concho River by way of Brady's Creek, and thence due west to the Horsehead Crossing of the Pecos River and to El Paso, where they arrived on May 2. The return journey, begun four days later, was through the

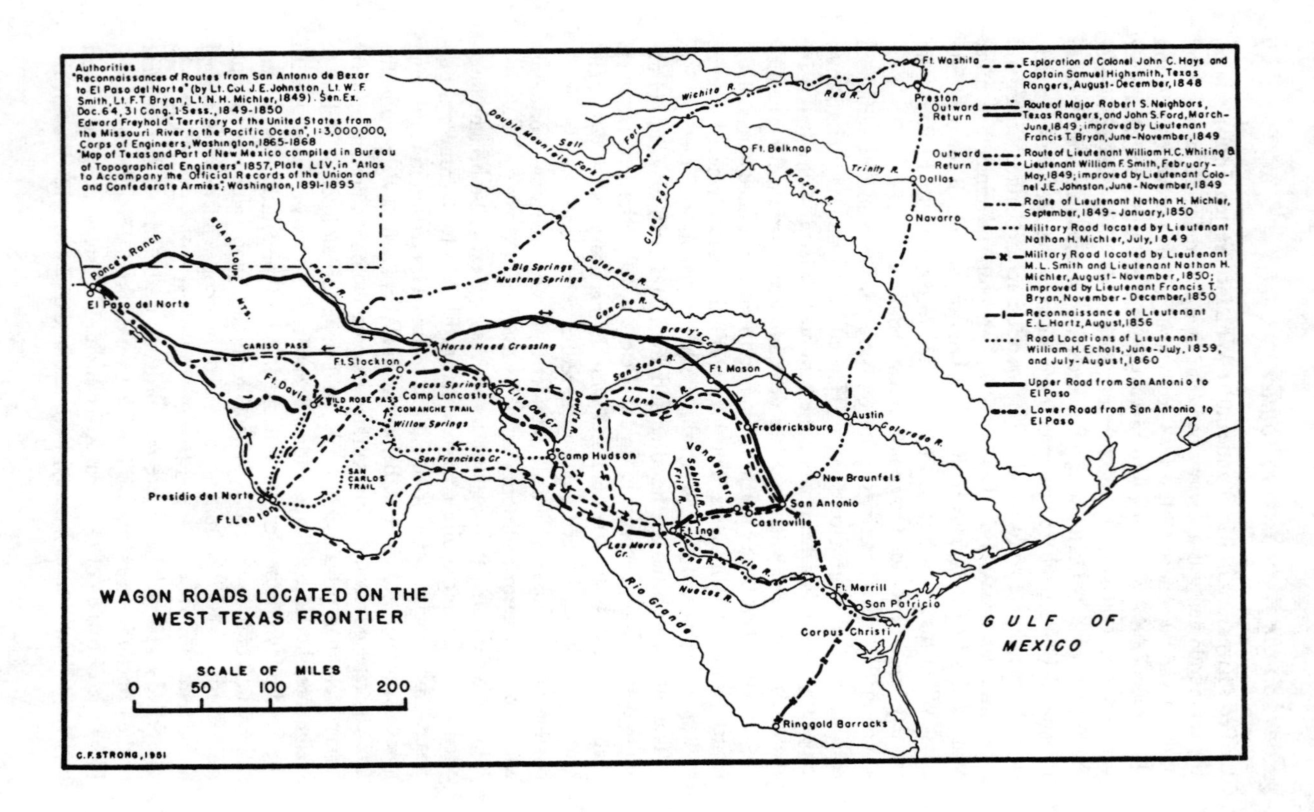

Authorities
"Reconnaissances of Routes from San Antonio de Bexar to El Paso del Norte" (by Lt. Col. J. E. Johnston, Lt. W. F. Smith, Lt. F. T Bryan, Lt. N. H. Michler, 1849). Sen. Ex. Doc. 64, 31 Cong. 1 Sess., 1849-1850
Edward Freyhold "Territory of the United States from the Missouri River to the Pacific Ocean", 1:3,000,000, Corps of Engineers, Washington, 1865-1868
"Map of Texas and Part of New Mexico compiled in Bureau of Topographical Engineers" 1857, Plate LIV, in "Atlas to Accompany the Official Records of the Union and and Confederate Armies", Washington, 1891-1895
Exploration of Colonel John C. Hays and Captain Samuel Highsmith, Texas Rangers, August-December, 1848
Outward
Return
Route of Major Robert S. Neighbors, Texas Rangers, and John S. Ford, March-June, 1849; improved by Lieutenant Francis T. Bryan, June-November, 1849
Outward
Return
Route of Lieutenant William H. C. Whiting & Lieutenant William F. Smith, February-May, 1849; improved by Lieutenant Colonel J. E. Johnston, June-November, 1849
Route of Lieutenant Nathan H. Michler, September, 1849-January, 1850
Military Road located by Lieutenant Nathan H. Michler, July, 1849
Military Road located by Lieutenant M. L. Smith and Lieutenant Nathan H. Michler, August-November, 1850; improved by Lieutenant Francis T. Bryan, November-December, 1850
Reconnaissance of Lieutenant E. L. Hartz, August, 1856
Road Locations of Lieutenant William H. Echols, June-July, 1859, and July-August, 1860
Upper Road from San Antonio to El Paso
Lower Road from San Antonio to El Paso
Ft. Washita
Preston
Red R.
Wichita R.
Double Mountain Fork
Salt Fork
Ft. Belknap
Trinity R.
Dallas
Brazos R.
Clear Fork
Navarro
Ponce's Ranch
GUADALOUPE MTS.
Pecos R.
Big Springs
Mustang Springs
Colorado R.
El Paso del Norte
Concho R.
Brady's Cr.
Horse Head Crossing
CARISO PASS
Ft. Stockton
San Saba R.
Ft. Mason
Pecos Springs
Camp Lancaster
Live Oak Cr.
Devil's R.
Llano R.
Ft. Davis
WILD ROSE PASS
COMANCHE TRAIL
Willow Springs
Austin
Colorado R.
Fredericksburg
Camp Hudson
San Francisco Cr.
Vandenburg
SAN CARLOS TRAIL
Frio R.
Sabinal R.
New Braunfels
Presidio del Norte
Ft. Leaton
San Antonio
Castroville
Ft. Inge
Las Moras Cr.
Leona R.
Frio R.
Rio Grande
Nueces R.
Ft. Merrill
San Patricio
Corpus Christi
GULF OF MEXICO
Ringgold Barracks
WAGON ROADS LOCATED ON THE WEST TEXAS FRONTIER
SCALE OF MILES
0 50 100 200
C. F. STRONG, 1961

Guadalupe Mountains to the Pecos, thence eastward via the Concho's headwaters, and finally southeast across the San Saba and Llano rivers to San Antonio. Both Neighbors and Ford were convinced that they had located an excellent road to El Paso, and without doubt they had opened up an area hitherto inaccessible to the soldier and merchant.[2] The exploration provides an excellent example of the coöperation between the United States Army and local interests desirous of improving commercial transportation.

Major General Worth, after reviewing the reports of Hays and Highsmith, ordered another survey from San Antonio to El Paso under Lieutenant William H. C. Whiting of the Engineering Corps and Lieutenant William F. Smith of the Topographical Engineers. This was the first wagon-road survey in West Texas to be sponsored exclusively by the Army and was carried out simultaneously with the Neighbors-Ford expedition during the winter of 1849. The reconnoitering party was instructed to retrace the route from San Antonio to Presidio del Norte to confirm its practicability for wagons. If, however, the route was found unsatisfactory, the men were to return from El Paso by a more direct way across the Pecos and San Saba rivers. General Worth preferred the longer and more devious way by Presidio because it paralleled more closely the international frontier and would more quickly attract settlers who could provide the military posts with supplies.[3]

Leaving San Antonio on February 12, the expedition moved northwestward to the German settlement of Fredericksburg and then to the headwaters of the south fork of the San Saba River. During the halt in Fredericksburg, Whiting, who was only twenty-four years of age and had no experience of frontier conditions, wisely secured the services of a guide who had been with the Hays' expedition the previous year. The party now totaled sixteen men, including an escort of nine trained frontiersmen. The route along the San Saba was well wooded and watered. Whiting recorded, "I have never yet seen, so far, a finer natural road."[4] However, in traversing the tableland west of the San Saba, the explorers encountered grave difficulty because of lack of water. Once they had arrived at Live Oak Creek, flowing southwest into the Pecos, they concluded the route passed over was impracticable for a wagon road unless wells could be dug to provide water. After crossing the Pecos, the party traveled southwest toward the settlement at Presidio del Norte. In this Apache country they unfortunately met one band of Indians under the control of Gómez, "the terror of Chihuahua," but managed to escape with their lives.[5] At Fort Leaton on March 24, the members of the exploration were warmly welcomed and offered the kind hospitality characteristic of that trading post. In five days the journey was resumed up the east bank of the Rio Grande. The practicability of the Presidio–El Paso road was thoroughly discussed by Lieutenants Whiting and Smith with their guides, and occasionally detachments were sent out to locate more acceptable short cuts or to determine what improvements were necessary for wagons to get through.[6] On April 12 they reached Ponce's Ranch, on the American side of the Rio Grande, opposite El Paso del Norte. Whiting wrote in his journal:

> The question of our homeward route has been much discussed among us. We all agree that, as far as known, the valley of the Rio Grande presents great difficulties, only to be overcome by time and ample means, to the passage of trains. A march for the column destined for El Paso, which should combine as little labor as possible with abundance of water and grass, is now the main object of the reconnaissance. As we take more northerly routes, the country is more open, and little or no labor is required for the immediate passage of trains; but as we go north, we lose the living water . . . General Worth was earnest in urging the great importance of the road between El Paso and Presidio, going so far as to direct me, should I find a practicable route, tc -eturn by it without crossing the country direct from El Paso . . . This not being found the case, I am in doubt. Strong reasons obtain to return via Presidio, equally strong to cross from El Paso to the Pecos; and this party, too small of itself, cannot be divided.[7]

Whiting and Smith resolved their dilemma by returning down the Rio Grande for 120 miles, along the outward route from Presidio, and then moved due east to the Pecos River. The explorers descended this stream for 60 miles in a southerly direction before crossing over to the Devil's River which they likewise descended to within a few miles of its mouth. From that point they again traveled east across the Nueces, Leona, and Frio rivers to San Antonio.[8] The distance from El Paso del Norte to San Antonio by this return, or "lower," route was 645 miles. The explorers had been gone one hundred and four days. The Neighbors-Ford return journey, accomplished during the same months, was along a route known as the "upper road" or the San Saba route. Lieutenant Smith insisted that the lower route would make a better wagon road for military and commercial purposes because water was everywhere available in sufficient quantities.[9]

In the absence of the Whiting-Smith party an epidemic of cholera had broken out in San Antonio, claiming as one of its victims Major General Worth.[10] The new departmental commander, Brigadier General William S. Harney, ordered the organization of two topographical parties for official surveys and improvements along the reconnaissances to the west. Brevet Major Jefferson Van Horn, at the time in San Antonio, was preparing to move his command of six companies of the Third Infantry to a new station in El Paso. An elaborate train of 275 wagons and 2,500 animals transporting government stores and properties of the battalion was also in his charge. Van Horn had planned to move west by way of Fredericksburg, but upon the return and favorable report of Whiting and Smith, General Harney ordered the battalion to follow the lower road.[11] The chief topographical officer of the Texas department, Lieutenant Colonel Joseph E. Johnston, was attached to the command with instructions to direct its march and to make a practical road for the provision wagons of the Army. He was accompanied by Lieutenant Smith, a civilian surveyor, and a working party of twenty laborers.[12]

The transport wagons were sent west to Fort Inge on the Leona River late in May, 1849, to await the arrival of troops. During the first week of June, the command traveled west along the well-known "Wool road" through the settlements of Castroville and Vandenburg and across the Sabinal River to the Frio. At that point the Whiting-Smith survey to El Paso had left the established road, and the topographical engineering party, following their trace, proceeded in advance of the main detachment to open the road. Captain Samuel G. French, assistant quartermaster, diligently represented his corps' interest in seeing that the route was improved so that Army supply wagons of this and future trains could travel without mishap. When improvements became necessary, large working parties from the battalion of infantry were assigned to the task. The road marked out for the Van Horn command diverged little from that of the Whiting-Smith survey. The route was changed only twice to avoid delay—once between the Devil's River and the Pecos, and again in approaching the Rio Grande. The expedition arrived in El Paso on September 8 after traveling approximately 650 miles in one hundred days.[13] This train had moved at half the speed of the earlier reconnaissances.

Captain French reported:

> El Paso, from its geographical position, presents itself as a resting place on one of the great overland routes between the seaports of the Atlantic on one side and those of the Pacific on the other.
>
> ... There can easily be established between the Atlantic States and those that have so suddenly sprung into existence in the west—and which are destined to change, perhaps, the political institutions and commercial relations of half the world—a connexion [*sic*] that will strengthen the bonds of union by free and constant intercourse. The government has been the pioneer in the enterprise ...[14]

Having completed his escort duty, Lieutenant Colonel Johnston set out on the return march along the northern route with the hope of examining the country between the Rio Grande and the headwaters of the Colorado and Brazos rivers. A train of twenty-five wagons bound for San Antonio accompanied his party of twenty-five men. Severe weather, however, forced the detachment to turn south when the Pecos Valley was reached, and the men returned by the lower road to San Antonio, arriving on November 23. Johnston was dissatisfied with the section of the lower road along the Devil's River. At the river's head he had decided, therefore, to travel east with the civilian engineer and ten men to seek out a smoother and shorter road, hoping to strike the well-traveled Wool road near Vandenburg. The 60-mile stretch between the Devil's River and the Nueces River was found to be entirely without water though the terrain was practicable for wagons. Johnston reported that the lower road, as originally located, was preferable. There was an abundance of grass and fuel, and water was available, though in limited quantities. Travelers had only to plan their itineraries carefully, he contended, and they would have no difficulty in going west.[15]

While the chief topographical officer of the department tested the practicability of the southern road for the wagons of Major Van Horn's command, Lieutenant Francis T. Bryan, another topographical engineer, made a similar inspection of the upper route. With thirty men and several Army supply wagons, he left San Antonio for Fredericksburg. From that settlement his route was across the Llano, the San Saba, and the Concho rivers to the headwaters of the latter stream. Up to this point the detachment made no improvements, other than cutting down and smoothing the river banks at stream crossings. Relative to the country between the San Saba and the Concho, Bryan reported: "A road can be easily made here by removing the loose stones from before the wagons—no other labor being necessary, except clearing away weeds and bushes whenever a stream is to be passed. Spades to cut away the banks have not been used since we crossed the San Saba."[16] Between the head of the Concho and the Pecos River, Bryan's men came upon a well-marked trail made by the wagon wheels of California emigrant parties, and so the Army wagons came through with little difficulty. The Pecos was forded at Horsehead Crossing. As they moved westward through the Guadalupe Mountains, the road surveyors utilized several water wells dug by California parties. El Paso was reached on July 29, after a six-weeks' march. Bryan reported that the upper route presented no obstructions to the movement of wagon trains. Essential water and grass could be located at least every 25 miles, except the stretch between the head of the Concho and the Pecos.[17]

Within only a few months in 1849 the United States Army had confirmed the desirability and practicability of two new wagon roads across west

Texas that were to remain main arteries of travel until the present time. The value of each was emphasized by its immediate use by the California Argonauts, headed for the gold fields. Soldiers and wagon trains of the United States Army traversed both highways. They were used by Texas cattle drovers pushing herds westward and by settlers migrating to New Mexico, Arizona, and California with their wagons.[18] Present paved highways into trans-Pecos Texas mainly follow the two surveys improved by the Army. The tracks of the Texas Pacific Railroad parallel in many places the northern wagon road and the Southern Pacific follows, in part, the lower road.

An additional major interest of the Army in the Southwest was the establishment of wagon-road communications from the western boundary of Arkansas to the Rio Grande Valley below Santa Fe. Nathan H. Michler, a young second lieutenant of the Topographical Corps stationed in San Antonio, was sent north to Fort Washita on the False Washita River, 30 miles above its junction with the Red River. Here he was to begin a survey to the Pecos River. With fourteen civilian laborers, Michler moved his wagon train to the north along the western line of Texas settlements, passing through Austin, New Braunsfels, Navarro, Dallas, and Preston. The country was well settled for 100 miles north of San Antonio, but after that farms were encountered only at intervals of ten to fifteen miles. No major difficulty was encountered in getting the wagons through. Lieutenant Michler, like other young officers of the Corps, thought this the best test of the practicability of a road.[19] His way mainly coincided with route 81 of the present federal highway system.

Constant rumors of Indian hostility to the west had been reported to this survey party on its northern march. At the outset the command had been ordered to retrace its steps if the serious hostility of the Indian tribes endangered the lives of the detachment. Under the circumstances, at Fort Washita Michler increased his party to twenty-one men, prepared for any eventuality, and loaded four wagons with provisions for two and one-half months. Captain Marcy arrived at the fort two days before Michler's departure and provided detailed information about the country just crossed. The lieutenant unsuccessfully attempted to secure the services of Marcy's Delaware Indian guide, Black Beaver. Leaving Fort Washita on November 9, the reconnaissance party turned southwest, cutting a road for their wagons through the Upper Cross Timbers as far as the Red River. They continued along the north bank of the river, not crossing until they reached the Wichita River near the present Wichita Falls, Texas. The expedition next moved southwesterly across the Brazos and along the double mountain fork of that stream. Forty-five miles beyond the Brazos they came upon Marcy's Trail which they followed across the divide to the Colorado River. Passing the Big Springs of the Colorado and Mustang Springs, the group struck the Pecos River about 40 miles above the Horsehead Crossing on December 30. Here the reconnaissance ended, 492 miles from Fort Washita. Although the engineering party had encountered several bands of Shawnee, Delaware, and Comanche, they had not found them hostile. Lieutenant Michler reported: "From this examination we may conclude that, for the distance passed over, a more advantageous country for roads of any kind cannot be found..."[20] The most difficult part was the sandhill section between Big Springs and the Pecos, where the wagons tended to get stuck and where water and wood were scarce. The wagons had gone through, however, with unusually little trouble. This reconnaissance served to confirm the report of Captain Marcy who insisted that a satisfactory wagon road could be found from the Arkansas border to the Rio Grande below Santa Fe. So a more direct route connecting with Cooke's road to California at the Rio Grande was a second time traversed by an Army detachment. It was now a certainty that emigrants, traders, and soldiers going through to California no longer needed to follow the long road westward along the Canadian River to Santa Fe and thence southward along the Rio Grande Valley to the trails through southern New Mexico.

Lieutenant Michler's command experienced a severe snow storm in the Pecos Valley which resulted in the loss of several mules and necessitated leaving two wagons behind. The return march was along the upper road surveyed by Lieutenant Bryan, and which was now used extensively by emigrant trains. Once more a topographical officer reported that the only undesirable stretch was the unwatered section from the Pecos to the Concho headwaters.[21]

Another important phase of the Army road-building program in Texas was the construction of military roads connecting the frontier posts, or one of these posts with near-by settlements. The primary and sometimes exclusive purpose of these roads was the movement of troops and military supplies. In the 1850's Texas had within her borders the greatest number of forts and military camps in the entire southwest. The most active military frontier in the West was in the state. During the summer of 1849, Lieutenant Michler was ordered to make a reconnaissance from Corpus Christi on the coast of the Gulf of Mexico to Fort Inge on the Leona River to determine the practicability of opening a wagon road. His route northwest from Corpus Christi followed the west bank of the Nueces to the mouth of the Frio. Here he crossed over and followed the west bank of the Frio to the mouth of the Leona, and thence up the east side of that stream to the fort. The engineer was convinced that a satisfactory road for all modes of conveyance could be constructed along the 217-mile reconnaissance. Timber cutting along the waterways would be the only labor required.[22]

A year later the Army officers in Texas decided to establish communications between San Antonio and the Ringgold Barracks on the Rio Grande River near Rio Grande City. Lieutenant Michler with the aid of another topographical officer, Martin L. Smith, examined the route between these military posts, via Fort Merrill, from October to November and made favorable reports on the road-building possibilities. Upon their return Lieutenant Bryan made a survey of the road usually traveled by Army trains between San Antonio and Fort Merrill. Discovering that the road was very crooked and swung too far to the east, he returned to San Antonio over a more direct and westerly course and reduced the travel time by 20 per cent. He reported that with a few days of labor, this new road could be made practicable for wagons.[23] Among the many young officers of the Topographical Corps assigned to the Texas department, Bryan was recognized by his Army superiors as one of the most successful road surveyors. He improved the important military road northwest of Austin to Fort Mason and was engaged on many reconnaissances from San Antonio to Fort Belknap on the Brazos River in the northern part of the state.[24]

The Army's military road-building program in Texas continued until the eve of the Civil War. As the decade of the 'fifties advanced, the essential communication network for central Texas was completed, and emphasis was transferred to the improvements in far west Texas. Fort Davis became the primary military establishment in the trans-Pecos country. In the summer of 1856, Lieutenant Edward L. Hartz of the Eighth Infantry was dispatched from the fort to seek a shorter and more satisfactory road to El Paso. With two wagons and a party of twenty-seven enlisted men, he planned to strike out due west, but soon found that his route had to be a winding one to locate water. Only with difficulties did his wagons scale the mountain passes, and when he struck the El Paso road in the Rio Grande Valley there was some debate as to whether the more direct route was truly practicable.[25] The advantages over the lower road between the fort and the river were negligible.

During the spring and summer of 1859, the United States Army conducted a camel experiment in west Texas.[26] Pursuant to orders from the Secretary of War, Major General David M. Twiggs of the Texas Military Department, ordered a thorough reconnaissance of the country between the Pecos and Rio Grande rivers for the purpose of locating another fort on the Rio Grande somewhere in the Big Bend country. After choosing the site, an expedition, directed by the department's topographical officers, was to seek the most satisfactory routes for supplying this future post, and by using camels as carriers to determine their success for military transport. Lieutenant William H. Echols organized the expedition at Camp Hudson on the Devil's River. Lieutenant Hartz, an experienced explorer, commanded the escort and served as quartermaster. Twenty-four government camels were brought west from Camp Verde on the Guadalupe River for the trial. To replace the usual supply wagons, the animals, together with a few mules, were loaded with packs varying from 300 to 500 pounds. The caravan's route from Camp Hudson was westward across the Pecos to Fort Davis. Here an infantry company joined the expedition which headed eastward for Camp Stockton, a supply base. On the march efforts were made to improve the wagon road between the two military outposts. Turning south from Camp Stockton, this command of fifty men spent a month in a comparatively unsuccessful reconnaissance of the Big Bend area.[27]

The next summer Lieutenant Echols again conducted a survey to ascertain the practicability of opening roads in the far west part of the state. His engineering detachment had an escort of thirty-one infantrymen and a pack train of twenty camels and fifteen mules. The starting point was again at Camp Hudson. Departing on June 24, this large expedition was to seek a shorter and more desirable route from the Pecos River to Fort Davis. Their route from the Pecos was westward for about 120 miles to San Francisco Creek over an unusually rugged, desolate, and dry area, unsatisfactory for any type of communication. Without the water supply carried by the camels, the explorers would have known untold suffering and possible death. At San Francisco Creek, the route turned northwest toward Fort Davis and though the terrain was smoother, permanent water supplies were no more available. Having failed in their road-locating assignment, the Army explorers turned south from Fort Davis once again to find a satisfactory site for a new fort on the Rio Grande. From Presidio del Norte, the command followed the Rio Grande to Comanche Crossing in the Big Bend of the river where the most widely used Indian trail crossed into Mexico. Twenty miles down stream the desirable site for the fort was found. The topographers returned to Presidio. From here they sought out a route northward to Camp Stockton, chiefly along the established Indian traces which, with a moderate amount of work, would constitute a good wagon road whereby the proposed fort might be supplied. The survey for another military road was thus completed.[28]

The United States Army made a major contribution toward opening west Texas between the Mexican and Civil wars. By three major wagon roads the way was prepared for the emigrant headed to the Pacific, the land-hungry settler desiring to take up land on the frontier, and the commercially minded merchant transporting his supplies. By forming a junction in the Rio Grande Valley with the routes through the Mexican cession, a continuous line of communication was established across the southern part of the trans-Mississippi West.

CHAPTER IV : *Minnesota Territory, an Experimental Laboratory, 1850-1861*

WHEN MINNESOTA became a territory in 1849, the population of four to five thousand was concentrated in its east central part at the junctions of the Minnesota, Mississippi, and St. Croix rivers. The most flourishing community, St. Paul, boasted eight hundred and forty inhabitants; next in size were Stillwater and St. Anthony. A sizable settlement known as Pembina was on the Red River of the North near the international boundary, but elsewhere west of the Mississippi settlements were few and small because the Sioux Indians had not yet ceded the lands.[1]

This new territory, large in area but sparse in population, had a framework of roads and trails upon which to build, but it was far from impressive. In the early years of the century, a road from Fort William in Canada had been built to Grand Portage on Lake Superior in what became the northeastern tip of the territory. The first roads serving the river towns ran from St. Paul to Mendota, where a ferry crossed the river to Fort Snelling; from St. Paul to Stillwater; and from St. Paul to St. Anthony and up the east bank of the Mississippi to Sauk Rapids and Crow Wing.[2] The road from St. Paul to Stillwater had been improved in 1848 and extended southward along the Mississippi Valley to Prairie du Chien, Wisconsin, and to Galena, Illinois. At these points, connections could be made with other roads, built by the United States Army, leading to Green Bay, Milwaukee, and Chicago.[3]

The roads of preterritorial Minnesota were, for the most part, trails which had been followed by traders, trappers, and Indians. The only work done upon them was removing the timber and brush, and they were quickly abandoned if they became rutted or full of mudholes or when a shorter route was discovered. The most important trails ran from the Red River Valley to the Minnesota–Mississippi River region. Two of these trails from the northwest joined the St. Paul–Crow Wing road along the east bank of the Mississippi at the Sauk Rapids and Crow Wing crossings.

Another followed the Minnesota River to its headwaters and thence northward up the Red River Valley.[4] Below the Mississippi and Minnesota rivers the only penetration of the prairie was a rough trace worn by the carts of Alexander Faribault in his traffic of furs and goods with the Indians.[5] The transportation system was inadequate even for a thinly settled area in which population was increasing. Most of the trails and roads were impossible for wagon traffic; few, if any, bridges and drainage culverts had been constructed.

An important reason in the agitation for the creation of Minnesota Territory was the hope that the federal government might be persuaded to launch a road-building program under the guise of frontier defense.[6] The navigable rivers, which had played the most significant role in early settlement and communications, were usually icebound four or five months during the winter and too low in summer for steamboat navigation. New roads were needed to improve mail facilities and to insure the arrival of supplies from the south.[7] The lumbering interests sought an outlet for their timber by road, and some insisted that immigration would be facilitated by more adequate transportation west.[8] When Henry Hastings Sibley was sent to Congress in 1848 his first task was to obtain territorial government for Minnesota, his second to secure congressional appropriations for road construction.

Sibley returned to Minnesota in the spring of 1849, having succeeded in the establishment of the territory but without gaining federal support for roads. Through his efforts a bill had been introduced in both the Senate and the House of Representatives to appropriate $12,000 for a road from St. Paul and Point Douglas, at the junction of the St. Croix and the Mississippi, northward through Marine Falls and Stillwater to the St. Louis River. Pressure of other business prevented its serious consideration.[9] At the first territorial election in August, Sibley was unanimously reëlected to return to Congress, and Alexander Ramsey, a newcomer to the territory, was chosen governor. The following month when the legislature assembled, Governor Ramsey's address on the essential needs of the territory emphasized the importance of road improvements by the central government. He proposed specific constructions and the legislature, in turn, drafted a series of memorials to Congress elaborating the recommendations of the governor.[10]

At the 1849–1850 session of Congress, Sibley was more fully prepared to press for congressional aid. Supported by the memorials of the legislature and encouraged by letters from home, he introduced a bill for the construction of five Minnesota roads.[11] The proposal received a favorable report from the House committee on roads.

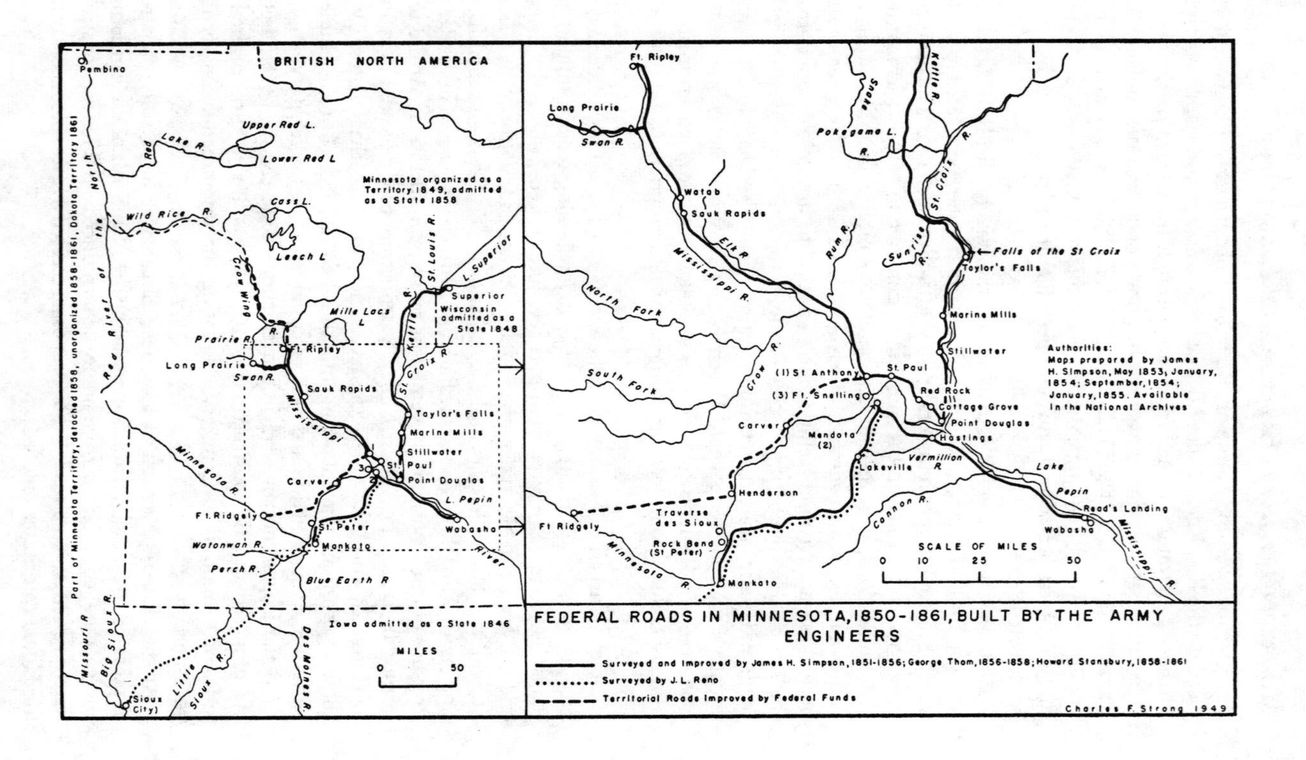
BRITISH NORTH AMERICA
Pembino
Part of Minnesota Territory, detached 1858, unorganized 1858-1861, Dakota Territory 1861
Upper Red L.
Lower Red L
Red Lake R.
North
Red River of the
Minnesota organized as a Territory 1849, admitted as a State 1858
Cass L.
Leech L
Wild Rice R.
Crow Wing R.
Mille Lacs L
St. Louis R.
L. Superior
Superior
Wisconsin admitted as a State 1848
Kettle R.
St. Croix R.
Prairie R.
Ft. Ripley
Long Prairie
Swan R.
Sauk Rapids
Mississippi
Taylor's Falls
Marine Mills
Stillwater
St. Paul
Point Douglas
L. Pepin
Wabasha
River
Carver
Minnesota R.
Ft. Ridgely
St. Peter
Watonwan R.
Mankato
Perch R.
Blue Earth R
Iowa admitted as a State 1846
MILES
0
50
Missouri R.
Big Sioux R.
Little Sioux R.
Des Moines R.
Sioux City)
Ft. Ripley
Long Prairie
Swan R.
Watab
Sauk Rapids
Elk R
Mississippi R.
Snake
Pokegama L.
R.
Kettle R.
St. Croix R.
Sunrise R.
Falls of the St Croix
Taylor's Falls
Marine Mills
Stillwater
Rum R.
North Fork
South Fork
Crow R.
(1) St Anthony
St. Paul
(3) Ft. Snelling
Red Rock
Cottage Grove
Point Douglas
Carver
Mendota (2)
Hastings
Vermillion R.
Lakeville
Lake Pepin
Henderson
Ft Ridgely
Traverse des Sioux
Rock Bend (St Peter)
Minnesota R.
Mankato
Cannon R.
Read's Landing
Wabasha
Mississippi R.
Authorities: Maps prepared by James H. Simpson, May 1853, January, 1854; September, 1854; January, 1855. Available in the National Archives
SCALE OF MILES
0 10 25 50
FEDERAL ROADS IN MINNESOTA, 1850-1861, BUILT BY THE ARMY ENGINEERS
Surveyed and Improved by James H. Simpson, 1851-1856; George Thom, 1856-1858; Howard Stansbury, 1858-1861
Surveyed by J. L. Reno
Territorial Roads Improved by Federal Funds
Charles F. Strong 1949

That same fostering care which has always been extended to the new Territories of the country may, in the opinion of the committee, well be manifested towards Minnesota, in opening and improving such thoroughfares as may be necessary for her protection, and useful in advancing her settlements. Such a policy will not only conduce to the general interest and welfare of the settlers, but will increase the value and sale of public lands to the benefit of the government.[12]

When the House discussed the proposed legislation, George W. Jones of Tennessee, an opponent of internal improvements at federal expense, suggested that the people in the territories could lay out and make their own roads as well as the people of the states. Sibley appealed to precedent:

...It has been the uniform practice of Congress to aid the organized territories, by appropriations of money for the construction of roads.... The roads asked for are to be the great thoroughfares of the country, some of them lead to your military posts and your Indian agencies, and this Government, by constructing them, will, in a very few years, save more than the sums asked for, by the consequent diminution of the cost of transporting military stores and supplies, and goods and provisions for annuities under treaty stipulations.

The House was reminded that Wisconsin had received $104,000 for roads and harbors between 1835 and 1845, and asked if Minnesota was to be less liberally treated. The House passed the bill on May 29.[13] Sibley reported developments to Ramsey in St. Paul mentioning that by "long and persevering *electioneering*" he had overcome the opposition of a "small but determined minority composed of the ultra-strict constructionists."[14] In the Senate, the bill was strongly defended by Stephen A. Douglas of Illinois and Augustus C. Dodge of Iowa against those who questioned the legislation because no War Department sanction accompanied it. Douglas realized that the initial appropriation of $40,000 was pitifully small and just a beginning. To minimize the fears of the opposition, he suggested that the appropriations were only "...to cut the timber through the woods, so as to mark the course of the roads. Once in a while a bridge will have to be built across a stream. It is not expected to make good roads at all."[15]

The Minnesota Road Act as it was approved on July 18, 1850, called for the construction of four roads and the survey of a fifth under the direction of the Secretary of War, pursuant to contracts that he might make. The sum of $15,000 was granted for one road which extended northward about 175 miles from Point Douglas, at the junction of the St. Croix and Mississippi rivers, through the St. Croix Valley by way of its major settlements to the falls or rapids of the St. Louis River. Under an appropriation of $10,000 a second road was to be constructed from Point Douglas northwest along the Mississippi for a distance of 145 miles, via Cottage Grove, Red Rock, St. Paul, and St. Anthony to Fort Gaines—later named Fort Ripley. With $5,000 a branch was to be located westward from this road to the Winnebago Indian Agency at Long Prairie. A fourth grant provided $5,000 for a road along the west side of the Mississippi River between Mendota and Wabasha. Finally, an allotment of $5,000 was made for surveying a military road from Mendota to the mouth of the Big Sioux River on the Missouri.[16] The roads were patterned wheellike, with the present Twin Cities area as the hub. Arthur J. Larsen has suggested that "one spoke reached northward to the Great Lakes; another, northwestward toward the Red River settlements; one extended southwest toward Missouri; and another to the southeast to Iowa."[17] Upon his return to the territory, delegate Sibley summarized his accomplishments before his constituents listing this road legislation as a major success.

> The sums allowed for the construction of roads between important and distant points in our Territory, although perhaps not sufficient to complete them, will go far towards opening the country to immigrants, and will prove of incalculable benefit, even on that score alone. And we may reasonably rely upon the liberality of Congress to supply any deficiency hereafter, which may operate to prevent the immediate completion of these great thoroughfares. . . .[18]

Colonel Abert, Topographical Engineering Chief, likewise concluded that "these several appropriations can, of course, have contemplated only the necessary preliminary operations and a limited portion of work."[19] Since the law did not prescribe the type of road required, the colonel assumed, for convenience, that a "country road" was intended. Bridges would be built where required, swamps and marshes made passable, trees felled and undergrowth removed, and drainage ditches constructed at the roadside in flat areas. The first duty of the Corps, he reported to the Secretary of War, would be to make preliminary surveys to determine the nature and size of contracts to be made. To continue operations through June, 1852, he recommended an additional appropriation of $70,000 for road construction and $5,000 for the survey of the military road to the Missouri River.[20]

Unable to assign an officer of the Topographical Corps to immediate duty in Minnesota, Abert employed John S. Potter, a civilian engineer, to take charge of all surveying and construction. Potter was directed first to survey the route from Mendota to Wabasha in the belief that it was all that could be accomplished before winter came. During November and December, 1850, the job was done and maps and estimates forwarded to the Bureau.[21] Many settlers in the St. Croix Valley, believing an early construction of the Point Douglas to St. Louis River route to be far more imperative than the Wabasha-Mendota road through unsettled Indian country, protested to Sibley against Potter's initial action. Sibley notified Colonel Abert of the agitation, and in March, 1851, Potter was ordered to direct his attention to a survey of the Point Douglas–Lake Superior route.[22]

During the winter session of Congress, Sibley worked to get additional funds for the Minnesota roads. In agreement with Colonel Abert's estimate of $75,000 necessary to continue the road program through June, 1852, this sum was included in the general appropriation bill of the Army by the House Ways and Means Committee. The Minnesota delegate urged its passage, but the item was subsequently cut out during House moves to reduce appropriations for that branch of service. In the meantime, the Minnesota legislature adopted a resolution, forwarded to the Secretary of War, requesting the employment of additional engineers to speed the federal road improvements. Governor Ramsey and Sibley both joined the pressure group to spur the War Department to greater effort.[23] Lieutenant James H. Simpson, who had just completed his Fort Smith–Santa Fe reconnaissance, was ordered to the territory to organize a second surveying party and, at the same time, to supervise all the surveys and construction.[24]

After a personal reconnaissance of the roads he deemed most important, Lieutenant Simpson dispatched surveying parties to make the actual location along the lines of his reconnaissance. During the summer of 1851 this preliminary work was completed on the roads from Point Douglas to the St. Louis River, from Point Douglas to Fort Gaines, and from the Mississippi River to the Winnebago Agency at Long Prairie.[25] The anxiety of Minnesota settlers over the construction of the roads, however, was not alleviated by Lieutenant Simpson's initial activities. The *St. Anthony Express* of June 7, 1851, denounced the failure of the Army to complete any construction on the Fort Gaines road and expressed the hope that the federal appropriations would not be exhausted in paying Army officers and their assistants for useless surveys. The federal government's policy of sending out Army officers unfamiliar with the terrain to make surveys when there were men in the territory who could build roads without such preliminaries was condemned as a wasteful expenditure.[26] The federal project was simultaneously attacked in a grand jury report after an investigation of the activities of the Army road builders.

In his first annual report to the Bureau, Simpson presented a general picture of the country through which the roads north of Point Douglas were to pass, the probable length of each road, the estimate of construction costs, and recommendations for future work. He was not greatly interested in the immediate need of constructing the Mendota-Wabasha road or the surveying of the military road because they passed through Indian country as yet unceded and not heavily occupied. About the remainder he wrote:

> The other three roads . . . are now of the utmost consequence, and should be made available at the earliest possible moment. The Point Douglas and Fort Gaines road runs through a portion of the territory to which emigrants are flocking in great numbers. It is the great highway by which the Government supplies reach the Indians in the Winnebago territory and in the Chippewa district. It is also the road by which the Government supplies are all transported to the troops at Fort Gaines (Fort Ripley). The Mississippi and Long Prairie road is the branch road by which the Government supplies are conveyed to the Winnebago agency at Long Prairie. The Point Douglas and St. Louis River road is of utmost consequence in the accommodation it will afford to the lumbering interests, high upon the St. Croix, upon the Snake River and Kettle River; this road being the only avenue, especially in the winter, by which supplies can be transported to these points for the maintenance of those engaged in the trade.[27]

While Lieutenant Simpson and his civilian assistants utilized the cold winter months, when no field work was possible, to prepare maps and estimates for use the following season, delegate Sibley renewed his campaign in Washington for additional appropriations. Through the adoption of a House resolution, the War Department was requested to reveal the progress of road construction in Minnesota, the amounts of the previous appropriation used, and the sums necessary to complete the roads.[28] Colonel Abert reported that no construction had been made because the necessary preliminary surveys had cost $12,089. He now presented an estimate of $100,000 to carry the work forward through June, 1853.[29]

Sibley's bill called for an appropriation of $45,000. When taken to task for an explanation of this variance from the Army's recommendation, the delegate frankly took full responsibility for the reduction and intimated that he was guided by practical politics. A larger sum, he did not feel, could be expended beneficially during a single year. Moreover, the Army estimates were based upon experience in Wisconsin where the country was more heavily timbered. Under additional questioning, Sibley finally admitted the $45,000 was insufficient to complete the roads but would render them passable between the points where they were most needed.[30]

Much opposition to the measure developed in the House. George S. Houston of Alabama led the attack by announcing that it was both premature and inadvisable to make new appropriations before the balance of the sums allotted in the 1850 act had been exhausted. Sibley's reply that contracts had been let to the limit of the appropriation and that the Army officer in charge had inquired whether more funds would be available only served to provoke Houston into a tirade against internal improvements in the territories at federal expense. Sibley assured the House that the Army considered these expenditures necessary for the transportation needs of the federal government. He also expressed frontier philosophy:

> The Government, being the sole great land proprietor in the Territories, is bound, by every consideration of equity and justice, to make its domain accessible to the settler, by means of roads. To do otherwise, would be to abandon the policy hitherto pursued towards all your Territories. How, sir, can your lands be sold if the immigrant can not reach them?

James Brooks of New York interrupted to insist that the only constitutional justification for federal road construction was military necessity, primarily for the movement of troops to combat the enemy. If justifications other than common defense were accepted, the government should build roads in the states as well as the territories. Abraham W. Venable of North Carolina supported this view and scoffingly inquired if it was good policy, after giving land to the landless, to build roads to the land for them. The majority of the House agreed with the contention of David L. Seymour of New York that the territories had every right to look to Congress for financial aid. Although the bill passed the House of Representatives, the Senate failed to reach a vote before adjournment. Twice Sibley's efforts to obtain a second appropriation for Minnesota roads had failed.[31]

In the meantime, Lieutenant Simpson continued to make maps, cost estimates, and construction specifications with the expectation of resuming spring operations under a new federal grant. His winter's work was complicated by protests from local groups about the location of the roads. The engineer had intended to improve an Indian trail along the north side of the Swan River to the Winnebago Agency. Settlers in the Sauk Rapids vicinity explored and blazed a shorter route to the agency, and requested construction along it. A winter reconnaissance only served to confirm Simpson's original judgment.[32] St. Croix River residents forwarded a petition to Congress, through Sibley, asking that Simpson be allowed to run the St. Louis River road through Bowle's Mill rather than through Cottage Grove as specified in the 1850 law.[33] Another group petitioned the War Department against the lieutenant's location of the road just north of Stillwater and persuaded Abert to suggest the postponement of all work on this section until a resurvey had been made.[34] No local opposition was raised to the survey of the Fort Gaines road, and during the winter construction bids were invited for several sections. Simpson considered the possibility of initiating construction on the Mendota-Wabasha road, but the remaining funds were too limited. His sudden interest in this road resulted from the ratification of the Treaty of Traverse des Sioux whereby the Sioux yielded lands west of the Mississippi to the United States. He urged Sibley to seek congressional funds to connect St. Paul with the road's terminus at Mendota by a bridge across the Mississippi.[35] When spring arrived, Simpson became impatient because of the limited work he could do with the remaining funds at his disposal and sought authorization from the Topographical Bureau to advertise for construction bids in anticipation of congressional appropriations. This he was forbidden to do.[36] When Congress adjourned without making money available, Abert announced in his annual report that "as the estimates of further progress on these roads have not met the approbation of Congress, now twice submitted to its consideration, work on these roads will be closed, as soon as the existing small appropriations are exhausted."[37] Plans were made for the transfer of Lieutenant Simpson.

The Army officer made his position additionally difficult by becoming involved in Minnesota politics. Simpson had a great deal of admiration for Sibley, the leader of one political faction in the territory, and the two men worked in complete harmony when Sibley was congressional delegate. An opposition faction, organized as the so-called Democratic machine, was led by Henry M. Rice, a New Englander who lobbied informally in Washington for territorial interests as successfully as the delegate. Early Minnesota politics were largely personal, and major issues involved the promotion of competitive economic interests. Both leaders had early associations with the American Fur Company. Sibley became procompany, but Rice bitterly opposed it, and their local parties soon assumed the titles of "fur" and "anti-fur." In March, 1852, at a meeting in St. Paul to discuss the congressional delay in ratifying the Sioux treaties, one William Hollinshead, a brother-in-law of Rice, accused Sibley of working for the defeat of these treaties to further his own promotions. Lieutenant Simpson, in attendance, seized this opportunity to defend Sibley and attack the motives of Hollinshead, whom he believed to be the author of the grand jury report denouncing the surveys by federal agents "as a humbug, and a gross misapplication of the funds of the government." This incident marked the beginning of a quarrel between Rice and Simpson that was to last as long as Simpson remained in the territory.[38]

When the new session of Congress convened in December, 1852, the Senate renewed consideration of the appropriation bill for Minnesota roads and speedily voted approval. On January 7, 1853, President Fillmore signed the bill allotting $45,000 for these public works, and the position of the

Army engineer and the future federal road picture immediately improved. Specifically the law designated $5,000 for surveying and laying out the military road, $20,000 for construction on the St. Louis River road, $10,000 for the Fort Ripley road, $5,000 for the Long Prairie road, and $5,000 for the Mendota-Wabasha road.[39]

Under the stimulus of new funds, contracts were awarded in the spring to local settlers. Realizing that the new appropriations were still inadequate to complete the highways, Simpson continued to follow the established pattern of using the appropriations in areas most urgently needing improved transportation. This meant that the work was piecemeal and highly selective. Instead of hiring one contractor for each road, several men were employed for special jobs, and at one time there were apt to be two or three independent construction groups working on each road. At no time was a contract made when it was believed that the work would continue beyond the fall. By the middle of May, 1853, nine contracts were awarded on such a basis, and construction was underway on the St. Louis River road, the Fort Ripley road, and the Long Prairie road. At the close of the working season the road from Point Douglas to the St. Louis River was completed and in traveling condition from Stillwater to a point 12 miles beyond the falls of the St. Croix, a total distance of 43 miles. The road to Fort Ripley was passable its entire length. Greater energies had also been directed to the Mendota-Wabasha route by its resurvey and the granting of construction contracts at its northern and southern extremities.[40]

While the road program progressed during the spring and summer of 1853, Simpson's political enemies were not idle. In April Rice and others wrote to Colonel Abert stating that they understood Simpson was to be removed for duty elsewhere and therefore recommended the appointment of Charles L. Emerson, a civilian engineer, as superintendent of government road construction. Emerson, Simpson's brother-in-law, had been brought to the territory by the topographical officer, but friction developed between the two men. The discord was heightened by Emerson's growing friendship with Rice. Members of the Rice faction began circulating petitions for Simpson's removal on the basis of incompetence and forwarded them to Washington. Abert steadfastly supported his junior officer who had been promoted to the rank of captain for fourteen years of meritorious service. During the summer, Rice's followers continued the attack upon Simpson in the columns of the *Minnesota Democrat.* The captain became further embroiled in politics by defending his record and berating the Rice group in the St. Paul *Minnesotian.*[41]

In his annual report for 1853, Simpson had made no mention of the

Mendota–Big Sioux military route. The original appropriation had been made in 1850, and when two working seasons passed without any survey, the settlers along the route began to manifest great anxiety. Some twenty thousand emigrants had moved to lands west of the Mississippi before the ratification of the Sioux treaties ceding this land. In the spring of 1852 several of the squatters decided to take the initiative in building a road to St. Paul if the agents of the federal government remained dilatory. Captain William B. Dodd of Traverse des Sioux and Auguste L. Larpenteur of St. Paul solicited funds to organize a road-building party with one surveyor and ten laborers which, during the spring of 1853, cut out a road through the woods from Rock Bend, or St. Peter as it is known today, to St. Paul along the ridge separating the drainage basin of the Minnesota and Cannon rivers. This was very little more than a rough trail, 65 miles long, but it made possible communication with St. Paul throughout the year. The efforts of these pioneers anticipated the beginning of government work by a few weeks. Captain Simpson had made inquiries about the possibility of obtaining supplies at the mouth of the Big Sioux River. Colonel Abert recognized the importance of making the military survey as soon as possible but knew also that Simpson had his hands full with work on the other four Minnesota roads. In May, 1853, he ordered Captain Jesse L. Reno to begin an immediate survey of the route.[42]

Reno's party, organized in St. Louis, ascended the Missouri by steamboat to Council Bluffs and then marched overland to the mouth of the Big Sioux. From the present site of Sioux City, Iowa, the surveying party traveled northeasterly across the Little Sioux and Des Moines rivers. After crossing the present Iowa-Minnesota boundary near Elm Creek, a tributary of the Des Moines, the detachment moved northward to Watonwan River. Descending this stream to its junction with the Blue Earth and thence up the valley of the latter stream to the Minnesota, they reached Mankato on July 25. From this settlement Reno led his party down the eastern bank of the Minnesota River for about 11 miles until he struck the Dodd road. With minor exceptions, this newly cut road through the "Big Woods" was followed to Mendota where the detachment arrived August 20. Reno's men had made only the improvements necessary to get the Army wagon train through. The captain computed the distance of his survey at 279 miles: 224 through prairie country, 40 through thick woods, and 15 through oak openings. He estimated the cost of constructing this road at approximately $52,000. Improvements on the Mendota-Mankato section were urgently recommended because the Minnesota River provided the only outlet to the southwestern part of the territory and it was not navigable for several months in the year. Not only was the road needed for the prosperity of the farmers, but it would provide more direct communication from Fort Snelling on the Mississippi to the newly established Fort Ridgely on the Minnesota.[43]

The year 1853 marked a turning point in the history of military roads in Minnesota. Important construction work had been accomplished on four roads contemplated in the program Sibley submitted to Congress in 1849, and the fifth project, the military road to the Big Sioux's mouth, had been surveyed so that the Bureau of Topographical Engineers could make plans for its construction. The year also marked an important change in the Minnesota political situation that would adversely affect the road work of the Topographical Engineers and particularly the efforts of Captain Simpson. Henry H. Sibley retired as congressional delegate, and Henry M. Rice assumed the position in December, 1853. He served as delegate throughout the remainder of the territorial period. At no time during the years in which Simpson was in charge of the government roads in the territory did Rice lessen his attack on the captain. Early in 1854 Simpson dismissed Charles L. Emerson, ostensibly for economy reasons, and the latter was immediately named editor of Rice's political organ, the *Minnesota Democrat*.

In spite of the increasing animosity between Rice and Simpson, both worked incessantly to secure additional congressional funds. On January 16, 1854, Rice introduced a bill allocating $50,000 for Minnesota roads in agreement with the War Department estimates, and when notifying Abert of his action, stated his approval of Simpson's construction plans. However, he urged the appointment of Emerson as superintendent of the St. Louis River road because Simpson would be occupied with other duties.[44] Rice's appropriation measure was approved by the Committee on Territories. Representative E. Wilder Farley of Maine, chairman of the committee, pointed out that the War Department estimates coincided with the territorial requests and drew the House's attention to the Simpson report of 1853, telling the progress that had been made. He explained that the funds were for a continuation and not a completion of the program. John Letcher of Virginia was disturbed by this suggestion and inquired how much money would be required for the future. Farley replied that in addition to the contemplated appropriation of $50,000, $64,000 would be needed to complete the four major roads. No estimates for the military road were under discussion. Farley also proposed an important amendment to the bill changing the terminus of the St. Louis River road from the falls to the mouth of that river. This change, the representative urged, would place the road in contact with the navigable waters of Lake Superior and also connect it with the military reservation at the head of the lake. A letter from Charles L. Emerson to Rice in support of this change was read into the *Congressional Globe.* By July 17, 1854, the amended bill had become a law.[45] There was no appropriation for the Big Sioux military road in this measure, but the matter was rectified by including an item of $25,000 in the Army Appropriation Bill for the completion of the road, a sum approximately half that recommended by Captain Reno.[46]

The road program in Minnesota had progressed steadily, if slowly, during the spring of 1854. After constructing only 12 additional miles of the St. Louis River road, Captain Simpson concluded that the work was so expensive and time consuming that the road should be cut through the forest to enable the passage of wagons, and then any remaining funds used to build bridges, culverts, and causeways. The funds approved by Congress in the summer did not reach him in time to be used during the working season.[47]

In the fall of 1854 the government roads in Minnesota were involved in a struggle over a federal land grant to a railroad from Lake Superior to Iowa. The railroad promoters, chiefly St. Paul men including Rice and his brother, were the proprietors of the lands in and around the settlement at Superior, Wisconsin, the proposed terminus of the railroad. The original bill authorizing the land grant was smothered in Congress, so the backers of the project drafted a second bill to eliminate the objectionable features of the proposal. In the interval between the failure of the first measure and the introduction of the second, the Minnesota road appropriation bill had been amended so that the northern terminus of the St. Louis River road would be at Superior, the river's mouth. Soon after a fraudulent scheme was exposed whereby a few individuals planned to secure control of the land grant for economic gain. Many persons doubted the motives of Rice, an advocate of the railroad charter, and were convinced that he was using his political office for private profit. The wrath of Minnesotans was aroused because the military road terminus had been given to Wisconsin, and the community at that point would have economic advantages over all others at the head of the lake.[48]

This controversy focused public attention upon the government road. From some quarters Simpson gained greater support in his defense against the attacks of Rice and his followers, but at the same time a stronger alignment against him was formed. The territory was split into two warring political factions. When Simpson delayed in expending the new appropriations in the fall, the editor of the *St. Croix Union* of Stillwater accused him of misappropriating funds allotted to the St. Louis River road and lending it to local bankers for investment. The *Minnesota Democrat* charged the engineer with indolence and incompetence. Rice presented charges against him to the Secretary of War, claiming that he had accomplished so little on the roads in three years that they could not be completed within the next ten. He reported that the St. Paul and St. Anthony road, in daily use, had not been improved, but a new one parallel to it was being surveyed. The delegate urged Simpson's replacement and once again asked for Emerson's appointment. A jury investigation was instigated to embarrass the engineer, but after a hearing he was acquitted of the charges of misappropriation by a vote of nine to three.[49] Captain Simpson continued in the confidence of the War Department.

During the winter of 1854–1855, Rice renewed his efforts in Washington to obtain more funds for Minnesota roads, and this short session of Congress was unusually liberal in making appropriations for internal improvements everywhere. Minnesota received more than $126,000 for road building. In the session, delegate Rice also made a single departure from his usual custom of seeking road appropriations to be expended under the Secretary of War and began to work with the Department of the Interior. Undoubtedly he was growing tired of the continual conflict with the Topographical Engineers and was convinced that the War Department intended to ignore his opinions. On the other hand, his relations with the Commismissioner of Indian Affairs over treaty negotiations had been amicable. As bands of Chippewa Indians relinquished titles to their lands by treaty in February, 1855, they were promised, among other concessions, appropriations for wagon roads. A sum of $5,000 was set aside to build from the mouth of the Rum River to Mille Lacs and $15,000 from Crow Wing to Leech Lake. Both sums were to be expended under the direction of the Commissioner of Indian Affairs.[50]

In response to local demands, Congress allotted $5,000 to cut timber along a territorial road between the head of navigation on the Mississippi at St. Anthony Falls and the new military fort on the Minnesota, Fort Ridgely. Twice the amount was approved for timber-cutting on the road from Fort Ripley via the Crow Wing River to its intersection with the main road leading to the Red River of the North. Increasing trade with the Red River country had justified this last request.[51] Funds for the five roads begun in 1850 were included in the Army Appropriation Bill approved March 3, 1855. The amounts for each road were, to the exact penny, estimated by the Corps of Topographical Engineers as necessary for their completion and, in all, amounted to more than $90,000.[52]

With these generous appropriations, Captain Simpson was able to push federal road constructions rapidly to completion. His best efforts were exerted on the St. Louis River road which he and the majority of Minnesotans judged the most important thoroughfare in the territory. Progress had been delayed by the change in its northern terminus and the resurvey. Congress, in 1854, had also instructed the War Department to complete the northern part before continuing work on other sectors. Simpson, who opposed this because of the sparse settlement in the north and comparatively small need for the road, succeeded in getting permission from the War Department to put the road in as good condition as possible throughout its entire length. During the winter of 1855, he advertised for bids to clear the roadway from the St. Croix Valley through to Lake Superior for a width of 25 feet and to grub out all the stumps and roots along a center strip of 18 feet. The plan was to construct a "practical" or passable wagon road through an area which not even a man on horseback had previously been able to travel. On the southern section of the road connecting the populous settlements of the St. Croix Valley, construction of a more permanent nature continued.

Approximately 130 of the 146 miles of the Point Douglas–Fort Ripley road were open to wagon travel, but at the same time the road remained in broken segments, chiefly due to the refusal of landowners in St. Paul and near St. Anthony to grant the right of way. Simpson was at a loss to know how to handle the situation. From the Attorney General in Washington, Caleb Cushing, to whom Colonel Abert had referred the matter, he learned that there was no way under existing laws in which private land could be taken for any government road in Minnesota without the owner's consent. The problem was referred to Congress. Stage coaches operating between St. Paul and Crow Wing were using the territorial road where it was completed and elsewhere followed an old Red River trail. The Swan River–Winnebago Agency road, an offshoot of the Fort Ripley highway, was entirely finished. The Mendota-Wabasha road was placed in traveling condition throughout its entire length. During the spring season when navigation of the Mississippi was prevented by the sluggish breaking of ice in Lake Pepin, this route was of inestimable value in transporting supplies to St. Paul. In the spring of 1855 Simpson initiated construction on the Big Sioux military road by making contracts for cutting out trees for a width of 66 feet and grubbing out a 25-foot roadbed between Mendota and Mankato.[53]

Timber-cutting on the two new roads, authorized in February, was at a standstill in spite of Simpson's anxiety to begin work. The St. Anthony–Fort Ridgely route proposed by the legislature had not been established by its commissioners because of the competition of various towns trying to influence the road's location. In September, 1855, the captain notified Colonel Abert he could not cut any timber until the road was surveyed and that the federal appropriation must remain unexpended. Early in June, he had directed one of his assistant engineers to make a reconnaissance of the Fort Ripley–Red River trail. The route crossed many swamps and mud-holes, and was so crooked that cutting out the timber would fail to improve markedly its traveling condition. Simpson concluded that all work should be postponed until Congress authorized the selection and survey of a better way. The road gave the captain great concern since he considered it second only to the St. Louis River road in military and commercial importance. Believing that he had an obligation to Minnesota as well as to the federal government, he addressed a circular letter, inclosing the assistant's report, to the leading men in the territory—Sibley, Rice, Governor Willis A. Gorman, and Norman W. Kittson, a former fur trader and state legislator from Pembina—asking their opinion. In general, they supported a postponement, but Rice was eager to have the current appropriation economically expended and estimates made for additional needs before the next Congress met. The War Department reluctantly accepted the postponement.[54]

Realizing that Congress had intended the 1855 appropriations to complete the original federal road system in Minnesota and that his tour of duty was undoubtedly coming to a close, Captain Simpson, in his annual report, included a summary of the knowledge he had gained from his experience in building government roads. Minnesota had been an excellent experimental laboratory for the Army. The captain assumed that the cardinal principle behind congressional appropriations for constructing roads in the territory, either military or territorial, was to establish thoroughfares serving the wants of the people only until they were numerous enough to care for their own needs. Assuming that this was the extent of federal support, he pointed out certain conditions that each road should satisfy. Experience had proved that to open a road through dense woods, trees should be felled for a width of at least 66 feet, and in some cases even 100 feet, to let the sun and wind in to dry the road and to prevent fallen trees from forming an obstruction. In areas where trees were sparse and very low, he considered only 33 feet sufficient. A satisfactory roadbed could be made by thoroughly grubbing and removing stumps for a width of 25 feet. This center strip should be leveled by removing all knolls and hillocks so that the remaining grade would not exceed ten feet in 100 and preferably not more than eight in every 100 feet. Whenever the roadbed was then found to be soft and muddy, the section should be corduroyed with logs covered by gravel and earth and the width of the road narrowed to 18 feet. Side ditches for drainage were most effective if they were four feet wide at the top, three at the bottom, and two and one-half feet deep. The engineer also had specific recommendations for the structure and dimensions of various types of bridges and culverts.

Simpson drew attention to his system of awarding contracts. When he first came to the territory, he followed the customary practice of letting contracts on a mileage basis. This plan was a temptation for contractors to slight their work to cover more mileage, so he developed a system of paying each contractor for the actual work completed. Contracts were awarded specifying definite rates for each unit of work, which could be measured by the acre, cubic yard, or rod. Although this procedure meant more labor for the engineer because of the necessity of measuring the work, he felt it was eminently successful and recommended that other officers of the Topographical Corps give it a trial.[55]

Simpson's experience in Minnesota also demonstrated the ease with which an Army officer could become involved in territorial politics, and the resulting folly. The election for congressional delegate in 1855 brought matters to a climax. Rice, who was seeking reëlection on the basis of his achievements in Congress, claimed, among other things, to have secured $200,000 for territorial roads. The Sibley faction challenged Rice's assertions that the delegate was responsible for these appropriations, and a war of words followed in the local press. Personal attacks on Sibley, claiming that his road proposals were introduced only to promote a town site at Mendota, invoked the ire of Simpson. In an open letter to the *Daily Minnesota Pioneer* he credited Sibley with initiating the federal road program and securing the first funds. Rice's personal contribution since then was declared nil because all the allotments granted during his term were the result of War Department requests, based upon the engineer's personal estimates of need. Rice was credited with securing two timber-cutting appropriations along the St. Anthony Falls–Fort Ridgely road and the Fort Ripley–Red River road. Yet the legislation was so worded by the delegate that no work could be accomplished on either project.[56]

Newspapers opposed to Rice's candidacy advertised the arguments of Simpson. In the process, the old claim that Rice had lost a railroad for the territory because of his attempt at fraud was republicized. The story of the relocation of the federal road to Superior for the benefit of land promoters there was likewise elaborated in the press. The delegate's role in securing land offices and Indian cessions, through treaties, was minimized. Rice was infuriated and wrote the Secretary of War, inclosing Simpson's articles to show how he wasted his time and neglected the roads. The captain had kept his Bureau chief informed of developments, however, and Abert took no action. Learning of Rice's protest to the Bureau, the engineer again wrote to the *Daily Minnesota Pioneer* in defense of his activities and questioned the delegate's right to try to drive him from the territory without a hearing.[57] Rice's political ally, the *Minnesota Democrat,* retaliated:

> We do not like to say that Capt. Simpson is an ass. Indeed, we do not think that he is—quite; if he were he would have shown less ears . . . An idiot would scarcely be so silly as to suppose that because the delegate asked for the construction of certain roads, and obtained a small appropriation therefore, and *because* Capt. Simpson *recommended* further appropriations, Congress would therefore grant them, *as a matter of course* . . . Let us have the money promptly expended, Captain Simpson, without reference to who is running for Congress, or which side you are on and we will speak a good word for you the first opportunity we have.[58]

In spite of Simpson's ability as a road engineer, his conscientiousness, and willingness to work hard, he obviously had become such a controversial political figure that his effectiveness in the territory had been undermined. For this reason or because the War Department felt he had enjoyed an assignment in a given place for longer than was customary for officers of the Topographical Corps, Simpson was transferred, and Captain George Thom, working on the Mexican boundary survey, was ordered to Minnesota in May, 1856. Simpson left the territory rightly convinced that he had laid the basis of Minnesota's road system.

The flow of requests from the Minnesota legislature for road appropriations, through congressional memorials, was ceaseless.[59] In 1854 a proposal had been made to secure federal aid in the construction of a comprehensive system of roads in the settled southern parts of the territory. This suggested legislation was never reported to the floor.[60] After Rice's success the next year in incorporating road projects into Indian treaty negotiations, he decided to return once more to the War Department. The Minnesota legislature in its 1856 session had asked for $30,000 to construct a military road from Winona to Fort Ridgely. The War Department reported that the road was not necessary or particularly useful for military purposes and so killed the bill in the Military Affairs Committee.[61] Minnesota politicians were aware that the territorial growth of the mid-'fifties would soon result in a statehood movement, and in the 1856 session of the legislature they bombarded Congress with memorials for road construction, knowing full well that no grants would be forthcoming after the end of territorial status. The following requests were made: a road along the west bank of the Mississippi from Fort Snelling to Pembina (the military road already existed along the east bank to Crow Wing); a road northward from St. Paul to Kettle River there to intersect the Point Douglas–Lake Superior road; two east-west roads, one from Wabasha to Fort Ridgely, the other from Brownsville to the Mankato territorial road; an extension of the Mendota-Wabasha road to the Iowa line; and numerous bridges and culvert constructions on local southern Minnesota roads.[62]

When a bill for the bridging of streams and opening roads in southern Minnesota came up for debate in the House of Representatives, John Letcher of Virginia requested a departmental recommendation on the measure.[63] Rice read into the *Globe* a letter of endorsement from George W. Manypenny, Commissioner of Indian Affairs. Letcher challenged this action.

> This bill comes in, it seems, upon a new platform. Heretofore roads were constitutional if they happened to be military; but now they are constitutional if they happen to be recommended by the Commissioner of Indian Affairs. Then they were constitutional if they were carrying munitions of war for the purpose of defending the country; now they are constitutional if they are for carrying goods to be distributed amongst the Indians.[64]

This statement precipitated a discussion in which several representatives asserted that Minnesota had received her share of federal subsidies for transportation. Rice charged these members with raising a great scarecrow each time legislation for the benefit of the territories was discussed. He also stated:

> The Indians that occupied Illinois, Iowa, and Wisconsin have been removed into the Territory of Minnesota. There are some thirty or forty thousand Indians there.... The laws of the United States prohibit a white man from cutting down a tree or marking out a road in the Indian country.... Yet these reservations are so placed between our settlements that the citizens, in many instances, cannot get from one hamlet to another without going a great distance around.... The men going to that territory are poor. They go there for the purpose of benefitting their condition; and, while they are standing there with the hoe in one hand and the rifle in the other, the gentleman from Virginia calls upon them to make roads.[65]

The Minnesota delegate, in further debate, imprudently criticized Virginians who held extensive territorial lands by military land warrants. This aroused Letcher's ire and brought forth his sectional bias. He launched a relentless attack on the bill. Asserting that $481,000 had been appropriated for Minnesota in seven years, he exclaimed:

> Talk about unkind treatment to a Territory when it has received appropriations like that!... It seems to me we have been liberal with this Territory. She has

gotten more than any other Territory that has ever been formed under this Government.... We have been liberal with her in making military roads, for the construction of which she has received $200,000. And when no other military roads are to be made, we are now called upon to do—what? To make Indian roads.[66]

Rice was unable to get a vote on the measure before Congress adjourned and not until January, 1857, did it pass the House.[67] When Senator Douglas submitted the bill for consideration in the upper house, it was immediately evident that many senators agreed with House colleagues that Minnesota had been well treated. Although Douglas announced, "this is the last appropriation for Minnesota," Robert W. Johnson of Arkansas stated:

> I rise to protest against the passage of this bill, and I believe I can do it with as clear a conscience as any man who ever opposed any measure on this floor. I have never cast a vote against the interests of this Territory... There seems to have been no stint in the favors extended by legislation to Minnesota. They have literally been showered on her. Now, after we have made provision for her admission into the Union as a State, we are asked to continue appropriations for the construction of her roads and bridges, and thus inducing population to flow into Minnesota. This seems to me not proper.[68]

Douglas denied and resented all suggestions of favoritism. He remarked, "I have never shown any sectional feeling in regard to a Territory. A southern Territory has fared as well at my hands as a northern Territory." The projection of the sectional issue into the debate, together with the previously announced opposition, was sufficient to insure the bill's defeat.[69]

Although this Congress was unwilling to launch new road projects in Minnesota with statehood imminent, it did provide a final appropriation to continue the original military road system by so designating $38,000 in the Army Appropriation Bill of 1857. The bulk of the appropriation was to be used on the Point Douglas–St. Louis River road, more than $4,000 for the completion of the Fort Ripley road, and $2,000 to build a bridge over the Cannon River on the Wabasha-Mendota road.[70] When Captain Thom arrived in Minnesota he learned that the majority of Minnesotans were very skeptical about the continued success of the Army road-building program. The Rice political faction had, through party newspapers, praised the Interior Department's construction of roads from Crow Wing to Leech Lake and from Rum River to Mille Lacs. The Commissioner of Indian Affairs had named William McAboy as civilian superintendent in April, 1855. By the summer of 1856, the routes were pronounced passable and all contractors paid off. The contrast between this rapidity of construction and the relatively slower progress of Army engineers gave hostile newspapers material with which to attack the officers of the Corps.[71]

During his first months in the territory Captain Thom examined the various federal roads in Minnesota and reported that the 1857 appropriation, and unexpended balances totaling $44,500, would make possible the completion of the roads from Point Douglas to Fort Ripley, from Wabasha to Mendota, and from St. Anthony to Fort Ridgely. One road from Swan River to Long Prairie had been finished and remained in excellent condition. Although several sections of the Fort Ripley road were yet to be built, particularly the segment between St. Paul and Point Douglas, Thom was certain the contractors could finish the job. As soon as the territorial road commissioners established the location of the road from St. Anthony to Fort Ridgely to run through Henderson, contracts were let for the removal of timber as provided by federal law. Work begun in July, 1857, was finished in the fall of that year.

Thom was dissatisfied with the state of construction on the other three roads and estimated that approximately $100,000 would be essential for their completion: $45,200 for the Lake Superior road, $36,000 for the Big Sioux military road, and $25,000 for the Red River road. Up to 1857, Congress had appropriated $120,000 for construction on the Lake Superior road, yet it was in usable condition for wagons only from Point Douglas to Taylor's Falls. North of that point it was passable in the dry season, but only for men on foot or horseback. Settlers at the lake's head had built a winter stage road from Fond du Lac to a point on the St. Croix near the mouth of the Yellow River. Here a connection was made with a lumbering trail that joined the military road at Taylor's Falls. Captain Thom spent the available funds during the summer of 1857 in rebuilding and repairing the most essential bridges, in corduroying, ditching, and grading, but the tasks remaining were insurmountable with the limited money and time. The captain confirmed Simpson's statement that the Mendota–Big Sioux road was improved to Mankato, but even its best sections were barely passable. No work had been attempted on the 178-mile stretch from Mankato to the mouth of the Big Sioux. In 1857 Congress amended the law authorizing the Crow Wing–Red River road to permit the Secretary of War to use his judgment in locating the road. The original appropriation was to be applied to construction as well as to timber-cutting. The original $10,000 appropriation, still unexpended, would serve only to initiate the work.[72]

Captain Thom's labors had scarcely begun in Minnesota when he was replaced by Captain Howard Stansbury in May, 1858. No more appropriations for military roads in the territory were made after 1857. In expending the small sums remaining, Stansbury concentrated his efforts on the Lake Superior road and the Red River road. The former was extended northward from Taylor's Falls to the mouth of the Kettle River. The engineer surveyed the Red River route as far as Otter Tail Lake and made a tour of inspection to its northern terminus. He recommended to the Bureau that this proposed highway be relocated to give Fort Abercrombie, west of the lake, a direct connection with Fort Ripley. The funds remaining after Stansbury's survey were sufficient to build 29 miles of road from Fort Ripley northward up the Crow Wing Valley.[73]

With the exhaustion of federal funds, road building in Minnesota by the national government reached a standstill. The engineering officers of the Army continued to make cost estimates for contemplated construction in each annual report from 1857 to 1861. The Bureau retained Stansbury in Minnesota for nearly two years after the funds were gone, partly in anticipation of future appropriations, but primarily to care for the vast amount of federal property accumulated for road construction over the eight territorial years. During two years in the field, Stansbury and his assistants prepared a series of military road maps for the Bureau. Orders finally came in June, 1861, to dispose of the remaining public property and close the United States Army road office. Thus the comprehensive program of road making in Minnesota, which the federal government had begun in 1850, came to an end, and the responsibility for maintaining the roads was assumed by the communities through which they passed.[74]

The federal government had indeed been generous to Minnesota Territory. According to a summary of appropriations prepared in 1882, $467,500 was granted for road and bridge construction: $87,500 spent by the Department of the Interior, $312,500 by the War Department on military roads, and $67,000 granted to aid the construction of bridges depending upon Army approval. These appropriations exceeded the combined total for Iowa and Wisconsin. Only aid to Michigan, with total appropriations approximating $400,000, approached that of Minnesota. Arthur J. Larsen has summarized the contribution of the federal government to Minnesota transportation development:

> The roads built by the federal government differed as widely as the purposes they served. Some were simply temporary trails through the wilderness, and others were so designed that they became arterial highways for the region through which they were built. In general, the roads built by the interior department were constructed solely to facilitate the business of the government. On the other hand, those built by the Army engineers were designed for a dual

purpose:—that of facilitating the transaction of government business, and that of opening and constructing "the great thoroughfares sufficiently to answer the wants of the people until they erect themselves into a State, or, at any rate, until they are populous enough and efficient enough to make and foster their roads themselves."[75]

CHAPTER V : *Federal Military Roads in Oregon Territory, 1853-1859*[1]

AMONG THE EARNEST PLEAS addressed to the United States Congress by the first legislature of Oregon Territory was a request for federal aid to improve transportation. Samuel R. Thurston, elected congressional delegate in 1849, was entrusted with the delivery of this memorial, but when his baggage and papers were lost on the way to Washington he was forced to draft his own statement for the House Committee on Roads and Canals.[2] Appropriations were asked for territorial roads on the west and east sides of the Willamette Valley, and from Astoria, at the mouth of the Columbia River, to the Tualatin Plains. Oregon residents in the valley agreed to be responsible for the labor, but a federal grant-in-aid was needed to speed their improvements for defense and exchange of goods. Thurston's words proved effective. The committee was easily convinced that in this new country of the Pacific Northwest "containing but a small population, scattered over a large territory" the need for the appropriation was obvious.[3] No federal funds for Oregon roads were appropriated, however, at this session of Congress.

Long before the creation of territorial government the transportation question had been of primary interest to Oregon's officials. Among the records of the provisional government, 1843–1849, six to seven hundred documents, including bills, memorials, and petitions, are related to the problem of road improvement.[4] Yet the Oregon roads did not meet the basic needs of the sparse population at the beginning of the territorial era. The surfacing of routes through forested areas to make them all-season roads was a primary concern of the citizens. During the first five years of the decade, a great deal of the road work was left to private initiative.[5] After 1849 appropriations from the territorial treasury, though small, were occasionally approved for the survey of new ways or the improvement of old wagon trails. The lawmakers instructed the territorial road commissioners to indicate the general route of projected roads and to require the traditional day's labor on the construction each year from every adult male. Both residents and legislators remained bitterly dissatisfied with the situation and earnestly hoped to secure federal support.[6] Historian Oscar O. Winther has rightly suggested that in the Oregon of that day "much of the political thinking and maneuvering evolved from the matrix of transportation."[7]

While local residents, working as individuals or in groups under authorizations of the legislature, were engaged in various types of road improvements, the committees of Congress considered the requests of the territorial delegate and the numerous memorials of the Oregon legislature, 1849–1852. Each session followed the pattern set by the first in memorializing Congress for assistance, and delegate Thurston and his successor, Joseph Lane, devoted much of their time to pleading with Congress for funds.[8] Lane was proud of his achievements in obtaining federal money for his territory, but political enemies accused him of deliberately wasting public funds on road construction to advance his own political position.[9] Nevertheless, the territorial press continuously reminded the politicians of the inadequacies of transportation and spurred them on to further action.

When delegate Lane requested his colleagues in the Thirty-second Congress to approve appropriations for two Oregon roads—from the Umpqua River to the Rogue River valley and from Steilacoom on Puget Sound to Fort Walla Walla—historical precedent justified his hope for a favorable vote. Congress had accepted the principle of federal responsibility in building roads in the territories for which a military justification could be established. At the opening of the 'fifties, Minnesota Territory was the chief laboratory for Army experiments in road construction.[10] Minnesota and Oregon, both beginning their territorial careers, were often associated in the laws of Congress and granted identical privileges and responsibilities. Hence in 1852, Congress found difficulty in justifying any failure to extend the federal road-building program to the Pacific Northwest.

No significant objection to the measure introduced by Lane was raised by the legislators during the first session, but House delays prevented its consideration by the Senate. As one of the first bills on the Senate calendar of the second session, it was quickly approved and signed by President Fillmore on January 7, 1853.[11] Twenty thousand dollars were appropriated for both the Steilacoom–Walla Walla road and for the route from Camp Stuart, near Jacksonville in the Rogue River Valley, to Myrtle Creek, a tributary of the Umpqua River.[12] At the request of the Oregon legislature, the Thirty-third Congress made an additional allotment of $20,000 to extend this military road from the mouth of Myrtle Creek northwestward to the settlement of Scottsburg, on the Umpqua.[13]

In southern Oregon, a route between Rogue River and the Umpqua Valley was first surveyed by Major Benjamin Alvord, assisted by Jesse Applegate, during the fall of 1853. Approximately $5,000 was spent on this reconnaissance. The primary purpose was to find a passage avoiding the Umpqua Canyon. Failing in this, Alvord decided to spend the remainder of the appropriation in improving the road through the canyon and the near-by Grave Creek hills. Construction contracts were made with the Applegate brothers, Jesse and Lindsay, and with Jesse Roberts, all local residents.[14] When the extension of the road to Scottsburg was authorized by Congress, Jefferson Davis selected a young infantry lieutenant stationed at Fort Vancouver, John Withers, to direct the expanded project, and stated in his orders:

> Your object will be first to secure a practicable wagon road between the points indicated, and then to devote the remainder of the funds at your disposal to the improvement of the most difficult places, aiming to make the road uniformly good throughout its length. You are authorized to have the work done by contract or to employ hands for the purpose. The former is believed to be the preferable mode, particularly if persons residing along the line, and thus interested in the success of the work, are willing to undertake it at modest rates.[15]

Lieutenant Withers surveyed the route from Myrtle Creek to Scottsburg, divided the construction work into sixteen divisions and advertised for bids in the Scottsburg *Umpqua Weekly Gazette.* Handbills describing the labor to be placed under contract were also circulated in the settlements along the road. Upon reviewing the bids, Withers discovered that both Jesse Applegate, as spokesman for his neighbors at Yoncalla, and a second group of citizens from Elk Creek had submitted proposals to build along different routes from his survey. Although these bids were lower than those accepted, the lieutenant rejected them for not being on the shortest and best route in agreement with his descriptive advertisements.[16] Angered by this treatment, the residents sought aid from an Elkton lawyer, W. W. Chapman, who protested directly to the Secretary of War and obtained an injunction to stay operations on two segments of the road.[17] Withers secured the services of Stephen F. Chadwick, an outstanding lawyer with experience as a member of Oregon's constitutional convention. Chadwick appeared before the courts in behalf of the federal government and had the injunction set aside. Chapman retaliated by delaying the improvements upon the section for which he personally had been granted a contract until Withers and Chadwick threatened to have the courts declare his bonds forfeited.[18] In the spring of 1855 Withers made a tour of inspection along the military road and discovered remarkable progress by the contractors. All road work was virtually

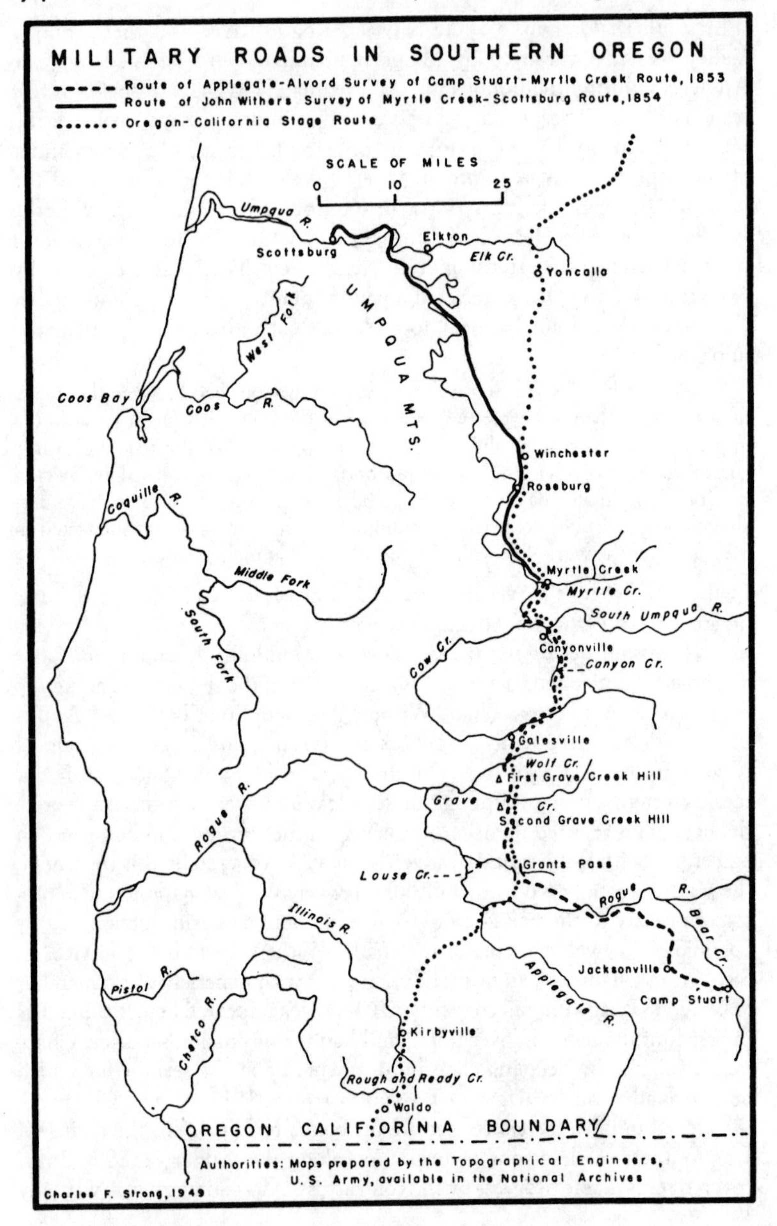
MILITARY ROADS IN SOUTHERN OREGON
Route of Applegate-Alvord Survey of Camp Stuart-Myrtle Creek Route, 1853
Route of John Withers Survey of Myrtle Creek-Scottsburg Route, 1854
Oregon-California Stage Route
SCALE OF MILES
0
10
25
Umpqua R.
Scottsburg
Elkton
Elk Cr.
Yoncalla
West Fork
UMPQUA MTS.
Coos Bay
Coos R.
Winchester
Roseburg
Coquille R.
Middle Fork
South Fork
Myrtle Creek
Myrtle Cr.
South Umpqua R.
Canyonville
Canyon Cr.
Cow Cr.
Galesville
Wolf Cr.
First Grave Creek Hill
Grave Cr.
Second Grave Creek Hill
Rogue R.
Grants Pass
Louse Cr.
Rogue R.
Bear Cr.
Illinois R.
Jacksonville
Camp Stuart
Applegate R.
Pistol R.
Chetco R.
Kirbyville
Rough and Ready Cr.
Waldo
OREGON CALIFORNIA BOUNDARY
Authorities: Maps prepared by the Topographical Engineers, U. S. Army, available in the National Archives
Charles F. Strong, 1949

complete. Two bridge builders, however, had to be given an extension of ninety days to complete constructions delayed by the inability to secure iron. By early summer he reported to the War Department that "the road, I am confident, is as good a wagon road now as any in the country and will greatly facilitate the transportation of any government supplies to Forts Lane and Jones, by way of Scottsburg." Withers considered the Umpqua River the only serious impediment to constant passage of wagons and urged that bridges be constructed at the three crossings of the military road.[19] Secretary Davis added a note to this report, "The course pursued by Lt. Withers is entirely satisfactory and his success gratifying."[20]

Parts of the southern Oregon military road built during 1853 and 1854 became a part of the Oregon and California Stage Road, for several years the chief north-south artery of travel between California and the Willamette Valley settlements. This stage road crossed the Oregon-California boundary near Waldo and continued northeastward to join the military road just south of the Grave Creek hills. Through the canyon of these hills northward to Myrtle Creek the stage used the military road surveyed by Alvord and Applegate. From Myrtle Creek, the route laid out by Withers was followed to the settlement at Winchester. At this village the stage continued northward to the Willamette Valley whereas the military road ran toward the northwest to the depot on the Umpqua at Scottsburg.

A new territorial road from Astoria to Salem in northwestern Oregon was authorized by the Thirty-third Congress with a $30,000 allotment for construction.[21] This Congress had approved so many federal road projects in Washington and Oregon territories under War Department supervision that Jefferson Davis notified Colonel Abert of the Topographical Corps to establish a superintendency of Pacific Roads with headquarters in San Francisco. Major Hartman Bache was placed in charge of the new office and ordered to assume direct responsibility for junior officers in the field.[22]

In the early summer of 1855 delegate Lane, who had returned to Oregon, wrote Davis urging his early attention to the construction between Astoria and the Willamette Valley. His constituents classified the road as a military necessity and he personally was certain that the completion of the route would "do much to develope [*sic*] the resourses [*sic*] and encourage the settlement of an important portion of our Territory now for want of a road inaacessable [*sic*]." Since the initial appropriation was inadequate to complete the road, the delegate urged the War Department to prepare a cost estimate that could be used in requesting the next Congress for funds.[23]

Lieutenant George H. Derby of the Topographical Engineers was ordered into the field by Bache with instructions to survey the Astoria-Salem road before those in Washington. "As the appropriations for the roads are small," wrote Bache, "the instrumental surveys should be confined to such portions of the roads as may be absolutely necessary to determine the trace which in every instance should be marked off in miles and quarter miles, for the convenience of letting for construction."[24] The lieutenant thought that the survey could be completed within two months at a cost of $4,000. He left San Francisco in mid-July aboard the mail steamer for Astoria and within a week had submitted requests for mules, instruments, and field equipment to the quartermaster at Fort Vancouver. Returning to Astoria to select the terminus of the road, he became involved in a squabble among the residents of the town. Two settlements claimed the terminus right. The principal village with one hundred inhabitants was the rendezvous for river pilots and the loading place for the mail steamers and Columbia River vessels; the second, farther upstream, was the site of the customhouse. The back country between the two was mountainous and covered with thick timber; the settlers used the water or beach as a route of communication since no road connected the villages. After careful surveys, Derby reported the easiest passes through the mountains were those immediately adjoining the customhouse. To start at the larger village would extend the road two miles at a total cost of $6,000; to connect the villages would mean an expenditure of only $3,000. No action was taken until the Secretary of War reviewed the case because the residents of the lower village, who had repeatedly petitioned the Treasury Department to move the customhouse, had threatened to protest Derby's action in Washington, D.C. In time the lieutenant was guaranteed support of the War Department in carrying out his decisions. This incident delaying construction was typical of many experiences that plagued the topographical engineers and which unsympathetic congressmen cited as examples of Army inefficiency and delay.[25]

From the outset, Lieutenant Derby knew the survey would not be easy. He informed the Department:

> The woods are very thick, with a dense growth of underbrush, and the mountains are represented as almost impassable. I can see no object in making these roads *one hundred feet* wide, as directed by instructions from the Bureau. If the idea is, that this width will prevent the road from being encumbered by falling timber it is a mistaken one, the growth being generally over a hundred feet in height. . . . I would respectfully suggest that sixteen feet is quite wide enough for all practical purposes . . .[26]

By the end of July Derby had succeeded in organizing a small party of eight men and procuring six pack animals. The excitement over the gold discoveries at Fort Colville had increased the price of labor, and good men were difficult to obtain at any price. The party cut a trail into the forest from Astoria to the Tualatin Plains, just west and south of the present city of Portland, and Derby relinquished supervision of the group to his assistant, a civilian engineer, with instructions to make an instrumental survey on the return to Astoria. The length of the Astoria-Salem road was 113 miles, but beyond the Tualatin Plains to Salem there was a satisfactory country road connecting the settlements so that the appropriation could be applied to the first 60 miles out of Astoria to the settlement of Harper on the middle branch of the Tualatin River.[27]

On August 12 the Army engineer proceeded to Vancouver for preliminary examinations of roads from that fort to The Dalles and Fort Steilacoom. News reached him there that the surveying party under his assistant had returned to Astoria the first week in October with animals broken and the men dissatisfied and exhausted. A new party of laborers was employed to chain the last 20 miles; however, the rains set in before the job was half completed and Derby was convinced that the survey should not be made until the road was constructed. This, he claimed, was the local custom, and a surveying party that could average only two miles a day was an unnecessary expense and delay. Before leaving the Northwest to return to San Francisco for the winter, the lieutenant concluded that the appropriation would suffice to construct only the first 20 miles, from Astoria to the Nehalem River, and he placed advertisements in the Oregon newspapers for contractors. He proposed to cut through the forest at a width of two rods, or 33 feet, and to build a carriage way one half of that width. Derby notified Bache:

> This road I intend when finished to be practicable for a loaded wagon throughout its whole extent, but I do not expect it will be by any means an excellent carriage road. The amount of the appropriation would not suffice to make five miles of good road, regularly graded, with ditches and culverts, through this country; and I presume the object of the appropriation is merely to get a road through at small expense, and trust the settlers to improve it at their convenience.[28]

Before Lieutenant Derby's departure, delegate Lane had conferred with him about the funds needed to complete the road and had obtained copies of his maps and specifications for the use of Oregon legislators memorializing Congress for additional support.[29] The lieutenant estimated the cost of constructing a satisfactory military road at $1,500 a mile, but after he reported in person to Major Bache in San Francisco the latter officer wrote to the Bureau that his own estimate was $2,000 a mile for 60 miles, thus

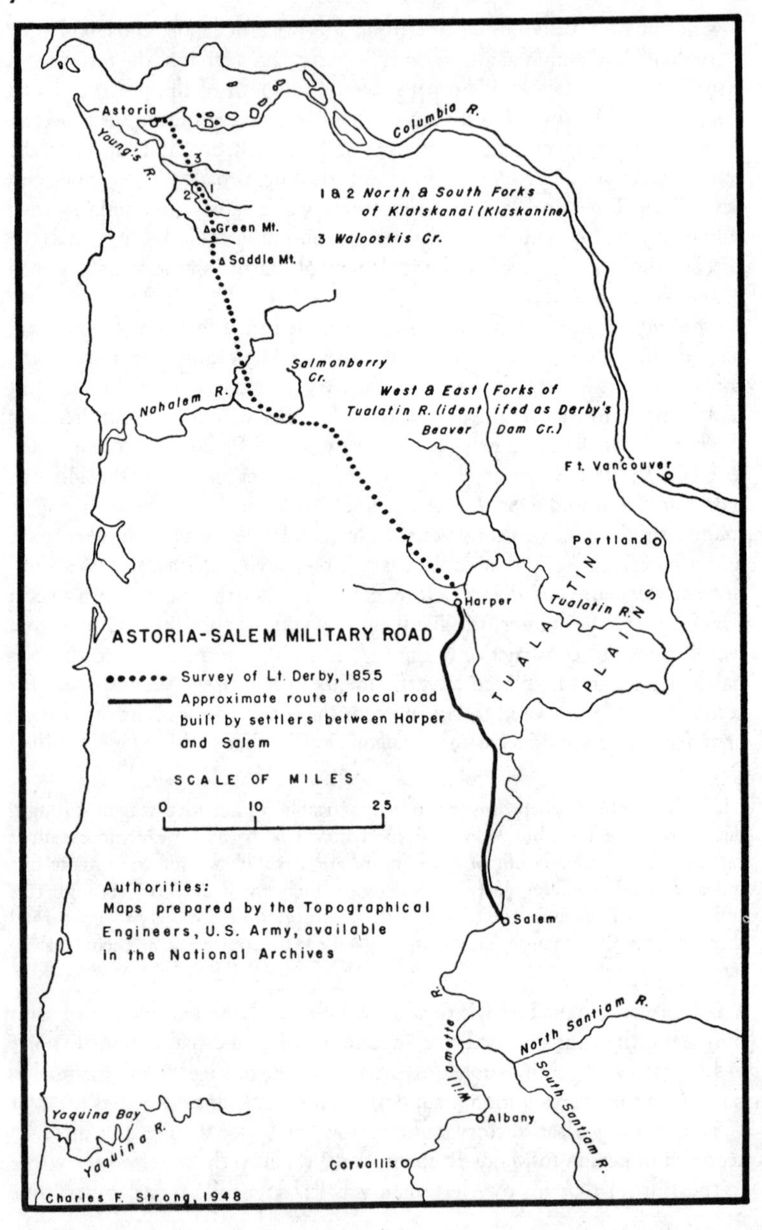
Astoria
Young's R.
Columbia R.
3
2
1
1 & 2 North & South Forks of Klatskanai (Klaskanine)
Green Mt.
3 Walooskis Cr.
Saddle Mt.
Salmonberry Cr.
Nahalem R.
West & East Forks of Tualatin R. (identified as Derby's Beaver Dam Cr.)
Ft. Vancouver
Portland
Harper
Tualatin R.
TUALATIN PLAINS
ASTORIA-SALEM MILITARY ROAD
Survey of Lt. Derby, 1855
Approximate route of local road built by settlers between Harper and Salem
SCALE OF MILES
0
10
25
Authorities:
Maps prepared by the Topographical Engineers, U.S. Army, available in the National Archives
Salem
Willamette R.
North Santiam R.
South Santiam R.
Albany
Corvallis
Yaquina Bay
Yaquina R.
Charles F. Strong, 1948

requiring an additional $96,000. This decision about the continuance of the project, should, in his opinion, be made in Washington.[30]

Lieutenant Derby experienced the usual difficulty in letting government contracts for construction. Few bids arrived before the deadline, and when one bidder on the first section out of Astoria was notified that his proposal of $1,500 a mile was the lowest, he declined to enter into a contract. The upshot of these negotiations was the granting of contracts for the construction of the entire proposed road from Astoria to the Nehalem River to Daniel Wright at $2,000 a mile. By March these contracts were in the hands of Jefferson Davis who, incensed that the funds were being spent on such a short segment of the road, wrote to Abert:

> Not approved. If contracts cannot be made in accordance with the original instruction, the officer in charge will hire men to execute the work conformably to the design heretofore communicated.[31]

Lieutenant Derby returned to the Northwest in the spring of 1856 and abrogated the contracts made with Wright. For two months he personally directed the work of laborers improving the route. When the funds were exhausted, one bridge and 20 miles of road were completed.[32] The War Department relieved him of his duties in the early fall and when he forwarded his annual report to Bache, the major penned a notation, "The report leaves nothing wanting in his efforts to try to carry out the law." The major was intensely annoyed at the attitude of the Bureau and the Secretary of War for canceling contracts for reasons "unknown to us on the Pacific Coast." He wrote Abert:

> The annual memoirs descriptive of this country were duly drawn up and submitted. Though quite full, these failed to convey to the department a just conception of the extraordinary features of the country.... From these natural causes and the prices which rule on this coast, the ordinary appropriations for like measures on the Atlantic, prove wholly insufficient to carry them out on the Pacific.... If the rule prevails that expenditures can be made only when the proposed object can be attained with the means then in hand, either the appropriations for improvements on this coast must be increased or important auxiliaries of defense and in the settlement of a new country must be abandoned.[33]

Lane, having returned to Washington for the convening of the Thirty-fourth Congress, introduced a measure requesting $55,000 for the completion of the Astoria road, and $30,000 for each of the two sections of the Camp Stuart–Scottsburg route. When J. A. Quitman, chairman of the Military Affairs Committee, asked Davis for the War Department's official position on the legislation, the Secretary's office ordered Colonel Abert to prepare a statement for Congress. His estimates were those prepared by Bache, Derby, and Withers, but Secretary Davis in forwarding the correspondence notified Quitman that he could not endorse the $96,000 for the Astoria-Salem road. Since the Camp Stuart (now Fort Lane)–Myrtle Creek road was part of a major highway between California and the Columbia River region, $30,000 might be justified; for the extension to Scottsburg, $10,000 should be sufficient. Apparently the Secretary was overly cautious in recommending large expenditures for the Northwest.[34] Quitman's committee, in time, returned the bill, reducing the amount for the Astoria road to $10,000, but leaving Lane's original request for the other two in spite of the Secretary's objections.

This House had several members who were prepared to question the legality and desirability of federal appropriations for roads in the territories. Intense party and sectional feeling was evident. Debate on the Oregon measure was opened in the House when George W. Jones of Tennessee, spokesman for the strict constructionists and continuous opponent of the federal road program, announced that he could not permit the bill to pass without protesting against the exercise of such a power. He inquired:

> Is it based upon the military power of this Government to construct these roads in the Territory? If so, then I appeal to my friends who claim to be strict constructionists, what more power has this Government—under the Constitution—under the military power—in the Territory than it has in the State?[35]

At this point the representative was interrupted by a member of his party to inquire if he supported the Democratic platform adopted at the Cincinnati convention urging the construction of wagon roads and a railroad to the Pacific Coast. "I repudiated it entirely and utterly," announced Jones. "The party may adopt it, but it is not a Democratic principle." The representative was further angered when his colleague expressed particular interest in this repudiation of the party. Jones quickly answered his challenge:

> I do not repudiate the party. . . . And, sir, I expect to remain with it, to battle with it, and to act with it; to contend for whatever I may think right, and to oppose what I believe to be wrong, although brought forward by that party.

Quitman, speaking for the committee, inquired if Jones was opposed to all appropriations for opening ways of transportation for military supplies in the common territory of the United States. Jones replied: ". . . I utterly deny the power of the government to make roads anywhere." Quitman then suggested that as the Constitution authorized the movement of the Army from one place to another, was not the government empowered to remove obstructions to the transfer of troops and military stores? There ensued a prolonged discussion of the constitutional issues involved. Quitman finally summarized his position: "I hold that where there is a delegated power under the Constitution, there is power to do everything necessary to the exercise of that delegated power." But Jones reiterated his view that road building was not "absolutely necessary" and, if justified under the military power, appropriations could be made for states as well as territories.[36] The attack on the Oregon bill was resumed the following day when William Smith of Virginia stormed at the generosity of the Military Affairs Committee to Oregon when the Committee on Territories had just inaugurated a policy of "starving out" Kansas by denying her a federal penitentiary. John Letcher, his colleague from the same state, moved to strike out the appropriation for the Astoria road; since the committee saw fit to reduce the amount from $55,000 to $10,000, perhaps nothing was justified. "If this is a military road, what military engineer surveyed it?" asked the Virginian. "Under what order of the military department of the Government was it surveyed?"[37] Lane rose to defend the legislation. He attempted to read the report from Jefferson Davis, but Letcher demanded the document and proceeded to read extracts. Concluding, he asserted that the Oregon Territory in the eight years of its existence had received $90,000 for roads, in his opinion a generous amount. He addressed Lane:

> Now, sir, it seems to me, from the very declarations contained in that report, that, so far from this being a military road, it is to be a road for the convenience of the Territory of Oregon. It is to be for a carriage way and a wagon way; to be clear of roots; thirty-three feet in width, and all that sort of thing—showing that it is to be a permanent character, for the use of the people of the Territory, rather than for any military purpose whatever. This is my objection to it.[38]

The Oregon delegate now spoke of the military necessity of the Astoria road to connect the mouth of the Columbia River with the settlements in the Willamette Valley, particularly in the winter months when the streams might be frozen. "There has never been a military road contemplated in any of the Territories, so essential to the protection of the Territory as that road," he stated. He informed the representatives:

> Up to the present moment the Government has not taken any steps to establish any kind of fortification or defense at Astoria. Not one dollar has been appropriated for the defense of that coast. The only way in which we can at present defend Astoria, either from an Indian or foreign foe, is by sending down there the people of the Willamette Valley, and the only way to do that with any sort of expedition, is by means of this proposed road.[39]

The recommendation of the War Department for the other roads was, in his opinion, explicit and justified on the basis of reduced costs for the transportation of Army freight between Scottsburg and the military establishments near the California border. At the close of his speech he called upon his colleagues to vote down the committee's amendment to the bill and restore the larger appropriation, but the amendment carried.[40] The result of this controversy was a postponement of the final action on Oregon road appropriations. Not until the third session of the Thirty-fourth Congress, January, 1857, did this bill pass both houses.[41]

The expenditure of the new grant was entrusted to Lieutenant George H. Mendell, the successor of Lieutenant Derby. Work on the Astoria-Salem road was resumed in April, 1857,[42] when a civilian labor gang was dispatched to each end of the road with instructions "to open a track through the forest along which a wagon will be able to travel, bestowing as little labor as will accomplish this end." By September, 40 miles had been opened; 24 made by the workers on the northern end, 16 by those laboring northward from the Tualatin Plains. When the fall rains set in, 16 miles remained to be opened. Lieutenant Mendell notified the War Department that the standard of construction had been greatly reduced from that advocated by Derby so that the route could be opened its entire length with the limited funds available in accordance with the Secretary's orders. He warned that the road was available for pack trains and stock driving, but before wagons could travel over it much grading would be necessary. Mendell concurred in all that Derby had written about construction problems due to the terrain and the forest, and suggested an additional $30,000 for a fair wagon road from the coast to the Tualatin Plains.[43]

During the summer of 1857, Lieutenant Mendell was busy with construction of roads in Washington and northern Oregon. Plans were made in the winter of 1857–1858 to concentrate the next seasons' road-building endeavors in the Northwest on the roads in southern Oregon. These were under the immediate supervision of Colonel Joseph Hooker who had resolved to accomplish improvements by hired labor, with special contracts let to civilians for bridge building.[44] Two labor parties, organized in San Francisco, were dispatched to the mouth of the Umpqua by steamer the first week in April. On the Camp Stuart road, from twenty to forty workers concentrated their efforts at Umpqua Canyon, eleven miles long. The road through this canyon, the artery of travel between the Pacific Northwest and California, had remained through the years impracticable for wagons and unfavorable for horsemen. Much of the route was relocated; vertical walls were blasted away to make a roadbed at places where the banks of the stream had previously served. After the removal of large mud deposits that had accumulated during the rainy season, permanent improvement of the drainage in the lower part of the canyon was provided by culverts. Contractors were, in the meantime, building bridges and improving the most southern section of the road out of Jacksonville northwestward toward Cow Creek. This route was now described as an excellent road, 16 feet wide, with timber cleared from 30 to 60 feet; it was practicable for a six-mule team.[45]

On the Scottsburg road a party of twenty men had completed about 15 miles of construction by August 1, when their enthusiasm over the Fraser River gold discoveries led most of them to leave their work. However, Hooker was able to keep two new parties employed, one near Roseburg and the other at Myrtle Creek, and with the aid of several civilian contractors who were building substantial bridges, he completed the necessary repairs by the close of the season. The topographical engineers were convinced that the permanent improvements made during the season could not have been accomplished in years by the sparse population of the region without federal financial aid.[46]

Meanwhile, a new Congress had convened, the Thirty-fifth, and again a request for federal funds to complete the Astoria road was introduced. The Military Affairs Committee of the House approached the War Department for its official standing and through the chain of command Lieutenant Mendell was asked for a statement. He wrote from San Francisco reiterating the reasons for his $30,000 request.[47] In May, 1858, the measure was debated on the floor of the House and the strict constructionist bloc, led by Smith of Virginia and supported by Alexander H. Stephens of Georgia, restated its views insisting that appropriations must be recommended by the War Department with a specific statement of the military purpose of the road. The spokesman of the Military Affairs Committee announced once more that the road had twice been recognized by Congress as a *military* road, and that the appropriation was to finish the job according to the original intent of that body.[48] When the bill went to the Senate, Jefferson Davis, who now represented the state of Mississippi, spoke for its passage:

> ... The money [previously appropriated] only sufficed to make it a good wagon road for part of the way, and for the rest of the route only a bridle path.... This road is direct from one point to the other, cutting off the great elbow made by the two rivers, and is deemed of great importance for military and territorial purposes, whether we look to defense against the Indians or a foreign foe on the exterior.[49]

The appropriation was immediately approved by the Senate, and the bill was signed at the White House two days later.[50]

In April, 1858, the War Department, under the leadership of John B.

Floyd, ordered the Office of Pacific Road Constructions transferred from San Francisco to Fort Vancouver. Major Bache was relieved of his duty and Lieutenant Mendell ordered east to participate in the Great Lakes survey. Captain George Thom, who had served the two previous years as superintendent of Minnesota roads, was given the responsibility of completing the Washington and Oregon roads, but upon his arrival in San Francisco an attack of Panama fever incapacitated him so that he could not work actively in the field. By the spring of 1859 direct supervision of the Astoria-Salem road was entrusted to Lieutenant Junius B. Wheeler. Two working parties of laborers and woodsmen were once more organized to repair the section of road built by Derby just out of Astoria and to widen the Mendell-Hooker trail toward the Tualatin Plains so that it could be used by freighters and settlers' wagons. When the season closed, 33 miles had been built across the Green and Saddle mountains of the coast range. Almost half of the appropriation remained,[51] but in September Wheeler was ordered to West Point.[52] Two months later the Secretary of War sent instructions to discharge all hands engaged on the construction and cease operations until further orders.[53] In March, 1860, the Chief of the Topographical Bureau requested the Secretary to authorize the renewal of construction for the coming season, and he was told to proceed with the work and to instruct Captain Thom to employ T. P. Powers of Astoria as civilian superintendent.[54] When the Civil War broke out, the Astoria-Salem road was still not opened throughout its entire length for the use of wagons. No engineer had been able to overcome the physical and political obstacles interfering with its completion.[55]

The establishment of a shorter emigrant road than the Oregon Trail between the Great Salt Lake and the Columbia or Willamette Valley settlements had continuously interested western Army officers. Captain Rufus Ingalls, quartermaster of the military department of Oregon, was more specifically interested in locating such a permanent wagon road to the west and south of the Blue Mountains that might connect with the Salt Lake–California Trail, thence to the Mormon capital. During the Mormon War of 1857, the War Department suddenly realized the imperative importance of speedy transportation of supplies and men from the Pacific Northwest to General Albert S. Johnston's Army, or any army, in the Great Basin. Emigrant, military, and possibly railroad uses justified a thorough exploration to ascertain the possibility of a shorter practical line of communication. From well-known guides in the Northwest, Ingalls collected information about the various passable routes to Utah. Submitting this evidence to Brigadier General William S. Harney, the quartermaster noted the comparatively longer distance to Salt Lake from Fort Leavenworth on the Missouri than from Fort Vancouver on the Columbia. General Harney, always an ardent supporter for the exploration of new communication routes, was convinced the best wagon-road connections should be established and fortified. In November, 1858, he reported to General Winfield Scott that he would order a thorough exploration in the spring.[56]

Captain Henry D. Wallen, of the Fourth Infantry, was placed in charge of the wagon-road expedition from The Dalles of the Columbia to Great Salt Lake. He was ordered to proceed southward along the John Day River to its headwaters in the Blue Mountains, across the highest ridge to the Malheur River, and to descend that stream to its junction with the Snake. From the Snake Valley the well-established emigrant and Army trails could be used to Camp Floyd in the basin. Placards were to be left with the inhabitants of Utah notifying emigrants to Oregon and Washington of the new route and providing the information necessary to travel over it without mishap.[57] Wallen's command consisted of two companies of dragoons, one infantry company, and a detachment of engineers. Captain Thom, now the ranking topographical officer in the Northwest, selected a young second lieutenant, Joseph Dixon, to accompany the expedition as surveyor and map maker and provided $2,500 to meet the expenses of his detachment.[58] Captain Ingalls, with whom Wallen had been ordered to confer freely, procured the services of Louis Scholl as guide. Four months' supplies from the subsistance, ordnance, and medical departments were allocated to maintain nine officers and 184 enlisted men. The dragoons' horses numbered 116; and in the quartermaster's department of the expedition there were 38 horses, 344 mules, 121 oxen, 30 wagons, and an ambulance.[59] This party was undoubtedly the largest and best equipped of any engaged in wagon-road survey and construction for the United States Army in the trans-Mississippi West.

Learning at the outset of the impracticability of constructing a wagon road along the John Day River, Captain Wallen decided to march southward from Fort Dalles to the Des Chutes River and to cross that stream near the mouth of its tributary, Warm Springs River. This first section of the new route, 70 miles long, was abundantly watered and usable for heavily loaded wagons. From the Des Chutes the expedition moved in a southeasterly direction for 250 miles to the western base of the Blue Mountains. In this long sector of the journey, the explorers had crossed a western spur of these mountains and followed the meanderings of the main stream of Crooked River to its headwaters.[60]

Here Captain Wallen established a military depot and divided his command. The large wagon train had proved burdensome and caused delay. From this encampment most of the wagons were to be returned to Fort Dalles under the direction of Lieutenant John C. Bonnycastle. A small detachment of cavalry was likewise assigned him to explore all possible return routes along the western slopes of the mountains and improve a road back to the Columbia River. Bonnycastle remained with the ox train while Lieutenant Robert Johnson and fifteen dragoons were sent forth into the country northwest of Camp Division along the tributaries of the John Day River. No passage for wagons was discovered, however, so Bonnycastle's reunited group moved westward over the same route traveled on the outward journey. At the crest of the mountains the detachment was temporarily halted while the dragoon commander took six of his company and rode over the mountain heights eastward to the John Day River. Nowhere was there a road over which it was advisable to bring wagons. So Bonnycastle with his wagons—seventeen pulled by oxen and six by a span of six mules—continued along the outward route. Arriving at Fort Dalles, he reported that his return had taken only twelve traveling days and that the wagons had averaged 17 miles a day. General Harney announced that a satisfactory wagon road, generally over level country and with grass and water in sufficient quantities for large trains, could be used from the Columbia River to south of the Blue Mountains, a distance of 200 miles.[61]

With his reduced wagon train, Wallen and eighty-five dragoons had moved more rapidly to a large lake, named Lake Harney in honor of the department commander. Wallen described this area as the best stock-raising country in Oregon. Lieutenant Dixon, with youthful enthusiasm, observed that along this first part of the reconnaissance there was permanent water, good grass in abundance, sufficient timber for fuel and wagon repairs, and, most important, a good natural road. He commended it both to emigrants and to troops on scouting expeditions.

Turning northeast from Harney Lake, the explorers entered the mountains expecting to find the source of the Malheur River and to follow it. Difficulties were immediately encountered since it was impossible for men on horseback to follow the chasms made by this river, perpendicular cliffs at times 1,000 feet high. Wallen recorded:

> I was somewhat disappointed in our route, as I expected to find it better than it really is. A wagon road cannot be constructed over a chain of mountains such as these before us without having hills to pull over; all the science of the engineer cannot change the general features of the country.

From an encampment on the Malheur, Lieutenant Dixon, as topographical officer, was sent forward to explore the canyon to make certain it was impassable. After a three-day trip during which he was forced to dismount and travel by foot most of the way, he was convinced that the route was not only impossible but also dangerous for heavily loaded wagons. So the expedition, both wagon train and cavalrymen, worked their way through the Blue Mountains by a circuitous and precipitous route in rugged and wild country beyond the description of the Army officers. Appraising the journey through the mountains, Wallen frankly admitted "...forty-four and a half miles may, in truth, be called a bad road requiring the labor of a couple of hundred men for one season to put it in order..."[62]

The wagon-road expedition crossed over to the Owyhee River near Fort Boise, the old Hudson Bay Company post abandoned in 1855. Before ascending the valley of the Snake River from the Owyhee's mouth, Wallen dispatched his guide, Louis Scholl, with an escort of dragoons to ascend the Owyhee and explore the country west and south of that stream and Goose Creek Mountains for a possible wagon route. Upon reaching the headwaters of the Owyhee, the Scholl party traveled in a southeasterly direction, somewhat parallel to the Snake but 30 to 40 miles south of it. Joining the command nineteen days later, the guide reported the location of a much superior route for wagons than that usually followed along the Snake River.[63] Both Wallen and Dixon found the Oregon Trail, along which the main command moved, to be dusty and desolate through sagebrush country, with no timber and a great scarcity of grass.

The captain ordered his expedition to go into camp at Raft Creek, a tributary of the Snake. One detachment was ordered to make a reconnaissance northeast along the Snake River to old Fort Hall while the animals were recruited and preparations made for the return march. Wallen, with an escort of twenty dragoons, explored Raft Creek to the southwest as far as City Rocks, a well-known landmark on the Salt Lake–California emigrant route. This much-traveled highway was then followed to the Bear River crossing, along the eastern shore of Salt Lake, and around its southern side to the military headquarters at Camp Floyd. The wagon-road exploration had taken ten weeks.

On the return trip, Captain Wallen's men escorted several emigrant parties on their way to Oregon. Many of these families had started from the Missouri frontier with the impression that they could renew supplies at Fort Hall and Fort Boise, but since these posts had been abandoned for some years, they found themselves starving and miles from relief. Three families, consisting of seven men, three women, and fifteen children, were not only without food, but their wagons had broken down. These and other destitute parties, later encountered, were provided with both food and transportation. Wallen's command moved slowly down the Snake Valley toward the site of old Fort Boise, allowing emigrant groups, who wanted security from Indian attack, to join the expedition. Detachments were periodically sent out to round up those in distress. From the ford of the Malheur, the expedition returned by way of the Oregon Trail across the Grand Rond to the Umatilla River and down this stream to the Columbia. In the valley they struck the military road from Fort Walla Walla to Fort Dalles which they followed to their destination. The 532 miles from the Salt Lake Valley to the Columbia had been traversed in twenty-nine days with loaded wagons.[64] In all, the wagon-road expedition had marched 1,900 miles in four months and sixteen days.

Captain Wallen recognized his failure to locate a new route to or from the Pacific Northwest. The shortest and most direct route remained that of the Oregon Trail along the Umatilla River, across the Blue Mountains at Grand Rond, but not touching the Snake River until within 40 miles of old Fort Boise. Wallen had succeeded in saving approximately 100 miles by a cutoff from Swamp Creek on the Snake to Raft Creek valley and directly up that valley to the California emigrant road. Lieutenant Dixon was unwilling to admit defeat. He proposed the survey of a new route that would follow the outward journey of the Wallen expedition as far as Lake Harney. From this point the country could be explored toward the southeast for about 65 miles to the headwaters of the Owyhee River. Here the route laid out by Louis Scholl, and pronounced excellent for wagons, would lead to the California–Salt Lake Trail at City Rocks. Such a route would be central and more desirable than the Oregon Trail to the north or along the Humboldt to the south.[65] After reviewing the many reports on the season's exploration, General Harney decided to send another command to Salt Lake during the summer of 1860 to open up a wagon road east of Harney Lake along the route suggested by Lieutenant Dixon. The Army detachment would escort the season's emigration back over the new road. On the return by way of Lake Harney, the expedition was to travel due west toward Diamond Peak and from there follow the road down the middle fork of the Willamette River to Eugene City. This grandiose scheme for road construction, saving hundreds of miles of emigrant travel, failed of fruition because of the outbreak of the Civil War.[66]

CHAPTER VI : *Federal Aid to Transportation in Washington Territory, 1853-1860*

AMPLE EVIDENCE is available to suggest that the settlers north of the Columbia River, in petitioning for a separate territory from Oregon, were prompted more by a desire to improve transportation than by existing economic and social differences. In this region the streams and trappers' trails provided the only means of travel. The Cowlitz River, a tributary of the Columbia, was the chief avenue of migration to Puget Sound as long as it flowed in a north-south direction. From the "landing" where the stream turned abruptly to the east, no wagon road led to the earliest settlement at Tumwater. The distance was negotiated on horseback or on foot. Residents north of the Columbia, assembled in the Cowlitz Convention in August, 1851, memorialized the Congress of the United States for separation from Oregon. Every paragraph presenting the reasons for this desired division of the territory stressed either geographic isolation or inconvenience of travel.[1]

In answer to the pleas of these pioneers and the concurrent appeal of the Oregon legislature, the Thirty-second Congress established the Washington Territory, March 2, 1853. Joseph Lane sponsored in this same body a bill, which was ultimately approved, appropriating $20,000 for the construction of a military road from Fort Walla Walla to Steilacoom on Puget Sound.[2] For the governorship of the new territory, President Franklin Pierce selected a former engineer, Major Isaac I. Stevens, who resigned his Army commission to accept the assignment. Stevens was intensely interested in the proposals for the construction of a railroad to the Pacific. The Congress had allotted $150,000 in the Army Appropriation Bill for the survey of several prospective routes, and Governor Stevens accepted an appointment from the Secretary of War to survey the northernmost of these on his way to the Pacific Northwest. If a practical route was to be found between the forty-seventh and the forty-ninth parallels, Stevens was compelled to find a suitable pass through the Cascades above the Columbia River over which a railroad could be run. Without it the Cascades would prove a blockade defeating all attempts to construct a northern transcontinental line and would infinitely retard the new territory's development. Stevens planned to organize his expedition in two main divisions. The eastern and larger party was under his personal command en route from St. Paul, Minnesota Territory, to Washington. Captain George B. McClellan was given a semi-independent command over a western surveying party to work eastward from Puget Sound to join the governor's force. McClellan was Stevens' personal choice, and the new governor was delighted with the spirit and enthusiasm with which the captain accepted responsibility and entered upon the exploration.[3]

Jefferson Davis, as Secretary of War, also charged Stevens with the task of choosing some Army officer or civilian engineer to construct the military wagon road from Walla Walla to Fort Steilacoom. Stevens notified McClellan that he desired this road work performed in conjunction with the assignment of the western survey party because it was in the line of the railroad explorations and would facilitate rather than hinder general operations.[4] The primary purpose of Congress in authorizing this road, according to the Secretary of War, had been to enable emigrants with their wagons to go directly from Walla Walla, across the Cascades, to the Sound. McClellan was urged to use every exertion to open the road before the fall migration. If this became impossible, the line of the road was to be fixed, particularly through the Cascade Mountains, and the most difficult parts improved to such an extent that the emigrants could make the road passable by their own efforts. McClellan was to dispatch a suitable guide to meet them at Walla Walla. The officer was given wide latitude in the matter of granting construction contracts and in securing the necessary supply of labor and tools. "In any event," admonished Davis, "you will so arrange your operations as, first, to secure a practicable wagon road between the extremities of the road; devoting the remainder of the funds at your disposal to the improvement of the more important points, always endeavoring to make the whole road a good one."[5] McClellan thus came to the Northwest with a dual responsibility. The Secretary of War apparently thought road construction the principal part of his undertaking, but Stevens simultaneously informed him that "...the first and most important point at which your attention is to be directed will be the exploration of the Cascade mountains. You will thoroughly explore this range from the Columbia river to the forty-ninth parallel of north latitude, making detailed examination of the passes...Pending this examination you will endeavor to examine the line of the proposed road from Wallah Wallah [*sic*] to Steilacoom, and to start its construction."[6] From the first there was confusion and conflicting emphasis in the orders issued to McClellan.

The captain traveled by sea to the Columbia City Barracks (Fort Vancouver), arriving June 27. Three weeks were consumed in organizing a party of sixty-five men, including three junior officers who assumed the duties of topographer, quartermaster, and meteorologist, and four civilians who served as assistant engineer, geologist, ethnologist, and naturalist-surgeon. Three noncommissioned officers and seventeen enlisted men made up the escort and working party; packers, hunters, and herders completed the personnel of the outfit. Men and supplies for one hundred days were mounted on one hundred and sixty horses and mules. The expedition headed northeast toward the Lewis River in a leisurely fashion with the men taking time to pan for gold along the mountain streams and to hunt marten and beaver. During the first month they averaged about five miles a day. The young commander seized every opportunity to distribute presents among the Indians, to inform them of the pending construction of the military road, and to issue warnings against interference with the emigrants. The party's route from the forks of the Lewis River was eastward across Klickitat River, thence northward across the many tributaries of the Yakima River flowing easterly out of the Cascade range. A major supply base was established on this stream near present-day Ellensburg. McClellan ascended the Naches River from its junction with the Yakima as far as Naches Pass, and here his party was engaged for several days in making observations. The captain was quickly convinced that it was the best pass over the Cascades for both railroad and wagon road.

A junior officer, earlier dispatched through the pass to procure additional mules at Fort Steilacoom, reported back to his commander that no surplus animals were available. Although he assured McClellan that his travels proved the White River valley on the west side of Naches Pass to be better than that of the Naches River on the east for a military road, no attempt was made to run the surveying line west of the divide. After ordering the escort's return and reducing the size of his party because of dwindling supplies, McClellan ascended the Yakima River to the pass of that name. Although he penetrated to the west of the pass only two miles, his men found time to fish in Lake Kachess, and he personally recorded his regret at being unable to procure a canoe for a sail.[7] In spite of Indian entreaties to explore the divide northward, no effort was made and therefore Snoqualmie Pass was missed.

Apparently the officer had not ignored his road-building instructions entirely. Immediately upon his arrival at Columbia City Barracks and before any survey, he had decided to make construction or improvement contracts. This was in direct violation of his orders from the Secretary of War. A. W. Moore, who was named superintendent of the Steilacoom and Walla Walla road, conferred with McClellan a second time at the main supply depot of the expedition on the Yakima River. Here he secured "a little assistance" for the road cutters and signed final contracts. Stevens, in time, questioned the wisdom of making these contracts and ordered McClellan to submit them to the Secretary of War for approval.[8]

From the supply depot the surveying party traveled northward to the Columbia River near present-day Wenatchee. The march continued along the west bank of the Columbia to the mouth of the Methou River, which the men crossed to reach Fort Okanogan. Several detachments went westward from the fort in half-hearted attempts to examine the mountain passes.

The entire command eventually ascended the Okanogan River, crossed over to the Kettle River, and followed a trail down that stream to Fort Colville, where the chief factor of the Hudson's Bay Company, Angus McDonald, greeted them. Here an unexpected meeting took place with Stevens who had arrived from his overland journey. Although relations between Stevens and McClellan remained cordial long enough to have a riotous celebration, Stevens was determined that McClellan should go across the mountains even though it was now late October. The captain was given orders to "go by the passes to the Sound" and some preparations were made when Stevens apparently was convinced that the western division of the exploring party had neither animals nor equipment left to succeed in the task. The two commands separated to go to Walla Walla by different routes. From this fort Stevens once more wrote McClellan urging him to go west by Naches Pass, stating he had information that emigrant parties had passed over with forty-eight wagons in September, and suggesting further that he could forward twenty additional horses for the trip. Two days later McClellan arrived in Walla Walla. The "old proposition," to use the captain's language, was revived, and he objected strongly, no doubt with a picturesque vocabulary, to the governor's repeated demands.

Stevens finally determined to send Frederick W. Lander, a civilian engineer with his eastern command, over the Cascades to do the job McClellan refused to do. Instructions were issued before Stevens' departure for The Dalles so that Lander was given the responsibility for the survey to Steilacoom. Within a few hours after Stevens' detachment left, an Indian reported in Walla Walla that several emigrant wagons had been abandoned in Naches Pass due to heavy snows. All animals had been lost. Lander, on hearing the news, sought the advice of McClellan, who in Stevens' absence had assumed command. The captain declined to issue orders, but gave his personal opinion that the endeavor should be abandoned. He recorded in his journal: "After the action of the Governor in this matter I have determined to give no orders to any persons out of my own party—& to allow no interference in it—I have done my last service (when I have finished this expedition by going to the Snoqualmi [Yakima] Pass from the Sound) under civilians and politicians."[9] McClellan's party headed for Fort Vancouver with Lander close behind. From The Dalles, the command took a steamer to the Cascades of the Columbia, where the baggage was loaded on a tramway by-passing this obstacle in the river. The men walked to a boat below the Cascades, which took them to Fort Vancouver where the expedition was disbanded on November 20.

McClellan remained at Vancouver until December 12, displaying no enthusiasm about reporting to Stevens in Olympia and not attempting to suppress his feelings about the governor. By the time McClellan finally arrived in the new territorial capital, the governor was determined that a survey should be run from the Naches and Yakima passes westward to the Sound. Stevens now knew that the Indian tale about abandoned emigrant wagons was false, that no snow fell in the Naches Pass until November, and then only a few inches. Instructions were issued to Abiel W. Tinkam, an assistant engineer on the eastern division who had not yet arrived at Walla Walla, to come west by the Yakima Pass. At last Stevens had found a man equal to the task. On January 7, 1854, Tinkam began the ascent of the Columbia with two Walla Walla Indians. At the mouth of the Yakima he turned up that stream, crossed the Pass, and reported in Seattle on January 27.

In contrast McClellan reluctantly set forth from Olympia on December 23 to make a survey eastward to the Yakima Pass. His detachment traveled by canoe to Steilacoom and, unable to engage guides, proceeded on out the Sound to the Snohomish River. They paddled up this stream to the vicinity of Snoqualmie Falls. Indians once more provided the welcome news that deep snows would make an entry into the mountains impossible. The explorers therefore descended the stream by canoe and returned up the Sound to Steilacoom. Professor Philip H. Overmeyer has summarized the captain's role in this endeavor:

> ...McClellan did not have those traits of character which Stevens had believed he possessed, and which were so necessary for the successful completion of

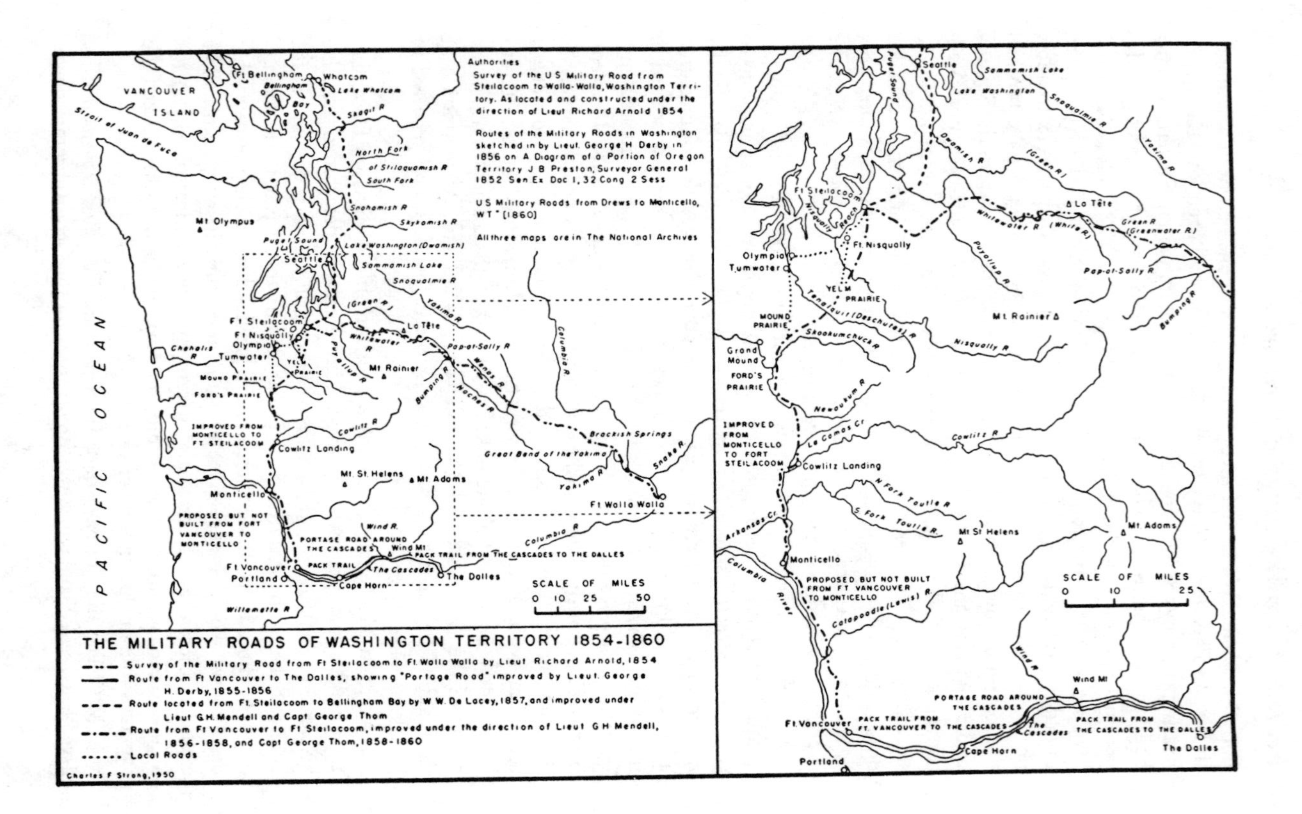

THE MILITARY ROADS OF WASHINGTON TERRITORY 1854-1860
Survey of the Military Road from Ft Steilacoom to Ft. Walla Walla by Lieut Richard Arnold, 1854
Route from Ft Vancouver to The Dalles, showing "Portage Road" improved by Lieut. George H. Derby, 1855-1856
Route located from Ft. Steilacoom to Bellingham Bay by W. W. De Lacey, 1857, and improved under Lieut G.H. Mendell and Capt George Thom
Route from Ft Vancouver to Ft Steilacoom, improved under the direction of Lieut G H Mendell, 1856-1858, and Capt George Thom, 1858-1860
Local Roads
Charles F Strong, 1950
Authorities
Survey of the U S Military Road from Steilacoom to Walla-Walla, Washington Territory. As located and constructed under the direction of Lieut Richard Arnold 1854
Routes of the Military Roads in Washington sketched in by Lieut. George H Derby in 1856 on A Diagram of a Portion of Oregon Territory J B Preston, Surveyor General 1852 Sen Ex Doc 1, 32 Cong 2 Sess
US Military Roads from Drews to Monticello, WT" [1860]
All three maps are in The National Archives
SCALE OF MILES
0 10 25 50
PACIFIC OCEAN
VANCOUVER ISLAND
Strait of Juan de Fuca
Ft Bellingham
Whatcom
Bellingham Bay
Lake Whatcom
Skagit R
North Fork of Stiloquamish R
South Fork
Snohomish R
Skykomish R
Mt. Olympus
Puget Sound
Lake Washington (Dwamish)
Seattle
Sammamish Lake
Snoqualmie R
(Green R)
Yakima R
Ft Steilacoom
Ft Nisqually
Olympia
La Tête
Chehalis R
Tumwater
Whitewater R
Pap-at-Sally R
Puyallup R
Mt Rainier
Mound Prairie
Ford's Prairie
Bumping R
Wenas R
Naches R
Columbia R
Improved from Monticello to Ft Steilacoom
Cowlitz R
Brackish Springs
Cowlitz Landing
Great Bend of the Yakima
Snake R
Mt. St. Helens
Mt Adams
Yakima R
Monticello
Ft Walla Walla
Proposed but not built from Fort Vancouver to Monticello
Wind R.
Portage Road around the Cascades
Wind Mt
Pack trail from the Cascades to the Dalles
Ft Vancouver
Pack trail
Portland
The Cascades
The Dalles
Cape Horn
Willamette R
Seattle
Sammamish Lake
Puget Sound
Lake Washington
Snoqualmie R
Dwamish R
(Green R)
Yakima R
Ft Steilacoom
Nisqually Reach
La Tête
Whitewater R
(White R)
Green R (Greenwater R)
Ft. Nisqually
Olympia
Tumwater
Puyallup R
Pap-at-Sally R
Yelm Prairie
Tenalquit (Deschutes) R
Bumping R
Mound Prairie
Mt. Rainier
Skookumchuck R
Nisqually R
Grand Mound
Ford's Prairie
Newaukum R
Improved from Monticello to Fort Steilacoom
La Camas Cr
Cowlitz R
Cowlitz Landing
N Fork Toutle R
S Fork Toutle R.
Mt St Helens
Mt Adams
Arkansas Cr
Monticello
Columbia River
Proposed but not built from Ft Vancouver to Monticello
Catapoodle (Lewis) R.
SCALE OF MILES
0 10 25
Wind R
Wind Mt
Portage road around the Cascades
Ft. Vancouver
Pack trail from Ft. Vancouver to the Cascades
The Cascades
Pack trail from the Cascades to the Dalles
The Dalles
Cape Horn
Portland

the objects of the survey. McClellan's slowness in organizing and moving his command, his timidity when approaching the mountains, and his fright at the thought of snow all combined to defeat the purpose of his expedition. But more than this, his actions as a leader in charge of his first command and his willingness to compromise and to magnify the difficulties of his task indicated that if given a command sometime in the future he would be tried and found wanting.[10]

Discouraged by McClellan's delay in initiating the road survey, Washington residents, early in June, had organized a volunteer party to search for a route across the Cascades. All were anxious to direct Oregon bound emigrants to the settlements on the Sound during the 1853 season. *The Olympia's Columbian,* Washington's first newspaper, sponsored and advertised the activity.[11] When it became apparent that McClellan had failed in his duty, a subscription fund of $1,200 was raised by Washington citizens to organize a road-working party. Under the leadership of Edward J. Allen, a party of road workers, many of whom contributed animals, tools, and provisions as well as labor, went into the forest to clear a path.[12] By October 15 the task was complete and scores of emigrant wagons followed immediately behind the work crews.

On March 6, 1854, Governor Stevens instructed McClellan to turn over to Lieutenant Richard Arnold the properties and moneys belonging to the military road. The lieutenant left Steilacoom on May 23 with a working party to make a reconnaissance to Walla Walla. Where passable, he traveled along the road opened by settlers the previous season, to determine what sections could be permanently improved. Beyond the Naches Pass he followed that river some distance before crossing over to the Wenas, thence by the Yakima Valley and the Columbia to Walla Walla.[13] The distance was measured at 234 miles. Arnold was far from satisfied with the route. East of the Cascades, emigrants were forced to cross the Naches River forty-four times to avoid perpendicular embankments; to the west, along Allen's road, the hills bordering the Puyallup River were dangerous. An additional appropriation of $10,000 to examine the country north of the road was recommended. The Army officer also suggested that the settlers be reimbursed for funds spent in 1853 because the greater part of the road cut by them from Steilacoom to the mountains had been adopted.[14]

The road across Naches Pass was never popular with emigrants. The population on the Sound shifted quickly toward Seattle after 1853, and residents there urged the construction of a more northerly route across the mountains at Snoqualmie Pass. No matter what route was followed across the Cascades, the eastern terminus invariably was at Walla Walla. Here emigrants on the Oregon Trail could be met and urged to take up residence in Washington. This was the immediate purpose, whether the improvements were federal military roads or those constructed by local residents.[15]

The government of Washington Territory was officially organized in February, 1854. Governor Stevens in his message to the first legislature revealed his particular interest in transportation by emphasizing the urgent need for roads. The lawmakers not only authorized the construction of many territorial roads, but memorialized Congress for federal aid through appropriations for military routes. The generous Thirty-third Congress complied in February, 1855, by allocating $25,000 to be spent between Fort Vancouver and The Dalles of the Columbia, and $30,000 from Fort Vancouver to Fort Steilacoom on Puget Sound.[16] The supervision of the roads was placed under the newly created Pacific Coast Office of Military Roads in San Francisco with Major Hartman Bache in command. Lieutenant George H. Derby was ordered to Fort Vancouver to superintend the field work in Washington and Oregon. The initial instructions issued by Jefferson Davis suggested that the general course of the roads would be indicated by the Columbia and Cowlitz rivers. Surveys and preliminary examinations should, as usual, first be made to ascertain the best and most economical routes. Contracts for construction were to be made with local residents if possible; otherwise the military should handle the arrangements of procuring materials and the necessary labor force. To hasten the work, the Army officer need not await War Department confirmation of the contracts as required by law. Should the Department later disapprove of any of the written agreements, the contractors would be paid for the work already performed at the rate originally agreed upon. Davis concluded his instructions with the insistence that the major object was to open each road through its entire length; utilizing the appropriations only in limited areas would be contrary to the intent of Congress.[17]

When Lieutenant Derby was delayed during the summer of 1855 with the Oregon roads,[18] Washington residents pressed for some progress there before the winter. Finally in September, Derby dispatched his civilian engineering assistant, George Gibbs, to survey a trail along the Columbia and Cowlitz rivers and on to Steilacoom. Another civilian, Robert Whiting, was to assist on the road to The Dalles.[19] Not until October was the lieutenant personally able to examine the 95-mile route from Fort Vancouver to The Dalles. The road along the north bank of the Columbia was found good for the first 15 or 20 miles until the Cape Horn Mountains, a part of the Cascade Range, were reached. Here all prospect for a wagon road terminated. The range could not be avoided and its descent was impossible for wagons. Then came the Columbia bottoms, often flooded by the river rise of 12 to 15 feet in the spring, and a road could not be raised above the flood level because the mountains came down to the river. Halfway to The Dalles, the traveler encountered the Cascades, a 30-foot fall in the river within five miles. Around these rapids the Portage Road had been built, but for more than a mile it was impassable, and speculators had built a wooden trainway three feet wide to transport freight. The United States Army paid twenty cents for each pound handled here. Above the portage the trail ascended the mountains, crossing the Columbia at Wind Mountain and continuing to The Dalles by a circuitous and rugged route along the south bank. Lieutenant Derby suggested that it would cost about a million dollars to make a good wagon road from Vancouver to The Dalles. He proposed to concentrate on improving the Portage Road and eliminate the necessity of using the trainway. The United States would save the amount of the appropriation in a single year. Derby insisted: "Good steamboat navigation from Vancouver to the Cascades, a good road across 'The Portage' and a continuation of steamboat navigation thence to the Dalles, certainly fulfills all the conditions of a 'Military Road' from the Dalles to Columbia Barracks and is moreover the only practicable route."[20] Colonel Abert agreed with the proposal and secured the approval of Secretary Davis.[21]

Because the soil at the Cascades became tenacious mud during the rainy season, Derby suggested the construction of a plank road along the entire distance of the portage. He reasoned that an ordinary dirt road would require repairs after each rainy season and without constant federal appropriations would become useless. A plank road could be expected to last ten years. The present appropriation would suffice to construct a road only 16 feet wide, graded and prepared so that one half of its width could later be planked. The lieutenant proposed to purchase supplies and hire labor to work on the project under his direct supervision. During the season of 1856, he also planned to improve the trail from Vancouver to The Dalles by widening, straightening, and reducing grades so that it could be used by dragoons, for a pack trail, and for driving stock.[22]

By May 1, he had a working party of fifty men whom he intended to employ for four months at the portage site. Two hospital tents were used to accommodate these laborers; a cook was hired to prepare meals. Difficulties immediately arose. Torrential rains fell three or four days out of each week making it impossible for the men to work. The expense of their board and the delay in construction perturbed Derby; the loss of wages on rainy days annoyed the laborers. The near-by Indian tribes, who had been on the warpath the previous season, burning down all but one building at the Cascades, threatened attack constantly, and Derby pleaded for a protective guard of soldiers. Planning and persuasion succeeded in keeping eighteen to twenty-five men on the job, but after each payday departures reduced the force. Recruits were secured periodically in San Francisco and sent by steamer to Portland. The Irish and Germans, according to Derby, made the best workers, and those coming from California remained on the project much longer than local laborers. When the harvest season approached, farmers attempted to outbid the Army and forced a wage increase to $52 a month.[23]

The crisis finally came July 10, when civilian guards stationed to protect the sleeping crew discovered three men hiding in the forest at midnight. The road workers, fearing an Indian attack, remained awake with rifles poised the rest of the night. At daybreak it was learned that the intruders were drunken soldiers who had deserted, but the entire force quit work. Derby refused their wages, claiming he did not possess sufficient funds, and secured a promise from the steamboat captain that he would not take them down the river. The men returned to the job in disgruntled fashion. The officer renewed his request for a military guard of ten men from the company of soldiers stationed at the Cascades. He protested:

> I can not see why even with the small force at the Cascades a guard should not be detailed for me, but possibly it is preferable to have an important public work obstructed and the property belonging to it lost and destroyed rather than the fixed and settled routine of the military duty be in the slightest degree disturbed.

If United States soldiers were not made available, he had planned to request Governor Stevens for a company of volunteers. On July 24 an Army officer and twenty men were assigned to guard duty.[24]

The Portage Road was half completed by August 1. A month later three and one-half miles were in excellent condition and all anticipated its completion within six weeks. The initial appropriation had served to cut out timber, grade the roadbed, and provide drainage but, as had been expected, no funds were left for planking. In the difficult places, mountain sides had been cut back and cribwork with rock foundations placed to prohibit slides. Derby reported the existence of a wagon road good in dry summer weather, but muddy and impassable for seven months out of the year during the rains. A sum of $18,800 was needed for planking and he proposed that the work be done by the quartermaster's department at Vancouver.[25]

Upon the return in November of George Gibbs' surveying party from the route to Fort Steilacoom, Derby planned construction for the summer of 1856. The route was divided into seven sections. The first two, along the east bank of the Columbia between Fort Vancouver and Monticello, at the mouth of the Cowlitz, were 55 miles long. He estimated the construction cost of this distance at $96,000, a somewhat needless expenditure because of steamboat transportation on the rivers. A road built along the five remaining sections from Monticello to Steilacoom, 84 miles distance, would cost $163,500. From Monticello to Cowlitz Landing transportation was by row and pole boats on the river and by a road along the west bank. The combined river and land routes temporarily afforded sufficient facilities for military and civilian travel. The lieutenant was convinced that the sector between the Cowlitz Landing and Ford's Prairie was in greatest need of improvement, and that the $25,500 remaining of the federal appropriation would make a road for wagons. Here the valleys of the tributaries of the Chehalis were often flooded and impassable so that wagons had to be abandoned. From Ford's Prairie a road had been cut through the timber by local residents to Olympia and Nisqually on the Sound. No road existed beyond this point, but no obstacles, other than a few narrow belts of wood, prevented construction to Steilacoom.[26] Major Bache personally concurred with Derby's proposal to concentrate the season's efforts on one sector, but referred the proposal to the Bureau for approval. Colonel Abert, in reporting to the Secretary of War, noticed that the plan did not coincide with Davis' instructions to make the roadway practicable for its entire distance, rather than perfecting one part at the expense of the remainder. He recommended that Major Bache be instructed to cease all work not necessary to make the whole route practicable.[27]

In the meantime Lieutenant Derby advertised for bids to construct the road and bridges just beyond Cowlitz Landing. The lowest proposals were in excess of the available funds, and he inquired of Bache for instructions. Several contracts on the Astoria-Salem road had just been rejected by the War Department on the grounds that they were not in harmony with the original instructions to the Pacific Coast Office of Military Roads, so Bache advised postponement. He forwarded a memorandum to Derby from the Secretary of War which stated in part:

> [The instructions] had in view the opening of the road the entire distance between these points, and the character of the road was to be such that its cost should not exceed the amount appropriated though this should limit the work done merely to the removal of the trees and underbrush from the way through the forests between the points named.[28]

Then followed his own interpretation:

> The question for you to consider, if I construe the wishes of the Department right, are [*sic*] whether the opening of the road for the passage of wagons, from

Cowlitz Landing to Fords Prairie will in effect open the entire route from Vancouver to Steilacoom and if not whether sufficient of the appropriation will remain after the completion of the above to make the rest of the road alike passable. If not sufficient, it might be inferred . . . that no expenditure whatever for the object could be made.[29]

Territorial residents, provoked by the delay, began improvements upon a route northward from Cowlitz Landing. Derby dispatched George Gibbs to survey the new route being improved, and on the basis of his report, the lieutenant was convinced that the settlers' line was perhaps better than the original located by Gibbs. Major Bache expressed understanding of the earlier mistake made by the civilian assistant:

Everyone similarly employed in a wild country is well aware that he must in some measure at least rely upon the accounts he receives from others of the physical features of the country, and that it is not always possible for him to know whether they are communicated to him in good faith or from intended motives to lead him astray. Considering the shortness of the season and the troubled state of the Indians, it is not surprising then that if attempts were made to mislead Mr. Gibbs he should have fallen into error. Of course, I cannot withhold my approval of the correction or as a consequence of the adoption of the new route.[30]

The settlers along the first survey were alarmed over the possible change in location. Two rival factions emerged, both urging construction along a route beneficial to their own interests, guaranteeing that contract bids would be within the limits of the appropriation, and volunteering services and supplies to the Army project. Derby attempted to resolve the conflict by advertising for bids without specifying the route.[31] Bache rejected such procedure as beyond his discretion and notified the lieutenant: "It is your duty to select the best route, and if you are satisfied upon the point without any question of doubt, advertise it and advertise it alone."[32]

The road from Cowlitz Landing to Ford's Prairie, along the new route, was accordingly divided into five subsections, specifications for each improvement were drawn up in detailed fashion, and construction bids called for. The Army procedure of publishing notices in the territorial press and posting placards in public places was followed.[33] As a counter move, a large group of Washington citizens presented Derby with a petition on September 15:

The undersigned citizens of Lewis and Thurston counties respectfully request that proposals be issued anew for the construction of the U. S. Military Road from Cowlitz Landing to Ford's Prairie . . . as surveyed and advertised last year and that in advertising the same the route be divided into smaller sections, say five miles each, and that payments be allowed on account of work actually done to the amount of fifty percent of the contract price. Owing to the disturbed state of the country last winter [Indian depredations], and the terms offered, . . . many persons were prevented from taking contracts who would otherwise have been glad to do so, and we are satisfied that if re-advertised in the manner above mentioned as favorable lettings can be made on this as on any other route.[34]

Meanwhile, Governor Stevens urged the War Department to action by reminding Jefferson Davis that Congress had made an appropriation for the *completion* of a military road from Vancouver to Steilacoom.[35] The climax was the removal of Lieutenant Derby. Major Bache was incensed at this action and in forwarding the lieutenant's defense to the Bureau of Topographical Engineers announced that he was "constrained to endorse the views of Derby and express regret that the Secretary of War had not, before implying a censure of the course of action taken, conferred with gentlemen in Washington relative to the country of the Pacific Northwest."[36]

Lieutenant George H. Mendell assumed the responsibilities as superintendent of military roads in Oregon and Washington in October, 1856. Operations on the Portage Road, virtually completed by Lieutenant Derby, were suspended the following month and laborers discharged. However, as a result of winter rains, slides from the embankments crashed down on the cribwork protecting the roads, and in places where the soil was soft these timbers gave way, falling into the roadway and blocking travel. In the spring of 1857 the funds remaining in the federal appropriation were used to rebuild the cribwork, drain the surface of the road, and to gravel, corduroy with logs, or plank those segments most difficult for wagon travel. Mendell, like Derby, considered it an excellent summer road. The quartermaster's department of the Army used it continuously. He further observed: "A six-mule team can haul two tons over it; and as the rate of transportation of the private company over the portage is $15 per ton, and as a team can easily make two trips per day, it will readily be seen that the public interests are much advanced by the construction of this road."[37]

In viewing the status of the Vancouver-Steilacoom road, Lieutenant Mendell confirmed all that Derby had reported. The farmers had formed an association to build a road for $15,000 along the latest survey from Cowlitz Landing to Skookum Creek, including a bridge over the Newankum River. Great excitement prevailed when this was made known and settlers along the older route subscribed several thousand dollars to aid any contractor willing to build for them. They insisted that the reason bids were so much higher when construction was first proposed was because a much finer road was contemplated. Moreover, no payments were to be made until completion, and men of moderate means were unable to bid.[38] Mendell advertised for bids to construct the road north of Cowlitz Landing. Those submitted by L. J. Tower and Louis Johnson, spokesmen for the old route, were disqualified because they submitted estimates for only a part of the advertised construction. The lowest bidder was Thomas J. Carter, representing those promoting the new survey, and Mendell entered into a contract with him.[39]

The losers protested directly to the Bureau of Topographical Engineers, and Colonel Abert referred the controversy to the Secretary of War. Davis disapproved of the Carter contract and declared: "Proposals will be invited to construct a road of certain standard and durability between these points, that is, between Cowlitz Landing and Ford's Prairie, without designating its location, and the contract awarded to the lowest responsible bidder."[40] Thus he reversed his original position and accepted the program of reconciliation urged by Lieutenant Derby several months previously. As a result of the new bidding, the Tower faction secured the contract for the controversial 25 miles. Later in that same season an additional agreement was made to build from Ford's Prairie on to Yelm Prairie, along the banks of the Nisqually River.[41] With the completion of these improvements in November, 1857, a usable road existed from Cowlitz Landing to Steilacoom.

With $10,000 remaining of the allotment, Lieutenant Mendell hoped to start construction southward to the Columbia River. Three possible routes ran from the Cowlitz Plains to Monticello, two on the east bank and one on the west. The Army engineer preferred the west bank because it was shorter, and the problems involved in maintaining a ferry to the east bank opposite the landing would be avoided. Settlements on Puget Sound received mail by this route, and also it was the chief avenue of travel to Oregon and California. The Cowlitz was a very rapid stream, at times becoming almost a torrent full of driftwood which rendered its navigation by canoes dangerous. These small craft were also unsatisfactory for moving any type of freight, and the charges of $40 a ton for the 30 miles between Monticello and Cowlitz Landing forced all Washington citizens to receive only the most essential goods. Farmers had no satisfactory outlet to the south. For these reasons, Lieutenant Mendell felt justified in recommending an additional appropriation of $15,000 to complete the road.[42]

Army officers were not alone in urging the federal government to provide additional funds for Washington military roads. The second and third sessions of the legislative assembly of Washington Territory had actively memorialized Congress for support. A favored project was the construction of a military road from Fort Steilacoom to Bellingham Bay in the northernmost part of the territory, where the Army had selected a site for a new military post. Washington lawmakers quickly suggested that such a road was a military necessity "... to say nothing of the development of the resources of our Territory, and establishing communication between our people."[43] The Thirty-fourth Congress considered an appropriation of $35,000 for this improvement. A bill was also under discussion to allocate $10,000 additional for a road from the Sound across the Cascades and to reimburse the settlers for their expenses on this endeavor during 1853, not in excess of $8,000. A bill was introduced at this session allocating $45,000 for a military road branching westward from the Vancouver-Steilacoom route to the mouth of the Columbia, a public work endorsed by the War Department.[44] Congressional approval was given only to one measure, the Steilacoom–Bellingham Bay construction.[45]

When Major Bache notified Mendell that Congress had launched this new project, as well as making funds available for the continued improvement of three Oregon roads, the lieutenant frankly admitted: "I cannot see how it will be possible for me to make the surveys necessary for the location of the Steilacoom and Bellingham Bay Road."[46] The country was unsettled and, except in the immediate neighborhood of Fort Steilacoom, the region to be traversed had never been explored. A search had to be made to find a capable civilian engineer. W. W. DeLacy, placed in charge, organized a small reconnaissance party of three white men and six Indians. Because of the dense timber and excessive number of fallen trees thought to blanket the area, pack animals were believed impracticable. The Indians were to carry supplies on their backs. Arrangements were made to provide the party with additional guides and canoes at the Snohomish and Skagit rivers.[47]

The surveying party left Fort Steilacoom on August 11 and pushed rapidly northward through the dense forest, yet following as closely as possible the coast of the Sound. Upon arrival at Bellingham Bay on September 20, DeLacy reported that the road was divided naturally into five sections by the Dwamish, Snohomish, Stilaquamish, and Skagit rivers. The surveyors were back at Fort Steilacoom by November 21, having run a line the entire distance of 137 miles. Although a territorial road existed from Steilacoom to the Puyallup River, beyond that point minimum improvements to open the route would call for an expenditure of $77,355, according to DeLacy's estimates.[48]

Convinced that a wagon would not pass over the proposed route for years and that any road built, but unused, would soon fill up with timber and become an impassable trail, Lieutenant Mendell proposed to spend the federal funds at each terminus, building a good road from Steilacoom to Seattle and from the military post at Bellingham Bay to the settlement at Whatcom, five miles distant, where there were coal mines and an excellent wharf. Any remaining funds would be used to cut a trail from Seattle to Whatcom. Early in the spring of 1858 contracts were made with William D. Vaugh and A. C. Lowell to extend the territorial road from its Puyallup River terminus on to Seattle, a distance of 27 miles. The Secretary of War rejected this contract, but not before 20 miles of road were completed. E. C. Fitzhugh and E. D. Warbass constructed five miles of road between Bellingham Fort and Whatcom. In agreement with the War Department's repeated instructions, Mendell now advertised for bids to construct a trail the entire distance from Steilacoom to Bellingham Bay. The one proposal submitted asked payment in excess of the $8,500 remaining from the original $35,000 appropriation. Mendell therefore recommended that the Department request Congress to appropriate additional funds to complete the road.[49]

During the spring and summer of 1858, work had progressed on the Vancouver and Steilacoom military road. The construction from Cowlitz Landing to Ford's Prairie was inspected by DeLacy upon his return from Bellingham Bay, and after several defects were removed, it was accepted by the government. George Drew and William James, local road builders, extended the road from Cowlitz Landing for 17 miles toward Monticello. Mendell repeated his request for $10,000 to finish the road and notified the War Department that $40,000 would be necessary to open the road from Cowlitz to Vancouver.[50] Operations in the Northwest had to be curtailed because no more federal funds were forthcoming. The Pacific Coast Office of Military Roads was transferred to Fort Vancouver, and both Major Bache and Lieutenant Mendell assigned to other duty. The federal road program became the responsibility of Captain George Thom.

When the Thirty-fifth Congress assembled, Delegate Stevens increased his efforts to secure federal appropriations. One measure introduced into the House of Representatives provided for the completion of those roads already begun: $10,000 for Steilacoom to Walla Walla with additional compensation to settlers for their 1853 efforts; $15,000 for Vancouver to Steilacoom to be expended between Cowlitz Landing and Monticello; and $60,000 for Vancouver to The Dalles, chiefly for planking the Portage Road. A bill also sponsored by Stevens was to launch three new projects: $60,000 to build from the mouth of the Columbia by way of Shoalwater Bay and Gray's Harbor to Olympia; $50,000 from Olympia to Port Townsend; and $50,000 from Seattle across the Cascade Mountains to Fort Colville. The

Washington delegate also asked $200,000 for a road to connect the headwaters of the Missouri with those of the Columbia. The Military Affairs Committee recommended the legislation to continue what had been started, but the other two bills were reported unfavorably. The committee then sought to be discharged from further consideration of five memorials presented by the legislative assembly of Washington Territory, requesting federal aid to transportation.[51]

In a desperate effort to offset the effects of the committee's adverse reports, Stevens insisted that the construction of these roads had been strongly urged by the War Department. Letters from Lieutenant Mendell and Colonel Abert were read into the *Congressional Globe*. The Chief Topographical Engineer believed all the newly proposed military roads essential: one would provide speedy access to the coast at the Columbia's mouth in case of Indian disturbances; the Port Townsend–Olympia construction would facilitate settlement to such an extent military protection would be unnecessary. Seattle was the natural port and outlet of the Yakima country east of the Cascades and of the mining region around Fort Colville, and this important connection traversed country where the Indian tribes were hostile and warlike. Then Stevens had his say:

> I introduced this bill, and urge the construction of these roads upon military grounds alone. They are required for the defense of the Territory. We have, west of the Cascade mountains, twelve thousand Indians, who, within the last two or three years, have been all disaffected, and in regard to whom disaffection may be their normal condition for years to come. It is a difficult country to communicate in. In the military service of 1855–56, the gist of the defenses consisted in cutting out trails and building roads. I therefore urge this measure as a measure of defense and a measure of economy.[52]

To all these pleas, the Thirty-fifth Congress turned a deaf ear.

Captain Thom served as liquidator of the federal road program from 1858 to 1860. The work of local contractors improving the road south of Cowlitz Landing for a distance of 17 miles was first inspected and approved. From its terminus, a trail along the west bank of the Cowlitz to Monticello was cleared of fallen timber so that pack trains, at least, could get through. Philip Keach was given a contract to open a trail six feet wide through the forest from Seattle to Whatcom, practicable for pack mules and stock driving. This exhausted the last federal appropriation.[53]

In each annual report, 1859–1861, the Bureau of Topographical Engineers recommended additional appropriations for these three military roads in Washington. A road from Seattle, across the mountains at Snoqualmie Pass and down the Yakima to join the Mullan road being built from Fort

Walla Walla to Fort Benton, was continuously presented as a military necessity.[54] The Thirty-sixth Congress ignored the recommendations, but approved $10,000 to be expended exclusively between "the Cowlitz River and Monticello" along the Fort Vancouver–Steilacoom road.[55] On examination of this section of the road, Captain Thom discovered that it was opened for the most part, and some improvements had been made by near-by residents. A party of laborers under foreman Ira P. Thraser completed construction during the spring of 1861. Elsewhere above Monticello, Washington citizens, working on a volunteer basis, had improved the route so mails were being delivered between Monticello, Olympia, and Steilacoom in horse-drawn coaches.[56]

No completely satisfactory explanation can be found for the determination of Congress to shut off appropriations for the military roads in Washington Territory so strongly recommended by the War Department. The McClellan failure taught national politicians the necessity of superlatives in describing the terrain of the Pacific Northwest. The difficulties encountered by Lieutenants Derby and Mendell further proved that comparatively large federal road appropriations were insufficient to do a creditable job. When funds were most urgently demanded, 1857 to 1861, the Thirty-fifth and Thirty-sixth Congresses adopted a basically more conservative policy in allotting funds for all territorial internal improvements. Perhaps Stevens was not as influential as his fellow delegates in securing patronage.

CHAPTER VII : *Federal Road Projects in New Mexico Territory, 1853-1860*

ROAD IMPROVEMENTS in New Mexico were neglected by the Mexican government. The colonists along the upper Rio Grande River were too few and too far away to expect much assistance of any type from a young republic struggling to solve the basic economic and political problems threatening its very existence. Moreover, the geography and climate of the New Mexico area tended to force the earliest settlers into the valleys of the larger rivers where a continuous water supply was available. The villages were often separated by uncompromising mountain ranges, and connections between these isolated communities soon became essential both for trade and for military defense. This latter consideration loomed large in the minds of inhabitants surrounded by Indians constantly threatening depredations. Yet the roads that existed were "... mere bridle-paths resembling more the trail of the Indian than the high-ways of a civilized people ..."[1]

The inadequacies of the New Mexico road system were obvious to the United States Army at the outbreak of the Mexican War. The troops, led by General Stephen W. Kearny, had more trouble with the roads than they had with the Mexicans. While Kearny and Cooke were exploring routes to California, two young engineering officers of the Topographical Corps, James W. Abert and William G. Peck, left behind because of illness, made a thorough reconnaissance of the upper Rio Grande. They charted the course of the river and its tributaries, determined the width of the valleys, and noted the position of towns and mountain peaks. The purpose was not exclusively geographic. Statistics were compiled on population, the available livestock, quantities of grain under cultivation, the mineral resources, and potential sources of water power. Information was likewise procured from this exploration to implement the Army's local road-building program in New Mexico. From their journal and map, the Bureau of

Topographical Engineers learned that roads radiated out of Santa Fe in four directions: northwest, northeast, east, and south. To the northwest, by way of Cañada and Abiquiu, was the beginning of the Old Spanish Trail to Los Angeles. To the northeast, the road to Taos was a part of the Santa Fe Trail. Easterly, communications wound through the mountains to reach the headwaters of the Pecos and Canadian rivers. Southern communications were along the Rio Grande Valley to El Paso, except for a divergence from Fra Cristobal where the main road swept onto the Jornada del Muerto. Local trails also extended to isolated settlements built along the various tributaries to the Rio Grande. Mountains flanking both sides of the Rio Grande Valley in northern New Mexico retarded expansion of the transportation system in that area; furthermore, passage through these mountain ranges was very difficult. Peck and Abert agreed that extensive labor on the roads was essential before they could become adequate arteries for trade and military defense.[2]

As usual, the officers of the Quartermaster Corps were perturbed at the condition of roads over which Army supplies had to be transported. Joseph H. Whittlesey, quartermaster officer of the First Dragoons, complained that the road northward from Santa Fe to Taos was cut up by arroyos and ditches. The hauling of forage and wood in wagons resulted in excessive wear and heavy breakage. As far as La Joya—present-day Velarde—the road, though inferior, was passable for wagons. Beyond La Joya the wagon road made a circuitous detour of 100 miles to Taos when the direct distance was only 35 miles. The route was so bad that wagons could carry only half loads. A shorter path existed through the mountains from La Joya to Taos, but one officer reported damage to the gun carriages he had forced over the trail. Obviously wagons could not be driven over it. Lieutenant Whittlesey recommended federal appropriations "to blast a good and direct road" over which he could haul supplies. From Taos, the roads to Las Vegas, El Moro, and El Rayado, settlements east of the mountains, were "mere bridle-trails along the beds of mountain torrents." The road from Taos to Ocaté was possible only for empty wagons.[3]

These observations of road conditions by Army officers stationed in New Mexico and their requests for alleviating the problems created were quickly reinforced by pleas from the residents of the territory. Between 1851 and 1861 the New Mexico legislative assembly presented, through its delegates in Congress, eleven petitions requesting federal assistance in improving transportation facilities, by building either roads or bridges.[4] The tenor of these memorials suggests that the settlers thought the necessary projects beyond the scope of individual or collective action of those in the territory.

Their pleas exhibit a tone of expectancy, a feeling that what they requested was a natural function of government. No evidence was shown of their interest or confidence in individualism.[5] In a mood of supplication, they pointed to their distress and need:

. . . Great necessity exists for public roads and high-ways; connecting the extremes of our Territory with its centre, where is the great mart of the trade and commerce of our country.

With frontier enthusiasm they also exaggerated the importance and potential of the community:

This want is now particularly felt in the region included within the county of Taos; a region rich in the most fertile vallies [*sic*], admirably adopted [*sic*] to agricultural pursuits, and already densely populated by an industrious and thriving community; but now measurably cut off from all access to a market for the products of the soil, from the absence of any road.[6]

In aiding New Mexico residents, the government would likewise provide for its own needs. The opening of such roads would speed military operations in the area as well as decrease the high cost of shipping military supplies over inadequate routes. This argument was recapitulated by the War Department in its statements to Congress.

The Thirty-second Congress ignored this general request for federal aid, so in January, 1853, the legislature repeated its demand. Although admitting the possibility of railroad construction sometime in the future, the lawmakers insisted that ". . . for the present we would be contented with Common Roads, to Missouri, Arkansas, Texas, California, and Utah." The petitioners reiterated:

The roads in the Territory, are in bad condition, and the great extent of these roads make the expense of a thorough repair of them altogether beyond the means of the people themselves.[7]

They agreed that the Army was the logical federal agency to be charged with the work, once Congress gave approval. Again in 1854 the territorial lawmakers requested money, this time limiting pleas to building bridges over the Rio Grande[8] and improving the roads from Santa Fe to Taos.[9]

Congress responded on July 17, 1854, by appropriating $20,000 to repair the road from Taos to Santa Fe, the route that had been a constant source of annoyance to Army officers and settlers alike; also $12,000 for improvements from Santa Fe to Doña Ana along the marches of Kearny and Cooke. On this last road any amount of the funds necessary were to be used in sinking experimental artesian wells.[10]

The debates in Congress over the passage of this first military road appropriation for New Mexico were overshadowed by the early discussions about the creation of Kansas and Nebraska territories, and no principle of constitutional importance arose in the arguments. Members of the national legislature, far removed from the New Mexico scene, relied heavily upon the testimony of federal officials and local residents. During discussion in the House of Representatives, Virginia's Charles J. Faulkner inquired if the Santa Fe–Taos project had War Department approval since the Secretary had notified the Committee on Roads and Canals of the previous Congress that he did not consider it a military necessity. Philip Phillips of Alabama, spokesman for the Committee on Territories, read into the *Globe* a deposition from Colonel Edwin V. Sumner, military commander in the territory, endorsing the construction of the road from Santa Fe to Taos. Relative to the road to Doña Ana, the colonel doubted the propriety of the grant unless public moneys were used to sink wells and build cabins on the Jornada del Muerto below Fra Cristobal to provide water and shelter for those who chose to travel over that desert. The road itself was generally level and satisfactory except for sandy sections which could only be improved by planking at an unjustifiable expense. As usual, strict constructionist John Letcher of Virginia supported his colleague Faulkner in questioning the propriety of legislation aiding the territories. He inquired if a survey had been made and estimates of construction costs prepared. Phillips assured the Virginians that the roads had been used sufficiently so that no survey to locate the line of the road was essential. Letcher, in turn, replied:

> The gentleman has had experience enough in legislation to know that when appropriations are asked for, even with estimates to back them, that the appropriations, although the whole amount given may be equal to the estimates, very rarely are sufficient to make the road complete. And I understand that the estimates formed here are formed by a gentleman who rode over the road, who never put an instrument upon it, who does not even describe the character of the ground over which this road is to be made . . .[11]

In an effort to change the pattern of discussion, Thomas S. Flagler of New York called for the opinion of the territorial delegate of New Mexico, Jose Manuel Gallegos. This highly respected priest, elected through the influence of the Spanish-speaking clergy who resided in New Mexico before the American occupation, could speak no English.[12] Phillips had to read his letter of endorsement. The final vote on the measure in the House was seventy-three in favor and fifty-nine opposed, and the Senate later passed it without a recorded vote.[13]

As the statute provided for the expenditure of funds under the direction and control of the War Department, Secretary Jefferson Davis directed the Bureau of Topographical Engineers to assume responsibility. Colonel Abert selected Captain Eliakim P. Scammon, an officer on duty with the Great Lakes surveys at Detroit, as superintendent of the New Mexico road building. Scammon expressed regret at being assigned duty in the Southwest for a prolonged period after seventeen years of service with the corps. His departure was delayed as long as possible.[14] After examining the roads, the captain estimated that the first year of operation would require $15,365. He organized on a grand scale. Funds were utilized to pay the salaries of one assistant and five engineers, who were to receive $4.00 a day, and the salaries and subsistence of twelve infantrymen workers, which amounted to $6,570. The cost of mules and labor was estimated at $7,000.[15]

Although funds were requisitioned, Scammon remained uncertain about the exact nature of his assignment and requested that the duties be specified.[16] Replying in November, 1854, Secretary Davis forwarded his standard letter of instructions to Army engineers engaged in military road constructions. Derby in Washington, Withers in Oregon, and Simpson in Minnesota had received the same message. The roads were first to be made practicable throughout the entire distance. Only then could any remaining funds be used on the most difficult sectors. The actual road work could best be performed by local citizens under contract because their interest in securing better roads for commerce and travel would inspire superior work. However, the New Mexico projects were in one sense unique. The lack of water on the Doña Ana route had to be overcome, and the officer was to direct his first attention to eliminating this difficulty if possible. The sinking of artesian wells was believed the easiest and cheapest means of procuring water, and when the apparatus used for that purpose in conjunction with the exploration of the railroad route to the Pacific was no longer required, it would be turned over to the road engineer in Santa Fe.[17]

In the first season, 1854–1855, Scammon spent $13,000 of the Santa Fe–Taos and Santa Fe–Doña Ana road appropriations. The complexity of bookkeeping and the procedure in granting government contracts defeated the captain, and his request for further advances of money was denied because of the irregularity of his accounts.[18] At the close of 1855 the surveys had not been started.[19] Scammon filed no detailed reports of his accomplishments, and there is no evidence of his success as a road builder. In 1856 he was dismissed from the service because he could not account for a $350 transportation charge held against him by Quartermaster General Thomas S. Jesup.[20]

In spite of Scammon's incompetence, Davis announced satisfactory progress on the military roads in all the western territories in his annual report for 1855.[21] In 1856, however, the Chief Topographical Engineer admitted that "... no further progress has been made since the last annual report toward the completion of the military roads appropriated for in New Mexico."[22]

On March 3, 1855, Congress had authorized three new military roads in New Mexico and appropriated $32,000 for their construction: from Santa Fe to Fort Union, Albuquerque to Tecoloté, Cañada to Abiquiu.[23] Responsibility for the five New Mexico roads passed to Captain John N. Macomb, who acknowledged his appointment in May, 1856.[24] Late in March of the following year, the captain arrived in Texas, and after a brief delay en route reached Santa Fe on May 27, 1857.[25]

Included among his inheritances from Scammon, Macomb found several ruined tents, harnesses, and pack saddles. Because the equipment had deteriorated beyond use, he requested that an Army survey board condemn it so that he could begin work with new gear. After reporting the fact that Scammon had ignored all roads except the Taos to Santa Fe route on which little had been accomplished, Macomb began an investigation of his predecessor's confused accounts.[26]

The new superintendent was eager to begin work. Although desirous of liquidating all claims as quickly as possible, Macomb proposed that work be suspended on the two roads under Scammon's control pending settlement, and that he work on the three new projects beginning with the Fort Union–Santa Fe road. His immediate and urgent need was an assistant who could speak Spanish. All attempts to secure the services of a junior officer who knew the language were unsuccessful, so Macomb asked permission to hire a civilian interpreter.[27] The final authorization from Washington to launch the new improvements was delayed by a cabinet change, but the new Secretary of War, John B. Floyd, approved his requests on August 20, 1857. While awaiting the new directives, Macomb had occupied himself for several weeks in surveying the boundaries of the Fort Stanton military reservation on the right bank of the Rio Bonito in present-day Lincoln County.

Upon completion of this assignment, the engineer returned to Santa Fe to organize his road-working parties. He soon learned that a major problem facing any officer dispensing money in New Mexico was that of obtaining specie. Much coin had left the New Mexico settlements because of a current premium on silver in meeting the demands of trade with Mexico.[28] This deficiency had meant that United States drafts were almost useless and accounted for some of Scammon's confusion. It was especially difficult for Macomb to obtain specie from local merchants when several of the old drafts amounting to $8,000 remained unpaid.[29] In January, 1858, the engineer secured $18,000 in specie from a willing merchant by writing an unauthorized check on the government for that amount. Although the procedure may have been questionable it was dictated by the local situation, and Macomb immediately informed the department of his action. These funds would enable him to begin work in March, 1858, almost two years after his appointment.[30]

The Army engineer's first preliminary survey was to locate the Santa Fe–Fort Union road which ran south to San José and in semicircular fashion turned northward through Tecoloté, Las Vegas, and on to Fort Union. While work by civilian laborers was progressing along this road, Macomb surveyed the Tecoloté-Albuquerque route. He next looked over the Cañada-Abiquiu road.[31] The improvements toward Fort Union consisted, for the most part, of widening the road from a single wagon track to 33 feet. At Tecoloté the road had to be macadamized for a half mile over an unusually steep hill, and also in Apache Canyon an 18-foot span was thrown across the arroyo. Elsewhere the road had been redirected so that difficult arroyo crossings were avoided. Those parts of the road that ran through mountain passes had their exposed sides reinforced with retaining walls of dry masonry. Frequent flash floods in the mountains made this type of construction imperative. During the working season, March to August, a total of 10,500 days of labor had been expended on the road by foremen, overseers, and laborers. Teams had been used for 131 days' time, based upon the eight-hour working day. The average size of a working party on this road was fifty-six men. The procurement of labor apparently was no problem in the Southwest in contrast to difficulties in the Pacific Northwest. This road provided communications between Fort Union and the headquarters of the Military Department of New Mexico in Santa Fe; it also was the best route between Fort Union and the principal Army depot at Albuquerque, pending the completion of the road from Tecoloté to Albuquerque. It crossed the southern Rocky Mountains at the most northern point practicable for a wagon route and was the major mail route from the Missouri River settlements to the New Mexico capital. For additional improvements on this principal entrance into the heart of New Mexico territory, Macomb recommended that Congress be asked for another $35,000.[32]

During April work was initiated on the Tecoloté-Albuquerque route. Sanctioning a $10,000 expenditure, Congress had provided that this road should run through Canyon Carnuel and Canyon Blanco. The first formi-

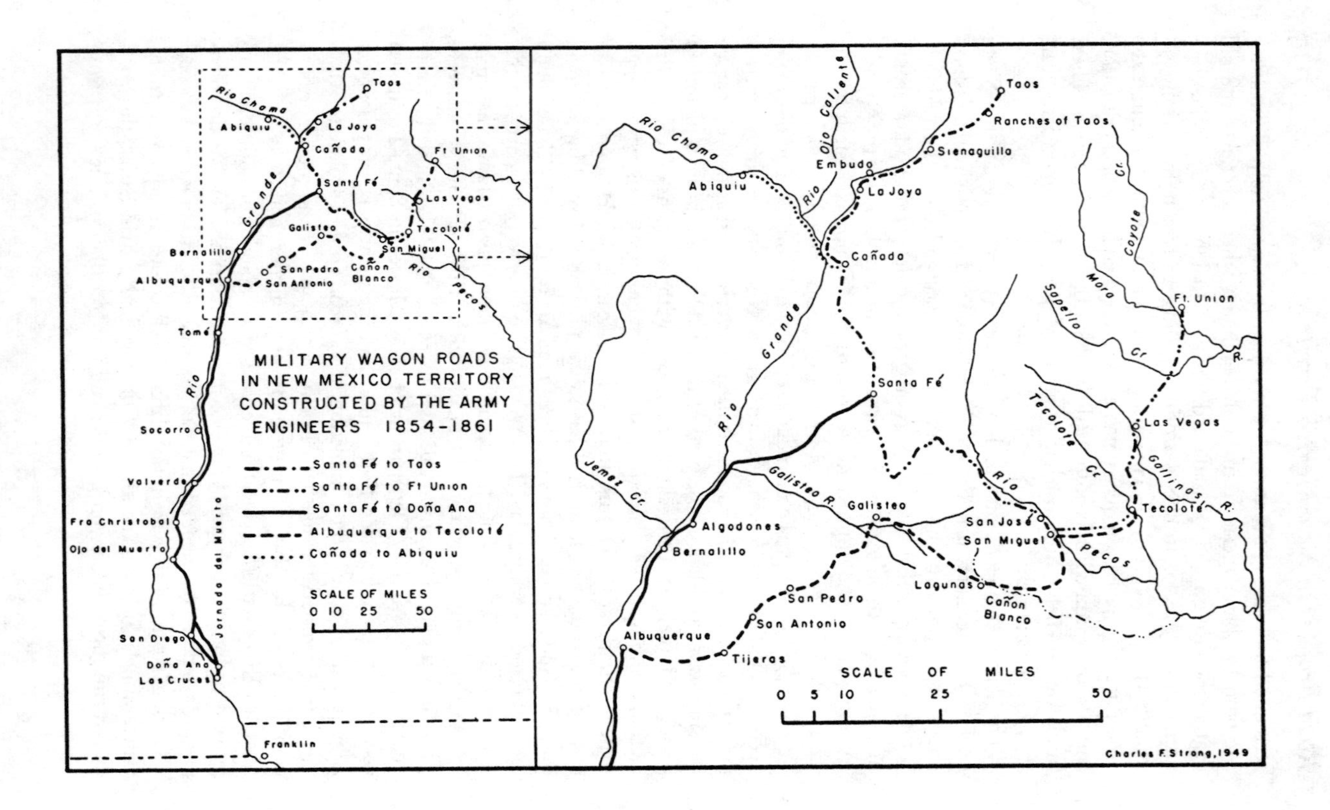
MILITARY WAGON ROADS
IN NEW MEXICO TERRITORY
CONSTRUCTED BY THE ARMY
ENGINEERS 1854-1861
Santa Fé to Taos
Santa Fé to Ft Union
Santa Fé to Doña Ana
Albuquerque to Tecoloté
Cañada to Abiquiu
SCALE OF MILES
0 10 25 50
Taos
Rio Chama
Abiquiu
La Joya
Cañada
Ft Union
Santa Fé
Las Vegas
Grande
Galisteo
Tecoloté
Bernalillo
San Miguel
San Pedro
Cañon Blanco
Rio Pecos
Albuquerque
San Antonio
Tomé
Rio
Socorro
Valverde
Fra Christobal
Ojo del Muerto
Jornada del Muerto
San Diego
Doña Ana
Las Cruces
Franklin
Ranches of Taos
Sienaguilla
Embudo
Ojo Caliente
Coyote Cr.
Mora
Sapello Cr
Ft. Union
R.
Tecolote Cr
Gallinas R.
Rio Grande
Jemez Cr.
Galisteo R.
Algodones
San José
Pecos
Lagunas
Tijeras
SCALE OF MILES
0 5 10 25 50
Charles F. Strong, 1949

dable obstacle was the pass through the lofty range of mountains running parallel to the Rio Grande River about ten miles east of that river. It was necessary to travel through this pass on the route between Albuquerque and Fort Stanton, as well as to get supplies of fuel and timber from the mountains so essential to the Albuquerque military depot. The road was also the last western section of the mail route along the Red River from Fort Smith to Albuquerque by way of Anton Chico. Progress in improving these canyon roads was slowed by the crystalline granite rock which stubbornly withstood blasting operations. Yielding slowly under constant dynamitings, the granite debris was used to pave the canyon road, thus furnishing a permanent passage. An appropriation of $23,000 was asked for its completion.[33]

The Cañada-Abiquiu road was a branch of the Santa Fe–Taos road, and ran along the Rio Chama for 22 miles. Abiquiu settlement, in the mountains, was of military importance as a former garrison site. This route was one of the principal avenues for entering the Navajo country besides being the eastern terminus of the Old Spanish Trail. Wagons passed over the entire distance with difficulty because of sand. Two major obstacles, a rocky cliff which projected to the river bank at one point and the steep hill on the outskirts of Abiquiu, received Macomb's attention. Making a practicable and safe road at these points consumed 1,138 days of labor and the $6,000 allotment. No request was made for additional funds.[34]

After paying the indebtedness incurred by Scammon, the appropriations for the first two roads were so reduced that it was impossible to accomplish what was originally expected. Macomb suggested that funds remaining for the Santa Fe–Taos road be spent in determining the best location and making estimates for construction. He inquired of the Bureau if such a survey could be warranted even though the law spoke only of "construction and repair." Approval was given and after examining three routes through the mountains between the two towns, the engineer proposed to use the existing road to Cañada in the Rio Grande Valley, and thence up the valley to La Joya, a distance of only 40 miles from Santa Fe. From this village he planned to cut a road through the canyon of the Rio del Norte to Sienaguilla. Leaving the valley at this point, the new survey ran northeast to Taos. The entire distance was 73 miles, 14 shorter than the circuitous route through the mountains. The commercial and military importance of this highway, known locally as "El Camino Militar," justified the engineer's recommendation that $113,000 be allocated for its completion.[35]

In September, 1858, operations were resumed on the Santa Fe–Doña Ana route. This principal north-south line of communication down the Rio

Grande Valley was 300 miles long, and the original $12,000 was inadequate for any permanent improvements for such a distance. As a result of Scammon's ineptitude, only $5,000 remained in the 1853 appropriation. Since the road in the valley below Albuquerque was passable, Macomb requested and received permission to spend the funds on the Santa Fe–Albuquerque sector. During the winter the trained crew of workmen that had just completed the Fort Union–Santa Fe route was transferred to this project and efforts concentrated at the crossing in the Galisteo Valley. Public reaction was favorable and Macomb reported, "Every traveller from the south is filled with praises of the route at or near the crossing of the Galisteo."[36]

In the spring of 1859, Macomb was assigned to a tour of survey duty at Fort Craig, eight miles below Fort Conrad on the right bank of the Rio Grande near Valverde.[37] In his absence, the final work on the military road was directed by Joad Houghton, a civilian assistant. His services had been invaluable because of his geographical knowledge of the territory and his acquaintance with its people and their language.[38] Macomb was pleased with the progress on all five federal roads and reported to the War Department:

> These improvements of the Roads in New Mexico are already attracting the favorable notice of the traveling public and have happened most opportunely to facilitate the very considerable extension of the mail service recently accorded the territory.[39]

From the first, local Army officers had brought pressure on the War Department for greater financial support. In the winter of 1857, Miguel A. Otero, New Mexico's congressional delegate, had asked General Garland to furnish him with information on desirable military roads in the territory. Realizing that the data would be used to encourage national legislation, Garland ordered Macomb to draft a statement. Five new military roads were needed, each of them necessitating funds from $10,000 to $20,000—Fort Union to Taos, Fort Massachusetts to Taos, Santa Fe to Fort Stanton, Albuquerque to Fort Stanton, and Albuquerque to Fort Defiance. The initial allotments were to be requested for exploration and surveys to ascertain feasible routes and make construction estimates; only then would improvements be made. A sum of $10,000 was needed to bridge the Rio Grande River at strategic points. All plans for digging artesian wells along the military roads in the territory should be postponed, according to Macomb, until the outcome of the boring experiment on the plains was conclusive. Finally, he proposed that the national government assume responsibility for improving the Fort Union–Santa Fe road, a part of the Missouri–New

Mexico mail route, to make it equal in facility for travel as the natural road across the plains.[40] Although delegate Otero presented to the Thirty-fifth Congress, in both the first and second sessions, bills providing for the road and bridge constructions in New Mexico Territory, no proposal reached the statute books.[41]

Much remained to be done, but the federal road funds were rapidly dwindling. Macomb sought available departmental money designated for related activities. As he traveled across the Jornada del Muerto in July, 1857, on the way to his assignment, he had noted the urgent need there for a permanent water supply. Congress had recently appropriated $100,000 for an experiment in sinking artesian wells on the public lands, and the captain urged his chief in the Topographical Bureau to suggest that one of these experiments take place on the Jornada.[42] Then in August, 1858, he asked that $10,000 be advanced to his road projects from the general appropriations for geographical explorations west of the Mississippi, but his request was not granted.[43] Finally in June, 1859, with the road allotments expended and no more subsidies in prospect, Macomb asked to be released from his New Mexico assignment.[44]

During 1859 Army explorations, inaugurated at the time of Kearny's invasion of the upper Rio Grande settlements, reached a most active stage in the New Mexico Military Department. Numerous reconnaissances were supported to find shorter and better routes of travel, to select sites for new military posts and Indian reservations, to sink artesian wells, and to construct roads.[45] Before Macomb was relieved, he was ordered to explore the San Juan River and the lower courses of the tributaries of the Colorado of the West to determine the most practicable route for a wagon road between Santa Fe and the settlements of southern Utah.[46] From the Office of Explorations and Surveys in the United States Army, $20,000 was granted to defray the expenses of his expedition.

Macomb employed five civilians as technical assistants to accompany him on this exploration. An infantry detachment under Lieutenant Milton Cogswell served as escort. Leaving Santa Fe on July 12, his party traveled in a northwesterly direction crossing the Rio Grande at the old Indian pueblo of San Juan, and followed up the Rio Chama, passed the settlement at Abiquiu to the river's head. Here they crossed the dividing ridge separating the waters flowing into the Gulf of Mexico and those into the Gulf of California. The expedition moved west for 70 miles in a rugged country broken by rapid mountain streams flowing into the San Juan. In this area they were accompanied by Albert H. Pfeiffer, a subagent for the Utah Indians, and his interpreter.[47] Turning northwest for 120 miles across desolate country, the party reached "Ojo Verde" or Green Spring, a well-known landmark on the Old Spanish Trail. Macomb noted:

> The greater part of our journey from Abiquiu to this point was by the Old Spanish trail, which has not heretofore been accurately laid down upon any map. This trail is much talked of as having been the route of commerce between California and New Mexico in the days of the old Spanish rule, but it seems to have been superseded by routes to the north and south of it, which have been opened by modern enterprise.[48]

At this point the expedition planned to leave the old trail which headed north. The main party remained at the spring and Macomb, with nine men, went westward 30 miles to seek the junction of the Grand and the Green rivers. Arduously they traversed this rough and dangerous terrain which prompted Macomb to remark, "I cannot conceive of a more worthless and impracticable region than the one we now found ourselves in." Completely baffled, the detachment stopped short of the goal and returned to Ojo Verde. Further plans for exploration were abandoned.

The combined expedition on its return veered southward for about 70 miles to the San Juan River which the men ascended along its north bank to a point opposite Canyon Largo. There a tributary of that name flowed into the main stream from the south. The San Juan was forded on September 15 and was followed in a southeasterly direction to San José Spring. From here the explorers moved down the Rio Puerco for 40 miles, crossed a spur of the Nacimento mountains to the old pueblo of Jemez, and on east to Santa Fe.[49] Macomb's exploration had proved that no route adequate for a wagon road existed from the settlements of New Mexico to those of Utah between the Old Spanish Trail on the north and the wagon route recently constructed along the thirty-fifth parallel.[50]

The captain was absent on this exploration from July until October. At the start of his assignment and after his return to Santa Fe he inspected his earlier work on the military roads to determine the effects of use and the weather. He wrote:

> I found, as I had expected, that the work in general had been improved by the rains, especially those parts of the new roadway which had been made in the dryest weather, as in such cases the first effect of travel is to make a deep dust by grinding the dry soil which may have been thrown up to form the roadway; but this, when once saturated by the rain, becomes cemented together, and in the course of the months of dry weather common in that climate, it hardens, and affords a smooth surface capable of resisting, in a great degree, the future rains to which it may be exposed; and it is for this reason that the common earthen road is remarkably well adapted to the greater part of New Mexico . . .[51]

In going over the Albuquerque-Tecoloté road, Macomb found that part running through the Canyon Carnuel endangered by swollen streams, and he set a crew to widening the road and protecting the shoulders by stone work. The captain regretted that no further funds had been provided by Congress because the improvements had definitely facilitated military operations and commercial transportation in the territory. Few officers of the Topographical Corps had so conscientiously and successfully completed a road-building assignment. Macomb was finally relieved of duty in New Mexico in June, 1860.[52]

While Macomb was exploring northwest from Santa Fe, Colonel Benjamin L. E. Bonneville, commander of the New Mexico Military Department, had sent out a series of reconnoitering expeditions under junior officers to examine the condition of roads and to seek new routes for wagons. With a detachment of forty infantrymen, Lieutenant Alexander E. Steen left Fort Garland, 85 miles north of Taos, to find a good road across the mountains to Bent's Fort on the Arkansas River. In moving eastward he followed the banks of permanent water courses and found abundant wood and grass. For one-half the distance to Bent's Fort the road was well beaten by wagon wheels and without obstacles. On his return to Fort Garland, a more direct and shorter route was followed but it was less satisfactory because of the scarcity of water. Later in the season Steen and his detachment located a second usable wagon road westward from Fort Garland to the Rio Grande River and up that stream to the San Juan Mountains.[53]

Less successful was Captain Thomas Claiborne who moved from Fort Stanton with an exploring party of one hundred and twenty men to seek a route directly east to the Pecos River. His detachment was mounted riflemen, but horses were so scarce that thirty-one of the men had to walk. They assisted eleven infantrymen in leading a string of forty-seven mules, which they apparently refused to ride in the absence of horses. This "ill-arranged command" to use the terms of its officer, moved very slowly in an area "where there was no road, much sand, and great heat." The explorer's course was generally southeast along the Rio Bonito to its junction with the Ruidoso and on down the Hondo Valley to the Pecos. Thence the command moved northward along the west bank of the Pecos to Hatch's Ranch where the reconnaissance ended. The attempt to find a practicable road eastward from Fort Stanton through the Capitan Mountains was a failure. The only possible route was that which they followed along the water courses.[54]

Several exploring parties marched from Anton Chico on the Pecos with the object of locating better roads. Lieutenant William H. Jackson took a command of fifty-four mounted rifles northeastward to the north fork of the Canadian River and ascended that stream to its head where they met the main trail from Independence, Missouri, to Santa Fe. On the return the command followed the Santa Fe Trail westward to Fort Union and Las Vegas and then struck out southward to Anton Chico. Thus two possible routes connecting Anton Chico with the Santa Fe Trail were known to exist. Lieutenant Henry M. Lazelle, Eighth Infantry, commanded the escort of the Texas–New Mexico Boundary Commission in its march eastward from Anton Chico to the Pecos and southward along the east bank of that stream to Fort Lancaster, Texas. The command was accompanied by a train of twenty-two wagons carrying 57,000 pounds of freight. The route down the Pecos was found to be excellent, with sufficient water and grass. By using this road rather than that down the Rio Grande, a distance of 200 miles could be saved between Santa Fe and Fort Lancaster.[55]

In the meantime the Topographical Bureau repeated its requests for additional federal funds for New Mexico roads,[56] but not until 1861 was any money granted. In that year the Thirty-sixth Congress appropriated $15,000 of the $113,000 requested for the Taos–Santa Fe road, and $35,000, the desired amount, was allotted for the completion of the Fort Union–Santa Fe route.[57] However, no further construction was attempted. In his annual report for 1861, Hartman Bache, new Chief of the Topographical Bureau, presented an explanation:

> Owing to the unsettled state of affairs, it was considered inexpedient to resume the work, and these appropriations are still applicable. The recommendation for further appropriations is now renewed, in the hope that at a more propitious period the required means for finishing these important lines of travel may be provided.[58]

Not until 1873 did Congress again grant money for specific road improvements in New Mexico.[59]

The routes of the pre-Civil War military roads pointed the way for modern lines of communication. For example, El Camino Militar from Santa Fe to Taos is today a main highway from the New Mexico capital northward into Colorado at Fort Garland. The tracks of the Atchison, Topeka, and Santa Fe Railroad follow in general the route of the old Fort Union military road from Las Vegas across the Sangre de Cristo Range into Santa Fe. This railroad's route southward to Albuquerque, on down the Rio Grande Valley, across the Jornada del Muerto to El Paso, follows the old Doña Ana road.

CHAPTER VIII : *Shortening and Improving Routes across the Plains of Kansas and Nebraska*[1]

IN 1854 the area between the western boundary of the prairie states and the Rocky Mountains and north of the Indian Territory was designated by Congress as the Kansas and Nebraska territories. Across these plains the great tide of migration had swept into Oregon, California, and the Great Basin. The valley of the Platte had been the greatest route of all, and since the beginning of the Great Migration in 1841, the Oregon Trail had been fixed upon its south bank. In 1847 the Mormons, leaving their Winter Quarters near Omaha, chose a new western route along the north bank of the stream. As a part of the expanding wagon-road program under the secretaryship of Jefferson Davis, the War Department proposed in 1854 to improve this Mormon Trail from Omaha as far as New Fort Kearny at the southern bend of the Platte. Military supplies could be more quickly and cheaply transported to that post by bringing them up the Missouri River, along the western Iowa boundary to the Council Bluffs–Omaha region and thence overland on a shorter land route than that from Fort Leavenworth. On February 17, 1855, Congress appropriated $50,000 for this public work.[2]

Fort Leavenworth, on the eastern boundary of Kansas, was at this time the principal depot from which the military stations along the routes to Utah, California, and Oregon were supplied. The contracts for the transportation of these supplies amounted to three or four hundred thousand dollars each year. One hundred and thirty miles west of Fort Leavenworth, at the forks of the Kansas (Kaw) River, a new fort, known as Fort Riley, was under construction in 1854–1855. This fort, built for the protection of the Kansas settlements and as a subordinate depot and an advanced rendezvous for troops, was connected with Leavenworth by a water route on the

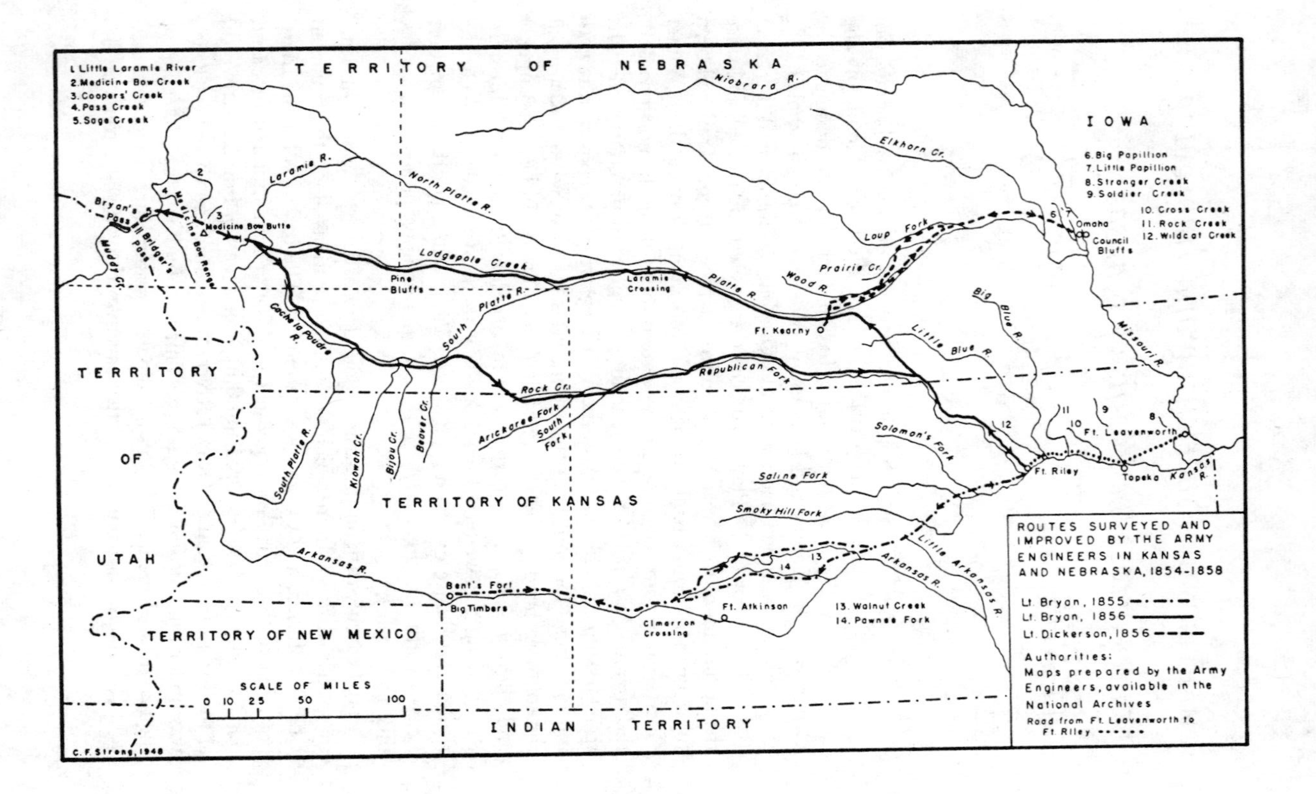

ROUTES SURVEYED AND IMPROVED BY THE ARMY ENGINEERS IN KANSAS AND NEBRASKA, 1854-1858
Lt. Bryan, 1855
Lt. Bryan, 1856
Lt. Dickerson, 1856
Authorities:
Maps prepared by the Army Engineers, available in the National Archives
Road from Ft. Leavenworth to Ft. Riley
1. Little Laramie River
2. Medicine Bow Creek
3. Coopers' Creek
4. Pass Creek
5. Sage Creek
6. Big Papillion
7. Little Papillion
8. Stranger Creek
9. Soldier Creek
10. Cross Creek
11. Rock Creek
12. Wildcat Creek
13. Walnut Creek
14. Pawnee Fork
TERRITORY OF NEBRASKA
IOWA
TERRITORY OF UTAH
TERRITORY OF KANSAS
TERRITORY OF NEW MEXICO
INDIAN TERRITORY
Niobrara R.
Elkhorn Cr.
Laramie R.
North Platte R.
Loup Fork
Prairie Cr.
Wood R.
Platte R.
Lodgepole Creek
Laramie Crossing
Pine Bluffs
South Platte R.
Omaha
Council Bluffs
Missouri R.
Big Blue R.
Little Blue R.
Ft. Kearny
Republican Fork
Rock Cr.
Arickaree Fork
South Fork
Bryan's Pass
Bridger's Pass
Muddy Cr.
Medicine Bow Range
Medicine Bow Butte
Cache la Poudre R.
South Platte R.
Kiowah Cr.
Bijou Cr.
Beaver Cr.
Solomon's Fork
Saline Fork
Smoky Hill Fork
Ft. Leavenworth
Ft. Riley
Topeka
Kansas R.
Little Arkansas R.
Arkansas R.
Bent's Fort
Big Timbers
Cimarron Crossing
Ft. Atkinson
SCALE OF MILES
0 10 25 50 100
C. F. Strong, 1948

Kansas and by a military road on its north bank.[3] On March 3, 1855, Congress appropriated $50,000 for the construction of a road from Fort Riley to the Arkansas River at any site the Secretary of War thought most desirable for military purposes. An equal sum was approved for a road from Fort Riley to Bridger's Pass in the Rocky Mountains.[4] The Army planned for the route to the Arkansas to strike that river either at the Cimarron Crossing or at Bent's Fort, so troops and supplies from the two Kansas forts, as well as emigrants, might then travel to the New Mexico settlements by the long-established Santa Fe trails. The road to the Rockies would provide a more direct route from the Missouri River towns and forts in Kansas to Utah and California than the Oregon Trail, shortening the distance to Great Salt Lake by 100 miles. The route was declared to be equally easy, and Bridger's Pass as accessible as the South Pass farther north.[5]

Lieutenant Francis T. Bryan, who was chosen to direct the construction of these three projects and to supervise the expenditure of $150,000 of federal funds, hastened to St. Louis where essential equipment for the surveys was purchased in the spring of 1855.[6] At Fort Leavenworth he first planned to travel the route to the Arkansas and hired several Delaware Indians, reported to be well acquainted with the country between Fort Riley and the Arkansas, to serve as guides to his party. An outbreak of cholera delayed his departure from Fort Riley until July 30, 1855. Accompanied by a military escort, the Bryan survey expedition traveled along the north bank of the Kansas River for approximately 50 miles, crossing Solomon's Fork about 35 miles from Fort Riley and the Saline ten miles farther west. At the Saline the party left the valley of the Kansas, crossing the plains in a southwest direction to avoid the bend in the Smoky Hill Fork. Immense herds of buffalo were observed here.

At their crossing of the Smoky Hill, the river was 220 feet between its banks, with the crests a full 22 feet above the bottom of the stream. Although the water was only a few inches deep at the time of crossing, some difficulty was experienced in keeping the wagon wheels from cutting too deeply and becoming stuck in the loose sand. In Bryan's opinion, a bridge was desirable but the thinly scattered cottonwoods on the banks of the stream near this crossing would be of little value in construction.[7]

Leaving the river, Bryan's men headed toward the southwest, crossing open country which they reported to be exceptionally level, covered with buffalo grass, and inhabited by prairie dogs. The lieutenant realized that though this would make an excellent dry weather road, it would be impassable in wet seasons. Crossing Walnut Creek, a tributary of the Arkansas, the surveyors reached the Pawnee Fork of that river and ascended it to the headwaters. They observed that the timber on the streams was more scattered and smaller, and the general appearance of the country indicated that they were approaching the dry region bordering the Rocky Mountains. In the short march from the Pawnee to the Arkansas, the country was destitute of timber and the party resorted to buffalo chips for fuel. At the Arkansas the detachment came upon the well-beaten road from Fort Atkinson to Bent's Fort.[8] From the Arkansas Crossing or these forts, trails provided through communication to the New Mexican villages on the upper Rio Grande.

At Bent's Fort, Bryan, learning that a direct route could be made from the Big Timbers at the fort to the head of Walnut Creek, attempted to employ competent guides who could direct his party there. Thus the timberless, desolate stretch between the Pawnee Fork and the Arkansas could be avoided. Bent recommended Cheyenne or Arapaho guides. However, these tribes strongly objected to the road-building activities of the government and would provide no aid. As a result the explorers returned to the camp where they first struck the Arkansas, gathered supplies of wood, and crossed directly to the head of the Pawnee.

Here the first norther of the season struck, bringing heavy rains and bitter cold. Having exhausted their fuel at this encampment, the men were forced to move quickly in search of firewood. The return route took the party down the Pawnee to a point close enough to the Walnut to cross over in a single day's march. The engineer decided it was unnecessary to bridge these streams unless a military post were established in the vicinity and the garrison could be convenienced thereby. On the trip down the Walnut and across to the Smoky Hill, bad weather continued to plague the party; it was now the third week in September. Once they reached the Smoky Hill, that stream was followed to Fort Riley along the outward track.[9] The total length of the road surveyed was 360 miles.[10]

Bryan reported to his superiors that as the route was, for the most part, over open prairie, and since there was no timber to cut out and none at sufficient intervals to provide stakes for the surveyors, there was no means of marking it except by the tracks of wagons. The trail which his few wagons had made was so dimly outlined that within six months it would be obliterated, and he urged the immediate passage of a large train over the road to mark it plainly. After the major streams were bridged, the only obstruction to wagon travel would be the small drains a few inches in depth that each pioneer party would be forced to make passable. Bryan recommended that a working party of twenty men should travel a day in advance of the next freighters and emigrant trains to prepare the way.

Bridges would be necessary at the crossings of Solomon's Fork, the Saline, and the Smoky Hill rivers. Oak could be found on the banks of the first two streams which would provide lumber for the 120-foot structures which were needed, but as no suitable timber could be found on the Smoky Hill and as the road crossing was all of 80 miles beyond Fort Riley's men and materials, cost of the 200-foot span would be greatly increased. Bryan requested the assignment of one company of infantry as an escort and patrol for the workmen employed upon these bridges.[11]

At the conclusion of the survey all camp and surveying equipment of the expedition was left with the quartermaster at Fort Leavenworth, and the animals that would be needed the next season were placed in the care of herders on the post. Bryan then returned to winter quarters in St. Louis where he opened an office and hired two draftsmen to assist in making maps and charts to accompany his report on the season's activities.[12] In February the contract for the building of five bridges on the Fort Riley–Big Timbers road was granted to James A. Sawyers, whose bid of $38,400 was the lower of the two submitted.[13] The Bureau refused Bryan's request for an escort for Sawyers' workmen, and the contractor, in desperation, wrote directly to Jefferson Davis:

> We have information of hostilities and depredations being commenced by the Cheyenne Indians, now in that region and as I have no protection . . . I should be provided with an escort as was verbally guaranteed to me by Lieut. Bryan and is really a part of the consideration of contract. . . . I am departing for the place of operation today. . . . I hope you will see the importance of granting me an escort, as any depredations, arising for want of protection, might prove disastrous to the government as well as seriously injurious to me.[14]

A detachment from the Second Dragoons at Fort Riley was finally ordered by the local commandant to join the laborers after they had been in the field for more than a month.

When the Army engineer left Leavenworth with his new exploring party to go to Bridger's Pass in May, 1856, he left a civilian engineer, Coote Lombard, to superintend Sawyers' construction of bridges on the road to the Arkansas. At Solomon's Fork the contractor worked from mid-June to mid-July hauling wood and building falsework. When he was ready to start the actual bridge, the stream began to rise as the result of freshets. In two days it was six feet above its earlier high-water mark, carrying off all the falsework. The contractor began again, but heavy rains in late August and September delayed the completion of the bridge, including the construction of icebreakers, until October. At the Saline Fork the river was also at flood stage most of the time and full of driftwood. The men continued to work, but several suffered from exposure and became ill, and the force was steadily reduced. One laborer died at this encampment. From here they moved up to the site of the Smoky Hill River bridge where the climate was drier; most of the men recovered, but a second laborer, who had been ill for several weeks, died soon after they arrived in the new camp. Lumber was hauled in from the two previous sites by ox teams, which, on at least one occasion, lost the road and had to be found and redirected by the mule wagons transporting rations for the crew.[15]

Sawyers had experienced a difficult season. Realizing that he was losing money on the contract, he appealed to Lombard, and the engineering agent permitted him to omit the construction of icebreakers on the Saline and Smoky Hill bridges because it had become necessary to build the Solomon's Fork bridge longer than the contract specified.[16] On his return from Bridger's Pass, Bryan proceeded to examine the work on the road and accepted the bridges for the United States government. At the beginning of 1857 Sawyers put in claims for what he termed "extra work" not in his contract. The Army engineer forwarded the claims to the Bureau with an evaluation of each and a recommendation that all be disallowed. His decision was sustained by the War Department. All admitted that the contractor had little profit to show for his season's work.[17]

Kansas settlers pushed westward as the road was built and the bridges erected. During the season of 1856 the civilian engineer observed:

> The bridging of this road has induced settlers to move out at least forty miles beyond the heretofore bounds of civilization, i.e., at and beyond the Saline Bridge. I expect that there will be settlers at the Kaw (Smoky Hill) River Bridge eighty-five miles west of Fort Riley by next Spring—the opening of this road has pushed the settlements beyond where they would be if the road had not been opened.[18]

Bryan notified the War Department early in 1857 that the road from Fort Riley to Bent's Fort was "passable for trains of any kind." His greatest worry was the section of the road beyond the Smoky Hill River bridge, which "would be very difficult to find except to persons who had once traversed it and knew it by landmarks, as the prairie grass of two summers has effaced the marks made by the surveying party of 1855."[19]

During the winter of 1856 in St. Louis, Lieutenant Bryan had notified the War Department that the survey to the Arkansas was his accomplishment of the previous season. He requested the appointment of a trained civilian engineer as agent to supervise the Nebraska road from Omaha to Fort Kearny in the spring when he would be engaged in locating the route to Bridger's Pass. An escort would be necessary for the safe conduct of both parties.[20] Colonel Abert reprimanded him for the failure to survey all three roads in Kansas and Nebraska territories in 1855 and requested an explanation that might be presented to the Secretary of War and possibly to Congress. Bryan reminded the colonel of the delay at Fort Riley because of cholera, the two months spent in traveling to and returning from the Arkansas, and explained that commerce over the plains stopped in October and did not begin until spring. An additional survey late in the season, he thought, would have meant a great loss of material and men from frost and starvation.[21]

Bryan wrote to the Bureau in April that his plans for the reconnaissance of the Fort Riley–Bridger's Pass road were nearing completion. Guides had been employed and he intended to start out in May as soon as the grass of the plains would support his animals. Officers from the West reported the Indians hostile to any attempt to make a road through their country and his guides thought an escort necessary in the western part of the territories. The lieutenant also restated his intention of placing the Omaha–Fort Kearny road under a civilian agent of the Army engineers, since his own time would be consumed in going to and returning from the Rockies.[22] On May 28 the Bureau notified him that Lieutenant John H. Dickerson had been assigned the responsibility of supervising the road in Nebraska Territory and Bryan replied by telegram, "I am prepared and wait only for Lt. Dickerson."[23]

When Bryan left Fort Riley on June 21 he was accompanied by several assistants: a topographer, John Lambert; a geologist, Henry Engelmann; a barometer expert; and two trained rodmen. They traveled northwest along the east bank of the Republican Fork for 100 miles in the direction of Fort Kearny, and then crossed over the prairie 35 miles to the Little Blue. After crossing this river, the party struck the established military road between Forts Leavenworth and Kearny. This was followed to the Platte about 15 miles east of Fort Kearny, and then up that stream to the fort. In the opinion of Lieutenant Bryan, a great amount of labor would be necessary on this first section of the route to the Rockies to make an acceptable wagon road. Many of the creeks needed bridging, and the approaches to practically all entailed grading to avoid the capsizing of heavily loaded freight and emigrant wagons.

Leaving Fort Kearny, the surveyor's route lay along the valley of the Platte, the usual way traveled by Oregon bound trains, to a point 16 miles beyond the much-used Laramie Crossing.[24] Here a new ford was located. The river was reported to be 610 yards wide, with a gravel bottom and water scarcely covering the axle trees of the wagons. Like all previous explorers, Bryan realized that bridging the Platte was out of the question, and trains must take their chances in finding a good ford. From the Platte crossing the party ascended the south fork of that stream and its tributary, Lodgepole Creek, to Pine Bluffs in what is now Wyoming, just across the western Nebraska boundary. This area was known as a favorite winter residence of the Sioux and Cheyenne Indians. The members of the expedition gathered dwarf pine for several days' use because fuel, even buffalo chips, was reportedly scarce at the headwaters of Lodgepole Creek.[25]

The party crossed the hills between this creek's source and the Laramie River in a single day and journeyed to the Little Laramie River on the next day. Here they struck an emigrant road along the foot of the Medicine Bow Range, which Captain Stansbury had used during his exploration of 1850, and followed it a few miles to an encampment on Cooper's Creek. The expedition had difficulty with the animals in this mountain country because of sore feet from worn out or lost shoes. Bryan recommended that trains traveling through the country should carry additional horse and mule shoes, a supply of shoe nails, and a forge. From Cooper's Creek they crossed rocky hills to the Medicine Bow in the vicinity of Medicine Bow Butte, a favorite rendezvous spot for beaver trappers in years past and at that time a council place used by the Sioux, Snakes, and Arapahoes.

From here their circuitous route toward the Continental Divide led to the headwaters of Pass Creek where, on August 9, they endured a mountain storm with the temperature dropping to freezing and leaving ice on the tents. From Pass Creek to the North Platte the route was so steep that ropes were used to hold the wagons in line and, in spite of precautions, two overturned. Leaving the North Platte the party traveled to Sage Creek, a tributary, which they assumed would lead to Bridger's Pass. None of the guides, who had spent years in the mountains, had been to the pass, and the appearance of the country did not coincide with Captain Stansbury's descriptions. The leaders agreed, however, that they could not be a great distance from Bridger's Pass, drawn on the map between the head of Sage Creek, flowing easterly to the North Platte, and Muddy Creek, flowing westerly into a branch of the Green. A consultation was held, and all concurred that the mission of the expedition was to find a practicable pass to the western slope and that they should not be concerned over its exact location. The party crossed the Divide and descended Muddy Creek to make certain that its waters flowed to the west. The reconnaissance was complete, and the pass over the Divide was named for Bryan.[26]

The engineering party returned to the North Platte, a few miles north of the outward route, but rejoined the old trace before reaching the Laramie River. Seeking a new way that might prove preferable to that traveled on the outward journey, they turned south, crossing the hills to the Cache la Poudre River in the vicinity of the present Wyoming-Colorado boundary. They descended this stream to its junction with the South Platte, forded the latter stream, and descended it in an easterly direction crossing Kiowah, Bijou, and Beaver creeks flowing from the south.

Fourteen miles beyond the mouth of the Beaver, Bryan decided to leave the South Platte and cross the open country to the Republican Fork of the Kansas. The party remained in camp, however, the following day, September 14, because of the illness and sudden death of Frederick Bortheaux, whom they buried on a ridge near the banks of the river. Resuming the march, they crossed the flat sandy prairie en route to Rock Creek, a tributary of the Arickaree Fork of the Republican. This proved the most desolate country of the entire trip and very fatiguing for the draft animals. A large party of Cheyenne Indians met the explorers on Rock Creek and gave evidence of preparing to attack, before they discovered the strength of the party's escort. Bryan's men went into camp immediately and the commander of the escort stationed sentinels to keep watch. A cold rain set in, and the party was greatly inconvenienced by lack of fuel, there being only buffalo chips which could not be used in wet weather.

The final section of the return route was down the Arickaree and the Republican Fork to Fort Riley. Bryan saw that the river bottoms furnished subsistence for large herds of buffalo and elk which made this valley a favorite hunting ground of the Cheyennes, Comanches, and Kiowas. These Indians intended to prevent the government from making a wagon road along the river. Leaving the main party in charge of John Lambert with instructions to proceed to Fort Riley, the lieutenant took a detachment across to Solomon's Fork for further reconnaissance. After inspecting the new bridges constructed on the Arkansas route in his absence, his party arrived at Fort Riley on October 24. Both groups disbanded at Leavenworth November 7, having been in the field four and a half months.

Bryan reported to the War Department that, in view of the limited funds remaining of the congressional appropriation, the route followed on the outward journey was the most advantageous. Running water was available over the entire distance and the section of the road along the Platte was already well established. The greatest obstacle was lack of fuel. From Fort Kearny to Pine Bluffs, a distance of 300 miles, only buffalo chips were to be found. In Bryan's opinion this absence of timber, and consequently fuel and shelter, would always make traveling along the Platte in the winter a hazardous and painful experience. However, the road between Fort Riley and Bridger's Pass could be considered "practicable," because thirty-three wagons had passed over it in the season of 1856. The engineer's only worry was the fact that his road led into the heart of the mountains with no definite terminus. To make it of some value the War Department was urged to connect it with the posts or stations west of the Divide, possibly the Salt Lake Basin.[27]

During the winter in St. Louis, Bryan and his associates prepared a comprehensive report on their season's work. The topographer, with two draftsmen, made an elaborate map of the road including near-by topographical features. Lambert also reported on several side-surveys made under instructions from the Army engineer; Engelmann, the geologist and mining engineer of the expedition, summarized his observations in a technical paper. The fossils he had collected on the government expedition were examined by B. F. Shumard of St. Louis who submitted a report on paleontology. In time these maps and reports were forwarded to the Bureau in Washington.[28]

In the spring of 1857 Bryan organized a party of laborers to pass over his road again to remove obstacles and to grade the banks of streams at crossings. Only with the assurance that an armed escort of cavalry would be provided could the engineer find men willing to leave the settlements for several months and to undertake the assignment.[29] The distance between Forts Riley and Kearny, 193 miles, was traveled in fourteen days and left in a "passable" state, so that the more distant parts of the road might be worked first. No improvements were thought necessary between Fort Kearny and the Laramie Crossing, a road distance of 168 miles. When the Bryan party arrived at the ford they had crossed the previous season, it was impassable because of high water, but four miles upstream a satisfactory crossing was found at the camping ground of the Cheyenne Indians.

Along the route from the Platte to the head of Lodgepole Creek, the crossings of streams were graded and in the timbered country at the headwaters of the creek, trees and stones were removed from the road. Crossings of the Laramie and Medicine Bow were improved, but Bryan noticed that the Medicine Bow was not susceptible to permanent improvement because boulders and gravel were brought down by mountain torrents each season when the snows melted. At several crossings of Sage Creek, small log bridges were constructed sufficiently wide for the passage of a single wagon. Bryan justified his cursory improvements by remarking, "In opening this road, I have endeavored to carry into effect the instructions of the Secretary of War, namely, not to expend an undue amount on any one section but to equalize as much as possible the expenditure, so as to make all parts practicable before any part was elaborated."[30]

The laborers returned to Fort Kearny by September 1 and then turned their attention to improving the eastern section of the road. At the crossing of the Little Blue, the banks were graded and the road opened through the timbered bottom. No bridge was believed necessary because the stream was usually fordable, but many of the smaller streams between the Little Blue and Fort Riley were so deep and narrow that bridges were required. Bryan did not have the requisite tools and mechanics to do the job, so he decided to discharge the party and sell the animals and property belonging to the project to secure additional funds for the construction.[31]

By March, 1858, drawings and specifications for ten small bridges on the road immediately north of Fort Riley had been prepared, and a construction contract given to Alfred Hebard for $12,500.[32] The unexpended funds for the road totaled $9,500, but Bryan assumed the mules, wagons, harnesses, and other equipment of the expedition would bring $3,000 at an auction. When this state of affairs was reported to the Secretary of War, Bryan was relieved of his command, and the Nebraska and Kansas roads were assigned to Captain Edward G. Beckwith.

On July 23 the public auction held at Fort Leavenworth was stopped by Beckwith's order because no reasonable bids were made by which a sufficient sum could be realized to cover the contract. Since the Secretary of War's approval of Hebard's contract was contingent upon raising $3,000 at the auction, Beckwith renegotiated the contract whereby Hebard would accept the balance of the funds for the road, plus income from property sales, even if less than $12,500. In recompense, an extension of time was granted to complete the bridges. The contractor was to be permitted also to use government mules for hauling supplies and for construction work. This arrangement was approved by the War Department.[33]

Hebard's laborers used the timber growing beside the Kansas streams to build several log bridges, but iron and flooring had to be brought in to construct a half-dozen frame bridges over the larger creeks. The first grading proved a simple problem, but the contractor noted that it was not permanent because once the sod was broken the dirt washed out on the slightest grades. Beckwith reported the road in good traveling condition 50 miles above Fort Riley in September. Should the season prove favorable for work during November, all bridges were to be completed by the month's close.[34] Beckwith then announced that the road was in excellent condition for travel of the heaviest trains across the plains, and hastened to Fort Leavenworth to report the close of his season's operations.[35]

While Lieutenant Bryan was engaged in locating the route west of Fort Riley to the Rockies in 1856, Lieutenant Dickerson concentrated his efforts on improving the eastern Nebraska military road. The fifteen months elapsing between the passage of the law authorizing this road and the assignment of Dickerson had been ample for Nebraska residents to evaluate the effects of the government project on the frontier communities. Residents south of the Platte were disappointed that all federal funds were to be spent on a road along the north bank and at least one, who described himself as "a resident of Nebraska interested in the development of the country," wrote the Chief Topographical Engineer urging the appointment of a surveyor to examine and report on the possibility of bridging the Platte near its mouth and building on the south bank to avoid the crossing of the Elkhorn, Loup Fork, and Wood rivers.[36] While the local debate continued, the Nebraska governor, Mark W. Izard, complained to officials in Washington that nothing had been done on the road in the season of 1856.[37] This communication had inaugurated the investigation of Bryan's activities that culminated in the appointment of Lieutenant Dickerson and the division of the Kansas-Nebraska road work.

Jefferson Davis, intensely interested in the pattern of the Army transportation system as well as emigrant travel to the West, forwarded his general circular of instructions to military road engineers. The instrumental survey was to establish the most direct and practicable line between the two termini after considering construction costs, the selection of points for river crossings, and an adequate supply of wood and water. The Secretary recognized the fact that road building on the plains presented divergent problems from those elsewhere in the West. Just as the heavy forests had proved a major obstacle in the Pacific Northwest and the lack of permanent water supplies in the Southwest, the principal difficulties to overcome here were the stream crossings.

Dickerson met Bryan in St. Louis on June 1 to receive the funds and instruments available for the survey, and within a week departed for Fort Leavenworth. Here five wagons with ox teams, twelve riding animals, camp equipage, and a forty-day ration for his party were provided by the commandant. Dickerson's command included two engineering assistants, hired in St. Louis, a wagon master and twenty teamsters and laborers.[38] From Leavenworth they crossed the Missouri River at Weston and marched through Missouri and Iowa to Council Bluffs where they recrossed the river to Omaha. The party remained in Omaha a few days, employing a guide and collecting information about the route. Out of Omaha the surveyors followed the "Winter Quarters' Trail" of the Mormons across the Big and Little Papillion rivers and struck the Elkhorn 18 miles above its junction with the Platte and 24 miles from Omaha. The broken country between the Missouri and the Elkhorn had made the route circuitous and would necessitate extensive grading on approaches to streams.

At the Elkhorn the party came into the valley of the Platte and continued upstream to the Loup Fork, which was crossed at the Mormon ferry. The Platte Valley west of this junction was reported to be too miry for wagon travel so the party followed the south bank of the Loup Fork for 57 miles. Turning in a southwesterly direction over sand hills, they reached Prairie Creek. Twenty miles upstream the men crossed over to Wood River, traveled six miles along its banks, then struck south to the Platte near Grand Island, and marched along this river to a camp opposite Fort Kearny.[39]

Dickerson's detachment saw no Indians along the route because the Pawnees who wintered in villages along the Platte had gone to the summer hunting grounds for buffalo. Having met hostile Sioux and Cheyenne, some three thousand five hundred Pawnees had returned to Fort Kearny for protection. Upon Dickerson's arrival, he was invited to attend a council of their chiefs who complained bitterly that the federal government was running a road through their country without approval and without purchase as had been the custom with other tribes. The Pawnees assured Dickerson, however, that they would offer no resistance to his party building the road, but they wanted to protest now lest it later be said that they had consented to the construction. The older chiefs observed that the roads always brought white men who chased away their game, and that emigrant roads involved them in many difficulties because the Pawnees received the blame when other tribes molested the trains and stole animals.

The Army engineer's outward route had coincided with that recently used by Mormon and California emigrant parties, but at the fort he learned that the earliest travelers along the north bank of the Platte had come directly up the valley along the stream without diverging to the north and going up the Loup Fork. He planned to return down the Platte Valley. First surveying a line due north of Fort Kearny for three miles, Dickerson turned east, striking the Wood River and following that stream to its junction with the Platte. Moving down the Platte, the party discovered excellent ground for a road with sufficient wood, water, and grass. By this new route the length of the march between Omaha and Kearny could be shortened 26 miles.

In his reports, Dickerson expressed an interest in the development of Nebraska along the route of his road. He had observed:

> This portion of the Territory is fast settling up with an industrious and enterprising class of pioneers. Pre-emption claims have already been located on all the timbered lands along the water courses as far west as the Loup Fork, above

which the Indian title has not yet been extinguished. But the scarcity of timber, stone, and coal, and the remoteness of the country from a market other than home consumption will operate against its ever becoming thickly settled.[40]

On the return trip the engineer was particularly observant of stream crossings to determine the nature and extent of bridge building required. The Platte, seldom confined to one channel, was too shallow to admit a ferry at Fort Kearny and reportedly too difficult to bridge. Dickerson was convinced the Indians would not allow any bridges on the route to stand long. The grass and tall weeds along the creeks were burned annually and a prairie fire would no doubt consume the bridges once the timbers had seasoned. The engineer therefore recommended an inexpensive construction at the lesser crossings by building corduroy flush with the beds of the stream and fastening the logs down so that they would not be washed away by freshets. The Loup Fork was 1,056 feet wide at the ferry, and he proposed to confine the channel of the ford by pilings. Bridging at any reasonable cost was impossible. At the Elkhorn, a stream 200 feet wide, a bridge would be constructed, and an embankment thrown up at its western approach for three-quarters of a mile.

Lieutenant Dickerson completed his season's survey August 14, stored his instruments and public property at Fort Leavenworth, and dismissed his party. In the winter months a contract for the bridges was made with Matthew J. Ragan who went immediately to Omaha intending to build some of the smaller structures before the spring. Dickerson recommended that the $4,500 remaining after the contract payment should be used for a laboring party under an Army engineer to improve the western sector of the road in the season of 1857. The congressional appropriation had made what Dickerson termed "a good wagon road for the greater part of the year." To render it passable at all seasons he urged the War Department to request another $25,000 from Congress.[41]

Captain Beckwith, who replaced Dickerson, supervised the actual bridge constructions at the eastern end of the road and, with a party of laborers, built small bridges over the creeks west of the Loup crossing. Deep trenches were dug beside each of these as a fireguard. Although this road was again reported as satisfactory in the dry season, it remained impassable along parts of the Platte after the spring freshets. In these months of April, May, and June the majority of emigrants using the north side of the Platte as a route to the west coast were delayed at the outset of their journey.

The $25,000 request that Dickerson recommended had been considered by Congress but no appropriation was granted.[42] Beckwith renewed the request to the War Department for additional funds at the season's close and suggested bridging the Loup Fork which he still judged the most difficult place on the road:

...Where it is most practicable to cross it with a ferry boat one day the boat grounds, the next, in the middle of the stream; compelling the discharge of loads into wagons, brought there across channels from the opposite shore.... And as it is impracticable for wagons or teams to stand still, even for a short time, anywhere in the river, without miring in the quicksands, the difficulties and labors and losses by emigrants are very great...[43]

Experience on the Elkhorn River had proved that piles driven 25 or 30 feet into the ground would be necessary to form the foundation work of any permanent bridge on the Loup. Cottonwood for the piles could be found near by, but hard timber for the substructure would have to be brought 80 miles overland from the Missouri. The estimated cost was $85,000.

These combined requests, totaling $110,000, repeatedly were included in the annual report of the Secretary of War to the Congress, but funds were not appropriated. In 1858 Captain Beckwith notified the Department that, had the appropriation been made in time to complete the contemplated improvements that season, the cost of transporting supplies overland to the Army of Utah would have been greatly reduced.[44] With a bridge across the Loup Fork, the fertile lands on that stream and the Platte would be taken up by settlers who could soon furnish supplies for Fort Kearny at reduced prices. Even these practical considerations failed to influence Congress.[45]

Between 1855 and 1857 another officer of the Corps of Topographical Engineers, Gouverneur K. Warren, was busily engaged in explorations in the northern part of Nebraska Territory, which would provide additional information for wagon routes west. In the spring of 1855 General W. S. Harney was ordered to ascend the Platte River to Fort Laramie and ultimately return overland to the newly acquired post on the Missouri River, Fort Pierre. Warren, assigned to this command known as the "Sioux Expedition," was temporarily detached and dispatched up the Missouri in June to make a preliminary survey of the military reservation. Once this area, embracing 310 square miles, had been mapped, Warren and seven companions traveled overland from Fort Pierre to Fort Kearny with hope of rejoining the main expedition to Fort Laramie. They crossed the domain of the Brule Sioux, against whom General Harney was waging a relentless war, and thereby displayed either unusual bravery or foolhardiness. Moving rapidly, using no tents, and lighting no fires at night, the little band halted only in daylight for Warren to jot down topographical details.

Arriving at Fort Kearny on August 24, Warren was instructed to continue as engineering officer with Harney's expedition to Fort Laramie. The topographer then traversed a new route from Laramie to Pierre along the valleys of the White and Teton rivers. He left the new outpost on the Missouri October 27 and established a new overland trace to the mouth of the Big Sioux River. This led down the Missouri River to Crow Creek, thence eastward to present-day Wessington Springs, South Dakota, down the Firesteel Creek to its mouth near Mitchell, thence southeasterly across the James and Vermillion rivers to the site of Sioux City. Within a three-month period Lieutenant Warren had examined a wide segment of the part of the Nebraska Territory that was eventually to become the state of South Dakota. Throughout he recorded observations on the topography, climate, soil, flora, and fauna and scientifically mapped his routes of travel, a procedure for which he was later to gain national recognition.[46]

In 1856 General Harney instructed Warren to ascend the Missouri River as far as the mouth of the Yellowstone River and to select suitable sites for military posts. Near Fort Randall, established this same year, the steamer on which his party had traveled from St. Louis was stranded on a sand bar. Warren and his assistants, therefore, proceeded overland to Fort Pierre. At this fort the scientists were given an escort of seventeen soldiers, and the enlarged detachment continued up the river aboard the *St. Mary,* an American Fur Company steamer. On July 10 they reached Fort Union at the Yellowstone mouth, where some stores were deposited, and then proceeded on up the Missouri to a point 60 miles beyond, to discharge the balance.[47]

While Mackinac boats were being constructed for Warren's return downstream, the lieutenant examined the Yellowstone Valley. From Sir George Gore, returning from a hunting excursion in the valley, the necessary equipment for a land expedition was procured. Leaving the Yellowstone mouth on July 25, the explorers ascended the stream almost 100 miles before the terrain blocked the passage of wagons. The expedition encamped at this point while Warren and seven companions proceeded with pack animals to the mouth of the Powder River, 30 miles farther. On returning to the wagon camp, Warren decided to divide his forces. An overland party retraced the route with the wagons to the Yellowstone mouth. A second detachment constructed a boat, 18 feet long, by stretching the skins of three buffalo bulls over a frame of cottonwood and willow. With this craft they navigated the Yellowstone, carefully mapping the islands and bends in the river. The expedition left Fort Union on September 1 and moved slowly down the Missouri, continuously mapping the river's course, taking notes on geological formations, and collecting zoölogical specimens. A detachment of herders drove the expedition's animals along the shore and encamped each evening with the boat crew. Fort Pierre was reached October

2. Here the animals were sold, and the soldiers of the escort returned to their companies. Warren proceeded on to St. Louis, thence to Washington, D.C. to report to the War Department and deposit his maps and scientific data. The season's work had cost the federal government $10,000. In two years of field experience, Lieutenant Warren, at twenty-six years of age, had established his reputation as one of the foremost topographers and scientists in the Army.[48]

When the War Department's Office of Explorations and Surveys determined in 1857 to explore the country west of the Missouri River to select the best routes for travel, Captain A. A. Humphreys, a topographical engineer in charge of that Bureau, recommended the appointment of Lieutenant Warren as the Army's best-informed officer on the Nebraska Territory. With $25,000 from the appropriation "for surveys for military defences, geographical explorations, and reconnaissances for military purposes," an expedition was organized ostensibly to continue the old military road leading westward from Mendota, Minnesota, to the Big Sioux mouth. This had been authorized by Congress in July, 1850, and surveyed by Captain Jesse L. Reno three years later.[49] The Secretary of War ordered Warren to the mouth of the Big Sioux River, then to head westward by way of the Loup Fork of the Platte to Fort Laramie and the South Pass, afterward to explore the Black Hills, and, if time and resources permitted, to return along the Niobrara River.[50] Much of the territory to be traversed was previously unexplored. True to the purposes of the Office of Explorations and Surveys, the expedition was not only to gain information on the character and resources of the country and its adaptability to settlement, but also to examine and report on the topographical features facilitating or hindering the construction of wagon or rail roads.

Leaving Sioux City July 6, Warren's party marched directly to the mouth of the Loup Fork, covering the 110-mile stretch in eleven days. On their westward journey along the Loup the expedition encountered no difficulties until within 50 miles of the stream's source. Here a gorge of the river proved impassable for the wagons and forced the expedition onto the sand hills. This sandy section extended southward to the Platte and northward to the Niobrara. Once the source of the Loup Fork was reached, the men did not venture westward through the desolate and dry country, but moved northward to the Niobrara. The surveyors struggled with the onerous task of getting the wagons through the sand hills. Throughout the journey most of the men suffered from fevers, the weather was inclement, buffalo and other game were unusually scarce causing the provisions to run low, and myriads of mosquitos so annoyed and exhausted the animals that a protective coat of grease and tar had to be applied.

At Fort Laramie the party was entirely refitted, an exceptionally slow process because supplies were in great demand for the Army of Utah moving against the Mormons. Due to the lateness of the season, the objects of the expedition could be accomplished only by dividing the command. Half of the wagons were left at the fort, and Lieutenant McMillan, the escort commander, was ordered to move down the Niobrara with the remainder. J. H. Snowden, a civilian topographer, accompanied him to make reconnaissances of near-by terrain. Warren, with F. V. Hayden, naturalist, W. P. C. Carrington, meteorologist, and P. M. Engle, topographer, organized a pack train to explore the Black Hills. With a guide-interpreter and seventeen packers, the explorers reached the peak of Inyan Kara before being intercepted by a large force of Brule Sioux. The Indians remonstrated with the scientists, threatening an attack if they marched farther into the hills. These Indians had made a treaty with General Harney in 1856 which gave the whites the privilege of navigating the Missouri and of traveling beside the Platte and along the White River between Forts Pierre and Laramie. They had been guaranteed, in turn, that the white men would not travel elsewhere in their country and frighten away the buffalo. They protested that Warren was examining their domain to see if it had value for roads and military posts. The Army officer sent for the warrior, Bear Rib, designated first chief of the Brule Sioux by General Harney's treaty, who reiterated the arguments of the lesser chiefs. In the company of this chief, the explorers journeyed eastward and completed their reconnaissance along the eastern parts of the mountains to Bear Peak. They had agreed not to penetrate to the heart of the Black Hills. The march was now southeast, striking the south fork of the Cheyenne at Sage Creek. After descending the south fork to French Creek, the party again turned southeast through the Bad Lands to White River and on to the Niobrara. Lieutenant McMillan's detachment was encountered about 40 miles below the point where Warren struck the river and 80 miles below the point where they had reached the river on the westward journey in August. Little Thunder, another Brule Sioux chief, protested each mile of the trip down the Niobrara.

The precipices and ravines shutting in the Niobrara River prevented the wagons from passing within five to eight miles of the stream. Forced out upon the sandy terrain, the train, following a devious route, found the going slow. Fort Randall was reached November 1, and a week later Warren's party began a survey of the route to Sioux City. With the third year's assignment completed, the topographer returned to St. Louis and again was ordered to the national capital to prepare maps and reports.[51]

Lieutenant Warren now assumed responsibility for comparing the advantages of the various routes from the Missouri Valley to Fort Laramie. In his opinion the broad and grass-covered valley of the Platte, leading to the west, furnished the best wagon road of its length in the United States, far superior to that of other river valleys. The route up the Loup Fork was impracticable and less direct than the Platte. By all means travelers were to avoid the desolate sand hills at its source, a miserable region impassable for ordinary wagon trains. The Niobrara route was not impracticable for wagons, but emigrants from Omaha and Nebraska City would have additional river transportation on the Missouri to the Niobrara mouth plus greater physical obstacles on the overland trip than along the Platte. A road from Sioux City along the Niobrara to Fort Laramie would be 40 miles shorter than moving directly southward to the Platte Valley at the mouth of the Loup Fork, thence westward to the fort. In Warren's opinion, however, the difficulties of constructing it would outweigh the advantages of the shorter distance. Farther north the military road from Fort Pierre to Fort Laramie along the White River was practicable and superior to the Niobrara route. Nevertheless, the increased river transportation along the Missouri to the White River mouth, in addition to the absence of settlements which necessitated higher prices for supplies, combined to make it less desirable than the Platte route. Above Fort Pierre, along the navigable part of the Missouri, there were no known advantageous routes leading to Fort Laramie or the South Pass.

In conclusion, Warren recommended the White River route as the logical continuation of the survey made by representatives of the Interior Department from Fort Ridgely to the Missouri River.[52] Although the Sioux City–Fort Randall road and the Niobrara River route appeared to be practical extensions of the military road from Mendota, Minnesota, to the mouth of the Big Sioux, the topographical officer advised a location direct from Sioux City to the mouth of the Loup Fork and along the Platte to the South Pass. All evidence pointed to the selection of the Platte Valley as best adapted for a national route for a Pacific wagon road or railroad leading from the Missouri River to the Rockies at the South Pass or Bridger's Pass.[53] Warren concluded:

> The same routes now most used and best adapted to the wants of the military occupation were long before used by the trader, the Indian, and the buffalo, as best adapted to their wants; and when future requirements shall demand increased facilities of transportation and locomotion and railroads shall be built, then they, too, will be found near the main routes now travelled by the trains of emigrants and the army.[54]

CHAPTER IX : *Road Locations in Utah Territory by the United States Army, 1854-1859*

WHILE THE ARMY ENGINEERS were engaged in shortening the routes across Kansas and Nebraska, the Mormons in the Great Basin, like other western groups, actively sought federal aid in building wagon roads. A foremost desire was to secure an extension of the ways across the plains, from Bridger's or Bryan's passes at the crest of the Rockies, on to Salt Lake along the shortest route practicable for emigrant wagons. A second objective was to improve the connections with the Pacific Coast by routes other than the emigrant trail north of Great Salt Lake and along the Humboldt River.

As a part of the Mormons' ambitious program to expand their domain, plans had been made early to secure an outlet to the Pacific Ocean. Brigham Young and his advisers devised the establishment of a chain of settlements in southern Utah and proposed the creation of others in present-day southern Nevada and California which would form a corridor to the sea in the San Diego region. Early in 1854 John M. Bernhisel, the Utah delegate to the Thirty-third Congress, introduced a measure in the House of Representatives authorizing a $25,000 appropriation to improve this so-called southern route to California.[1] The bill specifically provided for a military road beginning at Salt Lake City and running by way of Provo, Fillmore, Parowan, and Cedar City in the Territory of Utah, through New Mexico Territory to the eastern boundary of California in the direction of Cajon Pass. Fortunately for the interests of the Utah citizens, the bill was considered by the House of Representatives before that body became hopelessly deadlocked over the location of the Pacific railroad.[2] Even so, those in the House opposed to federal subsidies for road building moved to table the bill, and urgent pleas from the Utah delegate were necessary to defeat the motion. The demand for a roll-call vote did not block the passage of the bill. In the Senate the proposal received the endorsement of the Committee on

Military Affairs, was approved, and became law on July 17.[3] This appropriation was to be expended under the direction of the Secretary of War.

Lieutenant Colonel Edward J. Steptoe, with two companies of artillery and eighty-five dragoon recruits en route to California, arrived in Salt Lake City, August 31, 1854. Secretary of War Jefferson Davis, notified of the congressional appropriation and realizing that immediate action was expected of his department, decided that Steptoe was the logical federal officer to supervise the wagon-road construction. The lieutenant colonel's long-heralded arrival in Utah caused great consternation in the Mormon community. Delegate Bernhisel sent word in advance that Steptoe was President Pierce's candidate to succeed Brigham Young as governor. Before leaving Fort Leavenworth, the lieutenant colonel received additional instructions to assist in the capture of the murderers of Captain John W. Gunnison who had been slain by Indians just north of Lake Sevier on October 24, 1853, when engaged in surveying a railroad route along the thirty-ninth parallel. The assignment made Steptoe's task of establishing good relations doubly difficult.

Upon his appearance in the Great Basin, the military officer's relations with the Mormon leaders were outwardly cordial on most occasions, particularly when he joined with other federal officials, civil and military, in urging the reappointment of Governor Young. However, at the time a Mormon jury virtually freed the murderers of Gunnison, Steptoe's junior officers were highly incensed over church interference, and their reports revealed a strong feeling against the Latter-day Saints.[4]

Because of the smallness of the appropriation only the principal difficulties along the route could be removed. The War Department had received information that the most unsatisfactory stretches of the road would be found at the *jornada* south of Cedar City and in the mountains at the headwaters of the Santa Clara River. Davis enjoined Steptoe to inform himself as fully as possible and then make contracts "...with responsible parties for the construction of a practicable wagon road at these and all other difficult places, bearing in mind that the object is to obtain the best road that the money will make over the whole route." In anticipation of unusual difficulty in obtaining contracts, Steptoe was authorized to work out an arrangement whereby contractors could receive payments as the work progressed, rather than following the usual Army practice of payment only when the contract was fulfilled and the work inspected and approved. The lieutenant colonel was warned, however, that no money could be legally advanced to civilian contractors on the basis of promises to build.[5]

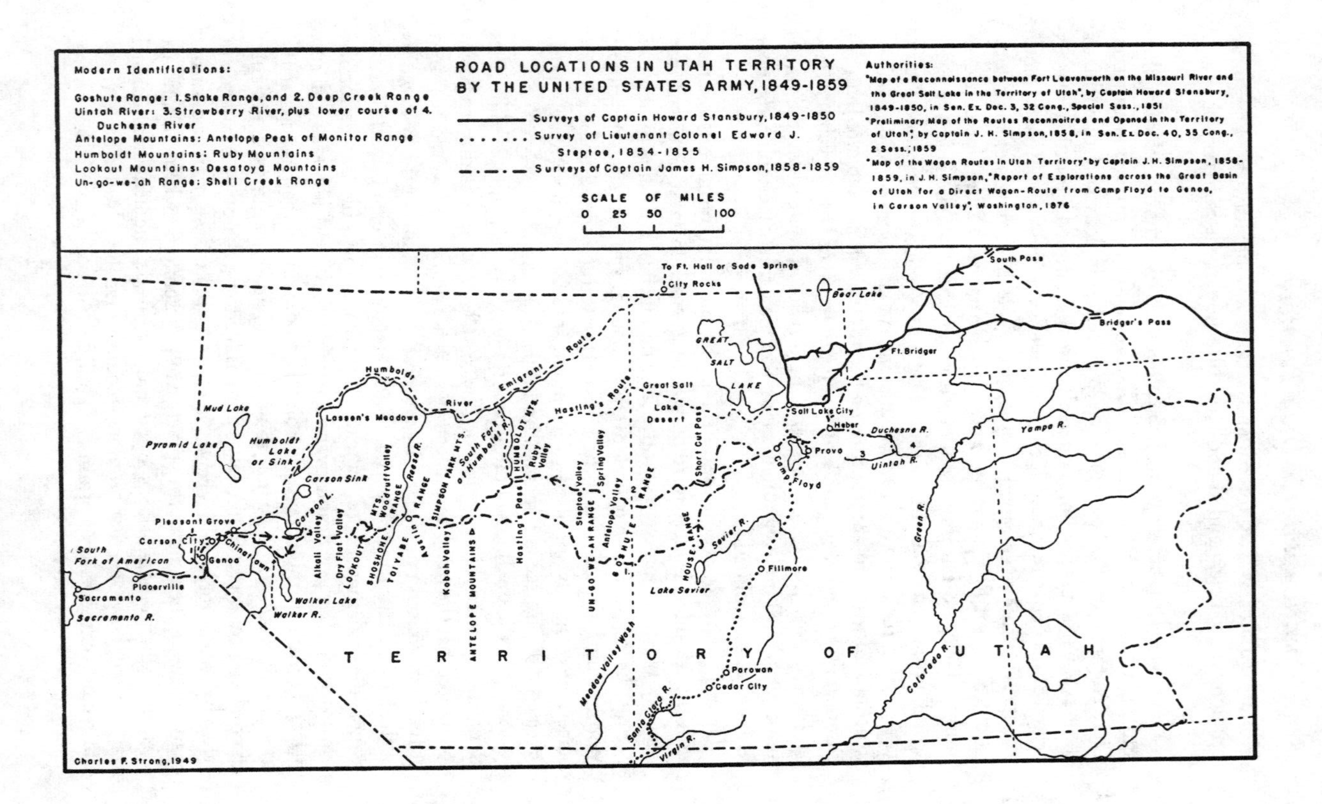
ROAD LOCATIONS IN UTAH TERRITORY
BY THE UNITED STATES ARMY, 1849-1859
Modern Identifications:
Goshute Range: 1. Snake Range, and 2. Deep Creek Range
Uintah River: 3. Strawberry River, plus lower course of 4. Duchesne River
Antelope Mountains: Antelope Peak of Monitor Range
Humboldt Mountains: Ruby Mountains
Lookout Mountains: Desatoya Mountains
Un-go-we-ah Range: Shell Creek Range
Surveys of Captain Howard Stansbury, 1849-1850
Survey of Lieutenant Colonel Edward J. Steptoe, 1854-1855
Surveys of Captain James H. Simpson, 1858-1859
SCALE OF MILES
0 25 50 100
Authorities:
"Map of a Reconnoissance between Fort Leavenworth on the Missouri River and the Great Salt Lake in the Territory of Utah", by Captain Howard Stansbury, 1849-1850, in Sen. Ex. Doc. 3, 32 Cong., Special Sess., 1851
"Preliminary Map of the Routes Reconnoitred and Opened in the Territory of Utah", by Captain J. H. Simpson, 1858, in Sen. Ex. Doc. 40, 35 Cong., 2 Sess., 1859
"Map of the Wagon Routes in Utah Territory" by Captain J. H. Simpson, 1858-1859, in J. H. Simpson, "Report of Explorations across the Great Basin of Utah for a Direct Wagon-Route from Camp Floyd to Genoa, in Carson Valley", Washington, 1876
To Ft. Hall or Soda Springs
City Rocks
South Pass
Bear Lake
Bridger's Pass
Ft. Bridger
GREAT SALT LAKE
Humboldt
Emigrant Route
River
Great Salt Lake Desert
Hasting's Route
Mud Lake
Lassen's Meadows
Salt Lake City
Heber
Duchesne R.
Yampa R.
Pyramid Lake
Humboldt Lake or Sink
Provo
Uintah R.
Carson Sink
Camp Floyd
Short Cut Pass
Carson L.
Reese R.
SIMPSON PARK MTS.
South Fork of Humboldt R.
HUMBOLDT MTS.
Ruby Valley
Steptoe Valley
Spring Valley
Antelope Valley
GOSHUTE RANGE
Green R.
Pleasant Grove
Carson City
Genoa
Chinatown
South Fork of American
Placerville
Sacramento
Sacramento R.
Alkali Valley
Dry Flat Valley
LOOKOUT MTS.
Woodruff Valley
SHOSHONE RANGE
TOIYABE RANGE
Austin
Kobah Valley
ANTELOPE MOUNTAINS
Hasting's Pass
UN-GO-WE-AH RANGE
HOUSE RANGE
Sevier R.
Fillmore
Lake Sevier
Walker Lake
Walker R.
TERRITORY OF UTAH
Meadow Valley Wash
Parowan
Cedar City
Colorado R.
Santa Clara R.
Virgin R.
Charles F. Strong, 1949

In discussing the expenditure of federal funds with local residents, Steptoe learned all were in agreement that the road was improperly located by Congress. No two concurred, however, about the areas in greatest need of improvement along the established route. Governor Young even brought pressure on Steptoe to expend the entire appropriation on the road between Salt Lake City and Parowan, which the officer refused to do. To obtain more reliable information about the route beyond the rim of the basin, the lieutenant colonel sent a party of quartermaster's men on a reconnaissance. In the meantime advertisements for construction bids were published in the Salt Lake paper.[6]

The first contract, made with William J. Hawley for the construction of a bridge over the Timpanogos River near Provo and the grading and widening of the road in that river bottom, provided for a payment of $8,400. The major contractor, James B. Leach, received $15,000 for cleaning, grading, ditching, and bridging the entire road "so as to render it practicable for general use and travel" before July 15, 1855. A third contract for $1,000 was signed with Lewis Robinson for widening the road to 25 feet and improving the grade across the mountains 20 miles south of the city of the Saints. The appropriation was thus virtually exhausted.[7]

In forwarding these contracts to the department, Steptoe informed Davis that at least $100,000 would be necessary to make this highway "what would be called in the east a tolerably good road."[8] The officer was impressed with the importance of the newly explored cutoff between the head of the Santa Clara and the Muddy because it would eliminate the difficult stretch of the old road down the Virgin River and shorten the southern route to California by 60 or 70 miles.[9]

Evidence is not available to determine how conscientiously the contractors fulfilled either the letter or the spirit of their contracts, but Utah residents generally believed the federal funds squandered. Lieutenant Colonel Steptoe, untrained as a road engineer and chiefly occupied with the political implications of his Utah visit, obviously considered the road-building program of secondary importance. Arriving in the late summer—he had been detained during the winter to settle the Gunnison affair—he did not plan to delay his California trip by remaining in Utah through the 1855 season for road building.[10]

Attached to the Steptoe command was a group of 130 civilian teamsters and herders employed by the Quartermaster Corps under Captain Rufus Ingalls. They were delivering 450 mules, 300 horses, and 70 wagons to forts on the Pacific Coast. According to Ingalls' reports, Colonel Steptoe was greatly disturbed by the circuitous routes and rough terrain of the known wagon roads west of Salt Lake City. Determined to explore the country to find a more direct route, Steptoe hired two Mormons to locate a road south of the Lake and along the Beckwith route to Carson Valley.[11] The Mormon party, after spending from September to November, 1854, in the field, reported the discovery of an excellent wagon road that would shorten the distance to Carson Lake by 150 to 200 miles. Steptoe planned to take his command and the large wagon train over this route and sought to employ one of the Mormons, O. B. Huntington, as guide. When spring approached and Huntington was unwilling to go, the lieutenant colonel, with his confidence shaken, decided to send out another Mormon expedition under Peter Rockwell to check the nature of the proposed new route. Rockwell found the desert tracts southwest of the lake impassable at that season, but suggested the possibility of locating a good road through the mountains somewhat south of the desert. Realizing how nearly he had led his expedition to disaster, Steptoe had no further interest in road exploration and decided to take his command around the north end of Salt Lake and along the emigrant road to the Humboldt. Before departure, the quartermaster's party was divided: one group of recruits was ordered to proceed over the new southern military road with horses and mules for Fort Tejon; a second command, under Ingalls, accompanied Steptoe to Lassen's Meadow on the Humboldt where it was detached to deliver wagons and animals to Fort Lane, on the Rogue River in Oregon.[12]

Utah's Bernhisel likewise sought congressional support for a shorter wagon route from the Rocky Mountains to Great Salt Lake. He introduced a bill in the Thirty-third Congress for a $3,000 federal grant to improve a military wagon road between Bridger's Pass and the Mormon capital. A copy of the measure was forwarded to Stephen A. Douglas in the Senate, and his influential support invited.[13] The emigrant route along the North Platte and the Sweetwater to the South Pass, thence southwest by way of Bridger's Fort, had been thought, since the time of Stansbury's exploration, to lie much too far to the north. The captain had proposed a more direct route from the South Platte Valley along its tributary, Lodgepole Creek, to Bridger's Pass, thence due west to Fort Bridger and on to Salt Lake City. Although this Congress approved an appropriation to survey and improve a road across Kansas and Nebraska to Bridger's Pass in accord with Stansbury's ideas, no authorization was given for its western extension into Utah Territory.

Efforts were renewed when the Thirty-fourth Congress convened. Captain Stansbury's suggestion about the importance of opening a trace due east of Fort Bridger, thereby avoiding the northern detour by Fort Laramie, was considered by the Military Affairs Committee to whom the bill was referred. His proposal prompted a favorable report. When the legislation was discussed in the committee of the whole, John Letcher of Virginia, desiring to block the bill's passage, ridiculed the idea that $3,000 could build 200 miles of road when $2,000 a mile was estimated as normal expenditure. Ignoring the Virginian's remarks, Bernhisel reminded the House of its $50,000 appropriation of the last session, authorizing a survey from Fort Riley in Kansas to Bridger's Pass in the Rocky Mountains. Without the extension now proposed, the earlier construction was useless. Lewis D. Campbell, Republican of Ohio, inquired if Letcher objected to economy, and the Virginian replied that he believed the appropriation an opening wedge to obtain additional support. John A. Quitman, of Mississippi, who usually supported federal construction of "military roads," rose at this point to state:

> ... This is probably one [road] in which it is ascertained by survey of examination that there are but very slight obstructions to the transportation of military supplies.... A considerable part of the line of the proposed road may be practicable for the conveyance of military stores and supplies, without further work or repairs.... I think that the sum of $3,000 would be sufficient to place an ordinary wagon road in a condition for the passage of troops and the transportation of munitions of war.[14]

Although Quitman had always insisted that the federal government could remove obstacles from the routes in the territories used by the United States troops or for the movement of materiel of war, he objected to a statement in the *Globe* earlier attributed to him, that in territorial affairs sovereignty was vested in the central government. At this juncture he began a digression on sovereignty which merged into a debate with Representative Campbell over the constitutional structure of the government. The lengthy argument was terminated by a reading of the War Department's statement supporting the appropriation, and the agreement that the bill should be recommended to the House for approval.[15]

The House failed to pass the bill during this session, and more than two years later in the second session of the Thirty-fifth Congress, Bernhisel introduced, and was successful in getting approved, a resolution instructing the Military Affairs Committee to inquire into the expediency of constructing the road from Bridger's Pass to Salt Lake City.[16] The Utah delegate wrote the Secretary of War for support on January 20, 1859. The Chief of Topographical Engineers was asked, in turn, for his opinion.

Colonel Abert reviewed for the Congress the history of Army road building across the central plains. The proposed Utah construction was admittedly a logical extension of the survey made by Lieutenant Francis T. Bryan between Fort Riley and Bridger's Pass in 1856 and hastily improved in 1857. Moreover, in 1858, Bryan, attached to the command of Colonel George Andrews en route to Utah, had seized the opportunity to open and improve a road along his survey from the Laramie Crossing of the Platte to Bridger's Pass and thence as far as the Green River. Andrews had then continued the march to Fort Bridger, following throughout the 1850 course of Stansbury and Bridger. Colonel Andrews was not as impressed with the route as Bridger, Stansbury, and Bryan and notified the Bureau of Topographical Engineers to that effect. As a result Colonel Abert stated that the Bryan route did not have equal advantages with the South Pass road for large government and emigrant trains because of the deficiency of water and grass. It was desirable as an express or mail route because it saved 60 miles.

The Bureau believed it quite possible that a better route might be found from Bridger's Pass to Salt Lake City, but no less than $20,000 would be needed for that exploration. Only the previous summer, James H. Simpson had opened a new road from Fort Bridger to Camp Floyd which was found to have superior advantages over that explored by Stansbury.[17] The colonel of the Corps recommended that any congressional legislation dealing with road construction between Bridger's Pass and Great Salt Lake should therefore authorize the Secretary of War to use his discretion in locating the best route. To Abert's report, John B. Floyd added a notation that Congress alone could decide whether the $20,000 necessary should be spent.[18] No grant was approved.

Therefore the memorials from Utah to three consecutive Congresses, presented over a period of five years, were disregarded. No specific endorsement was given to improving transportation from Salt Lake City to the Rockies. Due to the antagonism culminating in the "Mormon War" and the resulting bitterness, Congress had shown itself less generous with Utah than with other western territories in granting money for internal improvements.

The failure of Congress to support an improvement of eastern connections with Great Salt Lake did not, however, preclude Army activity. The program of exploration and road building launched in the Great Basin during the Army occupation of the Mormon War was one of the few positive aspects of that development in territorial-federal relations.

Soon after Albert Sydney Johnston established the headquarters of his Army of Utah at Camp Floyd in Cedar Valley, south of Salt Lake City, in the summer of 1858, he dispatched instructions to the chief officer of the topographical engineering corps attached to his command, urging him to hurry westward from Fort Kearny. His services were needed for wagon-road explorations and constructions. For this assignment the Bureau chief had chosen Captain James H. Simpson, the surveyor of the road from Fort Smith, Arkansas, to Santa Fe, New Mexico, in 1849 and the supervisor of the federal road constructions in Minnesota, 1851–1856. Simpson came to Utah with a record of outstanding success as a road engineer, and was generally recognized as one of the most able officers of the Corps. General Johnston first ordered the engineer to ascertain the feasibility of opening a wagon road between Camp Floyd and Fort Bridger by way of the Timpanogos River.[19] Eight years earlier Stansbury had suggested this route as desirable though he had not traversed it; Beckwith explored parts of the valley in making his railroad survey four years later. During the spring and summer of 1857, Mormon residents had built a wagon road for 12 miles up the river, northeast of Provo.

Simpson began his survey August 25, 1858, with an escort of twenty men under the immediate command of Lieutenant Samuel W. Ferguson. Isaac Bullock of Provo was employed as guide for the party. The route from Camp Floyd was east of north for eight miles out of the military reservation in Cedar Valley. The detachment gradually turned eastward toward the Jordan River to skirt the foot of Utah Lake. Crossing the river on a 60-foot toll bridge, Simpson traveled eastward through the settlements at Lehi, American Fork, and Pleasant Grove before turning south to the mouth of the Timpanogos River canyon.[20] The expedition ascended the Timpanogos along the Mormon road to its terminal point, thence on up the valley through Round Prairie for ten miles, and northward across the divide between the Timpanogos and Silver Creek, a tributary of the Weber.[21]

The engineer expressed great admiration for the Mormons who had built an excellent mountain road for wheeled carriages where only an Indian trail for pack horses had once been. Grades had been reduced on the rocky hillsides, embankments built up, and many promontories cut through to provide room for six-yoke ox teams. His only concern was about drainage improvements and widening the road so that teams might pass. To pay for their construction expenditures, the Mormons collected tolls at the bridge across the Timpanogos, one mile from the mouth of the canyon.[22]

From the divide between the Timpanogos and Silver Creek, the Simpson route was northward along the latter stream to a junction with the Parley's Park road from Salt Lake City.[23] Following this road, the engineering party crossed Silver Creek and the divide to the Weber River and journeyed down that stream for seven miles. Simpson left the old route by fording the

Weber and continuing to the mouth of its major tributary from the east, White Clay Creek. Ascending this creek for 30 miles, the explorers crossed the divide to the Bear River. From this stream they traveled northeast to Fort Bridger.[24]

The 155-mile reconnaissance had taken only nine days. Upon his return to Camp Floyd, Simpson organized a work party of fifty-five men under the command of Lieutenant Alfred T. A. Torbert, to open and construct the western section of the road so that it would be practicable for heavily loaded trains. Five wagons, road-working tools, and a twenty-five-day ration were given to these men. A similar detachment of thirty enlisted men under Lieutenant E. C. Jones, was simultaneously dispatched from Fort Bridger to improve the eastern end of the recently reconnoitered route. These workers carried their arms, twenty rounds of ammunition, camp equipage, and a twenty-day ration. Three mounted dragoons accompanied them for the purpose of carrying messages between the work crews and the forts. The roadway was to be marked by stakes, and guideposts to be erected at the crossings of other trails. The two parties were to work simultaneously and were expected to meet along White Clay Creek.

Simpson accompanied Torbert's party as it improved the river crossings, cut timber, built embankments, and widened the roadway. With the captain on his inspection tour was a representative of Russell, Majors & Waddell who was to determine if the company's freight trains headed for Camp Floyd should be directed to the new road. Simpson met Lieutenant Jones' party as anticipated in the valley of White Clay Creek. After leaving additional instructions, he rode on to Fort Bridger, inspecting the previous road work. On the return trip he observed the improvements completed by Lieutenant Torbert's men in his absence and arrived at Camp Floyd by mid-October. He was accompanied by Henry Engelmann, a geologist recently arrived in Fort Bridger who had previously been with Lieutenant Bryan on his surveys across the plains. This scientist was to be associated with Simpson on subsequent Utah explorations.

The engineer reported to General Johnston that his road was not nearly so rough as the usual emigrant, mail, and freight route up Echo Canyon and that more grass and water were available. He also believed the new way could be used for a longer season and was good for bringing in pack animals and herds of cattle early in the spring and late in the fall. All government trains began traveling over the route immediately, and Russell, Majors & Waddell discovered that its trains routed this way from Fort Bridger to Camp Floyd arrived more quickly than those traveling the old road. However, Simpson, the perfectionist, was far from satisfied with his road. The teams that traversed it the first months after his reconnaissance cut deep ruts, and the officer hoped he could supervise improvements the following spring.[25]

General Johnston, like his military colleague Steptoe who preceded him in the Great Basin, wished to open a route to California directly west of Lake Utah by way of the lower Humboldt or Carson Valley which would avoid the long detour north of Great Salt Lake. Immediately upon Simpson's return he was ordered to the west through Rush and Skull valleys, and instructed to explore as far westward as the late season and the condition of his animals would permit. The country was represented as a "vast clay flat, destitute of vegetation and water," that was "in the spring almost impassable for wagons." Johnston believed this reconnaissance a necessary preliminary to more extensive explorations the following spring. He was eager to learn of the grazing possibilities along the south rim of the valley and if it would be necessary to sink water wells in case a road could be put through. He was also debating the advisability of building a military post along the proposed route.[26]

Thirty-five men under Lieutenant Gurden Chapin, ten of whom were mounted dragoons, were assigned to Captain Simpson's exploration. Engelmann, the geologist, William Bean, guide, several laborers, and an interpreter completed the party's personnel. With their camp equipment packed in five wagons and the astronomical instruments placed in an ambulance, the expedition headed west on October 19, 1858. The first day's trip was to Meadow Creek in the center of Rush Valley. From here the party turned northwest to a small village on the mountain ridge near Reynold's Pass, known as Johnson's settlement.[27] There fifteen small log houses provided shelter for the one hundred inhabitants engaged in raising stock and hay. Simpson thought the westward descent from the pass into Skull Valley too steep to be practicable for loaded wagon trains and, should it be the only pass found, some blasting and heavy construction work would be required.

Turning southward through Skull Valley, the party traveled for two days to an encampment known as Pleasant Spring. En route they had crossed a "desert as level as a floor and in spots perfectly smooth and divested of every vestige of vegetation." From Pleasant Spring the explorers moved southwest to Short Cut Pass in the Thomas range. Here the captain determined to go no farther because the nearest water was reported to be 35 miles distant. He returned to his camp at Pleasant Spring and decided to march north-of-east to cross the divide between Skull and Rush valleys at a more southerly pass than that used by Reynold. From this new pass, named in honor of General Johnston, his route was northeast until it merged with the outward track just west of Camp Floyd.[28]

When Simpson's party returned, George Chorpenning, who held a government contract for carrying a weekly mail between Salt Lake City and Placerville,[29] took a small party over the new track to examine it before deciding to transfer his stock to this route. Chorpenning's outfit planned to use Simpson's road for winter travel because the snows in the Goose Creek Mountains, on the route north of Salt Lake, made it difficult for mail carriers to get through.

During the winter months, 1858–1859, Simpson was captivated with the idea of renewing his explorations to California in the coming spring. According to the official reports of the topographical engineers, he thought the distance between St. Louis and San Francisco could be substantially shortened by using Lieutenant Bryan's route from Fort Leavenworth to the Rocky Mountains, his new road from Fort Bridger to Camp Floyd, and a middle route that he hoped to locate through the central part of Utah Territory. He estimated this distance to be 2,265 miles between St. Louis and San Francisco. It was from 2,500 to 2,800 miles between these cities by other routes—by the southern Utah route via Salt Lake, the Santa Clara and Virgin rivers to Los Angeles and thence northward to San Francisco; by the southern mail route used by Butterfield from Fort Smith through Texas and New Mexico; or by the Oregon-California trail. Simpson obviously preferred the central route to the southern for mail deliveries to the Pacific and hoped to shorten the distance as a means of providing evidence to justify its support by congressional mail subsidies. He noticed with interest that the Californians had constructed a road from Placerville to Genoa across the Sierra Nevada. The telegraph was already in operation between these places, and its promoters were seeking a route to the East.[30]

In mid-winter Simpson requested the Secretary of War, through the Bureau in Washington, to authorize a program of exploration for the spring which included a wagon-road survey from Camp Floyd to California by way of Carson Lake, to be followed by an eastern trip to seek a shorter, better route from Camp Floyd to Fort Leavenworth. He wished to locate a new road to the headwaters of the Arkansas and thence along Bryan's survey of 1855 from Bent's Fort to Fort Leavenworth. This last way would be a more direct eastern connection than any through the Rockies at the South, Bridger's, or Bryan's passes. Secretary Floyd approved the project and through General Johnston gave Simpson a carte blanche to organize the expedition. Two junior officers, Lieutenant J. L. Kirby Smith and Lieutenant Haldiman L. Putnam, were named assistants, and Henry Engelmann continued as geologist. The party also included a taxidermist, a photographer, an artist, and two famous guides—a Mormon by name of John

Reese and Pete, a Ute Indian. An escort of twenty men, ten dragoons, and ten infantrymen under Lieutenant Alexander Murry completed the party. The expedition had twelve six-mule wagons for the transportation of three months' supplies. Two spring wagons carried surveying and meteorological instruments.[31]

In compliance with his instructions, Simpson left Camp Floyd May 3, 1859, and proceeded west along his return route of the previous fall from Short Cut Pass. From the pass the engineering party followed the mail route that Chorpenning's men had improved during the winter. This route was south of west to the Goshute Mountains on the present Utah-Nevada boundary. The expedition passed through these mountains in a westerly direction crossing Pleasant, Antelope, and Skull valleys.[32]

At Short Cut Pass and other places where Simpson thought the mail route inadequate for wagon travel, he ordered side-surveys and road relocations. The party next crossed the Un-go-we-ah Mountain range, passing through Spring Valley and emerging into Steptoe Valley. From here the route turned north of west, crossing several lesser mountain chains to Ruby's Valley. The agents of Chorpenning and Company had twice tried to turn southward from this valley to deliver mail to Genoa by a more southern route, but had failed. Simpson resolved to travel through Hastings' Pass in the Humboldt Mountains west of Ruby's Valley and from there to seek this southerly route.[33]

At Hastings' Pass, Simpson's travels coincided with those of Hastings, Beckwith, and Chorpenning. Just west of the pass was the valley of the south fork of the Humboldt, or Huntington's Creek, along which the Hastings' route to the north was located. It joined the regular emigrant trail to the Humboldt at the mouth of the south fork. Beckwith's survey turned northwest from the pass to join the old Humboldt road ten miles above Lassen's Meadows. Chorpenning delivered the mail northward from the pass to Gravelly Ford, thence along the same emigrant route used by the other two.[34]

After crossing the Diamond Mountains west of Hastings' Pass, Simpson's party turned southwest to the northern rim of the Antelope Mountains, thence westward through Kobah Valley across the Toiyabe Mountains by way of Simpson's Park into the Reese River valley. The engineer's route in crossing this valley was just north of the present site of Austin, Nevada.[35]

Many Indians were encountered by the engineering party. Simpson disregarded the Digger Indians, but was impressed by the conduct of Cho-Kup, chief of the Shoshones south of the Humboldt, whom he described as "respectful, intelligent, and well-behaved." In appreciation of his aid to the mail company, this chief had received a letter of approbation from Chorpenning. Simpson likewise gave him a paper stating that he believed him to be the friend of white travelers. However, their interview was disappointing because of the absence of Pete, the Ute scout and interpreter, who was accustomed to journeying a day in advance of the main group to explore the way.[36]

The explorers under Simpson crossed the Shoshone and Lookout (Desotoya) mountains west of the Reese River. The guides encountered difficulties in finding a route through the Lookout Mountains. Pete, Payte, and Sanchez were all lost for several hours on occasions, and John Reese's failure to return to camp one evening delayed the expedition until rescue groups were sent out for him. Reese's mule had given out 17 miles across the mountains from the main party. With no clothing but that on his back, no matches to make a fire, and without food, he trudged over the range. Kindly Digger Indians had offered him three fat rats that had been roasted with the entrails not removed, but Reese declined their generous hospitality. The next morning the guide was found in the sage brush by the roadside, his clothes nearly torn off, and in such a pitiable condition that Simpson first thought him to be a Digger.

Once the mountains were crossed, the expedition journeyed in a westerly direction across the desolate stretches of Dry Flat Valley and Alkali Valley to the north shore of Carson Lake. The entire party suffered from lack of water on the desert. The glare from the whitish clay flats that the expedition crossed east of Carson Lake was almost blinding, but Simpson, always optimistic, observed that the smooth arid land provided an excellent roadbed.[37]

From Carson Lake the engineer turned south to Walker's River, along which he traveled around its northern bend, thence overland to Carson River and down its south bank to Pleasant Grove. From the encampment at Walker's River, he sent Reese ahead to arrange for the construction of a raft to carry the expedition's wagons across the Carson River. The agent of the California Mail Company at Pleasant Grove pulled down a loghouse made of cottonwood to build this raft. Here the party crossed to the north bank and marched down the stream to Chinatown and Carson City.

In Chinatown the soldiers were impressed with the news of gold diggings in near-by Gold Canyon. Simpson observed the opium-smoking and gambling habits of the fifty male inhabitants. Carson City was described as a town of two stores and a dozen small frame houses. From here the detachment moved southward along the emigrant road between Carson River and the base of the Sierra Nevada to Genoa, arriving June 13.[38] The one hundred and fifty inhabitants of Genoa greeted the expedition with the raising of the national flag and a salute of thirteen guns, a fitting ceremony of appreciation for the successful exploration.[39] Simpson reported that he had shortened the old emigrant route from Salt Lake by the City Rocks and down the Humboldt by 288 miles and the Chorpenning mail route by 124 miles.

The topographical engineer left his party in Genoa and journeyed by stage to Placerville. Along the route he noticed the telegraph line strung on living trees and great coils of wire lying along the road. Colonel Fred A. Bee, president of the Placerville and St. Joseph Telegraph Company, conferred with Simpson about using his route immediately, but the engineer recommended that he wait for a report on his return exploration before deciding to use the outward trace.[40] Simpson traveled from Placerville to Sacramento by stage and then took a steamer to San Francisco where he reported to Army officials and procured supplies for the return journey.

During the two-day visit in San Francisco, Simpson was requested by the proprietors of the *Alta California* to allow one of their commercial correspondents, Walter Lowry, to accompany him to Camp Floyd. Lowry was suffering with a pulmonary disease and wished to go East immediately to visit relatives in Philadelphia before his expected death. Upon Simpson's return to Sacramento Colonel Bee introduced Lowry, whom the engineering officer tried to dissuade from making the journey, but with no success. From Placerville to Genoa Simpson endured one of the all-too-common night-stage rides in the mountains with a drunken driver holding the reins.[41] Partly as a result, he was convinced that this road over the Sierra Nevada was far more difficult for wagons than any part of his new route and urged the War Department to recommend a $40,000 appropriation to Congress for improvements.

Leaving Genoa on June 24, Simpson began his return reconnaissance by way of Carson City and Chinatown. At the last settlement his party left the old trace, crossed Carson River by ford, and proceeded along the south bank of the stream to Pleasant Grove. Experience proved the outward route on the north bank of Carson River far superior. From Pleasant Grove the expedition followed its former track for 25 miles, then struck due east to the south shore of Carson Lake, thus avoiding the Walker River valley too far to the south. East of Carson Lake, the men traveled through the flats to Lookout Mountains. A new route was worked out, slightly north of the outward trace, through these mountains that had proved so difficult for the guides. This new way led the explorers into Woodruff Valley where they rejoined the outward trail across the Reese River valley, through Simpson Park in the Toiyabe Mountains to Kobah Valley.

On this return journey, the captain had dispatched Reese, Pete, and four soldiers to go forward 150 miles in advance of the main party in the hope that they could locate a more southerly route than that through Hastings' Pass. Simpson left the Kobah Valley in a due easterly direction rather than turning northeast to return through Hastings' Pass, even though his guides had at first represented the intervening mountain range as impassable. The captain was overjoyed when a soldier from his advanced explorers reported the unexpected discovery of a practicable pass. He wrote in his journal, "The pass had been found by Ute Pete, who though he had been four days and three nights without food, except roots, yet had been the instrument of finding us a pass, and thus enabling us to keep on our course."[42]

The engineer's return road paralleled the outward trace some 40 or 50 miles to the south. In the Mon-tim Range Simpson turned abruptly south and entered Steptoe Valley at its most southern part. From this valley an eastward march was resumed across the Un-go-we-ah Mountains (Shell Creek Range), Antelope Valley, and the Goshute Range. Simpson crossed the present Utah-Nevada boundary just north of the thirty-ninth parallel. Here the expedition's direction of travel was changed to north of east.

In the House Mountains north of Lake Sevier, the reconnaissance party turned sharply north to enter Rush Valley at Oak Pass, five miles south of Johnston's Pass, used on the outward journey. From this point the route gradually converged with that followed two months earlier until a junction was made just west of Camp Floyd, where the party arrived August 4.[43]

In the region between Steptoe Valley and Lake Sevier, Simpson's party used a local road built by Mormons from Fillmore. Simpson surmised that it had been used as an avenue of escape to the west at the time of the "invasion" of Johnston's Army. After leaving his road in the House Mountains, the engineering party encountered the greatest difficulties of their entire exploration in crossing the Great Salt Lake Desert. With grass unavailable, the horses had to be fed with barley brought from Placerville; the last ox that had trudged all the way from Camp Floyd to Genoa and back was slaughtered for food; and small detachments constantly seeking water were handicapped by the sore feet of their overworked horses. Just when it seemed that the expedition would be demoralized by lack of water, Reese came upon a crippled Indian who offered to lead the party to water if he were lifted onto a mule's back. When the promised water was found, the wagons and mules left behind at various encampments were brought forward. The friendly Indian, paralyzed below the waist, traveled with the party for several days until hospitable valleys were reached.[44]

At Camp Floyd, Captain Simpson dispatched Lieutenant Kirby Smith with a small detachment to return over the last 100 miles of the route to straighten sections and mark them with stakes and guide posts. Wooden troughs were to be built at the springs in the desert to collect and save the water. Reese returned as guide to this party. An escort of dragoons was also assigned to Lieutenant Smith.[45]

Several days after Simpson's return, California emigrants started west over the new route. One party with seven wagons and another with thirty were supplied with an itinerary. The same information was given to Russell, Majors & Waddell who planned to drive a thousand head of cattle over the road to California.

Convinced that the season was too far advanced for Captain Simpson to organize an expedition to proceed to Fort Leavenworth by way of the headwaters of the Arkansas, General Johnston modified his earlier instructaions to authorize an exploration from Round Prairie in the Timpanogos Valley to the Green River by way of the Uinta.[46] Because of the imminent danger of being caught in the snows of the Colorado Rockies, Simpson was ordered to return from this exploration to the States via Fort Bridger.

Equipped with eight quartermaster's wagons, one large spring wagon, one light ambulance, and ninty-eight animals, a party of fifty-four left Camp Floyd August 9, traveling around Lake Utah and up the Timpanogos along the route Simpson had explored the previous year. The expedition arrived at Round Prairie the following day. A settlement of ten families, known as Heber City, had been built here since the reconnaissance of the last fall. From the Timpanogos they moved up the canyon of Coal Creek to the summit of the divide of the Uinta Mountains, down the valley of the Pott's Fork of the Duchesne, thence along that stream to its junction with the Strawberry River.[47] By this time the horses of the dragoons had been exhausted by the extraordinarily rough, steep, and stony character of the route along the mountain streams. Simpson considered this survey as difficult as any in which he had participated. Although the pass over the Uintas was an excellent one with only a slight grade, the route explored could not possibly be practicable for wagons or pack animals until the fallen and standing timber was removed. Luxuriant undergrowth and the boggy character of the soil also proved major obstacles.

Simpson, however, reported to General Johnston that, in his opinion, additional road explorations and constructions in this direction were highly important due to the opening of the Pike's Peak gold region. He hoped an extension of the road would be made eastward to connect Salt Lake City with Denver.[48]

To determine whether his road to Fort Bridger via the Timpanogos, Weber, and White Clay Creek could be shortened, Simpson organized a party of seven, including Engelmann, to travel by way of Kamas Prairie, the east fork of the Weber and one of its tributaries, across the head of White Clay Creek, and thence on to the Bear River. The main party under Lieutenant Alexander Murry, returned with the wagon train to Fort Bridger by the road located in 1858. Simpson's group on horseback explored two new routes, one shortening the distance of the Timpanogos route by twelve miles, another by seven.[49] Neither was worth improving, according to the engineer, because of the steep and rocky terrain and the fact that the road could not be used for more than the summer months because of early heavy snows in the mountain ranges. Simpson arrived at Fort Bridger April 26; three days later he started for Fort Leavenworth by way of the South Pass. When the party was encamped at Fort Laramie, Walter Lowry, the consumptive traveler from Genoa, died and was buried in the post cemetery.[50]

Simpson's accomplishments during his two years with the Army of Utah were recognized by his superiors as praiseworthy. The Pony Express which began running between San Francisco and Salt Lake City in April, 1860, used his northern route over the 300-mile course between Genoa and Hastings' Pass, and after continuing along the 175 miles of Chorpenning's extension of Simpson's route as far as Short Cut Pass, it traveled along the engineer's road to Camp Floyd on the way to the Mormon capital.[51] Simpson also asserted in 1861 that the telegraph line constructed eastward from California to join the western line from St. Joseph, Missouri, was following his route. This is entirely possible since the wires were strung from Virginia City through Ruby Valley near Hastings' Pass, Egan Canyon, and Deep Creek on to Salt Lake City.[52] According to General Johnston, emigrants passed daily over the new route to California, many driving large herds of stock, so that in a single season the road was well marked.

Simpson had pointed the way for future federal surveys. Through General Johnston, the War Department was notified of the importance of connecting the Utah and Colorado communities. The general urgently recommended that explorations be started on the eastern slope of the Rockies to find a road connecting with Simpson's Pass through the Uinta Range. Perhaps the pack trails of the gold seekers would disclose a route that could be converted into a wagon road. Whether the tide of migration passed through the Rocky Mountains by way of Fort Bridger or by the Uinta Pass, it was destined to pass down the Timpanogos. The general urged the government, therefore, to purchase the Mormon road improvements in that valley to relieve the emigrants from the tolls they could ill afford to pay.[53]

Captain Simpson, charged with the responsibility of furnishing the War Department with estimates of needed appropriations for Utah roads, requested W. H. Hooper, the territorial delegate, to learn what the Timpanogos River Turnpike Company would demand for its holdings.[54] Hooper replied that the canyon road had become the property of the Territory of Utah by action of the legislative assembly and that $20,000 was the price.[55] Simpson thought this a reasonable figure and included it in his recommendation of a $50,000 congressional appropriation for the improvement of the Camp Floyd–Fort Bridger road. He also suggested $50,000 as the sum needed to improve his road to California, along either way Congress chose to designate, with an additional $30,000 to carry the road across the first or most eastern range of the Sierra Nevada from Genoa to Lake Valley.[56] Twenty thousand dollars was needed to improve his road across the Uinta Mountains to the mouth of the Duchesne Fork and a like sum to continue explorations toward Denver. In this last recommendation he had the active support of B. D. Williams, the delegate from Jefferson Territory, who suggested that the Rockies could be crossed to the headwaters of the Blue River, where a small town of Breckenridge had been established as a result of gold discoveries in the fall of 1859. This settlement was 100 miles to the west of Denver, well into the Middle Park range, and was connected with the eastern slope by a road built by miners.[57]

The War Department's request for $170,000 for Utah road building involved one of the most comprehensive programs of construction yet sponsored by the Topographical Engineers. It was forwarded to Congress in 1860 when that body was perplexed over the impending division of the nation, and proposals for road construction appeared relatively insignificant. With the coming of the Civil War, no action was taken.

Part II

WAGON ROAD CONSTRUCTION BY THE DEPARTMENT OF THE INTERIOR, 1856–1861

CHAPTER X : *The Thirty-Fourth Congress and the Creation of the Pacific Wagon Road Office*

"CALIFORNIA, when she speaks, desires to be heard," announced Senator John B. Weller in presenting to his colleagues of the Thirty-fourth Congress a memorial signed by seventy-five thousand of his constituents. Weller, an Ohio Democrat, had gained legislative experience in the House of Representatives before the Mexican War. When the war was over, President Polk appointed him chairman of the commission to survey the boundary line between the United States and Mexico, under the Treaty of Guadalupe Hidalgo. He planned to continue his political career in California and, by 1851, was elected Frémont's successor to the Senate as a Union Democrat. Here he became an outstanding spokesman for better road and mail connections to the Pacific Coast.[1]

On May 19, 1856, Weller delivered to the Senate a resolution of the California legislature for federal construction of a wagon road connecting that state with the Mississippi Valley, and the establishment of military posts and watering stations along the route to protect emigrants and mail carriers.[2] A week later he laid two heavy folio volumes, handsomely bound in hand-tooled leather and with a title page elaborately lettered with red, blue, and gold leaf, upon the desk of the presiding officer of the Senate. In this book were the signatures of men in many villages and mining camps attached to a petition for congressional action to improve communications. No petition containing so many signatures had ever before been presented to the Senate.[3] Weller began to read this document:

> We are a population of five hundred thousand in number, occupying the western limits of American possessions upon the Pacific. Our State is the growth of little more than five years; yet within that brief period we have erected as an outpost upon the frontier of our republic, a municipal government faithfully reflecting the spirit of our common institutions.
>
> By energy and enterprise we have laid the foundation for a great State which,

properly protected and encouraged, shall add to the dignity and stability of the Union....

Our mines, not yet fairly opened for successful working, have realized by moderate estimate $300,000,000, which we have sent forth to the world; and thus, in no inconsiderable degree, contributed to the prosperity of our whole country.... And while we appreciate the benefits of all the appropriations of the Federal government for our debt, defense, and commerce, which have been so generously expended on our coast, and which give us reasonable assurance of the future, there is a pressing necessity, for the want of which our State of California and the whole Pacific Coast are now languishing and unprosperous. Our great necessity is increase of population; our requirement is an immigration of the working and producing classes.... Every inducement for immigration is now as strong as when, in 1849 and 1850, all approaches to our state were thronged with immigrants.... The best portion of our present population arrived here by passage across the Plains;... this journey was not without its inconveniences and injuries. The recent hostile demonstrations of the Indian tribes have so multiplied the hardships and increased the hazard of this our only land approach, as to render it nearly impracticable; and the result has been an almost entire suspension of travel by this route—a cause from which we are now laboring under embarrassment.[4]

To relieve this situation, the Californians petitioned for a wagon road from the Missouri Valley to the Great Salt Lake basin and thence to the eastern slope of the Sierra Nevadas at Carson Valley. This had been the route of the "Great Migration" along the North Platte, across the South Pass, and down the valley of the Humboldt; already three hundred thousand had crossed it, and the location was central. According to the Californians, a few small bridges, occasional excavations, the sinking of a limited number of wells, and the establishment of ferries over the larger streams would do the job. Equally important would be the creation of new military posts for emigrant protection and the construction of blacksmith's shops and other workshops along the route for the public. "We are now, as it were, a distant colony," complained the men of California.

Senator Weller insisted that road building was of immediate and pressing necessity. All of his colleagues did not agree, however, when he stated:

These memorialists do not ask you to stretch the Constitution to accommodate them. They ask you to make no works of internal improvement within the limits of a State; but they simply ask you to construct a good wagon road *through your own Territories*.... Here is no constitutional objection to be interposed. You have the absolute power to expend every dollar in the national Treasury, if you choose, in making roads through the Territories.[5]

Early in the session Weller had introduced a bill providing for a triweekly mail between the Mississippi and San Francisco. His primary object in introducing this bill had been to obtain the construction of a good wagon road. A provision had been included, therefore, to allow the contractor $150,000 for improving his route. Before reporting the bill to the Senate, the Committee on the Post Office and Post Roads had struck out this provision as well as a statement directing the President to see that military posts and stockades were placed along the road. Weller, incensed at this rebuff, took the occasion to criticize the work of the Topographical Engineers.

I desired to place the construction of the road under mail contractors. They are the best road makers in the world. They do not go out, as do the topographical engineers, with barometers and other instruments, to determine the altitude of mountains; nor do they care about the botany, mineralogy or geology of the country; they take no other instruments than the ax, the shovel, the spade, and the pick-ax. Their only object is to locate a good road.... My desire, then, was to place the expenditure of this money in the hands of practical civilians—men who would see that the road was made at once. Time has become important to us;... we want the road, and that at once. This accomplished, we have no objection to parties to examine the geology and mineralogy of the country.

The California senator reminded his associates of the $50,000 appropriation of the previous session for the Fort Riley–Bridger's Pass road and stated that when he inquired at the War Department about the status of the survey, he was told that parties were just starting work. With little understanding of the difficulties of disease and distance on the Plains, which had made it impossible for Lieutenant Francis T. Bryan to complete his Kansas and Nebraska surveys, Weller stormed at the fifteen months' delay. If Indians on the warpath had been the cause, why did not the Secretary send out a strong military force to protect the working parties?

The Californian announced his intention to sponsor three separate road construction bills: one, appropriating $50,000 for a route from Fort Ridgely in Minnesota to the South Pass of the Rocky Mountains; a second, allowing $300,000 for a central road westward from Missouri via the Great Salt Lake to Carson Valley; and a third, southern route between El Paso and Fort Yuma at the mouth of the Gila for which $200,000 was to be allotted. Having lost confidence in the Topographical Engineers, Weller included a provision transferring the direction of each project to the Secretary of the Interior rather than the War Department, with the understanding that construction contracts would go to civilians whom the senator described as "practical men."

Weller told the Senate that stages and mail carriers awaited the construction of his proposed roads, that the nation would have a telegraph immediately thereafter, and above all California would have "the population which we particularly desire, agriculturalists and laboring men." In concluding his address, the senator stated the position of his state clearly:

> California does not now ask you to make her roads. There never has been a single dollar of the Federal Treasury appropriated to the construction of any roads in the State of California.... We are not now asking it. We ask you merely to make roads through your own territory and we will take care to connect with it. Is there anything unreasonable in this? or do you still think that California is a "humbug," a bubble, and will soon explode? Let me say that my State has thus far been loyal to the Union, and will remain so if you deal justly and fairly by her. One would suppose that after you had extracted more than three hundred millions from her, you would concede she was no "humbug."[6]

The Weller peroration in Congress came as the climax of a movement for federal road-building aid that was well known to California residents. Many had been discouraged and chagrined by the policy of the Franklin Pierce administration, which was hostile to governmental appropriations for internal improvements, other than territorial military roads, on constitutional grounds. The more optimistic leaders, nevertheless, called a mass meeting in San Francisco on December 12, 1854, and organized a committee of twenty-five to secure more adequate means of overland communication. Colonel John C. Frémont and Captain William T. Sherman were active committeemen. Later in the month the executive committee of this group agreed that a wagon road and stage line from Missouri to California was imminently possible, but the construction of a Pacific railroad was an enterprise for future consideration. The emigrant road committee adopted resolutions on January 2, 1855, incorporating this conclusion and urging the state legislature to appropriate funds for the improvement of a wagon road to the eastern boundary of the state. Within two weeks another meeting was held, attended by Governor John Bigler and Sherman, at which the association was incorporated under the laws of the state of California as the "Pacific Emigrant Society."[7]

In his message to the California lawmakers on January 5, 1855, Governor Bigler reiterated the views of those agitating for road construction and stated that, in his opinion, years must elapse before the desirable and highly important Pacific railroad project could be consummated. The immediate protection of family groups, particularly women and children moving west to join their menfolk, was of paramount importance, and the creation of military posts along the overland route vital for security. The governor urged the state legislators to present the question before Congress and plead for federal action to "render travel over the road, with coaches, expeditious and safe."[8]

The emigrant road committee preferred a route to the eastern boundary by way of Sacramento, Placerville, and across the Sierra to Carson City. The governor also thought this route the most practicable since it was the way he had traveled to California as an emigrant. Sacramento and Placerville newspapers began a crusade for wagon-road legislation, local enthusiasts called mass meetings of the interested citizenry, and petitions poured in upon the legislature.[9]

In consideration of these petitions and the governor's message, the California legislators passed a law on April 28, 1855, authorizing the construction of a wagon road from the capital at Sacramento to the eastern boundary of the state. The law provided for a preliminary reconnaissance under the direction of the surveyor general to determine the most practicable and economical route. Five thousand dollars was thought an adequate expenditure for this. A commission composed of the governor, secretary of state, and surveyor general was to receive bids for the construction of the road, along the lines of the survey, across the Sierra to Carson City. The state was to contract with the lowest and most reliable bidder at a cost not to exceed $100,000.[10] The legislature failed to appropriate $5,000 for the reconnaissance and the surveyor general was forced to appeal to community-minded citizens of Placerville and Calaveras and to the El Dorado County officials for assistance to finance the survey.[11] Several explorations across the Sierra Nevada were made during the summer months. The first was conducted by Sherman Day, a California state senator; on his second trip he was accompanied by the surveyor general of California, S. H. Marlette. Upon the return of this group, George H. Goddard, a recognized engineer, was sent to mark the eastern boundary of the state and to make a profile of the route advocated by Day. Surveyor General Marlette seriously doubted that this was the most satisfactory route across the mountains, but the construction commission, headed by Governor Bigler, insisted on making contracts. Marlette finally entered a protest against the decision to advertise for sealed bids, stating that the road was not yet surveyed, nor were the plans and specifications for construction ready.[12] When the bids were opened, the commission discovered that one L. B. Leach of Stockton had agreed to complete the work for $58,000. As the lowest bidder he was granted the contract, subject to the approval of the surveyor general. However, upon further investigation Leach proved to be a fictitious individual. Various groups, including the steamship operators and rival communities to Sacramento and Placerville, were accused of entering the low bid to delay the construction.[13]

The constitutionality of the 1855 law had been questioned because of a provision authorizing the California treasurer to issue bonds for the amount of $100,000 to secure funds for the road. The attorney general brought action in the state courts to challenge the law as a violation of Article VIII of the California constitution, setting a maximum figure for the public debt and providing that the approval of the people was essential to any increase over that figure.[14] Some accused the "Bigler clique," who were well aware of the illegality of the legislation, of rushing the road program to completion before the new governor, J. Neely Johnson, was inaugurated. The state supreme court, during the summer of 1856, sustained a lower court's decision that the original wagon-road law was unconstitutional.[15]

The legal tangle never lessened popular enthusiasm for the road. The unofficial surveys of 1855, financed by private subscription and El Dorado County, were completed, although those who had covered the expense waited two years for reimbursement from the legislature. Four additional reconnaissances were made in the summer of 1856.[16] In view of the supreme court decision nullifying the 1855 statute, the voters of Sacramento and El Dorado counties persuaded the California legislature to provide for an election whereby they might authorize each county to appropriate $25,000 for the construction of the road to Carson Valley. The new law provided that should the voters approve, a property tax of one-fourth of 1 per cent would be levied to raise the funds.[17]

The newspapers also continued the campaign for federal aid. The editor of the *Stockton Argus* wrote:

> Let wagon roads be made through the State to Utah, and memorialize Congress to prosecute the work through the Territory to the bordering States of Missouri, Arkansas and Texas. We can by this means secure the hardy population of the Western and South-western States, bringing with them their stock of horses, cattle and sheep. It is very rare that an agriculturalist can be met with who did not migrate to this State by land, bringing with them [*sic*] stock. Railroads cannot benefit this class of emigrants as the cost of transportation could not be made by a company to meet their means. . . .[18]

The *Sacramento Daily Union* joined the plea:

> From the West we must obtain our permanent agricultural population, and in order to give them a chance to come with their wives and children, and their household goods, cattle, horses and sheep, a good emigrant road must be built across the continent and the immigrants protected from the murderous rifle and tomahawk of the Indians. The accomplishment of this national object the people of this state have a right to demand Congress.[19]

The state had not failed to use the limited means at its disposal to make demands on Congress. In 1855 the California legislature passed a joint resolution instructing the state's representatives in Washington to urge the establishment of three or more roads across the plains. The routes suggested were by way of the Humboldt Bend through Nobles' Pass to northern California; from the Great Salt Lake to Carson Valley, thence by way of Johnson's cutoff to the middle part of the state; and into southern California through Walker's Pass to Tulare Valley, or from the mouth of the Gila to San Diego.[20] The next session of the legislature insisted that though the Pacific Coast hoped for railroads, immediate advantages could be derived from "the construction and defense of good wagon roads across the territory of the United States, which separates California from the other states." The federal government was charged with the responsibility of using its constitutional power and sources of wealth in this way to strengthen the political bonds of the Union.[21]

When the legislature was in session, news came to California that a new stage line had been organized in St. Louis and incorporated by the state of Missouri. The Missouri Overland Mail and Transportation Company sought the coöperation of the west coast citizens and solicited subscriptions for stock. Mass meetings were held in San Francisco, Sacramento, Marysville, and Placerville in support of this venture, and a people's petition began to circulate throughout the state requesting Congress to build wagon roads for emigrant and stage travel.[22] A popular movement thus reinforced the legislative action, and with the unified support Weller could speak his mind in the Senate.

Simultaneous with the promotion work in California, was a drive for additional federal support for road building in the Minnesota Territory. Here the leadership was invested in Henry M. Rice, congressional delegate since 1853. The delegate had little respect for the ability of the Topographical Engineers of the War Department, and a declared enmity had existed between him and James H. Simpson, the Corps' representative in the territory, because the engineer opposed his reëlection to Congress in 1855. From that time on, Rice no longer sought road appropriations that would be administered by the War Department but began to work through the Secretary of the Interior. Successful negotiations with the Commissioner of Indian Affairs, George W. Manypenny, in handling reservation and land problems of the Minnesota tribes prompted him to attempt the enactment of treaties or laws authorizing the Indian Office to build federal roads in Minnesota.[23]

Second only to Rice as a promoter of a federal road program in the territory was William H. Nobles. As one of the gold seekers in 1850, he had sought a shorter and easier route to the Pacific than that followed down the Humboldt by the hordes going to the gold fields that season. He discovered an excellent pass of easy grade across the Sierra Nevada, which bore his name. For several years thereafter Nobles assumed the role of congressional lobbyist, attempting to interest the government in his new route. In 1853 he moved to Minnesota and in this new territory continued the campaign. Nobles was convinced that the frontier settlements in western Minnesota could be connected with central California by a direct road that would easily prove to be at least 100 miles shorter than that from the Missouri River towns. In the territorial press he urged that a trans-Mississippi West route should, therefore, begin in Minnesota. Moreover, on this northern route water and wood were more abundant than along the Platte River. Many emigrants from the region north of the Ohio would find the milder summer climate more satisfactory for traveling than the route that lay along the central plains.[24]

Minnesota boosters saw in Nobles' proposals an opportunity to bring a transcontinental road through their territory and hoped that this might prove a forerunner of the all-important railroad to the Pacific. The territorial press, led by the *Minnesota Democrat,* was enthusiastic. Early in 1854 the editor of that paper urged the legislature to memorialize Congress for a road from Minnesota to Fort Laramie where it might join the Oregon Trail from Independence. It was reported further that papers in Chicago and other western cities endorsed this route and surely Minnesota, whose benefit would be greatest, should join the campaign.[25]

The upshot of the newspaper publicity was the calling of a mass meeting in St. Paul on February 9, 1854, which was attended by most of the leading citizens of the territory and presided over by the associate justice of the territorial supreme court, Andrew G. Chatfield. The assemblage appointed a committee to draft resolutions for submission to Congress requesting the wagon road and selected Nobles, the chief spokesman, to go to Washington and work for the passage of a federal law appropriating funds for the project. Five hundred dollars was raised for his expenses.[26] The territorial legislature endorsed the popular proposal by locating a territorial road from St. Paul to the Missouri River via Fort Ridgely and adopted a memorial requesting funds from Congress to construct a military road from Minnesota to California and Oregon.[27]

Between 1854 and 1856 no territorial legislature was more aggressive than that of Minnesota in memorializing Congress for road-building support. Statehood was soon to be granted, and road appropriations would then be unattainable.[28] In 1856 the local lawmakers again adopted a resolution pleading for an emigrant route across the Rockies and reminded Congress that the territory had already appointed commissioners to locate the road to the Missouri. Funds were needed for improvements and the extension to the South Pass.[29] Once more Nobles, now a member of the legislature, hurried to Washington to lobby for the road.

The first of the proposed wagon-road bills to be considered by Congress was the Minnesota-Nebraska route between Fort Ridgely and the South Pass. The $50,000 appropriation measure passed the House, where the Republicans had a majority, with no noteworthy opposition, but when it reached the stage for final debate in the Senate on July 14, 1856, Asa Biggs of North Carolina, a perennial spokesman for economy, challenged the legislation because it was not recommended by an executive branch of the federal government. John Bell of Tennessee expressed doubts as to the necessity of the road or the justification of the expenditure. He elaborated his views:

> The difficulty has been to get communication to the Minnesota Territory, of which a very large portion is yet unsettled. It is a place which invites immigration. Now, is it supposed to be necessary that this road should be constructed in order that the immigration to Minnesota may continue on to California or Oregon? There is a great deal more direct road than this, if they propose to go to California or Oregon. . . . I have no objection to having any necessary or any expedient communication; but I object to making an appropriation, and authorizing the construction of a road like this, merely because it is suggested by the representative of that Territory in the other House, or in any other quarter that such a road would be useful.[30]

Lengthy debate appeared imminent, so action on the bill was postponed. Senator Weller opened his argument the following day by saying that the distance between Fort Snelling and the Missouri was nearly 1,800 miles by the route in use, but this road would shorten the distance to 200 miles. Jacob Collamer, Republican from Vermont, interrupted to request the reading of a memorial from the legislative assembly of Minnesota Territory in behalf of the proposed law. The Minnesotans claimed their route would be the cheapest, shortest, and most convenient for any emigrant reaching the Mississippi above Rock Island. The road would facilitate the transportation of troops from the upper Mississippi to the upper Missouri and provide an avenue for Indian supplies going west of the Missouri and as far south as Fort Laramie.[31]

After the legislation was reread at the request of Clement C. Clay, State-rightist from Alabama, the senator remarked, "If the purpose of the road were to provide for the military defense of the country, I should have no scruples in voting for it." He added:

. . . I have been surprised to find that the control and direction of the construction of this road is given to the Secretary of the Interior. This strikes me as a novel feature; and it changes, to some extent, the character of the work, as I have understood and am inclined to regard it. . . . I move to strike out "Secretary of the Interior" and insert "Secretary of War."

Weller countered, stating that the bill was in the form that it passed the House, and summarized all his previous arguments against the War Department's control of federal road building and his reasons for wanting civilian contractors. Moreover, he feared the House was so divided by partisan and sectional conflict that should the measure be returned with amendments, it would never be reported again to the Senate. Richard Brodhead of Pennsylvania, unequivocally opposed to internal improvements, inquired if Weller had any assurance that the Secretary of the Interior would not call upon the Topographical Engineers, as usual, for assistance. The California senator replied, "I take it for granted that the Secretary of the Interior will select civilians, if he thinks they can perform the work better than the topographical engineers of the War Department. He is a western man, and he is a practical man, and will make the best selection for the Territory."[32]

Bellicose Andrew P. Butler of South Carolina, the intellectual disciple of Calhoun and a man uncompromisingly against federal aid to roads, rose at this point in the debate to state, "I think that in some respects it is the most extraordinary bill that has ever come to the Senate, or passed the House of Representatives." This senator felt the legislation should have originated with a department of the federal government, and he spoke of the Minnesota legislature's attempt to dictate to Congress. Weller and Butler, in anger, exchanged harsh words; Brodhead quickly came to the support of Butler. He also joined Bell in questioning the advisability of locating a road, ostensibly to California, in the Minnesota Territory, and with Clay, he believed that military roads should be constructed by the Army Engineers. Michigan's Charles E. Stuart shattered the opposition's argument by pointing out that the road was through Indian country and that these Indians were supervised by the agents of the Interior Department. According to Stuart, the House had wisely seen that the Secretary of the Interior, more than any other executive officer, could administer the road work with less likelihood of interference by the Indians. By a vote of twenty-four to eighteen the amendment to transfer the construction to the War Department failed, and the original bill was passed with a majority of twenty-eight to seventeen.[33] The President signed the measure on July 22, 1856.[34]

The appropriation for the central overland route from Missouri via the Great Salt Lake to Carson Valley came before the Senate for final consideration on June 26. William H. Seward and Lewis Cass both spoke briefly to indicate their approval of the bill, but Kentucky's John Crittenden urged more consideration of such important legislation. Weller reported that the bill had been referred to the Military Affairs Committee, of which he was chairman, and had been reported favorably to the Senate. During the committee stage, the responsibility for construction had been reassigned to the War Department, in the hope of gaining the support of those who approved of "military roads" but no other. The measure passed the Senate without a recorded vote.[35] A week later Weller presented the appropriation bill for the southern route between El Paso and Fort Yuma. No opposition was raised and the bill quickly passed, because the Senate wished the wagon-road appropriations to go to the House as companion measures. In spite of his tirade against the Topographical Engineers a month earlier, Weller had permitted these roads to pass as "military roads."[36]

In the Republican-controlled House of Representatives there was objection to the bills because Jefferson Davis would direct the expenditure of the money. All efforts to get a vote upon the legislation failed.[37] Thus, the opposition of the majority party to a southern Secretary of War aided the Californian in his desire to transfer federal road building to the Interior Department and thereby assure construction by civilian engineers.[38]

During the presidential campaign of 1856, James Buchanan, the victor, and all parties in Congress had been pledged to sponsor measures for improved communication with the Pacific, even to constructing a railroad. The Republicans, in their platform, had urged that in the meantime there should be "the immediate construction of an emigrant road on the line of the railroad."[39] When the Thirty-fourth Congress reconvened after the election, the House once more considered the wagon-road question. Samuel A. Purviance of Pennsylvania reported on the floor the bill for the central route, now located from Fort Kearny via the Great Salt Lake to Honey Lake on the eastern California boundary, with the approval of the Committee on Territories over which he presided. Galusha A. Grow of the same state and Solomen G. Haven of New York immediately spoke for the passage of the measure. Grow, a free-soiler who had built a national reputation championing free homesteads in the West, pushed through an amendment locating the road through the South Pass, rather than the Salt Lake Basin, so it would connect and form an extension of the South Pass–Fort Ridgely road. Haven said:

> ... This is the right place to begin.... If a measure of this kind, making a grand road... across to the Pacific, had been entered on four or five years ago, we should now have a good communication, and have fully tested whether there would be any use in undertaking to make a railroad communication hereafter. Some practicable good old-fashioned wagon road through the Territories can be made for less money than the railroad surveys have cost....[40]

Purviance discussed his committee's proposal stating that the road ran fourteen hundred miles along the chief route of emigration to the West. No deep cutting or tunneling would be necessary, timber would be available for bridging, and all deserts would be avoided.

"I cannot remain silent when a proposition of so much importance to the people whom I have in part the honor to represent is pending," announced Philemon T. Herbert of California. After repeating the worn-out arguments about the federal government's neglect of California and the incessant and urgent demand in the state for a wagon road, as reflected in the petitions to Congress, the Representative reached the heights of oratory:

> The graves of wayworn travelers upon every single mile upon the routes lying between this and the Pacific Coast tell a tale of neglect, and should be sufficient argument, not only to make the donation asked for, but to double, yea, treble the amount if necessary. Give us the appropriations asked for in the bill, so that the poor man as well as the rich man can with his family visit our shores in safety, where the increase of the value of his wagon and team alone will enable him to purchase one hundred and sixty acres of land, unequaled in fertility upon the globe. Sir, it is this class which has built us up on the Pacific and made us what we are.[41]

William Smith of Virginia offered an amendment to include a $300,000 appropriation for the El Paso–Fort Yuma road and, since the committee had now placed the construction of the central route with the Interior Department, he proposed that the southern route also be supervised by that executive department. Smith represented the interests of both the South and West. After serving as governor of Virginia, he went to California in 1849, where he declined the nomination as United States senator to return to his native state. Grow, spokesman for the northern group, accepted the amendment quickly, and the legislative impasse was broken. The delegate from New Mexico, Miguel A. Otero, seized the occasion to present an amendment appropriating $50,000 for a wagon road from Fort Defiance, in the New Mexico Territory, to the Colorado River, near the mouth of the

Mohave. Admitting that he had been advised not to submit a claim for this road for fear of defeating all wagon-road appropriations, he requested it nevertheless because of its importance in extending westward into California a commercial route already established from Independence, Missouri, to Fort Defiance. His amendment was approved by a vote of seventy-three to fifty-seven.

Michigan's David S. Walbridge then moved an additional amendment providing $300,000 for a road from Lake Superior to Puget Sound. "If . . . we cannot carry a road through the central portion of the country without hitching on to it two southern roads, then I propose to put on at least one northern road," he remarked. John S. Phelps of Missouri insisted that the road requested by Otero was local in character to open trade with California. Walbridge called for a vote on his amendment which was defeated eighty-seven to thirty-six.

George W. Jones of Tennessee, unusually silent up to this point, showed his opposition to internal improvements by moving to strike out the appropriation figures. He asserted:

> . . . I make this motion to enable me here to enter my protest against all this doctrine of, and all these schemes of legislation for, making territorial, State, military, or any other roads, by the Government. My opinion is, that there is no authority or warrant in the Constitution of my country to justify the appropriations which are being voted by this committee for the purpose of constructing roads. I entirely repudiate the doctrine, come from what quarter it may. . . .[42]

Representative Haven admonished Jones for his strict interpretation of the Constitution and taunted him with the fact that the incoming President, a member of his party, was committed to the principle of federal aid for better communication to the Pacific. Haven stated that he personally had preferred to construct only one road as an experiment, that it could not approach in cost the railway survey expenditures of the previous Congress, and that the present appropriation was far less than the extra compensation paid to the mail carriers by the federal government in the preceding two years. He was willing, however, to accept the House's decision to build two roads, but cautioned that the bill should not be further amended for fear of ultimate defeat. When Jones' amendment to strike out the appropriations came to a vote, it failed ninety-nine to twenty-six.[43]

The bill, as finally approved by the House of Representatives, allotted $300,000 for the central route, $200,000 for the southern, and $50,000 for the Fort Defiance–Colorado River road in New Mexico. Upon Senator Weller's recommendation, the upper house accepted this combined measure on February 14 without debate, and three days later President Pierce gave his approval.[44] In this third session of Congress another measure, appropriating $30,000 for the construction of a wagon road between the Platte River and Running Water River (Niobrara River) in Nebraska by way of the Omaha Reserve and Dakota City, had quietly jumped every legislative hurdle. Perhaps this local project was to match that in New Mexico and preserve the delicate sectional balance. The Secretary of the Interior was likewise to supervise the project.[45]

Although the Department of the Interior had expressed concern over better highways for emigrant travel to the Pacific as early as 1850,[46] the Secretary of the Interior had never expected to assume direct responsibility for road construction. Federal government road building was the province of the War Department and the term "military" had been applied to these works even though emigrants, freighters, and mail carriers had continuously benefited. Obviously the majority of the Thirty-fourth Congress no longer believed an apparent military need necessary for the constitutional justification of federal aid in constructing roads.

Presented with this new assignment, Pierce's Secretary of the Interior, Robert McClelland, temporarily assigned the supervision of the Fort Ridgely–South Pass road to the Commissioner of Indian Affairs, presumably because the sponsors of the legislation in Congress had spoken of the aid that agents of the Indian Office could give the road builders. McClelland came from Michigan and as a friend of Lewis Cass represented his political faction in the Pierce cabinet. At the beginning of the administration he had favored federal aid to western transportation, even to the extent of granting land subsidies for a Pacific railroad project. When Pierce began to voice constitutional objections to federal grants for internal improvements, McClelland lost interest.[47] The Pierce administration in its closing months made little effort to push construction on the Minnesota wagon road.

Jacob Thompson of Mississippi, a spokesman for the interests of the South and who had served as chairman of Committees on Public Lands and on Indian Affairs in the House of Representatives, was chosen by President Buchanan as Secretary of the Interior.[48] Finding his department entrusted with the construction of four federal roads, the Secretary assumed a personal responsibility in selecting the superintendents of each project and drafting detailed instructions. These men were charged with the supervision of construction, the protection of all public property assigned to their undertaking, and the preparation of an itinerary recording the events of each day.

Immediately upon the approval of the appropriation for a wagon road to the Pacific from Fort Kearny by way of the South Pass, the Minnesota and

California lobbyists united forces to secure the superintendency of the western sector for Doctor O. M. Wozencraft, a San Francisco civic leader who had been chairman of the California Emigrant Road Committee, and the eastern section for William H. Nobles of Minnesota.[49] Nobles, the experienced explorer and lobbyist who had been placed in charge of the Fort Ridgely–South Pass road late in the summer of 1856 by Commissioner Manypenny, desired a transfer to the Honey Lake road, which had received the largest appropriation and was in many respects the most important route to the Pacific.[50] The political strength of these combined forces, with the aid of Weller and Rice, appeared sufficient to dictate the appointments, but these men had overlooked the potential candidacy of William M. F. Magraw, a personal friend of President Buchanan. Magraw, once a resident of Virginia who, in the early 'fifties, moved to Philadelphia, had received a mail contract for deliveries west of Kansas City in 1853, and at the same time worked as western representative of a Philadelphia mercantile firm, Smith-Murphy Company. Failing to secure a renewal of his mail contract in the fall of 1856, Magraw had returned to Washington with endorsements from the Democratic members of the Missouri legislature for a place in a Kansas land office. Learning of the available wagon-road superintendencies, he made application at the Interior Department and had his entire personnel folder, including extensive recommendations and testimonials of his value to Democracy in Virginia, Maryland, and Pennsylvania, transferred from the Post Office Department. Magraw was supported by General Persifer F. Smith, Attorney General J. S. Black, Senator Stephen A. Douglas and five of his senatorial colleagues. Moreover, James Buchanan, in endorsing him for a federal post in 1853, had thought him "a decided and active democrat" who had worked for the party in Pennsylvania.[51]

Nobles had the continuous support of the Minnesota legislature. In the memorial requesting the construction of the Fort Ridgely road, that body had urged some congressional subsidy for Nobles in recognition of his previous service to the nation as a pathfinder. The Minnesotan conferred with Buchanan and Douglas in the capital and, upon their advice, proposed a tentative route for the road in a statement addressed to Secretary Thompson and published in the *Washington Union,* April 14, 1854.[52]

The Secretary of the Interior, attempting to satisfy the divergent political pressure groups, worked out a compromise whereby the Fort Kearny–Honey Lake road was to be divided into three divisions for construction: from Fort Kearny to Independence Rock, thence to City Rocks, thence to the California boundary near Honey Lake. Magraw was named superintendent of the eastern and central sections, with the understanding that his work to Independence Rock would be cursory and the emphasis placed on the central sector.[53] Nobles' appointment on the Fort Ridgely road was continued for the season of 1857, with the understanding that he would return over the eastern division of the central road once he had completed the Minnesota branch.[54]

There remained the task of selecting a superintendent for the Honey Lake–City Rocks division. The assignment was not immediately awarded to Wozencraft, whose political support was sufficient, because of the rival candidacy of John Kirk from Placerville, California.[55] Kirk, an engineering contractor, had supervised the construction of several roads, mining canals, and quartz mills and was supported by Democratic party leaders on the west coast.[56] After reviewing the respective claims, Thompson appointed Kirk.[57]

Responsibility for the New Mexico project was given to James B. Leach, to whom Colonel Steptoe had granted the contract for improving the Salt Lake City–southern California route for the War Department in 1855. This candidate's early background included canal and railroad contracting in Michigan and Ohio. During the Mexican War he had been a quartermaster agent on Colonel A. W. Doniphan's expedition, and had performed the same duties on a march with a rifle regiment to Oregon in 1849.[58] Leach was recommended to the Department of the Interior by George Chorpenning, the mail contractor, and endorsed by Congressman John S. Phelps, the Missouri Democrat who had championed the cause of improved road and mail facilities for the West as a member of the Ways and Means Committee.[59] The Nebraska road from the Platte to the Niobrara River was placed in charge of George L. Sites, who was to prove one of the most effective supervisors of the group.[60] Annual compensation for these superintendents was usually $3,000, though Sites received only $2,500. Each of them was authorized to hire a clerk at a salary ranging from $600 to $1,200.

Second in command on each of these road building expeditions was a chief engineer. Frederick West Lander, Magraw's assistant, had an established reputation as an engineer, having aided Isaac I. Stevens in his survey of the northern railway route to the Pacific in 1853. Lander did not share Stevens' enthusiasm for a northern route, but thought Oregon and Washington could best be served by the construction of a branch road from the central route by way of the Snake and Bannock rivers to the South Pass.[61] He had returned by such a route and made a report of his reconnaissance to Congress.[62] On the west end of the road, Kirk was assisted by Francis A. Bishop.[63] N. Henry Hutton, chief engineer on the El Paso and Fort Yuma survey came to his new assignment with extensive experience, having worked as principal assistant engineer for A. W. Whipple on his railroad survey along the thirty-fifth parallel; with J. G. Parke of the Topographical Engineers on an exploration from Benicia, California, to Fort Fillmore, New Mexico, in 1854–1855; and with G. K. Warren in his reconnaissance of the Missouri and Yellowstone rivers in 1856.[64] Sites' engineer in Nebraska was Henry Smyth of Platte City, Missouri;[65] Nobles was assisted by Samuel A. Medary.[66]

In consultation with the superintendents, each engineer was to determine the exact location of the entire road. Surveys and improvements were then to be concentrated on the least known and most difficult sections. A record was to be made of the course of travel and the distances between watering stations, fuel, and grass, so that each road could be made immediately available to emigrants moving west.

In completing the organization of these parties, the Secretary of the Interior selected a physician and surgeon who was to pass on the physical fitness of each party's personnel before departure, to maintain the general health of the group, and to care for those sick or who became injured in line of duty. As men of science, the doctors were to prepare reports on the topography and collect specimens in natural history as well as to analyze the economic potential of the country, should their medical duties not be too arduous and the road construction not delayed thereby.[67] Disbursing agents were also attached to each group, with instructions to pay all claims for supplies, contract work, and the wages of the crew. Due to the distance from Washington and the poor communication facilities, the financial agents were to pay all vouchers signed and presented by the superintendents, even though they doubted their validity. All expenditures were to be reviewed by the Interior Department so agents were enjoined to present reasons for questioning any doubtful voucher. Detailed instructions relative to financial records were sent both to the superintendents and to the disbursing agents.[68]

Personnel records reveal that "service to the Democracy" in pivotal states was largely the certification required for an appointment to one of the dozens of lesser positions on the wagon-road expeditions.[69] The cabinet triumvirate of Howell Cobb in the Treasury, John Floyd in War, and Thompson in Interior controlled most of the patronage.[70] Occasionally the figure of Colonel Abert of the Topographical Engineers loomed rather prominently in the background in an advisory capacity. As a professional military man he was careful to channel all requests for personal endorsements through Secretary Floyd.[71]

All field parties were notified that they were not expected to engage in heavy grading or in bridge building, but to report such necessities to the Department with a cost estimate. The basic problem was opening a road over which emigrants with loaded wagons might pass without difficulty. No finished road was expected. Furthermore, Congress had given no indication that additional road appropriations would be forthcoming, so the entire route had to be located and surveyed with the single allotment. The Secretary repeatedly warned that he would tolerate no deficits.[72]

Fear was expressed in some quarters that the Indians might interfere with the prosecution of the various road projects. From the funds of the Commissioner of Indian Affairs, each superintendent received an allocation, varying from $200 to $500, to purchase gifts for the chief men of the tribes encountered.[73] The Indians were to be assured of the friendship and good will of the United States, but they were to be warned against molestation of travelers, which would bring extreme displeasure and severe chastisement by the federal troops. Superintendents were urged to thwart any Indian designs to embarrass operations by such acts as driving away animals or stealing instruments, and to use their discretion in handling emergencies such as an Indian attack.[74] Thompson wrote to Floyd for Sharp's rifles, Colt revolvers, and sufficient ammunition for each of his Pacific Wagon Road surveys.[75]

The Secretary drafted his instructions to the superintendents during April and May, 1857, and toward the close of the latter month he resolved to place the administrative responsibility in an official to be stationed in his own office, Albert H. Campbell. Campbell, who had served with Whipple and Parkes and crossed the continent at the thirty-second and the thirty-fifth parallels, was familiar with the southwest. He had the endorsement of Jefferson Davis, John B. Floyd, and Texas' Senator Thomas J. Rusk.[76] He assumed the title of General Superintendent of the Pacific Wagon Roads and explained to the field men:

> All communications of an official character from the various attaches of the several Wagon Road expeditions are expected to come to the Secretary through the Superintendent. Any inquiries or other matters will be directed to me & if such a nature as are proper will be submitted by me to the Secretary for advice.[77]

The two men were in constant consultation over administrative matters such as the review of reports and auditing of accounts. All correspondence to the fields of operation, whether signed by the Secretary or by Campbell, went out from the Pacific Wagon Road Office. Its records were preserved as an independent administrative unit within the Secretary's office.

CHAPTER XI : *From the Minnesota Frontier toward the South Pass*

As soon as the funds for the Fort Ridgely–South Pass road were made available by Congress and the Indian Office was selected by the Secretary of the Interior to administer its construction, Commissioner George W. Manypenny faced the problem of choosing a field superintendent. As has been mentioned, William H. Nobles was the obvious and inevitable choice. He was, therefore, instructed to survey and construct such an emigrant road as $50,000 would permit.[1]

Nobles gave up his lobbying activities in the national capital in September, 1856, and returned to St. Paul. During the fall he began a preliminary reconnaissance of the route with a crew of twenty men and seven mule teams, to gain additional information about the terrain over which the road would pass. Two possible ways west of the Minnesota River, along Cottonwood and Redwood creeks, were examined for comparative advantages. The engineering party reached Lake Benton on November 5, and Nobles ordered the construction of a storehouse to serve as a supply depot the next season. The superintendent then apparently rode on to the Big Sioux River where he was met by Indians who blocked his crossing. The tribes had burned the prairie, according to reports, all the way to the Missouri River. They insisted there would be no road survey without payment for the right of way. Nobles admitted to the Indian Commissioner that he had made no road because of his late start in the fall, but the way had been "located from Fort Ridgely to the Cotton Wood valley, *nearly in a direct line* to the crossing of the Big Sioux River, a distance of about one hundred and fifty miles."[2]

Early in the spring of 1857, Secretary of the Interior Jacob Thompson appointed Nobles for another season and forwarded new instructions.[3] With a working party of seventy-five men he was to go directly to the James River, crossing it between Redstone and Firesteel creeks, on to the Missouri in the vicinity of Bijou Hills, thence in a direct line to Independence Rock.

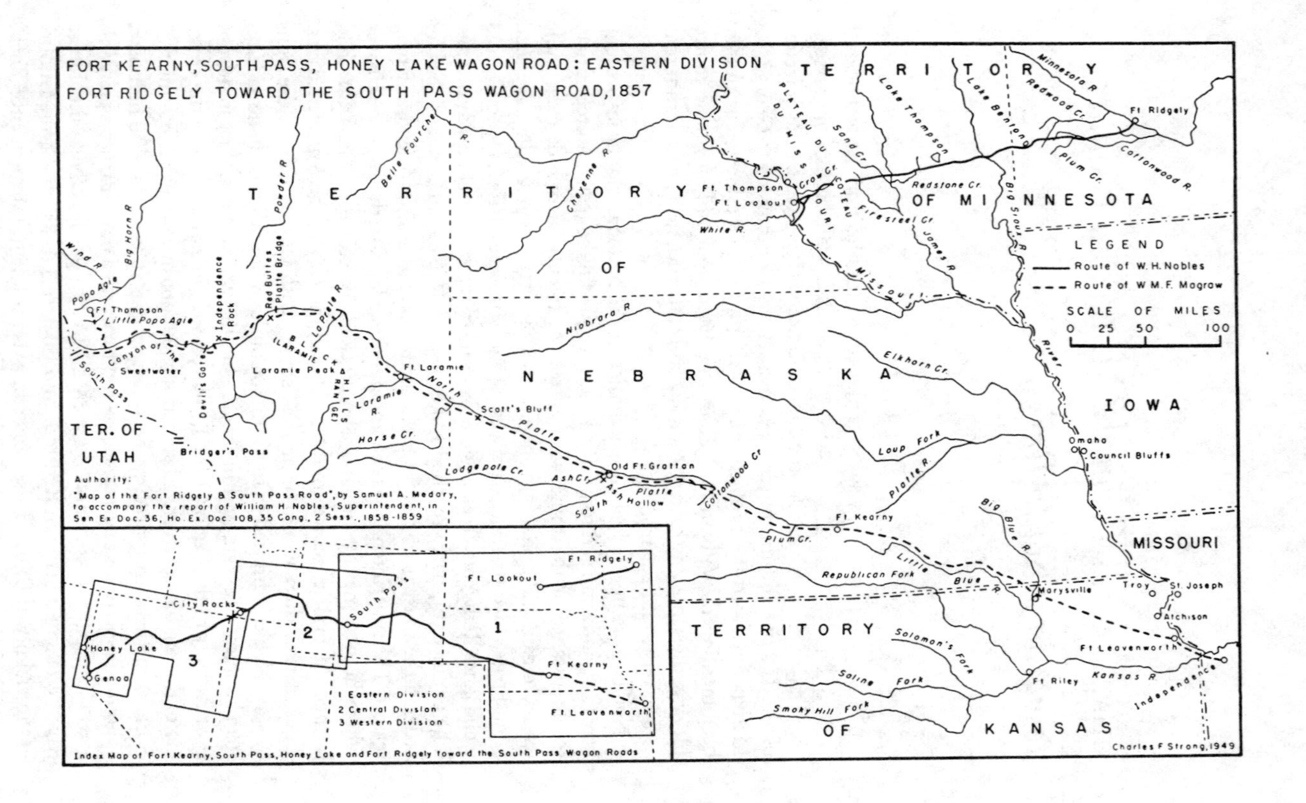
FORT KEARNY, SOUTH PASS, HONEY LAKE WAGON ROAD: EASTERN DIVISION
FORT RIDGELY TOWARD THE SOUTH PASS WAGON ROAD, 1857
TERRITORY OF NEBRASKA
TERRITORY OF MINNESOTA
TERRITORY OF KANSAS
TER. OF UTAH
IOWA
MISSOURI
LEGEND
Route of W. H. Nobles
Route of W. M. F. Magraw
SCALE OF MILES
0 25 50 100
Ft Ridgely
Minnesota R
Redwood Cr.
Lake Benton
Plum Cr.
Cottonwood R.
Lake Thompson
Redstone Cr.
Sand Cr.
Big Sioux R
Firesteel Cr.
James R
PLATEAU DU COTEAU DU MISSOURI
Crow Cr.
Ft. Thompson
Ft. Lookout
White R.
Missouri River
Cheyenne R
Belle Fourche R.
Powder R
Big Horn R
Wind R
Popo Agie
Ft Thompson
Little Popo Agie
Canyon of the Sweetwater
South Pass
Devil's Gate
Independence Rock
Red Buttes
Platte Bridge
La Prele R
BLACK HILLS (LARAMIE RANGE)
Laramie Peak
Ft Laramie
Laramie R.
Horse Cr.
North Platte
Scott's Bluff
Bridger's Pass
Lodgepole Cr.
Old Ft. Grattan
Ash Cr.
Ash Hollow
South Platte
Cottonwood Cr
Plum Cr.
Ft. Kearny
Niobrara R
Elkhorn Cr.
Loup Fork
Platte R.
Omaha
Council Bluffs
Big Blue R
Little Blue R
Republican Fork
Marysville
Troy
St Joseph
Atchison
Ft Leavenworth
Independence
Kansas R.
Ft Riley
Solomon's Fork
Saline Fork
Smoky Hill Fork
Charles F Strong, 1949
Authority:
"Map of the Fort Ridgely & South Pass Road", by Samuel A. Medary, to accompany the report of William H. Nobles, Superintendent, in Sen Ex Doc. 36, Ho. Ex. Doc. 108, 35 Cong., 2 Sess., 1858-1859
Ft Lookout
Ft Ridgely
City Rocks
South Pass
Honey Lake
Genoa
Ft Kearny
Ft Leavenworth
1
2
3
1 Eastern Division
2 Central Division
3 Western Division
Index Map of Fort Kearny, South Pass, Honey Lake and Fort Ridgely toward the South Pass Wagon Roads

At Fort Laramie additional instructions were to be received for his return from Independence Rock by way of Fort Kearny along the eastern sector of the Fort Kearny–Honey Lake road. The Secretary suggested that the party should be outfitted for six months in the field and authorized the purchase of twenty-five wagons to haul equipment and subsistence. The superintendent was to use his own discretion in building the road, keeping in mind that it was an "Emigrant Road" and should, therefore, be made passable the entire distance for the loaded wagons of those headed for the Pacific side of the Rocky Mountains.[4]

For two months Nobles delayed his departure from St. Paul; first, because of continued cold weather, and then with the excuse that funds were unavailable to purchase supplies because of the resignation of one disbursing agent and the incompetence of a successor. In late May, provisions were sent up the Minnesota River to Fort Ridgely, and on June 19 Nobles left the territorial capital for this frontier post two days behind his working party. Arriving at Fort Ridgely in mid-July, the superintendent found an excuse to return to St. Paul within two weeks. During this and subsequent absences, the chief engineer, Samuel A. Medary, performed the road work.[5]

From the Minnesota River ferry at Fort Ridgely the general course of the road laid out by the engineer was southwest, passing through a marshy region to the north fork of the Cottonwood, thence over rolling country to the Cottonwood, just below the mouth of Plum Creek. From here the route was west across the Plum, the headwaters of the Cottonwood, and the Redwood Creek. Skirting Lake Benton to the south, the road builders crossed the prairie to the Big Sioux River, where a near-by lake was named for general superintendent Albert H. Campbell and a small affluent of the Big Sioux for engineer Medary. In traveling south of west to the James River, the party passed slightly south of Lake Thompson. The country between the Big Sioux and the James rivers was described as "a vast sandy prairie, with no timber whatever." West of the James, the road was located across the Coteau du Missouri to the river.[6] At Fort Lookout on the Missouri, construction came to a standstill with no time or money left for its continuance.

Nobles reported that he had built a good wagon road throughout the 254 miles to the Missouri River. Heavily loaded wagons could traverse the route without once being forced to unload or to double the teams. To serve as markers, earthen mounds were erected from three to five feet in height at a distance of a quarter of a mile apart. The road was described as being 30 feet wide. In the grassy marshes just west of Fort Ridgely a timber road bed was laid 12 feet wide covered with earth. A temporary bridge was built at the first crossing of the Cottonwood. At the wide Big Sioux and James rivers, fords were made by paving the bed of the stream with boulders and gravel.[7]

From the outset of his appointment by Thompson, Nobles found his relations with the Interior Department difficult. Vexed by the delay in the appointment of a bonded disbursing officer, he purchased extensive supplies for the expedition and began to borrow funds without the authorization of the government.[8] Much of the resulting controversy was due to Nobles' belief that the federal government's Pacific Wagon Road program was an endorsement of his personal plans. He overemphasized his political influence in Washington and thought of himself as a dispenser of federal patronage. Secretary Thompson and general superintendent Campbell objected to his financial manipulations. Treasury Department regulations required that vouchers for expenditures should record the details and cost of purchases, be signed by a road superintendent, and countersigned by a disbursing agent. When Nobles refused to complete and sign purchase forms before demanding the certification of his disbursing agent, the latter resigned, notifying the Interior Department that he doubted the superintendent's integrity.[9]

Thompson wrote Nobles early in June inquiring about an excessively large bill for surveying instruments, since he assumed that sufficient equipment had been retained from the previous season's work. The subsistence expenditures were also challenged by the Secretary who thought certain food items were not "strictly necessary but belong to that class usually denominated the luxuries of life, and consequently cannot be paid for out of the appropriation for the construction of a wagon road, for reasons which will suggest themselves to you upon reflection."[10] From the reports on the outfitting of the party, the Secretary also noticed that his superintendent was purchasing supplies from members of the expedition and urged an explanation. On June 3 Thompson, in desperation, wrote Nobles that $30,000 had been spent without any commencement on the road, warned him not to exceed the appropriation, and ordered the expedition into the field immediately. Nobles apparently had interpreted the Interior Department's original instructions to return from Independence Rock by way of Fort Kearny as releasing sizable funds from the larger appropriation for his road work, but Thompson assured him that he would not get more than $20,000 additional.[11] Within the week Jerome R. Gorin, the new disbursing agent, was notified that "*no more funds*" would be deposited to his account until the total expenditures could be ascertained.

In fury, Nobles reminded Thompson of his labors to get the Pacific road built, and suggested that when *he* transferred control of construction into the hands of the Interior Department from the War Department, he had hoped for more cheerful coöperation. "If I am to be detained over a can of sardines and a few codfish," he inquired, "how can I bring honor to myself and to the department?" Once more he proclaimed the road to be the great object of his life.[12]

The member of the expedition from whom Nobles had apparently bought supplies was not on the public payroll as believed, but instead was a land speculator going west to examine the country in the valley of the Big Sioux River. A group of St. Paul promoters had organized the Dakota Land Company in the hope of selecting promising town sites along the route of Nobles' road. These Minnesotans had been the instigators of the federal appropriation, anticipating that the proposed highway would become an avenue of migration to the west and ultimately a railroad line, thus enhancing their holdings. Both the superintendent and his engineer, Medary, who was appointed territorial governor of Minnesota by President Buchanan in 1857, were among the incorporators of the land company. Henry M. Rice, likewise, was a sponsor and several congressmen, including Galusha A. Grow of Pennsylvania, chairman of the House Committee on Territories, and Charles E. Stuart of Michigan planned to accompany the Nobles' expedition to "look over the land."[13] These interests delivered a virtual ultimatum to general superintendent Campbell reminding him that too many eminent men were concerned with the success of this expedition to fail to remember the person who might cause disaster. Rice reported that the whole Northwest, both Republican and Democratic, was upset over the delay.[14]

The faction supporting Nobles launched an attack against Thompson in the Minnesota press, accusing him of deliberately interfering with the success of the expedition. The Secretary was described by the *Pioneer and Democrat* as "a narrow minded southern negro driver; a hater of the north and northern people, and ready at any moment to use his official position to cripple and retard northern public enterprises that came within his reach."[15] *The Daily Minnesotian* stormed about the financial arrangements, insisting that a political hack had been appointed to obstruct Nobles' lofty desires. "Had a southern fire-eater undertaken to thwart the will of Congress?"[16] The tension mounted daily. When Willis A. Gorman, former Minnesota governor, attacked the purposes and methods of Nobles' road-building promotions, the superintendent knocked him down in the streets of St. Paul and pummeled him before a public gathering.[17]

Nobles now officially accused Campbell of deliberately trying to embarrass his operations and suggested that the objectionable items purchased by his vouchers were in the accounts of other government officers, only in a disguised form. In reply, Thompson sent the Minnesotan a severe reprimand stating that his apprehensions were "ungenerous and unfounded." Moreover, the superintendent's declaration was "... a reflection, in the absence of specific charges, highly improper in itself, and does not increase the confidence of the Department in the honorable bearing of its officer."[18] The Secretary reminded Nobles that he had repeatedly been made aware of the amount of the appropriations and had been told to confine himself within these limits. Yet he had expended the funds without beginning road work, and appeared to be attempting to force the Department into a deficiency. This policy called forth the Secretary's unqualified disapprobation.[19]

The disbursing agent had early thrown his support to the Nobles' clique, bringing pressure on the Department for additional funds and praising the economy of the superintendent.[20] When no funds were placed to the agent's credit by the end of June, Nobles made a personal arrangement with Rice and the land company promoters to secure sufficient funds to meet the outstanding obligations. With a note of defiance, he wrote Thompson that he could not give up the exploration. Rather than lose six years of preparatory work and see his dream fail, he was calling on his friends for financial support.[21] Gorin was to remain in St. Paul in anticipation of receiving more of the appropriated funds.[22]

Once superintendent Nobles went west with his expedition, he had more interest in suppressing the Indian tribes than in road building. On the Big Sioux River, news came to him that one hundred lodges of Yankton Indians were again that year encamped on the James River, awaiting his arrival to demand payment for building the road through their country. Medary was sent back to Fort Ridgely on July 14 to secure a mountain howitzer which the Army refused to release. A small detachment then returned to St. Paul for additional guns and ammunition claiming that the materiel furnished by the ordnance department of the Army was defective.[23]

The Interior Department had, in the meantime, sent a federal inspector to Minnesota hoping to ascertain the true state of affairs. This official was quickly convinced that Nobles was seeking a fight with the Indians, that he and his confederates were primarily concerned with land speculation, and, therefore, their interests were hostile to the Indians. The inspector likewise forwarded the disappointing news that Nobles had announced his intention of going no farther than the Missouri River this season, and that he publicly and privately placed the blame for the lack of progress on the Department.[24]

Nobles was back in St. Paul again by the close of July under the pretext of securing presents for the Indians and ammunition to force their hands, if necessary.[25] In answer to his proposal that additional gifts be given the Indians to secure a passage through their lands, the Secretary suggested that the wagon-road funds could not be applied for Indian gifts, and that he personally disapproved of buying the way. Nobles was told to consult the United States Army and Indian Office representatives in Minnesota about the best methods of handling the Indian problem.[26]

Two weeks later the Secretary learned that Gorin had abandoned the Fort Ridgely expedition and had returned to his home in Decatur, Illinois.[27] The funds belonging to the road appropriation had been deposited with a St. Paul banker, recommended by Nobles, for safekeeping. In the agent's absence the bank failed, and officials promised him a partial payment only if he signed a release for the remainder. Gorin was forced to raise $6,000 among friends in Minnesota and Illinois, and in doing so, exhausted his credit to save his honor.[28] With confusion reigning supreme at the close of the working season, Nobles was ordered to place the property of the expedition in headquarters near the mouth of the Cottonwood. A substantial log-house with a storeroom and barracks, and stables were erected so that the property and stock could be placed under guard.[29]

Developments had reached the stage of a national scandal. The *Deseret News,* whose readers in the Great Basin of Utah were interested in the success of this road project, reprinted a news story from the *New York Daily Times* placing the blame squarely upon the shoulders of Thompson. The newspaper claimed that Nobles had not abandoned the Minnesota–South Pass road out of fear of Indian attacks, but because it had become necessary for him to return to St. Paul to expedite affairs deliberately held up by the Secretary of the Interior. That the "Mississippi fire-eater" did not want the northern road made was a well-known fact. The *Times* summarized events:

When Col. Noble[s] set out upon his duties he was sent from Washington without a dollar in money with which to fit out his expedition.—"Buy your supplies and equipment," said the Secretary, "and send the bills to me, when I will settle them!" This certainly was a novel way of transacting business ... Nevertheless, Col. Noble[s] proceeded to his rendezvous, called on the merchants, showed his instructions, and attempted to buy on the terms suggested therein. Of course he received for an answer a decided refusal. Parties having merchandise for sale were perfectly willing to let the colonel have whatever he wanted on his own security,—but they wouldn't think of trusting Uncle Samuel on any such loose terms, justly fearing that if they should venture the speculation, they would have to go to Congress for their pay, and then spend the greater share of their claims in buying votes to secure its allowance.[30]

The article recounted the story of Nobles' protest, the Department's agreement that he could certify his purchases and forward the vouchers for final approval in Washington, and the Secretary's ultimate refusal to honor his bills "because he had bought some dried codfish, which the Department esteemed one of the 'luxuries of life,' and could not therefore authorize it for use of the poor 'critters' who were mean enough to locate a Northern wagon road!" According to the *Times,* "the next move was to send a disbursing agent out to pay the expenses of the party and so prevent Col. Noble[s] from stealing any of it." No understanding or sympathy was shown by the newspaper for the administrative procedures of the federal government:

> ... It is difficult to see how the public money was rendered safer, in the least, by this appointment of a second individual who might be tempted to steal, and who had no discretion except to fork over what his chief ordered him to pay. When this disbursing agent reached the party he found it necessary to raise several thousand dollars at once to pay up the obligations and let the commission start. Secretary Thompson having instructed him that he would place the money to his credit at once, he drew against the funds, his draft being cashed by a banking house, and the main body of the party started for St. Peters, where Col. Noble[s] intended to overtake it in a few days. But on the very day on which he proposed starting, the draft aforesaid came back protested. These are the true causes for the failure of the expedition thus far.[31]

The *Times* warned: "Depend upon it, he [Thompson] does not intend that the Northern route shall have any fair chance for competing with the more southern ones. How else shall we account for his remarkable conduct, equally dishonorable and injurious to the interests of the government?"[32]

The *New York Tribune* published an article, signed by G. A. Grow, suggesting that the Buchanan Cabinet was notorious for the art of knowing how not to do something. An analysis of events in Minnesota proved, to Grow at least, that the "Honorable Jacob's job" was to keep the road from being built rather than constructing it. He thought that perhaps the Secretary was overly agitated because Minnesota was about to become a Republican state.[33] The *Washington Star* presented the Interior Department's version of events, claiming that Thompson was scrupulously fair but had been embarrassed by the appointment of an incompetent subordinate. A full scale congressional investigation was warranted and welcome.[34]

Reports from the federal inspector in Minnesota told another side of the story. The Minnesota legislature, in chartering the Dakota Land Company, had granted to Nobles and a single partner the right to keep and maintain ferries across the Sioux, James, and Missouri rivers where the government road from Fort Ridgely to the South Pass crossed. This was to be a twenty-year monopoly. From the outset, the federal road project had been conceived as a part of real estate promotions sponsored by several congressmen who had traveled west with Nobles. The Indian menace had been grossly exaggerated, for none were encountered other than a small hunting party of twenty. The inspector was convinced that a secret financial deal was made by Nobles and Gorin with the Minnesota bankers, though he was unable to get the facts. Gorin, he thought, was an honest man who had been trapped to serve as the scapegoat for the delinquency of others.[35]

Nobles' extravagance not only exhausted the appropriation but his accounts were $10,000 in arrears at the close of the season's operations. This deficit, he insisted, should be charged against the Fort Kearny–South Pass road. The Interior Department suspended Nobles and appointed as his successor, William McAboy, who had competently built roads in Minnesota for the Commissioner of Indian Affairs.[36] By January 12, 1858, McAboy arrived in St. Paul, having crossed the wilderness from Bayfield, Wisconsin, in midwinter snow. He had traveled 100 miles on snowshoes. With Nobles in Washington, he was unable to secure the road property and soon news came that his mission was useless.[37]

The antiadministration press launched another attack upon Thompson for his removal of Nobles. The *New York Times* led the way:

> ... Within the last few days it has been given out that Col. Noble[s] is to be removed. I have no doubt that this is true. He has had the manliness to look official delinquency on the part of his superiors, fairly in the eye. The chances are that he will be sacrificed to appease the wrath of the Secretary.... It remains to be seen whether Mr. Buchanan will acquiesce in the sacrifice of a faithful officer under such circumstances as these, and simply because he refused to be kicked and cuffed around the Northern Territories for the amusement of even a Cabinet member.[38]

Political pressure was strong enough that the Secretary of the Interior was forced to reinstate the road builder.[39]

In the winter, the Treasury Department informed Thompson that Gorin's incompetence in handling funds had made him liable to an indictment for embezzlement.[40] Upon the agent's return to Minnesota he was warned, for his own best interests, not to place money in superintendent Nobles' hands since the road work had been officially suspended and the appropriation exhausted.[41] While the auditors of the Treasury sought vainly to bring order to the chaotic financial accounts for the Fort Ridgely road, Nobles campaigned for an additional appropriation to continue the road west of the Missouri River. He mentioned $300,000 as the amount needed and succeeded in getting the Minnesota state legislature to draft a memorial to the Thirty-fifth Congress requesting $100,000. An additional $50,000 was asked for an eastern extension from Fort Ridgely to Lake Superior.[42]

When Congress adjourned without approving the additional funds, Secretary Thompson appointed a board of appraisers consisting of Nobles and two Minnesota Indian agents, W. J. Cullen and Charles H. Mix, to evaluate the government property on the Cottonwood and transfer to the Indian Office all that would be usable on the reservations. The mathematical and surveying instruments were to be given the surveyor general of Minnesota.[43] Nobles declined to comply with the Secretary's instructions and forwarded his resignation which Thompson quickly accepted. However, he warned the superintendent, "you certainly are not aware of the consequences of maintaining the position which you threaten to assume."[44] When Cullen and Mix forwarded their reports on the extent of missing property, Thompson turned all records of the road over to the solicitor of the Treasury with the request that he "take the most proper and necessary steps to protect the interests of the United States."[45]

On February 9, 1859, Nobles expressed a willingness to close the records of the Fort Ridgely road, but the Treasury by this time had turned his affairs over to the Attorney General's office. The superintendent had always been careful to insist that Campbell, and not Thompson, had mistreated him and he pled with the Interior Secretary to conduct a double investigation. However, he repeatedly refused to submit his own accounts.[46] In June, 1860, the United States district court of Minnesota handed down a decision in the case of the United States vs. William H. Nobles, awarding the government a judgment of $3,446 and court costs. The Interior Department had charged the road builder with $21,000 missing property. The United States attorney reported that Nobles' only testimony was that three-fourths of the missing property was wasted on the expedition and the remaining one-fourth decreased in value 50 per cent.[47]

After the court decision, Nobles' lawyers submitted his accounts and property returns, requesting an audit. Campbell replied that the vouchers were not in correct form, according to instructions issued by the Department, nor was the property return complete. After all legitimate expenditures were approved by the general superintendent, outstanding vouchers totaled $10,168.[48]

In the tense weeks just before the outbreak of the Civil War, Nobles took his case to Congress and petitioned for redress through the Committee on Claims of the Senate. In his memorial he recounted how the Department shut off funds when $20,000 was left in the appropriation. According to his tale, the laborers became angry about getting no wages, seized the property belonging to the expedition, and threatened his life. The workers had been promised, if they returned the property, that the superintendent would not surrender it until a final financial settlement was made. Arriving in the national capital to investigate his removal, Nobles heard rumors that he would be arrested. Thompson refused to see him and explain the reasons for his dismissal. In time the superintendent interviewed him at a night session in the Secretary's private rooms, reminding him of the many friends of the project in Congress. Temporarily reinstated, Nobles returned to Minnesota only to receive orders to surrender the property. After this recitation of all his grievances, the superintendent requested the money for his suspended claims and vouchers, the additional salary due him, and an indemnity for damages and court costs.[49]

With the disappearance of Jacob Thompson from the Washington scene, Moses Kelly became acting Secretary of the Interior and immediately set about making restitution to Nobles. In reply to Kelly's inquiry at the Treasury Department about procedure in having the judgment against the Minnesotan set aside, he was told that the decision of the court and jury must stand. The United States attorney who had presented the government's case admitted, to Kelly's satisfaction, that there were equities in the suit that could not be reached because of the laws the United States had written for its protection. Moreover, Nobles' lawyer prepared a brief raising objections or questioning the procedure of the former Secretary of the Interior. As a result of the judgment, the former superintendent was bankrupt. According to the lawyer the tragedy was caused either by a misapprehension of the facts or a basic prejudice.[50]

When pressed for information by Congress, Acting Secretary Kelly proclaimed that Nobles' difficulties were indeed due to conflicting instructions issued by the Interior Department, and that he was not furnished with the means adequate to the labor he was expected to perform. The Acting Secretary was convinced that Thompson and Campbell, southern sympathizers, had conspired against the success of Nobles' project. He informed the Senate committee that his convictions about the case could not be expressed "because the language would be objectionable."[51] Upon Kelly's recommendation an amendment was added to the Army Appropriation Bill of March 2, 1861, providing:

That the Secretary of the Interior be, and he is hereby, authorized and directed to audit and state the accounts of the late Superintendent of the Fort Ridgely and South Pass wagon road, up to the time when he was relieved from the care of the public property in his possession, allowing him all such sums as, in the opinion of the Secretary, may be fair, reasonable, just, and charging him with all such sums as in his opinion he ought to be charged with, and report the same to Congress.[52]

The friends of Nobles in Congress urged a settlement. One by one those who had earlier accused him of wrongdoing lined up to retract their charges.[53] On March 8, 1862, a joint resolution passed Congress authorizing Caleb Smith, the new Secretary of the Interior, to pay the former superintendent $8,199.99.[54]

But for the work of Sam Medary, the construction of the Fort Ridgely–South Pass road would have been a fiasco. Secretary Thompson undoubtedly had slight interest in the successful construction of this northern road. Moreover, he was notoriously guilty of partisan politics during his term of office, but no conclusive evidence was produced that he and general superintendent Campbell entered into a conspiracy to stop the construction of this specific road. Nobles, on the other hand, had at least proved himself utterly incapable of mastering the necessary technical details associated with the administration of a federal government project. At times he appeared not only unfit, but belligerent and insubordinate. There was little in the story of this endeavor to encourage those members of Congress who thought civilians were superior road builders to the Topographical Engineers of the Army.

CHAPTER XII : *Improvement of the Central Overland Route: Fort Kearny to Honey Lake via the South Pass*

THE CENTRAL ROUTE across the plains along the Platte to the South Pass, with branches leading to Oregon and California, was a well-known and thoroughly used trail. Thousands of homeseekers had trod this way to establish residence on the Pacific Coast. Looking back upon the experience, the majority rebelled at the hardships they endured and determined to alleviate the problems of future travelers. Although they had failed to gain congressional support for military posts to lessen the Indian danger and blacksmith shops to repair the emigrants' wagons along the route, they had obtained funds to remove physical obstacles to travel.

The Secretary of the Interior, a newcomer to the road-building business, was uncertain about the best procedure of improving most effectively the trail between Fort Kearny and Honey Lake, the new legal termini of the central overland route. Before appointing the superintendent and chief engineer for the sections east of City Rocks, northwest of the Great Salt Lake, he called upon the leading applicants to present proposals for the type, methods, and cost of construction desirable for a wagon road to the Pacific. Thus he could gain information, and on the basis of these documents appraise the professional training, judgment, and thrift of each candidate.

From William M. F. Magraw came many helpful suggestions for the organization and outfitting of a wagon-road expedition. Magraw was the first to point to the necessity of naming a chief engineer, physician, clerk, and disbursing agent for each road-working party.[1] When approached for his ideas, engineer Frederick W. Lander displayed great interest in the location of the route west of the South Pass. At the crest of the mountains, Oregon and California emigrants were forced to turn south to avoid the northern Wasatch Mountains, usually along the Mormon road. After reaching Fort Bridger, it was necessary for Oregonians to turn north again to reach Fort Hall. To avoid this detour, the more venturesome had attempted to use the unsatisfactory cutoffs between the two forts, leading directly to Soda Springs. The greatest obstacle on all these direct ways was, in Lander's opinion, the desert immediately west of the South Pass between Sandy and Green rivers. A good road slightly to the north of this inhospitable sector would shorten the distance to the Pacific by almost 200 miles and might prove to be a forerunner of a railroad. He proposed the exploration of the area first with a pack party of engineers, followed by laborers to construct the road along the survey.[2]

After Magraw's appointment as superintendent, his initial instructions prepared on May 1, 1857, authorized him to organize a party of one hundred men equipped for ten months in the field. The well-known route from Fort Kearny to Independence Rock was to be improved for emigrant travel, particularly at Ash Hollow between the crossing of the South Platte and the North Platte, but no major time-consuming construction was to be attempted. From Independence Rock, the Department was to be notified of the improvements needed to make this a "first-class road," originally with the intention that William H. Nobles would return this way from Independence Rock and supervise constructions.

Lander, somewhat apprehensive about Magraw's ability and fearful of his political prestige as a friend of the President, requested and received authorization for a separate engineering corps that might move rapidly to the South Pass and conduct his proposed reconnaissance in the summer of 1857. As an advanced command, the engineering corps was to examine the country as far as City Rocks. If a suitable new course west of South Pass was found it was to be improved immediately.[3] If not, Lander and Magraw, in consultation, were to adopt and report on a passage that would meet the desires of Congress. On reaching City Rocks, the combined party was to go over the western section of the road, which was being surveyed and improved simultaneously by superintendent John Kirk, and then disband the expedition.[4]

Magraw selected Independence, Missouri, as the outfitting headquarters for his party. While he devoted the first weeks of June negotiating for cash to purchase essential subsistence, Lander, vexed by the delay, spent idle hours enjoying the social festivities of the frontier community honoring members of the expedition.[5] On June 15 Lander resolved to wait no longer and with his detachment of fourteen men, six of them belonging to the engineering corps, left Independence. With no rations other than sugar and coffee for sixty days, and anticipating the procurement of additional supplies at Fort Leavenworth, the party was prepared for rapid movement on mule back. Two days later Lander was encamped just west of Fort Leavenworth, still more annoyed by discovering that the commandant was not authorized to grant him supplies. Magraw had assured the engineer that Secretary Thompson had made proper arrangements with Secretary of War Floyd. Requisitions of the Pacific wagon-road superintendents were to be honored by the commanders in the western forts pending a later financial accounting between the departments. Neither Thompson nor Magraw had reached an agreement with the War Department, however, and the superintendent's letters to the officers at Forts Leavenworth, Kearny, and Laramie authorizing requisitions were of little assistance.[6] Fortunately for the prosecution of the season's work, the difficulty was adjusted by the time Lander had reached Fort Kearny.[7]

With the entire party in the saddle, the engineer averaged fully 38 miles a day between the Missouri and the Rockies, despite the fact that the supply wagons were now overloaded. Lander, hurrying to the South Pass to begin work, did not fail to dispatch reports to the superintendent about necessary improvements. Near Fort Laramie, where the detachment encamped on July 6, he notified Magraw that his men had forded the South Platte nine miles below the usual emigrant crossing thereby shortening the distance between the South and North Plattes to a day's drive. He had examined Ash Hollow. The improvements needed there were so extensive that any attempt to make an adequate road would preclude the expedition's arrival in California by fall. The first assistant engineer, Henry K. Nichols, and another young surveyor, M. M. Long, were left behind to move westward with Magraw's train along the Platte Valley. Lander proposed that they make triangulations and take water soundings at all crossings located on the South Platte.[8]

Arriving at the South Pass July 15, Lander and his men spent the next sixty days thoroughly exploring the area between the South Pass and Soda Springs. The Interior Department had indicated that a good wagon route might exist from the summit of the South Pass running near the south end of the Wind River Mountains westward to the Green River, near the New Fork, and thence to Soda Springs.[9] As fabulous tales existed about the desert from the headwaters of the Big Sandy to the New Fork of the Green, Lander's first trip was to explore this desolate waste. On returning to the South Pass, he decided to divide his corps into several detachments for simultaneous explorations. John F. Mullowney was dispatched with one party to examine various cutoffs along the "old emigrant road," chiefly that located by Milton Sublette, the fur trader. Particular attention was to be

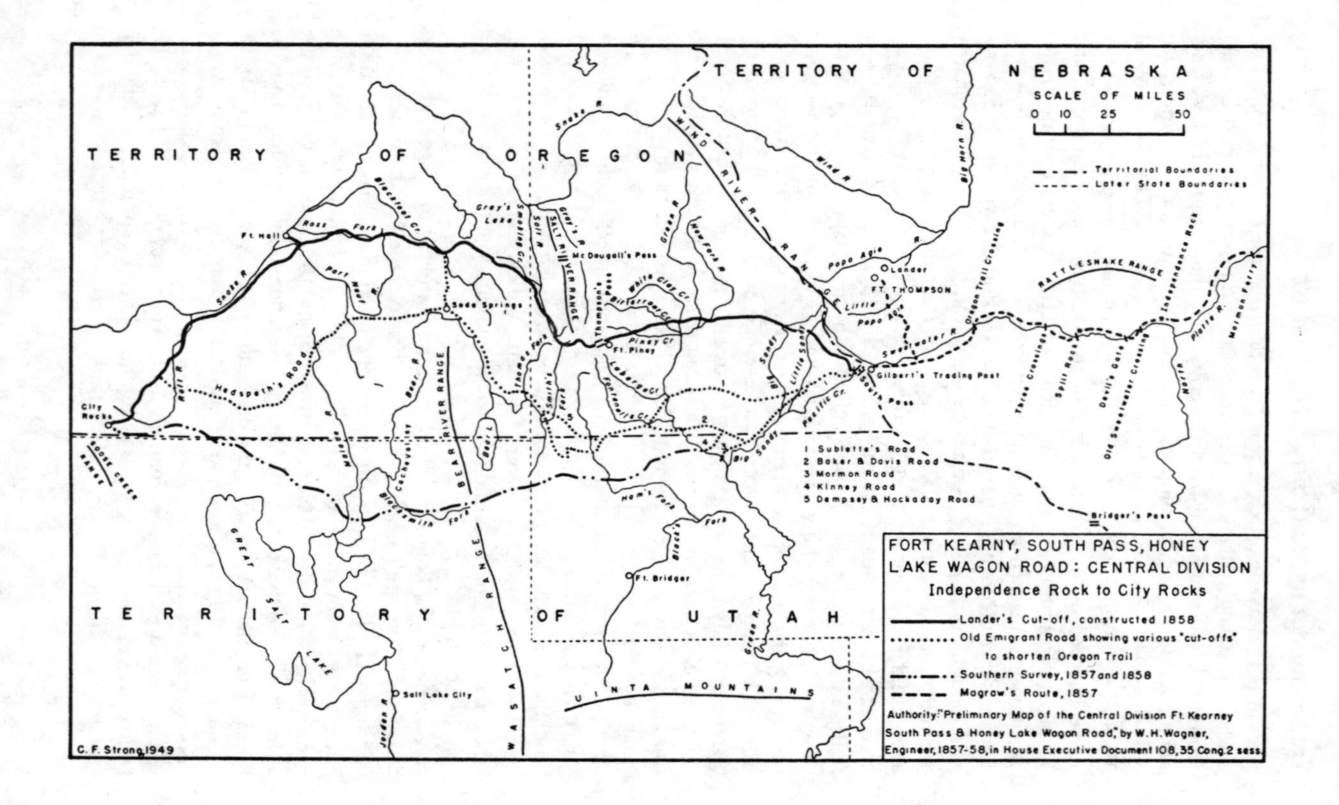
TERRITORY OF NEBRASKA
SCALE OF MILES
0 10 25 50
Territorial Boundaries
Later State Boundaries
TERRITORY OF OREGON
TERRITORY OF UTAH
FORT KEARNY, SOUTH PASS, HONEY
LAKE WAGON ROAD: CENTRAL DIVISION
Independence Rock to City Rocks
Lander's Cut-off, constructed 1858
Old Emigrant Road showing various "cut-offs" to shorten Oregon Trail
Southern Survey, 1857 and 1858
Magraw's Route, 1857
Authority: "Preliminary Map of the Central Division Ft. Kearney South Pass & Honey Lake Wagon Road," by W. H. Wagner, Engineer, 1857-58, in House Executive Document 108, 35 Cong. 2 sess.
1 Sublette's Road
2 Baker & Davis Road
3 Mormon Road
4 Kinney Road
5 Dempsey & Hockaday Road
City Rocks
Goose Creek Range
Ft. Hall
Snake R.
Ross Fork
Blackfoot Cr.
Port Neuf R.
Raft R.
Hedspeth's Road
Soda Springs
Grey's Lake
Smoking Cr.
Salt R.
SALT RIVER RANGE
Grey's R.
McDougall's Pass
Thompson's Pass
Bitterroot Cr.
White Clay Cr.
Green R.
New Fork R.
Piney Cr.
Ft. Piney
Thomas Fork
Smith's Fork
LabargeCr.
Fontenelle Cr.
Bear R.
BEAR RIVER RANGE
Bear L.
Cachevalley
Malade R.
Blacksmith Fork
WASATCH RANGE
GREAT SALT LAKE
Salt Lake City
Jordan R.
Ham's Fork
Black's Fork
Ft. Bridger
Green R.
UINTA MOUNTAINS
Big Sandy
Little Sandy
Pacific Cr.
South Pass
Gilbert's Trading Post
WIND RIVER RANGE
Wind R.
Big Horn R.
Popo Agie R.
Lander
FT. THOMPSON
Little Popo Agie
Sweetwater R.
Oregon Trail Crossing
Three Crossings
Split Rock
Devil's Gate
Old Sweetwater Crossing
RATTLESNAKE RANGE
Independence Rock
North Platte R.
Mormon Ferry
Bridger's Pass
C. F. Strong 1949

given the Dempsey and Hockaday road between the headwaters of Ham's Fork of the Green and Smith's Fork of the Bear River. At Soda Springs the assistant engineer was to prepare a map of his work and wait for Lander's arrival. If Lander should fail to show up, Mullowney was to bring his wagons eastward to the South Pass by the shortest possible line, report to Nichols, and consider him the chief engineer.[10] B. F. Ficklin, in charge of a second detachment, was ordered to examine the Big Sandy–Green River desert, deciding upon the best route between the Pass and the Green by August 1, and then go west from the New Fork through the Salt River Range to Soda Springs. Lander's personal examinations were continuous as he traveled by devious routes to keep the rendezvous with his men.

After joining forces, the engineering parties returned eastward exploring more thoroughly to find the most acceptable pass through the mountains.[11] Reaching South Pass on August 7, Lander met Thomas Adams who had been with him on the Stevens railway survey. He hired Adams to accompany Mullowney, newly returned from Soda Springs, on a second western trip to improve the various surveys by preliminary and cursory constructions. On August 26 J. H. Ingle was sent from the Green River northward through McDougall's Pass in the Salt River Range, and thence to City Rocks by the waters of the Portneuf River. From City Rocks he was to explore a southern route toward Salt Lake as far as the Bear River mouth and then ascend this stream to Soda Springs.[12] William H. Wagner, a fourth assistant, made valuable reconnaissances with James Baker as guide.

Lander considered the season's work a tremendous success. He reported to Magraw that he had covered 3,000 miles on horseback in ninety days, explored a much larger area than contemplated, discovered sixteen mountain passes, and defined the topography of the northern Wasatch Range (Bear River and Salt River ranges). Of the dozens of routes and byways explored, two continuous passages between the South Pass and City Rocks appeared to be "practicable" for a wagon road. He was perturbed, however, by the intent of the Washington administrators when they used the word "practicable" in their instructions. If it meant the shortest way possible, he proposed construction of a "southern route" to follow along the Sandy to the Green, across the plains to Ham's Fork, and thence to the headwaters of the Bear River, on across the southern end of the Bear River Mountains, and by way of Blacksmith's Fork of the Bear across Cache Valley. From here the proposed route could next cross the Bear and the Malade rivers near their outlets in the Salt Lake and proceed northwesterly to City Rocks. He felt that emigrants following this road could shorten their trip by 500 miles or seven days, but that, like all routes in the lower Green Basin, it had major disadvantages. The area through which it would run was comparatively dry, partly sandy, and with limited grass. Lander thought the route would be valuable for mail carriers, but doubted if it could be satisfactory for the great mass of population flooding the plains with covered wagons.

A northern survey, on the other hand, had temporarily located a possible route northwest from South Pass across the headwaters of the Sandy and westward across the Green to Thompson Pass. From the mountains the road could turn north along the Salt River valley for about 20 miles, thence westward to Fort Hall. From there the route would be southwest along the Snake River valley and across Raft River to City Rocks. This proposed road, through a region of ample water supply, grass, and timber, and with no ferry tolls to pay, was reported to be better adapted for ox-team migration than any route west of the South Pass.[13]

Superintendent Magraw, meanwhile, dawdled in Independence until Albert H. Campbell of the Interior Department notified him that the Secretary thought the season far advanced, that his projected work would cover many miles, and that he should, therefore, make every exertion to take the field.[14] James R. Annan, the disbursing agent, who arrived on the scene late, was told to remain behind, if necessary, to get the accounts straight, but to urge Magraw to begin operations.[15] Finally, on July 6, Magraw reported that the train was on the prairie and "*going through to California this season.*"[16] The Department acknowledged his departure, but pressed him to hurry westward to avoid the expense of a winter in Utah Valley. The expedition had already spent $80,000, and since there was no evidence that Congress intended to make annual appropriations for this work, the Secretary was anxious for the party to go through to Honey Lake without spending more than $110,000 for the year.[17] Not until the end of the month did Magraw arrive at Fort Kearny, having traveled the usual route from Fort Leavenworth across the Big Blue and along the Little Blue northwest to the Platte.[18]

The Interior Department continued to protest the superintendent's excessive expenditures and his failure to report or acknowledge receipt of communications.[19] In desperation, funds were suspended with the notification, "Your expenditures have far exceeded the expectations of the Department and an explanation in regards to them and the delay in getting on the march is expected of you. In the next communication . . . state the monthly expenditures of the entire command in order that a definite idea can be formed by the Department as to its future course in regard to your operations."[20] Magraw, in the meantime, had moved on to Ash Hollow where he delayed for a time to make repairs. From the vicinity of Scott's Bluff he sent word to the Interior Department that the long-awaited report would be written at Fort Laramie.[21]

On September 3 the engineering officers attached to Magraw's force, joined by the disbursing agent and the physician, wrote Lander enumerating grievances against the superintendent and suggesting that by misconduct he had forfeited his position. They urged Lander to assume command. Nichols, spokesman for the revolters, also notified Campbell that all hope of reaching California during the season was at an end. According to reports, days and weeks had been lost by the superintendent's neglect, due to chronic intoxication. Magraw was charged with using government wagons to carry personal supplies, including a 6,300 pound liquor shipment to Fort Laramie, where he and Tim Goodale, the guide and interpreter, were to divide the profits in a joint sutlership. Goodale and Magraw could not come to terms at the journey's end. Although the guide pulled the superintendent's beard and tromped on his feet to invoke a fight as a means of settling the matter in true mountain style, the dispute had to be adjudicated by the officers at Fort Laramie. Five days were lost as a result of the altercation.[22]

The engineers felt they were resented because of professional competence. Magraw was reported as saying he could build a road more effectively without them. Affairs reached a climax when the superintendent announced his intention to go into winter quarters, and the malcontents inquired about the extent of provisions for those months. Thereupon all personnel was called together, read the inquiry, and the laborers were told that these men had indirectly suggested the reduction of the force to guarantee their personal subsistence. The laborers were thus turned against the engineers and officers of the party. The professional men, who repeatedly had been threatened with a thrashing at the hands of the workers or with being put in irons, resolved to leave the expedition.[23]

When Lander joined Magraw at the South Pass he learned that Annan and J. C. Cooper, the physician, were in Salt Lake City, and that Nichols had started toward the Missouri frontier. Mullowney, having arrived at Magraw's camp a few days ahead of Lander, had been appointed Nichols' successor and ordered to make improvements below the Platte River bridge.[24] Lander informed Nichols that he could not assume command of the train even though asked to do so by the superintendent and the entire force. Should "small ambition or insane folly" prompt him to accept the office, his concept of official etiquette would be violated by doing so, and having given no bonds he could not expect the Department to honor his drafts. The whole affair would be presented to the Department, knowing that

Secretary Thompson would see full justice accorded to all. Lander declared his intention to request an honorable discharge, but only after his own work of the season was reported. He chided Magraw's officers: "To resign in the field, to forsake the expedition for the mere need of articles of personal comfort, or to avoid danger or privation, is not in accordance with the character I have hitherto maintained as an engineer and explorer."[25]

Magraw requested Lander to make a thorough investigation of the charges against him and the factors leading to the collapse of the expedition, and upon his arrival in Washington to ask a departmental inquiry into the recent difficulties.[26] The chief engineer, impatient to get to the East and report on his explorations, ordered Mullowney to stay behind to take depositions from the men about the fracas and the alleged misconduct of the superintendent. Ficklin was chosen from among the engineering assistants to stay in the field during the winter as a representative of the corps. Upon departure Lander notified Magraw, "You may rest assured so far as I am able to obtain it for them as Chief Engineer, personally representing them as head of the corps, as a private gentleman, the humblest member of my party shall have strict justice accorded to him by yourself, by the Department or by the country, that is if I am able to procure it."[27]

The first snow had fallen in the mountains by the time Magraw reached Independence Rock, and his mules were unfit for further service because the animals of emigrants who had preceded him had eaten the grass all along the way. Fearful of both Indians and Mormons, the superintendent had decided to travel with the Utah Army under Colonel E. B. Alexander. At South Pass, Lander urged Alexander to continue along his northern route to the Piney Valley and encamp there, north of the Utah boundary, to await Colonel Albert S. Johnston's forces. The engineer had been advised earlier by his mountain guides that the Mormons planned to scorch the earth and destroy all supplies in advance of the Army's arrival. Alexander and Magraw, nevertheless, followed the well-known route down the Sandy to the Ham's Fork of the Green before going into camp. Magraw, in consultation with Lander, finally decided to go into winter quarters north of South Pass on the Popo Agie. Ficklin, the engineer, James Saunders, mountaineer, and James Bromley, an old mail rider through the South Pass, went on a four-day reconnaissance to locate the site for a winter fort.[28]

The expedition moved into winter quarters on October 3 at Fort Thompson and began the construction of log cabins and a corral for the stock. Their site on the Popo Agie was in a wintering ground of the eastern Shoshone who would keep the Mormons away. Magraw sent Ficklin on several missions to buy flour and beef and to seek out Chief Washakie, the friend of the whites, for consultation. The assistant engineer found no flour in the Green River settlements.[29] While the laborers continued construction of the fort and cut hay for the winter, Magraw went to see Colonel Johnston, ostensibly about supplies, but in reality to offer the services of his sixty men and his eighteen wagons and mule teams to the Army. He returned to the fort with a letter from Johnston inviting the men to volunteer for military service. Forty-one enrolled with enthusiasm, took oaths of allegiance, and elected their superintendent captain of the volunteer company.[30] The expedition's clerk, who had been serving as disbursing agent since Annan's departure, was left in charge of the fifteen men remaining behind at Fort Thompson. During late October and November they labored in the snow and ice to complete the fort, equipped with a kitchen, storeroom, and dormitory.[31]

Unaware of these latest developments in the Rockies, the Interior Department launched a full-scale investigation of Magraw's conduct. Thompson wrote Lander upon the latter's arrival in Washington:

> I have learned through several official sources, of an unfortunate difficulty between certain officers and employees of the expedition, and the superintendent, which occurred near Fort Laramie, resulting in a partial dismemberment of the party, which has proven a serious detriment to the service upon which they were engaged and which has had the tendency to throw discredit upon this Department, under the direction of which, this expedition was employed.[32]

The chief engineer was requested to furnish a report of the matter from personal knowledge and the statements of his men. What was the true situation of the party, property, and animals at Wind River? Did he think the equipment could be preserved until spring and be usable on the road? In reply, Lander assured the Secretary that the property was reasonably safe and urged him to await the arrival of Annan, Mullowney, Long, and Cooper before launching the investigation. As there was no one in Washington to represent Magraw, any departmental decision against him would, in his opinion, be a reflection on the honor of the engineering corps.[33] Lander then called on Magraw's brother to urge the preparation of a defense against the serious charges.

At the close of December, Annan was called upon to present written testimony about Magraw.[34] Lander insisted that the depositions Mullowney had taken be considered. The evidence, pro and con, was compiled by general superintendent Campbell in a forty-page brief for the Secretary's consideration. Thompson, thereby convinced of Magraw's incompetence, offered his position to Lander,[35] and wrote to the superintendent that the Department had been notified of the charges of habitual drunkenness, unnecessary delays, brutality, overbearingness, and ungentlemanly conduct toward his officers. His failure to present a defense was unjustifiable, and the Department now assumed that he had abandoned his position as superintendent by joining another branch of the service. This action was accepted as his voluntary resignation. The Secretary was under the impression, however, that the property of the expedition was scattered from the frontier of Missouri to the Rocky Mountains, and he warned Magraw that an eventual accounting must be made for the property consumed, wasted, lost, sold, destroyed, or turned over to the Army of Utah.[36]

Lander talked with Magraw's political friends in Washington and notified Thompson, "The delicate relations existing between myself and that gentleman, in reference to his late difficulties with the engineer department, precludes my acceptance of the appointment unless requested to do so in writing by his representatives here." No written communication was received, so Lander declined the appointment.[37] Within the week, however, the delegates of Oregon and Nebraska territories presented Lander with written requests from several members of Congress that he assume the responsibility for building the road, and he forwarded an acceptance to Thompson.[38]

Magraw, in winter camp with the Army, reported to the Interior Department on military operations. Colonel Johnston had suggested the abandonment of Fort Thompson and dispatched several soldiers for the additional supplies there, contrary to Magraw's wishes. The superintendent considered it imprudent to resume road work before Mormonism could be rooted out, and advised a definite postponement of the completion of the road from South Pass to Honey Lake.[39]

News of the acceptance of his "voluntary resignation" by the Department of the Interior reached Magraw at Camp Scott on March 25, 1858. He registered amazement and surprise in a letter to Thompson, stating that Lander had been his deputy to present the case and that he had considered it improper to write the Department until charges were officially preferred and a notification sent him. Had the Secretary accepted only the views of his enemies?[40] Magraw also penned a note to "his excellency the President of the United States," apologizing for having requested an Army commission since Thompson had listened and acted upon charges of a group of "miserable and designing men" and interpreted his admission into the Army as a resignation of the wagon-road superintendency. Could Lander be guilty of perfidy? Magraw promised President Buchanan that if he lived to get back to Washington, the affair would be straightened out. "I owe my appointment to you, and to you, and you alone am I accountable for my conduct . . ." he wrote. He also informed the President that, in his opinion, the Secretary of the Interior had stopped one step short of his duty if he considered him guilty as charged. Why had he not been publicly dismissed by Thompson? He concluded, "I think Mr. Thompson has acted a bit hastily, how wisely future developments will determine. At all events, I am content to bide my time and wait the result."[41]

The struggle between the supporters of John Kirk and O. M. Wozencraft for the superintendency of the western subdivision of the Fort Kearny–Honey Lake road symbolized a basic conflict in the Far West over the location of the federal wagon road. Wozencraft, spokesman for northern California above the Bay district, sponsored the Honey Lake terminus whereas the Kirk faction hoped that the road might be located farther south through the Carson Valley settlements east of Placerville.[42]

After selecting Kirk for the superintendency, Thompson ordered him to conduct a reconnaissance from Honey Lake directly to the northern bend of the Humboldt by way of Mud Lake. From the Humboldt's headwaters, the road was to go toward City Rocks via Thousand Spring Valley. Kirk was advised to avoid the river as much as possible because of the deleterious character of its waters. The Secretary admitted that "the country is little known, but a good line is said to exist through this region avoiding the river for many miles." Thompson did not expect the road to be constructed in a single season, but hoped that a route over which loaded wagons could pass with ease and safety might be opened so emigrants would feel free to use it another season.[43]

Upon his return to California from the national capital, Kirk organized a party of seventy laborers, mechanics, and teamsters equipped with four months' provisions. Early in July this road-building crew started over the Sierra Nevada with six ox wagons and fifty-eight head of oxen, two mule wagons, and one spring wagon for surveying instruments.[44] The superintendent expected to arrive at Honey Lake in six or eight days and was certain his expenditures would not exceed $40,000.[45] The Department was gratified that he entered the field so quickly and with such little expense.[46]

Crossing the summit of the Sierra Nevada at Johnson Pass on July 10, Kirk's party descended the eastern slope to Lake Tahoe. From the camp on this lake, several exploring groups went out to examine the roads into Carson Valley and the laborers spent five days improving Carson Canyon. The expedition turned north through Eagle and Washoe valleys and encamped in the meadows along the banks of the Truckee River. Traveling west of north they struck Long Valley and went into camp at Beckwith

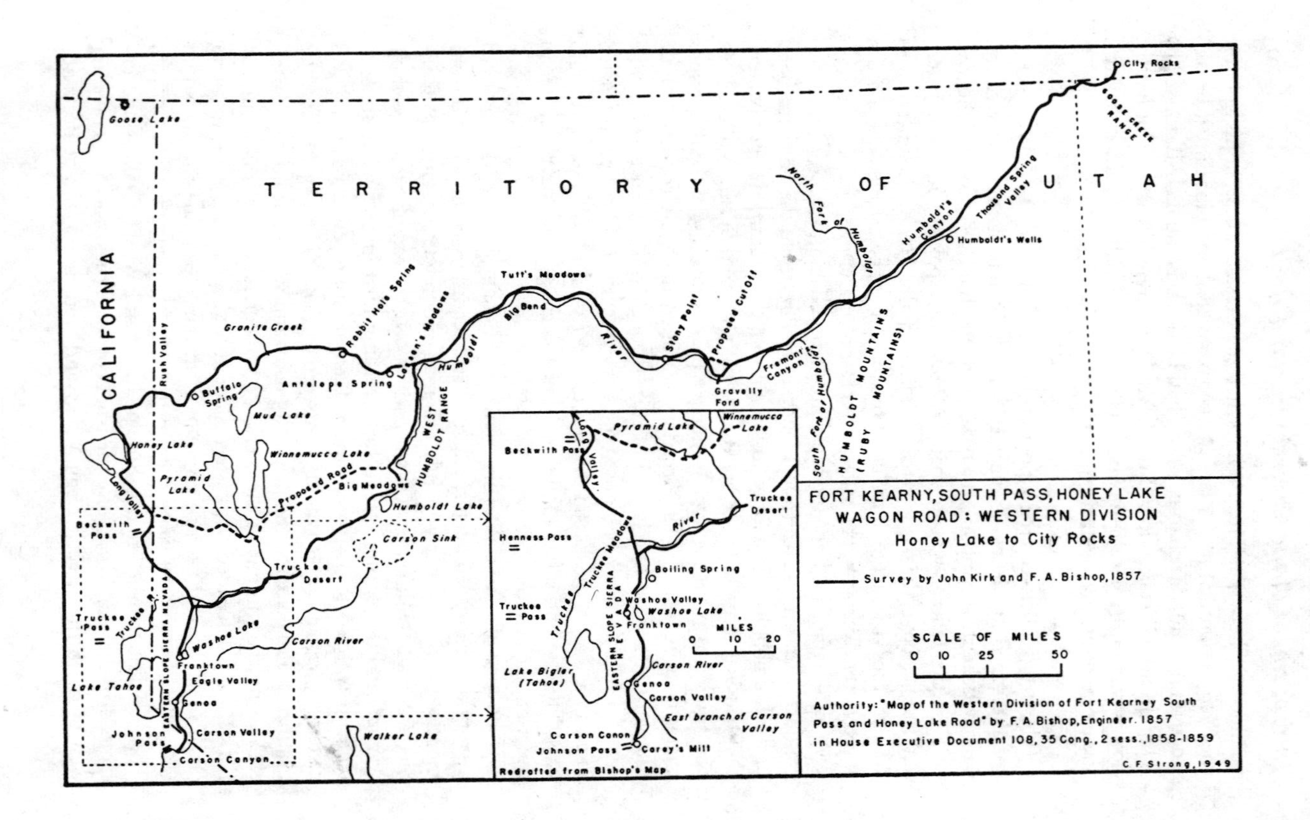
FORT KEARNY, SOUTH PASS, HONEY LAKE
WAGON ROAD: WESTERN DIVISION
Honey Lake to City Rocks
Survey by John Kirk and F. A. Bishop, 1857
SCALE OF MILES
0 10 25 50
Authority: "Map of the Western Division of Fort Kearney South Pass and Honey Lake Road" by F. A. Bishop, Engineer. 1857 in House Executive Document 108, 35 Cong., 2 sess., 1858-1859
C F Strong, 1949
TERRITORY OF UTAH
CALIFORNIA
Goose Lake
Rush Valley
City Rocks
GOOSE CREEK RANGE
Thousand Spring Valley
Humboldt's Wells
Humboldt's Canyon
North Fork of Humboldt
HUMBOLDT MOUNTAINS
(RUBY MOUNTAINS)
South Fork of Humboldt
Fremont's Canyon
Gravelly Ford
Proposed Cut Off
Stony Point
River
Tutt's Meadows
Big Bend
Humboldt
WEST HUMBOLDT RANGE
Lassen's Meadows
Rabbit Hole Spring
Antelope Spring
Granite Creek
Buffalo Spring
Mud Lake
Honey Lake
Winnemucca Lake
Pyramid Lake
Proposed Road
Big Meadows
Humboldt Lake
Long Valley
Beckwith Pass
Carson Sink
Truckee Desert
Truckee Pass
Truckee R.
EASTERN SLOPE SIERRA NEVADA
Washoe Lake
Carson River
Franktown
Eagle Valley
Lake Tahoe
Genoa
Johnson Pass
Carson Valley
Carson Canyon
Walker Lake
Pyramid Lake
Winnemucca Lake
Beckwith Pass
Long Valley
Henness Pass
Truckee Desert
River
Truckee Meadows
Boiling Spring
Truckee Pass
Truckee
Washoe Valley
Washoe Lake
Franktown
MILES
0 10 20
Lake Bigler (Tahoe)
EASTERN SLOPE SIERRA NEVADA
Carson River
Genoa
Carson Valley
East branch of Carson Valley
Carson Canon
Johnson Pass
Carey's Mill
Redrafted from Bishop's Map

Pass on July 23. Two days later they were at Thompson's Ranch on Honey Lake.

From local residents the road superintendent received opinions that a satisfactory wagon route could be found between the southeastern shore of Honey Lake around the north side of Pyramid Lake. Interested in locating a road that approached the Carson Valley settlements to the south, as near as possible and still adhere to the Honey Lake terminus established by Congress, Kirk and his engineer, Francis Bishop, accompanied by Peter Lassen and the disbursing agent, Frank Denver, spent several days exploring the vicinity.

On August 3 the expedition made a belated departure from Honey Lake in an easterly direction toward Lassen's Meadows at the Humboldt. Their route through the section immediately northeast of Honey Lake, known as Mud Lake after its most prominent physical feature, was by way of Rush Valley to Buffalo Spring, across Granite Creek, thence to Rabbit Hole Spring, Antelope Spring, and on to the Humboldt River.[47] The watering holes in the desert were so small and their waters so warm and sulphurous that Kirk was reluctant to describe them as springs. The expedition had gotten through the area only with discomfort and suffering for men and animals. An ox died each day along the line of reconnaissance. At Lassen's Meadows the animals were given a rest, much of the baggage was stored, and the party reduced to fifty men.

As soon as the road surveyors arrived on the Humboldt, daily reports of Indian attacks upon emigrant trains were received.[48] From Lassen's Meadows they followed the river around the Big Bend to Stony Point, where they had a brush with the Indians but succeeded in reaching Gravelly Ford. Several days were spent here in relocating the emigrant route between Stony Point and Fremont's Canyon to avoid the sharp southern bend of the Humboldt at Gravelly Ford. Continuing northeast in the river valley, the expedition passed through the Humboldt Canyon and out onto the plains at the river's source, known as the Humboldt Wells. Between here and Thousand Spring Valley no water was available. Kirk's party reached the difficult Goose Creek Range on September 4, and Bishop surveyed a line up the rocky slopes into City Rocks, which Kirk described as "an amphitheater-like assemblage of conical masses of Granite Rock."

> It is nearly circular with a diameter of about three miles.... The rocks adjacent to the valley rise in fantastic pinacles [*sic*] and as the sun falls on them their whitewashed sides gleam like monuments in a grave yard. Almost every person that has passed through this place has ambitiously left his name. The rocks near the road were literally covered with inscriptions.[49]

The road in the Goose Creek Mountains was unsatisfactory for emigrant travel, but the necessary improvements would take more time than the expedition could devote due to the lateness of the season. Both the superintendent and engineer were convinced that the wagon road must run along the Humboldt River between Thousand Spring Valley and the Bend because the water and grass there were far superior to any in the surrounding country.

On its return the expedition was accompanied by several emigrant trains, including forty wagons and one hundred men, who feared to proceed without an escort. Side surveys were made, and necessary improvements and desirable short cuts noted on the way to Stony Point. Here engineer Bishop succeeded in smoking the peace pipe with a small group of Indians caught in the act of trying to ambush two men out looking for stray cattle. At the Big Bend of the Humboldt, a party of California Rangers from Plumas County greeted the emigrants whom they had been dispatched to protect.

From Lassen's Meadows, where Kirk claimed his cached equipment, the main expedition traveled the most direct route to Placerville via the Humboldt, Truckee Desert, and the Carson Valley. Near the Humboldt Lake, or Sink, Kirk and Bishop met a large party of Mormons moving toward Great Salt Lake as a result of a "call" home by Brigham Young. Kirk wrote: "This train consisted of 133 wagons, between eight hundred and nine hundred animals, and 800 persons of all ages and sex. They had a large quantity of powder and lead and were very anxious to purchase our Pistols. They treated us with great friendship but were not at all communicative."[50] The Mormon village of Franktown in the Washoe Valley was found deserted and the population at Genoa greatly reduced. The party arrived in Placerville October 18.

The superintendent had reservations about the Honey Lake–Lassen's Meadows route. Although the terrain might have no peer for wagon travel during the dry season, the grass and water along the way were inadequate during the period of heaviest migration. Moreover, the distance saved by relocating a road from the Humboldt Bend to Thousand Spring Valley would not justify the expense.[51] Kirk presented a counter-proposal to establish a road west of the Big Meadows above the Humboldt Lake to the south end of Pyramid Lake. Here migration could be divided with those for northern California traveling to Long Valley and Honey Lake, and others by way of Carson Valley to Placerville and the Bay district.[52] If this construction was impossible under the act of Congress, the old emigrant route was preferable to that by Honey Lake, and federal funds should be spent on improving it.

All Californians were not pleased with Kirk's seasonal performance. Wozencraft complained to the Interior Department that he had disregarded the law, the public good, and the Secretary's instructions by taking a circuitous route to Honey Lake. The stock and most of the season were exhausted, according to Kirk's critics, before he reached the initial point of his reconnaissance. Wozencraft cited this evidence to support his accusations, made to Thompson in Washington the previous spring, that Kirk's only interest in the wagon road was speculative. Apparently the superintendent had property investments in El Dorado County and spent many days surveying and improving the road leading there, in an area not included in the bill.[53]

The *North Californian* at Oroville inquired into Kirk's failure to travel by way of Sacramento and Oroville. As his time at Honey Lake was spent in looking for a pass to aid emigrants to get through the Sierra Nevada to Placerville, the editor believed him guilty of improper conduct. Emigrant groups reported to the Marysville *Herald* that Kirk met trains at Antelope Spring and advised them to go to California by Carson Valley, saying that their cattle would surely die on the Honey Lake road. Others claimed the embankments built by Bishop were nothing more than rock piles in the center of the road, making travel more difficult than before.[54]

Sufficient evidence is available to prove Kirk's partisanship for Placerville and the Carson Valley route. He used practically as much time surveying from Johnson Pass in the Sierra Nevada to Honey Lake as on his reconnaissance to and from City Rocks. Moreover, he continuously sought a route as far south as possible that still might terminate at Honey Lake, even though a circuitous detour around Pyramid Lake became necessary. The Secretary of the Interior thought the criticism of Kirk on this score relatively insignificant. His accounts were closed in a businesslike manner, with far less confusion than those of the other superintendents of the cross-country wagon roads during the season of 1857.[55]

Having completed the authorized survey, engineer Bishop divided the western division of the central wagon road into four subdivisions: Rush Valley to Lassen's Meadows, 107 miles; Lassen's Meadows to Gravelly Ford, 133 miles; Gravelly Ford to Humboldt Wells, 104 miles; Humboldt Wells to City Rocks, 93 miles. A fifth section, along the proposed route from Lassen's Meadows southwest to Pyramid Lake was 141 miles in length. These units were recommended for the construction of a future season;

the location, nature, and cost of specific improvements within each were catalogued, with a total cost estimate of $187,905.[56] Upon discharging the officers of the western division, Secretary Thompson assured them their duties had been performed to his "entire satisfaction."[57]

In the winter of 1857–1858, Lander became Secretary Thompson's confidant and adviser on the Central Pacific Wagon Road. The engineer estimated the cost of construction along his northern survey at $70,000. If the mules and all major equipment had been transferred to the Army and only the hand tools left at Fort Thompson or Fort Bridger, no more than the grading of his newly surveyed route and the Dempsey-Hockaday cutoff could be attempted in 1858. Should the Mormon difficulty develop into a full-scale war, a military escort would be necessary to protect the working parties. Even more important, was the reëstablishment of the expedition's credit in the mountains. Lander suggested to the Secretary that he could dispatch a representative, as late as January 1, to go by steamer to San Francisco, then to Portland, The Dalles, and overland to Fort Colville. Here supplies could be purchased from the Hudson's Bay Company and taken to the camp on the Popo Agie by April 15. Although the trip might be classified by some as "dangerous," Lander was ready, and perhaps eager, to go,[58] but the Secretary did not sanction the proposal.

Asked to prepare a formal plan of organization for the construction between the South Pass and City Rocks immediately upon his acceptance of the superintendency,[59] Lander urged that steps be taken at once to recover the equipment from the Army of Utah. He now proposed to organize a small train of fifty men, mounted on Indian ponies or mules at the Missouri River, that could reach the South Pass by May 20. A forage and supply train, with a hundred-day ration for the party should move forward slowly to the scene of operations. When an advanced detachment went to Fort Thompson to obtain the remaining supplies, according to Lander's plans, a rapid reconnaissance could be made between the South Pass and the Green River on through the northern Wasatch Mountains to observe the damage of spring freshets. Both McDougall's and Thompson's passes needed reexamination to compare their respective advantages.

Lander knew the season's labor would be tough, and he informed Thompson that no one should come with him who did not know the dangers and contingencies of frontier life. "Men are required for this kind of service who will eat dried mule." As his cutoff would lose much of its value if a more direct road was not located west of City Rocks, the superintendent urged consideration of a new reconnaissance from Fort Hall to the north fork of the Humboldt.[60]

Thompson drafted Lander's instructions in the early spring, in accordance with the engineer's wishes. Operations were to continue either to City Rocks or the head of the north fork of the Humboldt as he thought advisable, and the way there might be through Thompson's or McDougall's Pass. The size of his work crew was to be determined by the superintendent; he was authorized to obtain any expedition property that had been transferred to the Army, and to attempt an adjustment of all claims against Magraw. The Secretary also stated, "In the construction of the wagon road, should it seem in your view the most economical and expedient method, you are authorized to contract with reliable parties to perform such jobs of work as may occur at various points along the route." No road superintendent had been given such sweeping authorizations and heavy responsibilities by the Interior Department; for the exercise of his judgment and discretion Lander was paid a maximum salary of $4,000.[61]

At Independence, Missouri, the superintendent learned that all the supplies taken over by the Army were to be retained. Since those remaining at Fort Thompson were known to be insufficient for his season's roadwork, he made extensive purchases on the frontier. The sutler at Fort Laramie, S. E. Ward, was contracted to bring the expedition's supplies forward to that fort. John Richards, a mountain trader who owned a bridge across the Platte, was in Independence engaging carpenters for repair work. He assured the road engineer that beef cattle and oxen for mountain work would be available at his post, 830 miles up the Platte, in exchange for some of the expedition's mules. News was received here of the close of the Mormon War, flooding the country with unemployed, so common laborers were easily obtained.

The expedition reached Fort Kearny on May 15, 1858, having averaged more than 38 miles a day.[62] At the close of the month Lander was at Fort Laramie. In expanding the working force here, he gave preference to the stranded members of the last year's expedition. Supplies were available from New Mexico, via the Salt Lake Valley, at prices which averaged a third lower than on goods from Leavenworth, so he purchased extensively when they were offered.[63]

Arriving at the South Pass in mid-June, the engineer directed the construction of a blockhouse for storing the expedition's equipment. One detachment returned to Fort Laramie to bring up the supplies transported by sutler Ward and the flour, meal, and *frijoles* obtained from the Mexican trains; a second group headed north to Fort Thompson to reclaim the remaining tools of the 1857 expedition; the wagon master went to Salt Lake City to procure Mormon laborers. With lumberjacks and bridge builders hired in Maine, and drifters attached to the train from the hordes of Army employees, superintendent Lander headed west over his northern survey.

Progress in road building was rapid to Piney Creek, where the cutting of heavy timber and removal of rock to reduce the grades through the Wasatch Mountains proved exceptionally hard and time consuming. The party from Salt Lake arrived with forty-seven destitute Mormons, mostly English, Norwegians, and Swedes, who gave timely assistance. By August 2, Lander had built the new highway to Smith's Fork of the Bear River and had made a new survey along the Salt and Blackfoot rivers where the laborers were to follow.[64]

As the road-building party started construction along Salt River, the supply depot was brought forward to a warehouse on the Piney Creek known as Fort Piney, and weekly shipments were sent to the point of farthest advance. Lander, with a mountain guide, remained in the front, engaged in a reconnaissance for the definitive location of the road, until he and his disbursing agent, John H. Ingle, were forced to attend to the business affairs of the expedition in Salt Lake. Wagner assumed the responsibility for pushing the work to the headwaters of the Ross Fork of the Snake. Lander, on his return, completed the reconnaissance to the Goose Creek Mountains with a party of his engineering assistants. The construction gangs continued to build down the valleys of the Snake and its tributary, Raft Creek, to City Rocks.[65] Lander's cutoff between the South Pass and City Rocks was established on the following route:

> Beginning at Gilbert's trading station, in the South Pass, it passes along the base of the Wind River mountains, heading Little and Big Sandy creeks; thence west, across the Green River basin, crossing the New Fork, Green River, and White Clay and Bitter-root creeks to the valley of Piney creek; thence up this valley through Thompson's Pass to the headwaters of Labarge creek; thence, *via* the head of Smith's fork of the Bear river to the valley of Salt river. The road continues down this fertile valley about twenty-one miles to Smoking creek; thence up the valley of this creek to the head of Blackfoot creek, and the valley of John Gray's lake to Blackfoot creek, lower down; thence over to Ross creek. Passing several miles down this creek the road crosses over to Snake river or Lewis' fork of the Columbia, near the mouth of Pannock river; thence down the valley of Snake river to the valley of Raft river; thence up this valley direct to City Rocks; a total distance of 345.54 miles from Gilbert station at the South Pass, and 950.54 miles from Fort Kearney. [*sic*][66]

William H. Wagner and J. C. Campbell, assistant engineers, simultaneously made a further exploration of the proposed southern road through Cache Valley. Lander asked the Department's authorization to maintain his engineering party in the mountains during the winter and an additional $35,000 to construct the Cache Valley road. Approval was granted these proposals, but had not been received by Lander before he resolved to return to Washington.[67] With the northern cutoff completed at the close of September, Lander reversed his decision to winter in the mountains because of the outbreak of war between the Crow and Shoshone Indians and an inability to find a suitable wintering area for the stock, not occupied by the Mormons, the Army, or transportation contractors. Sixty employees were discharged at City Rocks, the majority of whom returned home to Salt Lake, though thirteen of the wagon party chose to travel westward to California.[68] The main expedition headed for the States. Some of the mules and wagons were sold in Salt Lake, the tools and instruments stored at Fort Laramie, and an old mountaineer, Charles H. Miller, stationed at the South Pass to make weather observations and direct the earliest immigration to the new road in the spring of 1859. Miller was faithful in the performance of his assignment until killed in a gun fight in early March.[69] Lander arrived at St. Joseph, Missouri, on November 17, discharged the remaining employees, and wired Washington for instructions. Campbell ordered him to the capital for the winter to prepare reports and maps for Congress.[70]

The superintendent knew that the two groups upon whom the success of his new road chiefly depended were the Indians and the Mormons. For the eastern Shoshone, under Chief Washakie, he had words of praise and urged Secretary Thompson to give the tribe financial assistance. Since the wagon road traversed their hunting and root-gathering grounds their subsistence was greatly diminished, and without aid they would likely take the warpath or become marauders along the trail. Lander believed the Bannocks, near Fort Hall, more recalcitrant because they had burned Fort Thompson to the ground. The superintendent, always judiciously appreciative of the Mormons, found them excellent laborers, trained in ledge work and masonry. Some had gained experience in Cornwall, others in excavating the irrigation ditches in the Great Basin. Lander had been assured by former Governor Brigham Young that his people were grateful for the income from working on a federal government project.[71]

During the winter Lander prepared his *Emigrant's Guide* for those who might use the government road. In an introductory note he advised the overland traveler:

> You must remember that this new road has been recently graded, and is not yet trodden down; and, with the exception of grass, water, wood, shortened distance, no tolls, fewer hard pulls and descents, and avoiding the desert, will not

be the first season as easy for heavily loaded trains as the old road, and not until a large migration has passed over it.

All stock drivers should take it at once. All parties whose stock is in bad order should take it, and I believe the emigration should take it, and will be much better satisfied with it, even the first season, than with the old road.[72]

Deposed superintendent Magraw had conferred with Lander in Salt Lake City in the fall of 1858, in the presence of General Johnston, and assured him that the difficulty over the expedition's property would be settled in Washington.[73] Magraw arrived in the national capital with his Army discharge during the winter, convinced that Lander had misrepresented events of the 1857 season. Defamatory accusations were made against the new superintendent. On learning this, Lander addressed a billet doux to Magraw accusing him of libel and issuing a challenge to settle their difficulties in a duel or by physical combat. The two men met in Washington's Willard Hotel soon thereafter, and Lander inquired why Magraw had ignored his challenge. Magraw countered by striking Lander with a billy held in his hand, knocking him to the floor. When Lander gained his feet he returned full measure. Both men, though physical giants, were evenly matched and came out of the affray badly cut. Their encounter, disrupting the Willard's guests, was publicized by the Washington press.[74]

Magraw's bondsmen pressed him to settle his accounts with the Interior Department. Sympathetic with his bewilderment, Secretary Thompson requested Lander to enumerate the property that he, as chief engineer, had received from Magraw in the season of 1857, to state how it was liquidated, and what equipment of the "old expedition" he did not see fit to use when he went west in 1858. The Secretary frankly admitted that Magraw had difficulty in accounting for extensive amounts of property, yet the Department wanted to make an intelligent adjustment of his records, if possible.[75] The Secretary of War was requested at the same time to check a schedule of expedition property presumably taken by the Army and to ask if he would accept the figures of the Interior Department setting an evaluation on the equipment.[76] The auditors struggled to balance the accounts of Magraw and his disbursing agent, but the former superintendent was more interested in settling his quarrel with Lander.

When the wagon-road expedition was in the field for the third season, 1859, Magraw presented his first written defense to the Interior Department. Character references were used to refute the personal charges.[77] Lander was accused of not "acting as a man of truth," and a full investigation was once more requested. Secretary Thompson tried to squelch the feud—a source of increasing embarrassment to the administration—by assuring Magraw that the Department had no intention of going into the charges because of his voluntary retirement. Moreover, no assumption had been made that he was evading charges and no judgment in his prejudice had been handed down. His papers were officially filed as a denial.[78]

In Lander's absence from the city, Magraw continued to circulate stories about his former assistant's misconduct. He boasted to Army acquaintances, including Major Edward L. Yates, that he had thrashed Lander the previous winter for his insults. In the spring of 1860 Lander and Yates, walking in front of the Kirkwood Hotel, met Magraw and friends alighting from a carriage, and Lander seized Magraw, asking him to repeat his remarks made to Yates. According to a newspaper report, Lander told Magraw to "speak up, and speak up loud." When Magraw asked to step inside the hotel and then sought to depart, Lander grabbed him again and demanded, "Turn around, sir, and face me and answer me." Kirkwood stepped in to stop the altercation, giving Magraw an opportunity to draw his pistol and threaten Lander by saying, "Approach me again, sir; and you are a dead man." Lander hurdled a chair and grappled with Magraw shouting, "I am unarmed, you scoundrel, but no matter." The scuffle was once more terminated by the management. Upon Magraw's refusal to leave the hotel, Lander called him a liar, coward, thief, and blackguard in the hope of invoking a fight. Failing to do so, he inquired, in great anger, if his adversary had a friend among the forty present who would come to his defense. None stepped forward so Lander and Yates departed after extending apologies to the hotel management and the ladies present.[79]

Secretary Thompson exercised far greater patience in attempting to adjust Magraw's deficiencies than in settling those of other superintendents like Nobles of Minnesota or Leach in New Mexico. Presumably this was out of respect for President Buchanan's wishes. Dozens of letters continued to be received in the Department from men who had failed to receive wages, business houses with unpaid claims, and Army representatives who had discovered Magraw's vouchers to be invalid. The auditors complained that they could not adjust the partial and confused accounts. After the Army estimated the value of its acquisitions from the expedition at $52,306,[80] the Treasury listed the missing property for which Magraw was liable, evaluating it at $14,000.[81] Once more, Thompson asked Magraw's intentions in accounting for the equipment so he might adopt a general policy for the settlement of claims against the Pacific Wagon Roads before his term came to a close.[82] With the inauguration of Abraham Lincoln, Caleb B. Smith assumed the duties of Secretary of the Interior and within a month authorized a solicitor of the Treasury Department to take legal action against

Magraw and his sureties "to secure according to the law what he owes the government."[83]

In the spring of 1859 Lander expressed an interest in returning once more to the mountains. Congress had recently debated the importance of improving the overland mail routes, and the engineer quickly reported that his new trace was not properly placed for a winter mail route across the continent. This road was "especially and emphatically" an emigrant road for ox-team migration. Three superior mail routes would be found: in the Salmon River Valley; along the "old emigrant road" between South Pass and Fort Hall; and by his Cache Valley reconnaissance.[84] Lander suggested that he should go west on a dual mission to improve his emigrant road and seek mail routes that he was certain existed through the mountains where no emigrant route could be made. He wanted a train to precede the season's migration. Once the stream of covered wagons began to roll through the South Pass, his road-working party could move on toward Honey Lake, exploring and mapping the area along and north of the Humboldt.[85]

Annual instructions from the Department again coincided with his proposals. Not only was he to improve the road built in the 1858 season but make a continuous survey to California to determine the advisability of opening another route from the Bend in the Humboldt directly west to Honey Lake. Improvement of the Cache Valley route was to be postponed. The Secretary of the Interior obtained $5,000 from the Commissioner of Indian Affairs to be spent on gifts for the Shoshone and Bannock tribes, and he urged Lander to hold conferences with them, maintain amicable relations, and secure pledges that travelers along the new road would not be molested. As funds for the wagon road were almost exhausted, expenditures for the season were limited to $25,000.[86]

Upon arrival at Troy, Kansas, Lander found his mules, pastured during the winter, unfit for travel, so the main party was delayed and William Wagner, with an advanced detachment, headed for Soda Springs to meet J. C. Campbell in charge of the expedition property at Salt Lake. Together they were to begin surveys on the western section beyond City Rocks.[87] On his way west Lander met the backwash from the Pike's Peak region who termed the gold rush a "humbug." At the South Platte crossing, the destitution of the Colorado miners was found to be so great that the road party voted to go on short rations into Fort Laramie to share their food with these men. Mountaineers assured Lander that the pitiable condition of those whom he met was insignificant compared with those on the Smoky Hill route.[88]

Wagner traveled rapidly to the South Pass and dispatched a messenger to Campbell with instructions to join the engineering party at Soda Springs. At the Pass, the advanced forces were divided. The professional engineers went west to City Rocks by Hedspeth's old road, and a construction gang of Mormons, supervised by C. C. Wrenshall, proceeded to the western end of the new highway and commenced repairs working toward the South Pass. Wagner's reconnaissance to Honey Lake confirmed John Kirk's findings and recommendations of 1857. After a careful survey north of the Humboldt, he was likewise convinced that the cost of building a practicable road away from the river between Humboldt Wells and Lassen's Meadows was inadvisable; bad as it was, the Humboldt River road was still preferable. An *Emigrant's Guide* to the western part of the road was prepared as a supplement to that of Lander.[89]

Lander reached the South Pass at the close of June to discover that traders along the old routes to Soda Springs or the Salt Lake Valley were meeting emigrants and trying to divert them from his new road. Miller's murder of the previous winter was indirectly attributed to these men, so Lander stationed a former soldier of his party at Gilbert's Trading Post to inform travelers of the advantages of the federal wagon road and present them with the published guide. Fist fights became weekly occurrences in the bid for the emigrant's favor, so Lander decided it would be necessary to leave a blacksmith at the pass during the winter to ply his trade and explain the merits of his road.

Moving on to Green River, the road builders discovered its waters were so high that the emigrants had staked out a new ford, thought to be more desirable than that chosen by Lander's party. Here two wagons had been swept into the deep water by the current and one man drowned; from this time on, Lander pressed the Interior Department for $30,000 to build a permanent bridge. Laborers from the periodically passing emigrant trains were employed to improve the road, and at one time there were one hundred and fifty men engaged in heavy grading.[90] The main working crew met those under Wrenshall on August 1, along the Ross Fork of the Snake, and arrangements were made to break up the party. Those who did not wish to go on to California returned to the States with the wagonmaster, the public property was sent to Salt Lake for disposal, and Lander went west to join Wagner.[91]

The road builder's primary interest during this third season appears to have been promotional. A corps of artists sketched and prepared a set of stereoscopic views of emigrant trains, Indians, and camp scenes.[92] Lander's time was largely consumed in distributing route guides, and persuading emigrants to sign petitions in praise of his road. These petitions reveal that three out of every four emigrants were headed for California, only one for Oregon. The previous residence of most had been Illinois, Iowa, and Wisconsin; though many also came from Missouri, Minnesota, Michigan, Ohio, and Indiana. A train of ten wagons was large, but a few with nineteen wagons and one with twenty-seven passed by. The average group was composed of nine emigrants, young and old. Exceptionally large herds of stock were driven west over the new road. Quite often a drover had several hundred head, and some controlled as many as eight hundred to a thousand.

Upon his arrival in California, Lander read in the press the praises of Captain James H. Simpson's road between Camp Floyd and the Carson Valley, and was piqued to learn that he had a formidable rival as a road builder. Simpson's discovery was greatly stressed in the Salt Lake press because the entire Mormon community had been disappointed at being bypassed by the Fort Kearny–South Pass–Honey Lake road. The Mormons had found a spokesman in Chief Justice D. E. Eckles who wrote to Secretary Thompson stating that the Army road was the shortest by more than 300 miles and by far the best road to California, and "the public would soon follow it but for *interestedly false representations made to emigrants.*" Lander's annoyance was further increased after his arrival in Washington by reading this correspondence and the annual report of the Secretary of War, claiming that Simpson's surveys in conjunction with those of Captain Howard Stansbury and Lieutenant Francis T. Bryan had established the shortest emigrant road between the Missouri River and the Pacific Coast.

The road superintendent prepared an elaborate brief to show that Simpson's road, though good for the overland mail and perhaps for the telegraph, was unsatisfactory for emigrant travel. Moreover, in answer to Eckles' charges, he published his instructions to agents at the South Pass which said in part:

> . . . We cannot do more than simply inform the emigrants of the actual facts in regard to our road. Let them choose which of the two roads they wish to travel; it is nothing to us; we simply and plainly obey instructions from the department; therefore do not persuade anyone to take the road, although we know it to be the best.[93]

Testimony in praise of Lander's route was presented in two petitions, signed by over 9,000 of the 13,000 said to have used the road in 1859. The petition prepared at Fort Hall contained the following endorsement:

> This is to certify that we, the undersigned, have travelled over the Pacific wagon road, better known as Lander's cut-off, and find it a very acceptable road for emigrants. We think it preferable to any other road across the mountains in

many respects. Most of the way it is well worked, and with a bridge across Green River (the only stream at all troublesome) it would be as good a road as many now travelled in the States. It is some five days' travel shorter than any other route across the mountains; there is no desert to cross on this route, no alkali to kill your stock, but instead plenty of good water, abundance of grass, and wood enough to satisfy any reasonable man.

Many of the undersigned have crossed by other routes and give this the preference.[94]

Extensive clippings from the California newspapers, lauding the federal road program, were likewise filed with the Interior Department.

Lander wrote to Simpson, inquiring if he concurred in certain quoted statements praising the location, shortness, and utility of his road as published in the Secretary of War's report and in the West Coast newspapers, warning him "all these reports will cause the less experienced emigrants to take your road." "The question is," stated Lander, "are you ready to advocate it as suitable for ox-team emigration? If so, there is nothing more to be said upon the subject." Simpson assured Lander, "... it has been the farthest thing from my mind to do injustice to you, or any one else, in what I may have reported of the explorations I have recently made between the Rocky Mountains and the Sierra Nevada." The topographical engineer prepared a statement, sent to the editors of the Washington *Constitution* for publication, in which he admitted that whereas the data available in Utah had made his route appear much shorter than others to the Pacific, the great saving of distance was far less significant in view of Lander's recent improvements. He generously admitted, "... it is very possible that immigrants desiring to travel through to California without passing through Great Salt Lake City or Camp Floyd, for purposes of replenishing supplies, ... would do best to take the Lander cut-off at the South Pass and keep the old road along the Humboldt river." The Army officer hastened to point out that he had no intention of disparaging either of his two routes westward from Camp Floyd, but since the deficiency of water and grass at some points would create hardships for large herds of cattle or big trains, he could not take personal responsibility for diverting the thousands of western emigrants from the road to City Rocks, as improved by Lander. He was convinced that "Time can only settle which is the best route to the travelling public; and to that arbiter do I leave the decision; only feeling desirous that that route which furnishes the greatest facilities may, as it will, be eventually taken."[95]

Lander, somewhat reconciled, was determined to prove the superiority of his road and outlined a program for the Secretary of the Interior whereby the sufferings of the emigrants between the Humboldt and the eastern

Sierra Nevada might be alleviated. One of the three desert routes from Lassen's Meadows should be improved by the construction of aqueducts and reservoirs to transport and store the limited water. As the law would not permit the improvement of the Carson Valley road, he proposed expenditures on one of the two surveys made by Kirk and Bishop in 1857. To provide adequate water across the Truckee Desert would prove twice as expensive as improving the water holes on their outward trace between Honey Lake and Lassen's Meadows by way of Mud Lake.[96]

Thompson sent Lander west for the fourth consecutive season, this time to improve the Honey Lake–Humboldt trace. The Secretary explained the final mission, "This being the last portion of the road over which the emigrants to California and Oregon have to pass after a long and toilsome march across the continent it is the desire of the Department to render its passage as comfortable and easy as the nature of the country and the means at its disposal will admit of."[97]

Lander arrived in Sacramento in mid-May at the outbreak of the Paiute war and was unable to obtain a contractor willing to improve the springs and construct reservoirs in the domain of hostile Indians. He therefore prepared to mount and arm a party of forty men to build the water tanks under the supervision of engineer Wagner. Winnemucca, the Paiute chief, was said to have eleven hundred men west of the Humboldt, including his Shoshone and Bannock allies, three hundred ponies for his best warriors, and sufficient dried beef for several seasons. The Paiute had announced his determination to fight for ten snows until the whites gave up trespassing on his domain.[98] Lander notified the press of his intention to seek out the chief for peace talks.

The road builders reached Honey Lake to find the settlement virtually depopulated and the near-by residents in a state of panic. At the request of Major Isaac Roop, the provisional governor of Nevada Territory, Lander assumed command of sixty men, including his own staff and the local Honey Lake Rangers, and fought a running engagement with the Indians to drive them north of the emigrant roads.[99]

Returning to the first watering place northeast of Honey Lake at Mud Springs, the laborers stoned its bottom, then moved to Buffalo Spring where a fifteen-foot well was sunk. As the group moved eastward, new springs were located, and old watering holes cleaned out. At Hot Springs a large reservoir was built, and the water of the springs diverted so that it would cool in the process of running to the tank. The major construction job was at Rabbit Hole Spring, where the water supply was so inadequate the mules had to be sent forward to the Humboldt. A labor crew of fifteen worked for three weeks tapping the water supply, building a split-stone culvert set in cement to carry the water to a reservoir of solid masonry. Within a short time the tank held eighty thousand gallons, and a train of three hundred emigrants and one thousand animals, accommodated shortly after its completion, did not materially lower the surface of the water. At Antelope Spring close to the Humboldt, additional excavations and masonry made the water more accessible to the traveler.[100]

Upon his return Lander succeeded in interviewing Winnemucca and arranged a year's truce with no promise of concessions at the end of that period. The grateful Californians expressed their praise of the superintendent in the press, and his reputation as an Indian commissioner equaled that as a road builder.[101] During the ensuing winter in Washington, Lander requested authorization and funds from the Interior Department to build the Green River bridge. Although he and Thompson had earlier agreed that the traders owning ferries on the old roads would probably burn the government bridge, if erected, the superintendent now proposed the construction of a blockhouse and blacksmith's forge at the proposed bridge site, to be occupied by a mountain guard. Lander also announced his willingness to go west as a peace commissioner to the Paiutes or as a military agent to locate forts along the California road.[102] But the Democratic administration drew to a close; Lincoln had been elected, and the prospect of a civil war monopolized the conversations and plans of men in Washington. In submitting his resignation from road-building activities, Lander offered his services to the administration until the inauguration. The acting Secretary acknowledged his contribution:

> In accepting your resignation it affords me great pleasure to bear witness, on behalf of this Department, of the fidelity, skill and ability with which you have conducted the operations entrusted to your charge, and that your successful effort to render the march of the emigrant comfortable and easy across the broad domain of the United States entitles you to the approbation of the country.[103]

CHAPTER XIII : *Interior Department Improvements in New Mexico: El Paso and Fort Yuma Road*

SENATOR THOMAS J. RUSK of Texas was the primary motivating force in inaugurating federal wagon-road construction along the thirty-second parallel. After piloting the legislation through Congress, he demonstrated an intense interest in the details of its location and the selection of personnel for the project. His concern was the natural outgrowth of a desire to promote the development of his state by securing a western commercial outlet. Moreover, Rusk shared with Jacob Thompson, Secretary of the Interior, and with Jefferson Davis a strong desire to prove the advantages of a southern route to California.

With Thompson's approval the senator assumed leadership in requesting statements from men familiar with the Southwest, recommending, in detailed fashion, the exact route to be followed between El Paso and Fort Yuma, an estimate of construction costs, and the best type of contract to be made. Captain John Pope proposed that the wagon road should follow the route of Lieutenant John G. Parke's railroad survey westward from the Rio Grande River to the San Pedro, down that river to the Gila, and on to the California border at Yuma. He thought the road could be built for $125,000, with heaviest expenditures to locate and improve necessary water facilities, to cut out the mesquite brush through the 60-mile stretch along the San Pedro, and to remove the rocks from the boulder-studded 12 miles beyond the San Pedro–Gila junction. The Army officer reflected a familiarity with topographical engineering procedure for he proposed that the road work be done by contract, the superintendent and engineer assuming roles of overseers.[1]

Albert H. Campbell, who was at the time sponsored by Pope for the road superintendency, agreed that if Congress intended temporarily to substitute a wagon road for the needed railroad to transport emigrants and mails to California, the route should follow the railway survey, with slight deviations to locate water supplies. Campbell disagreed with Pope's plan to make the road by contract. The Army system of granting construction to the lowest bidder, who invariably ran short of funds before the task was complete and then sought relief from Congress, had proved disadvantages. If the road was to be built with the single appropriation, then the bonded superintendent and engineer, with fixed salaries, should assume responsibility for directing labor gangs.[2]

From a political associate in Tennessee who had traveled over the proposed route to California, Rusk received a third endorsement of Parke's survey directly west from the Rio Grande to the San Pedro River, avoiding the horseshoe detour to the south made by Cooke. In organizing the working force, according to this informant, the Department would also be well advised to employ Anglo-Americans at a ratio of one for each two Spanish-Americans. These men would ward off expected Indian attacks and bolster Mexican morale to maintain a high level of accomplishment. The recommended labor force of two hundred and fifty should be broken into three groups: the commissariat, or supply, department; the vigilance party; and the laborers used for road building or tank and well digging. Among the many other procedural details discussed was a suggestion that wagon trains constantly shuttle between Texas settlements and the various field locations to keep the road party continuously supplied. The politician reminded Rusk, at the same time, that Congress took the road-building program from the Army so that it "might not be baffled, or smothered by routine, and forms, *delays,* and ceremonies of *military etiquette,* and *parade.*" With the senator and the Secretary of the Interior, he was concerned over the interests of the southern states:

> It is the only little scrap of southern territory we have. It is the great link to the Pacific. I hope the road may be made. I know it could be done, and promptly. It is a thoroughly practicable route.[3]

A fourth proposal, that of James B. Leach, was sent directly to Secretary Thompson. He also recommended a route due west from the Ojo de Vaca (Cow Spring) to the Gila River rather than turning southwest through Guadalupe Pass along Cooke's road. Leach reported that Cooke's road was so desolate that Texas cattlemen had learned to avoid it and instead to drive their herds northward to Kansas and then west to California over the central emigrant trails. According to his plan, wells and reservoirs would be needed along the whole line of the southern road, ". . . for the great desideratum of this road is water, in the rainy season there are numerous springs, creeks, and other supplies of water which become dried up under intense heat of summer."[4] Bridges would have to be built across the Rio Grande, San Pedro, and Gila rivers.

In addition Leach advanced his candidacy for the road superintendency by presenting a grandiose scheme for the wagon-road expedition. The train would be outfitted on "the Atlantic side" and marched overland through Texas to El Paso, where the road work was to commence. Construction would be toward the west so the men could be discharged in San Diego or Los Angeles, California, thereby saving the expense of transporting them back to the settlements. He planned to hire one hundred laborers in the East, and two hundred Mexicans upon reaching El Paso, sharing with many frontiersmen the conviction that the Anglo-Americans would aid morale and serve as a protection against the Indians. The working force would be divided into six gangs of fifty men, each under a subordinate assistant, independently organized with its own carpenter, cook, blacksmith, herdsmen, and watchmen. All provisions, except fresh meat, could be purchased in the East rather than in the "erratic" El Paso market. Necessary commissary supplies would cost $28,500. An estimated $35,000 would be used for mules and wagons, $4,000 for tools, and $6,300 for camp equipage. After figuring the wages of the Mexicans, eastern laborers, and mechanics at one, two, and three dollars a day, respectively, Leach thought $155,000 would complete the wagon road. He concluded:

> I have estimated for a very large working party that the road might be finished as rapidly as possible consistent with economy, and made available to the emigration at the earliest moment; indeed the road should be finished and the announcement of such . . . made early in the coming winter that the Pacific emigration of 1858 may avail themselves of the Southern Road.[5]

Such detailed plans and resolution inevitably impressed the Department with the potential ability of the author.

Davis and Rusk were joined by John B. Floyd, Secretary of War, in prevailing upon Thompson to name Campbell general superintendent of the Pacific Wagon Road office in Washington, D.C. because of his knowledge and interest in the southern route.[6] The road superintendency, in turn, was assigned to Leach, and N. Henry Hutton, a man with greater professional competence than political influence, became chief engineer.[7] Other personnel appointed by the Secretary of the Interior included W. Drayton Cress, a Georgia friend of Campbell and Hutton, as assistant engineer; Dr. J. R. McCay, a protegé of Secretary Floyd, who also had excellent connections at the Smithsonian Institution, as physician and surgeon; and M. A.

McKinnon, of Oxford, Mississippi, as disbursing agent.[8] Leach, who was empowered to select his assistant, named a personal friend, D. Churchill Woods. The Secretary and superintendent later resolved to add N. P. Cook to the party as an assistant engineer. Cook, a resident of Tucson, had been elected by the people of western New Mexico to represent them as a delegate to Congress. Although his mission to get a new territory created had failed in 1856–1857, he had effectively aided the passage of the Pacific Wagon Road Act. By profession Cook was an explorer, surveyor, and engineer, and his congressional colleagues, headed by Douglas of Illinois and Rusk of Texas, secured the road-construction assignment.[9] All appointments to key positions were thus dictated by personal friendship, political expediency, or sectional interest.

After much consideration, discussion, and review, Thompson and Campbell settled upon a final plan of construction. Leach was enjoined to locate the road from Franklin, a town opposite El Paso, Mexico, north to Fort Fillmore along the eastern bank of the Rio Grande. Crossing the river between the fort and the town of Mesilla, the road was to ascend the Sierra Madre plateau to Cooke's Spring. Continuing westward across the Rio Mimbres, the survey was to run to Aqua Frio (Cold Water), a watering place six miles south of the Ojo de Vaca. Both springs were to be improved. From these oases an old wagon trace was to be followed westward to the Railroad Pass by way of a pass in the Piloncillo Range. Both the Mexican Boundary Commission and Parke's Pacific Railway survey had reported the existence of this trail. The area northwest of Railroad Pass was to be explored by the engineers in the hope of finding a shorter route to the San Pedro River. Campbell maintained that a route might be located down the Arivaypa, one of its tributaries. Should the investigation prove fruitless, the road would be run southwest by Croton Springs, through Nugent's Pass, and due west to the San Pedro. From this point, the valleys of the San Pedro and Gila rivers would be followed to Maricopa Wells. A direct westerly crossing to the small settlement at Tozotel, avoiding the Great Bend of the Gila, would be made, and the river followed on into Fort Yuma.[10]

In enumerating the places where Leach and his working parties were to conserve the existing water supplies or to search for new ones, the Department blueprinted the construction technique. Wells and pits were to be dug in the bottoms of canyons, to concentrate the water and reduce evaporation following rains. Wide earthen dams lined with stone across a dry arroyo might stop the runoff of water following downpours and provide a watering place in the wet season. If a well was dug behind these dams, accumulated water would be preserved even for the dry season. Where feasible, wells of small depth were to be sunk, but the absence of timber would preclude any attempt at deep drilling. The paramount importance of providing water was generally recognized.

According to the Department's design, no heavy grading jobs or bridging were to be attempted. The description and specifications for any such exigencies would be forwarded by the superintendent for the Department's consideration. Fully aware of the magnitude of the assignment, Thompson encouragingly wrote to Leach:

It is not expected by the Department that you will be able to construct a finished road over this route in one season; but it is confidently hoped that you will be enabled to report by the 1st of December that a road over which loaded wagons can with ease and safety pass has been opened, and that ample provision has been made for the collection and preservation of the rainfall on the plateau, and also that the permanent sources of supply have been improved; so that by another season the Emigrants shall feel no hesitation in adopting it.[11]

For closing out his operations, further instructions were to be sent to Leach at Fort Yuma. It was clearly anticipated that the road would be worked back again to Mesilla from that point. Together with the other superintendents, Leach was warned that the total appropriation, $200,000, was to be expended proportionately along the entire road, without relying on the possibility of future congressional grants. Accepting Leach's scheme for moving the wagons, men, and equipment from some point east of the Mississippi River, Thompson selected Memphis, Tennessee, as the depot. After assembling his group, the superintendent was to proceed by the most direct route to El Paso on the Rio Grande, the initial point of his operations.[12]

Soon after receiving his appointment, Leach and his inseparable assistant, Woods, made a tour of eastern markets to outfit the road expedition. Harness and leather goods were obtained in Philadelphia; rolling stock, including wagons and a spring ambulance, in Concord, New Hampshire; mechanical tools in Boston; and the necessary mechanics and foremen in Fall River, Massachusetts.[13] They purchased commissary stores on a lavish scale including 27,000 pounds of bacon, 19,225 pounds of sugar, 10,012 pounds of coffee, 2,675 pounds of soap, 234 bushels of beans, and 40 barrels of vinegar. Ploughs and scrapers, axes and ax handles, picks and shovels, tents and bake ovens were obtained by the dozen, and plates, cups, knives and forks were purchased by hundreds. The Ordnance Department of the Army furnished 75 percussion rifles, 20 Colt belt pistols, 11,250 rifle-ball cartridges, and 3,000 Colt pistol cartridges. In addition, the superintendent purchased $5,000 worth of ammunition and 35 kegs of blasting powder. Each wagon was equipped with two ten-gallon water kegs, and one traveling forge, complete blacksmith's tools, and a small quantity of iron and coke.[14] These supplies were shipped westward to Memphis where Leach and Woods, with the eastern mechanics and laborers, arrived in late May to join the other officers of the expedition.

On June 27, 1857, Leach informed the Interior Department that his party was organized. With bullwhips cracking, the caravan of forty wagons crossed the Mississippi River and headed for El Paso.[15] En route to Little Rock by way of Des Arc on the White River, the first difficulties became apparent. The forty wagons had begun their westward roll with fewer than the necessary oxen because Leach, unable to procure them in Memphis, was certain of their availability in Arkansas. Fifteen wagons were, according to original plan, drawn by mules; the other twenty-five wagons required oxen. Upon the arrival at Des Arc, Cook was dispatched, first to Doaksville in the Choctaw Nation lands and then to Fort Smith, to buy needed stock. Due to the extensive delay in the delivery of these animals, Leach moved forward with the mule train, leaving the ox train to follow.[16] The superintendent decided not to take his disbursing agent, McKinnon, with him because the man had brought his family, and travel with the slower moving ox train would be easier on McKinnon's pregnant wife.[17] Leach predicted that his party would reach El Paso in forty days.

Leaving Des Arc, the expedition moved forward to the Arkansas River and encamped opposite Little Rock. The rough country and poor roads broke several wagon tongues and wheels, and sickness laid many men low, thereby slowing down the advance. From Little Rock the mule train traveled to Ultima Thule on the western Arkansas boundary via Hot Springs and the settlements of Dallas and Mineral Hill. The trip across Arkansas took forty days, the length of time thought necessary to reach El Paso.[18]

The wagons were driven on across the Choctaw Nation lands in a westerly direction through Doaksville to Preston on the Red River. Turning toward the Southwest, their route was through Gainsville, Fort Belknap on the Brazos River, and Camp Cooper on a tributary of that stream. Pushing across the Brazos tributaries near the Comanche Indian Agency, they arrived at the site of old Fort Phantom Hill, built in 1851–1852, but by this time abandoned. From here the mule train continued to Fort Chadbourne where Leach, who had been ill since leaving Camp Cooper, was forced to remain behind. Impatient at the delay and anxious to reach the field to begin road building, the superintendent ordered his men westward along the Middle Concho River to its headwaters and across the dividing ridge to the Pecos. This river was forded at the Horsehead Crossing used

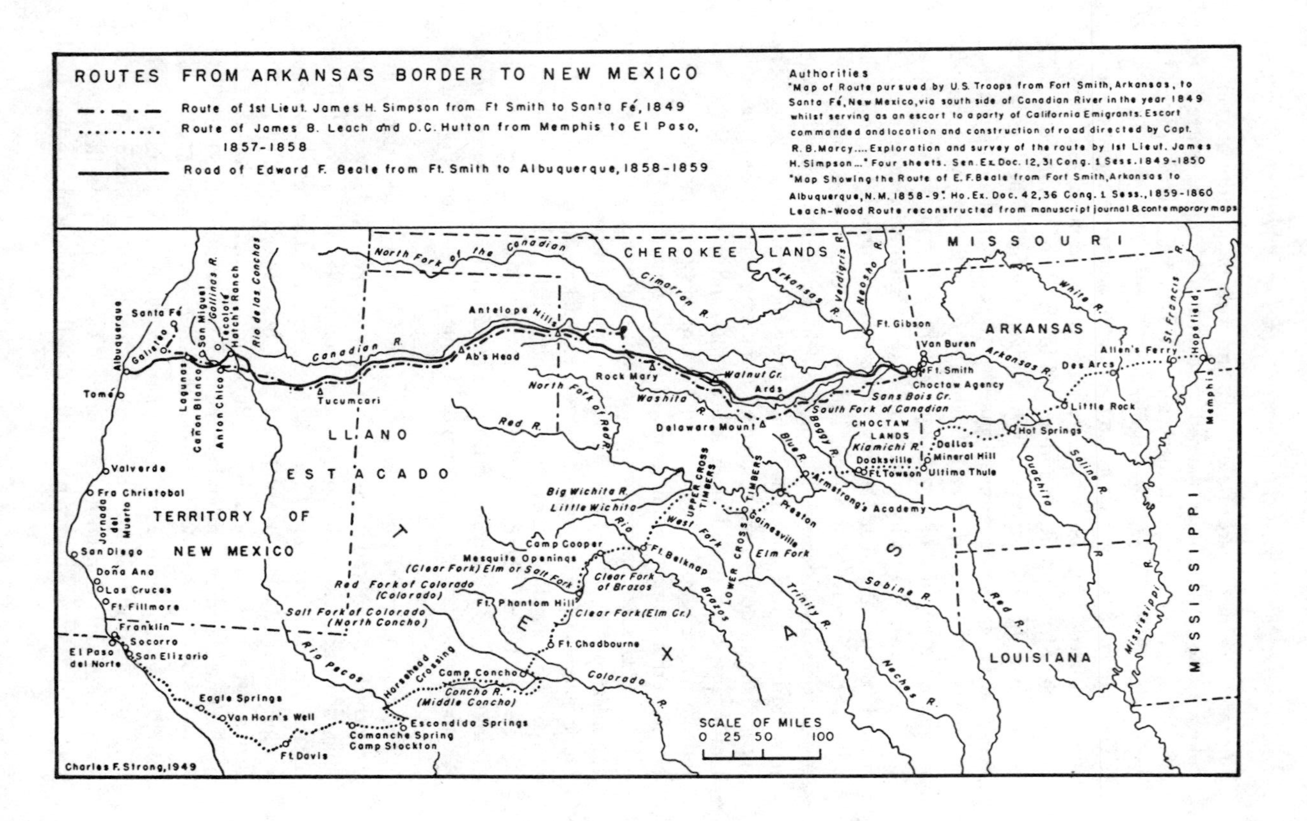
ROUTES FROM ARKANSAS BORDER TO NEW MEXICO
Route of 1st Lieut. James H. Simpson from Ft Smith to Santa Fé, 1849
Route of James B. Leach and D.C. Hutton from Memphis to El Paso, 1857-1858
Road of Edward F. Beale from Ft. Smith to Albuquerque, 1858-1859
Authorities
"Map of Route pursued by U.S. Troops from Fort Smith, Arkansas, to Santa Fé, New Mexico, via south side of Canadian River in the year 1849 whilst serving as an escort to a party of California Emigrants. Escort commanded and location and construction of road directed by Capt. R. B. Marcy.... Exploration and survey of the route by 1st Lieut. James H. Simpson..." Four sheets. Sen. Ex. Doc. 12, 31 Cong. 1 Sess. 1849-1850
"Map Showing the Route of E. F. Beale from Fort Smith, Arkansas to Albuquerque, N.M. 1858-9". Ho. Ex. Doc. 42, 36 Cong. 1 Sess., 1859-1860
Leach-Wood Route reconstructed from manuscript journal & contemporary maps
MISSOURI
CHEROKEE LANDS
ARKANSAS
MISSISSIPPI
LOUISIANA
TEXAS
LLANO ESTACADO
TERRITORY OF NEW MEXICO
CHOCTAW LANDS
North Fork of the Canadian
Cimarron R.
Arkansas R.
Verdigris R.
Neosho R.
White R.
St. Francis R.
Hopefield
Memphis
Allen's Ferry
Des Arcs
Little Rock
Hot Springs
Saline R.
Ouachita
Mississippi R.
Red R.
Ft. Gibson
Van Buren
Ft. Smith
Choctaw Agency
Sans Bois Cr.
South Fork of Canadian
Kiamichi R.
Dallas
Mineral Hill
Ultima Thule
Doaksville
Ft. Towson
Boggy R.
Blue R.
Armstrong's Academy
Preston
Gainesville
Elm Fork
Trinity R.
Sabine R.
Neches R.
Delaware Mount
Washita R.
Ards
Walnut Cr.
Rock Mary
Antelope Hills
Ab's Head
Canadian R.
Tucumcari
North Fork of Red R.
Red R.
Santa Fé
Albuquerque
Galisteo
Lagunas
Cañon Blanco
San Miguel
Gallinas R.
Anton Chico
Tecolote
Hatch's Ranch
Rio de las Conchas
Tomé
Valverde
Fra Christobal
Jornada del Muerto
San Diego
Doña Ana
Las Cruces
Ft. Fillmore
Franklin
El Paso del Norte
Socorro
San Elizario
Eagle Springs
Van Horn's Well
Ft. Davis
Comanche Spring
Camp Stockton
Escondido Springs
Rio Pecos
Horsehead Crossing
Camp Concho
Concho R.
(Middle Concho)
Colorado R.
Ft. Chadbourne
Clear Fork (Elm Cr.)
Ft. Phantom Hill
Salt Fork of Colorado
(North Concho)
Red Fork of Colorado
(Colorado)
Mesquite Openings
(Clear Fork) Elm or Salt Fork
Clear Fork of Brazos
Camp Cooper
Ft. Belknap
Brazos
Big Wichita R.
Little Wichita
Rio
West Fork
UPPER CROSS TIMBERS
LOWER CROSS TIMBERS
SCALE OF MILES
0 25 50 100
Charles F. Strong, 1949

by Marcy in 1849, by the Army road reconnaissances from east Texas, and by many Argonauts. Traveling along the San Antonio–El Paso mail route, the wagon train meandered through cactus-studded country to Fort Davis. Leach and his attendants in the meanwhile had rejoined the group. From Fort Davis they moved by way of Van Horn's Wells and Eagle Springs to the Rio Grande River, 80 miles below El Paso. The men drove the mule wagons hurriedly through San Elizario, Socorro, and Ysleta, three border towns, to arrive in El Paso on October 22, 1857.[19] The Memphis–El Paso trip had required one hundred and four days of travel and had covered 1,309 miles.[20]

Upon reaching El Paso, Leach's advance party numbered about eighty men, mostly mechanics and engineers. The remaining personnel, approximately thirty-five laborers, remained with McKinnon and the ox train, under Woods. From the initial delay in seeking oxen at Des Arc, the story of the ox train was one of misfortune. Following several hundred miles behind, in the path of the mule train, the rear detachment advanced uncertainly across Indian territory. Sickness incapacitated so many men that the entire train was immobile for days. Arriving at Fort Belknap, Texas, in November, almost a month after Leach had reached El Paso, Woods decided to winter there and discharged the men until the following spring. In addition to the fact that the oxen were exhausted, all who were consulted advised against attempting the plains during the approaching cold season.[21] After a six-month pause, the ox train moved out of Fort Belknap on May 6, 1858, finally pulling into Mesilla June 25, 1858, two days short of a full year from the time of the Memphis departure. Eleven of the twenty-five wagons had been left by the wayside.[22]

The confusion which attended the gathering of equipment, men, and supplies at the eastern terminus, the unexpected delay that the mule train under Leach experienced, and the absence of the supply-laden ox train with its men and needed wagons, precluded extensive road building immediately. However, the working season in the Southwest was the entire year, so Leach was able soon after his arrival to divide the available men into several working parties for preliminary improvements. Some went westward with Hutton to select the location for drilling an experimental water well.[23] A group of laborers under Cook's direction set up camp at the Pima villages on the Gila River, from which two parties worked toward Fort Yuma, and another one was dispatched to build eastward from the fort. Work on this western sector was completed first, to take advantage of the cool winter months.[24] Leach personally directed improvements on the line of the road running from Franklin to Mesilla. Cursory construction started here as soon as the wagons were unloaded. Another main camp was established on the San Pedro River out of which sixty men, mostly Mexicans, labored in cutting the brush in this valley.[25] Thus an attempt was made, in haphazard fashion, to prepare the road for an anticipated increase in emigration during the spring.

The road, as finally located, coincided in general with the route recommended by the Department of the Interior as far as Railroad Pass. Although the railroad route surveyed by Lieutenant Parke was thoroughly reëxamined and found to possess superiority in grade and distance over all others, it was finally abandoned between Mesilla and the pass because of insufficient water. The cost of necessary well digging and tank construction there would exceed the appropriation. Engineer Hutton's trace, like the railway survey, avoided Cooke's semicircular detour to the south, but left the Rio Grande farther north and passed through the established watering places, Cooke's Spring and Ojo de Vaca.[26]

In agreement with the Department's instruction, the engineering party attempted to locate a road from Railroad Pass down the Arivaypa to the San Pedro, a potential saving of distance. Exploration revealed that this stream ran down a canyon too narrow for a wagon road. So the engineer accepted the alternate route to the south via Nugent's Pass, striking the San Pedro River 13 miles below the point of Parke's survey. Following the right bank of the San Pedro to the mouth of the Arivaypa, the road crossed the river and ascended the Santa Catarina Mountains lying between the San Pedro and the Gila. The bank of the latter stream was reached 15 miles below the mouth of the San Pedro. In this way many rocky spurs abutting closely on the Gila River banks were avoided, the route shortened, and expenses in constructing a practical wagon road correspondingly reduced. Continuing down the south bank of the river to Maricopa Wells, the road retraced that followed by Kearny and Cooke in 1846–1847 into Fort Yuma. Engineer Hutton concluded that emigrant travel would save 47½ miles by adhering to his route, rather than following any combination of old traces or surveys. Moreover, there remained the advantage of excellent grass and running water in the San Pedro Valley, where emigrant parties could rest half way on their western journey.[27]

The road was opened 18 feet wide on the straight stretches and 25 feet at curves. Timber, brush, and rocks were cleared from the right of way along the entire distance to facilitate the passage of wagons. In making side-cuts or building up the roadbed, more than 50,000 cubic yards of earth and stone were excavated. Several irrigation ditches west of Mesilla, wide enough to block the crossing of a wagon wheel, were bridged. A major effort was made to reduce grades and to smooth the road surface, particularly in the Gila valley. Hutton reported to the Interior Department: "...from the improvement of surface consequent upon our labors (I am informed by freighters over the road,) two days' time is saved by loaded wagons between the Maricopa Wells and Fort Yuma."[28]

In accordance with the major assignment to locate new watering places, three wells, nine tanks, and one reservoir were sunk, increasing water potential by some 300,000 gallons.[29] In summarizing his accomplishments, Hutton boasted that future emigrants had been saved five days' traveling time, 70 miles of the road had been relocated to parallel supplies of running water, six new watering places reduced the greatest distance between camps to 27 miles, and the location of the road through the rich valleys of the Gila and San Pedro rivers now opened them for settlement.[30] The improvements on the southern overland road were equivalent to those made on other federal wagon routes in the 1850's, and apparently all that was anticipated, or thought necessary.

As customary, additional appropriations were requested from Congress. A bridge across the Rio Grande, costing about $50,000, was needed; $25,000 would line the earthen tanks with stone, making permanent the work already finished.[31]

Though the reports prepared for Congress and printed for public consumption covered the El Paso–Fort Yuma road project with an air of success, smoldering underneath was the threat of disgrace and humiliation for its leaders. Suspicion, investigation, and court hearings were to hound Leach for many years. The confusion began with the administration of financial affairs. The Secretary of the Interior laid down the rules to guide the wagon-road superintendents, requiring them to keep an elaborate set of accounts, in multicopies, and to forward quarterly reports to Washington. All purchases were made with a voucher system, signed by the seller and the superintendent, and countersigned by the disbursing agent at the time of payment. The Washington office assumed that the two officers would be inseparable in the field.[32]

McKinnon's certification was delayed by his procrastination in executing a bond and Leach, already bonded for $30,000, was granted federal funds directly to cover his eastern purchases.[33] This divergence from procedure was made because of Thompson's earnest desire to get the road ready for the spring migrations of 1858. Leach's long delay in the East and his decision to purchase items of questionable utility forced Thompson to withdraw the authority hastily.[34]

Soon after the expedition reached the field, it was apparent that the complicated and voluminous paper work was too great for either officer to master. Thompson found it necessary to warn Leach to state exactly the price of supplies per piece, pair, or pound and to make certain the seller signed the vouchers.[35] If illiterate frontiersmen substituted their marks, as had been done in many previous cases, the signatures of witnesses had to be procured.[36] Frontier merchants repeatedly made erroneous mathematical calculations in their bills, always to personal advantage; these were being detected by the auditors in the Treasury and charged against the reporting officers.[37]

Another factor upsetting all plans was the slowness of communication by the inadequate mail service. Two and a half months were needed to send instructions from the national capital to New Mexico and to receive a reply. Thus the expedition, for all practical purposes, was isolated from the Washington command. During the entire march from Memphis to El Paso, Thompson and Campbell lost contact with the road party. Not until Woods reported personally in Washington during the winter of 1857–1858, did the Department officials realize that the expedition had split into two sections at Des Arc and that the ox train was wintering at Fort Belknap.[38]

For many weeks Leach, in Mesilla, was at a loss to know where his ox train was. This division of forces increased the expenses of the expedition beyond all previous calculations. More important, it split Leach and McKinnon. Their jobs were complementary. Neither was adequate by himself. In fact, each was hamstrung by the other's absence. Leach bitterly complained, and in the end decided to send vouchers covering his expenditures directly to the Department without his agent's countersignature.[39] In other words, he was disbursing money not deposited to his credit by the Treasury but in the account of his financial agent. The superintendent raised cash by using his personal credit, sometimes writing a draft on the Department, and literally begging and borrowing to keep the work going. The Treasury Department was forced to establish a special temporary account for these disbursements.

Only adverse reports were received in the Pacific Wagon Road Office about the preferred southern road. Secretary Thompson was alarmed by the rapid influx of drafts and the slow presentation of properly verified vouchers. To cap the climax, there came to the Department a report from Fort Davis charging the superintendent with misconduct, "frisking, gambling, and frolicking," to the discredit of the expedition. From San Antonio news arrived that wagon-road equipment was being used by the San Antonio–San Diego mail contractor.[40]

The Interior Department concluded that its only recourse was an investigation. Because of the possibility of fraud in connection with the mails, Welcome B. Sayles, confidential agent of the Post Office Department, was chosen for the assignment. Thompson authorized him to review all charges made against Leach, to ascertain the nature of road improvements, and to probe into the financial records. He was authorized to interrogate any member of the expedition. The Secretary provided him with letters containing varying orders and instructions to the principal officers. They were to be delivered if Sayles considered it advisable to reshuffle assignments and responsibilities. In two of these, Leach was to be deposed and Hutton promoted to his position.[41]

En route to Mesilla, Sayles interviewed Leach's accuser at Fort Davis whom he was convinced was a gross exaggerator. The superintendent was addicted to gambling, had played a few hands with the Army officers at the fort, won and lost like a gentleman. What the informer reported about celebrations was based upon hearsay. After a brief investigation, the special inspector decided the charges of collusion with the mail contractors were likewise without foundation. Although he was unprepared to report on Leach's general ability, preliminary evidence indicated that every effort had been made to push the road work.[42]

Upon the arrival of the disbursing agent in Mesilla, a conflict between his statements and those of the superintendent convinced Sayles that he did not yet have the whole truth.[43] The investigator placed his assistant in a position to watch Leach's transactions. In time, the financial accounts were found to be not only unsatisfactory, but also highly irregular. Pending the inquiry, large groups of the working force remained in camp, becoming demoralized with drinking and gambling. Sayles resolved to break up the party, send Leach and Hutton over the road to San Diego to make final repairs, and to return to Washington by way of Santa Fe.[44]

After an interview with Secretary Thompson, Sayles carried written instructions to Theodore Sedgewick, United States Attorney in New York City, requesting the examination of certain vouchers which were doubtless fraudulent. The attorney was asked to bring the guilty to justice if the evidence so warranted.[45] The government moved against the accused parties immediately. Sayles located an associate of Woods, Charles F. Whitcomb, in Boston, and found him ready to make a full confession. In making purchases, Whitcomb had been instructed by the assistant superintendent to get various dealers to sign blank vouchers, to pencil the amounts paid for supplies in the margin, and forward them for Woods' completion. Whitcomb had kept a record of payments in a personal diary, and these entries proved that Woods had consistently claimed a greater amount than that expended. Woods was arrested on November 17, 1858, and bail placed at $10,000. For four weeks he was held *incommunicado* while the prosecution checked Whitcomb's story and gathered evidence.[46] Sayles left for Memphis to compare Woods' vouchers with the testimony of local merchants.

On the way the investigator stopped in New York long enough to discover that a voucher from Berford and Company, purporting to cover the freight and commissions for the shipment of wagons and other property to the Mississippi River rendezvous, was a forgery. The company had been paid $1,692 but the voucher in the inspector's possession claimed $3,406. More damaging evidence was uncovered in Memphis. Eleven dealers agreed that the amounts received for their animals or supplies did not coincide with the figures on the vouchers. By a systematic increase of funds claimed for each item, Woods had cleared $3,205.[47] Sayles also questioned two discharged laborers who stated that they had not received the amount recorded as paid.[48]

The mounting evidence appeared sufficient to convict Woods. Whitcomb insisted that Leach had not directly participated in any of the frauds. During the early stages of investigation, the superintendent's responsibility appeared to be only that of an executive for a dishonest subordinate. But he stood in great danger because of his ardent support of Woods. In Mesilla, the federal investigator had urged the assistant's dismissal along with other "worthless hangers-on" but Leach insisted on restoring him to the payroll. Sayles had to threaten to dismiss the superintendent to get rid of Woods.[49]

During Woods' confinement, Leach tried to cover up for him in the Berford and Company affair. An attempt was made to pay the company officials the money claimed to be outstanding, and Berford disappeared as a witness for the prosecution. In an interview with officers of the Abbot Company of Concord, famous wagon manufacturers, the federal investigators learned that Woods had taken the manager aside at the time of purchases for the road expedition to request that he add $25 to his price for each wagon, evidently with the idea of dividing the amount with the manufacturer. Woods stated that Leach knew this proposition would be made, but the manager was to say nothing to the superintendent who would have to deny it emphatically. In the closing moments of the sale, Leach was told of Woods' offer, and he disclaimed knowledge of any such arrangement.[50]

Support of Woods eventually resulted in Leach's indictment. On May 16, 1859, the foreman of a grand jury empanelled in New York City reported against Leach, who

> ... being an evil disposed person and contriving and intending to injure and defraud the said United States, and to cause and induce the said United States to pay to him large sums of money unlawfully, falsely, fraudulently and feloniously ... did ... present ... paper writings purporting to be vouchers of expenditures incurred.[51]

The prosecution submitted these vouchers, totaling over $10,000, that it claimed to be falsified or forged.

At the same time Woods was returned to Texas for trial. Most of his flagrant swindles had been committed while the ox train wintered in Fort Belknap, and indictment was thought to be easier in the federal judicial district where the illegal action had occurred. Charges before the grand jury were scheduled for June, 1859. Campbell, who had planned to testify, failed to put in an appearance, other scheduled witnesses were not on hand, and Woods escaped prosecution.[52]

Leach, in the meantime, returned to his California home. In preparing for his trial, the Interior Department wrote for the evidence used unsuccessfully against Woods in Texas. Because of the delay in receiving the necessary papers, the case, on the docket for the December session of the Criminal Court in the Washington District, had to be postponed until January, 1860.[53] Leach never stood trial. In March, 1861, Campbell wrote the district attorney handling the government's fraud suit against the superintendent that he was convinced a further continuance would not benefit the United States. The difficulty of procuring witnesses from the "remote parts" of Tennessee and Texas made an early trial impossible. The general superintendent concluded that "as this prosecution was made mainly at my instance, I unhesitatingly recommend a 'Nolle Pro' in the premises."[54] Campbell informed Leach that the Interior Department was notified, though not officially, that the Attorney General had secured a discontinuance of the suit.[55]

Clearance on the charges of fraud settled one problem for Leach. Shortages in the accounts presented another. In September, 1860, the auditors of the Treasury Department reported to Secretary Thompson that a final review of the superintendent's records revealed a shortage of $23,003.[56] A copy was forwarded to Leach in California, who immediately announced his intention to report in Washington, present explanations, and defend himself against all accusations.[57] In a desperate defense he forwarded numerous documents of explanation to the Interior Department, striving to reduce his financial liability. The Secretary wrote a memorandum on April 18, 1861, reviewing all disputed vouchers and following each entry recorded "allow," "disallow," or "explanation not satisfactory." The Department was prepared to compromise the matter. Leach and his bondsmen were to pay

$13,400 or risk a suit for the whole amount.[58] No evidence is available that the payment was ever made. Within a few weeks, the outbreak of the Civil War terminated all interest in the cash deficiency.

While the investigation of the financial chaos surrounding the El Paso–Fort Yuma wagon road expedition progressed, Secretary Thompson refrained from airing the scandal in public, as in the case of the central overland route and also that from the Minnesota frontier. In an attempt to salvage something from the threatening debacle, he contracted with Captain Charles P. Stone to improve and repair a section of the road between the Pima villages on the Gila and the mail station at Ojo Excavada. The contractor agreed to organize a working party of thirty men, provide them with tools, arms and ammunition, and provisions for five months' work in the field. The War Department agreed to furnish mule teams to haul provisions for the party and to dispatch an escort of soldiers for protection.[59] This was arranged in May, 1860. Stone spent $10,829 in repairs, primarily on the watering places. Thompson reported the task completed in his annual report. The federal appropriation was thus exhausted.[60]

James B. Leach joined Magraw and Nobles, superintendents of the other two overland wagon-road expeditions sponsored by the Interior Department, 1856–1860, as a failure. The appointments of each had been dictated by politics. All were frontiersmen with practical experience in exploration, road building, mail carrying, or freighting. In contrast with these superintendents, the engineers—Medary, Lander, Hutton—were comparatively successful. Each of them had constructed wagon roads with varying degrees of professional competence, and Lander emerged as one of the greatest road builders in the nation's history. The backgrounds of these men were similar to those of the superintendents, and the fact that they were chosen less as a result of political pressure does not explain their greater accomplishment. Apparently their task of road survey and improvement was an easier assignment than the management of a field party with its endless records and reports. Complexity also encouraged dishonesty among men so inclined.

Administration within the Department of the Interior was partly responsible for this dark chapter in the history of transportation. There was no tested organization to supervise these road projects; no experience upon which to rely. The War Department's Topographical Engineers, on the other hand, had training in both road building and bookkeeping, and through the Quartermaster Corps and other branches of service, established organizations existed to relieve them of the problem of logistics. This Department was somewhat freer from partisan political influence in handling personnel matters. Moreover, its officers respected the principle of the chain of command so necessary for field workers with the federal government.

CHAPTER XIV : *Interior Department Construction to Meet a Local Need: Road along the Missouri River, Nebraska Territory*

When the Pacific Wagon Road Office was established, a single local road project in Nebraska Territory, likewise assigned to the Department of the Interior, was placed under the jurisdiction of the general superintendent. Bird B. Chapman, Nebraska territorial delegate, had sponsored a bill through the Thirty-fourth Congress authorizing $30,000 to be spent on a road running from the mouth of "the Platte River, via the Omaha Reserve and Dakota City, to the Running Water [Niobrara] River."[1] Since both of these streams were tributaries of the Missouri River, and one side of the Indian reservation adjoined it, the route was certain to be located along the Missouri's banks in Nebraska.

Chapman had arrived in the territory from Ohio with the avowed purpose of building a political career for himself. Local politicians soon divided into two factions: "the Omaha clique," representing the interests of those north of the Platte River and the "Nebraska City clique," those to the south. In a spirited contest where blocs of votes were counted in, and then out, and in again, Chapman had obtained the delegate's seat in the Thirty-fourth Congress. The contested election had been settled in his favor by the national legislature.[2] The wagon-road appropriation was one reward for his constituents in Omaha and the Missouri River settlements just north of the Platte.

The Nebraska delegate wrote the Secretary of the Interior, Jacob Thompson, that the territory would benefit greatly if the road went along the west bank of the Missouri River for the 120 miles from the Platte Valley to Dakota City, connecting the various towns. Beyond this point it should run northwesterly, still paralleling, in general, the river valley for 130 miles to the Niobrara. Supplies and equipment would be available in the settled country as far as the Omaha Reserve; beyond that was Indian country. The chief improvements necessary, besides grading, would be bridges over the streams pouring into the Missouri River.[3]

The superintendency of this federal road was assigned to George L. Sites of Fort Wayne, Indiana, who had attended the Cincinnati convention of the Democratic party and worked for the nomination of James Buchanan of Pennsylvania. In the 1856 presidential campaign he had written a pamphlet entitled "Fremont and His Supporters—Their Record," which had circulated widely in Indiana, Illinois, Ohio, and Pennsylvania, aiding the election of Buchanan. Sites was not only recommended to the Interior Secretary as an ardent and devoted Democrat whose appointment would be "gratifying to the Democracy of Indiana," but also to the President as a man who helped put him in the chair.[4] The Democrats of Indiana, through their spokesmen Senator Graham N. Fitch and Thomas A. Hendricks, commissioner of the general land office, also claimed the chief engineer's assignment for the former principal clerk of the Indiana House of Representatives.[5] A post more suitable to the clerk's qualifications was found elsewhere, and delegate Chapman, who had been disappointed in failing to secure the superintendency for an Omaha friend, prevailed upon Thompson to name Henry B. Smyth, a Missouri resident, as chief engineer.[6] Smyth had earlier been endorsed by John B. Floyd in the War Department.[7] Political considerations again dominated the selection of personnel; patronage was inconsequential, however, due to the comparative smallness of the project and the appropriation.

In sending appointments and instructions to Sites and Smyth, Secretary Thompson ordered them to the Platte River for a rapid survey of the route to determine its proper location and to ascertain the number of bridges required. Economy was essential on the reconnaissance to conserve the appropriation for actual improvements. On this initial trip a few saddle horses and a single wagon to handle the surveying instruments were believed sufficient. Congress had not contemplated a thoroughly graded and bridged highway to be supported by future appropriations, but only a road to meet the immediate wants of the settlers. Once the survey and estimates were completed, the Interior Department would provide instructions for construction. Perhaps local labor could be relied upon, with teams and wagons hired for specific jobs. If it should prove desirable, the whole road might be placed under contract.[8]

The superintendent established headquarters at Bellevue, Nebraska, early in June, and during a two-weeks wait for the chief engineer, examined the country between Omaha and the Platte River mouth. For this job a transport wagon, surveying instruments, harness, and saddles had been purchased in St. Louis. Sites chose to hire labor at a monthly wage and to furnish subsistence for the crew. Rather than renting wagon teams for hauling supplies and doing road work, he preferred to buy mules outright and dispose of them when the project closed.[9] Having worked as a clerk and auditor in the Treasury Department, he knew the wisdom of obtaining specific authorization from Washington. General superintendent Campbell, somewhat impatient with inquiries about the details of organization, urged discretion in such matters, but warned against using funds for too much paraphernalia rather than for road work.[10]

By July 10, 1857, the reconnaissance party was at Dakota City, 105 miles from the Platte. The survey had been run through the settlements of Omaha, Florence, Fort Calhoun, De Soto, Cumming City, and Decatur, all along the Missouri River bottom lands. Engineer Smyth, who submitted a map of the route, reported the necessity of bridging twenty-four streams, with structures varying from ten to 75 feet in length, at a total cost of $11,725. Where streams were too deep to ford and excessively expensive to bridge, the Department proposed construction of flat boats to take travelers' wagons across the water.[11] Sites was perturbed over the proper location of a ferry across the Platte River.[12]

A month later the preliminary survey was completed between Dakota City and Niobrara, 130 additional miles. From the "Serpentine Bend" of the Missouri where the river changes its course from a direction south of east to east of south, the surveyors crossed the bluffs to Elk Creek. Although they followed in general the river's course, once they crossed the bluffs the trace was not directed by the meanderings of the stream. As the country was still Indian domain, no settlements had to be included on the march. Moving through the Lime Creek hills, across the various branches of Bow Creek, to the east branch of Bazille Creek, the engineers followed the latter stream to its junction with the Missouri and thence five miles along the bottoms to the mouth of the Niobrara.[13] Superintendent Sites reported: "... from the character of the country, the comparatively few bridges to construct, the feasibility of a good road without grading, and being almost direct, I cannot hesitate to respectfully recommend the location of the road upon or near the route passed over from Neobrara to Dacota City [*sic*]."[14] On this western end, sixteen bridges from ten to 100 feet, costing $4,850, were recommended. All cost estimates were based upon the assumption that laborers would make the improvements under superintendence. The esti-

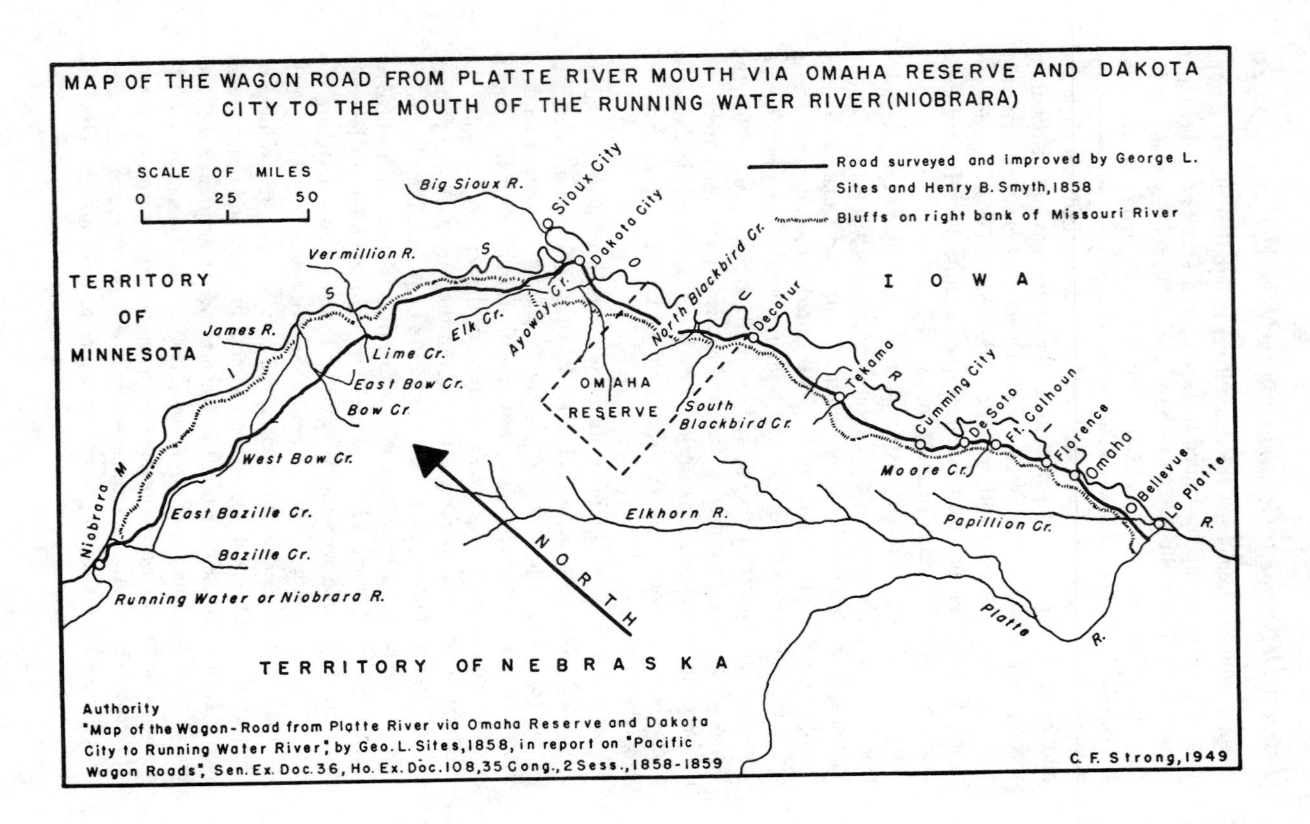
MAP OF THE WAGON ROAD FROM PLATTE RIVER MOUTH VIA OMAHA RESERVE AND DAKOTA CITY TO THE MOUTH OF THE RUNNING WATER RIVER (NIOBRARA)
SCALE OF MILES
0
25
50
Road surveyed and improved by George L. Sites and Henry B. Smyth, 1858
Bluffs on right bank of Missouri River
TERRITORY OF MINNESOTA
IOWA
TERRITORY OF NEBRASKA
NORTH
Big Sioux R.
Sioux City
Dakota City
Vermillion R.
James R.
Lime Cr.
East Bow Cr.
Bow Cr.
Elk Cr.
Ayoway Cr.
North Blackbird Cr.
Decatur
OMAHA RESERVE
South Blackbird Cr.
Tekama
Cumming City
De Soto
Ft. Calhoun
Florence
Omaha
Bellevue
La Platte
Moore Cr.
Papillion Cr.
Elkhorn R.
Platte R.
West Bow Cr.
East Bazille Cr.
Bazille Cr.
Niobrara
Running Water or Niobrara R.
Authority
"Map of the Wagon-Road from Platte River via Omaha Reserve and Dakota City to Running Water River," by Geo. L. Sites, 1858, in report on "Pacific Wagon Roads", Sen. Ex. Doc. 36, Ho. Ex. Doc. 108, 35 Cong., 2 Sess., 1858-1859
C. F. Strong, 1949

mates would need to be doubled to induce local contractors to undertake the work, "for in this country they expect to make a small fortune in every contract with the government."[15]

August and September were spent in building eleven bridges between the Platte and Dakota City and setting up the frames for thirteen more. When the October rains came, Sites asked for instructions for future operations, proposing at the same time to discharge the labor force for the winter, store the property at Bellevue, and send the horses and mules to southern Iowa or Missouri.[16] In answer to this inquiry about continuing the following spring, the Department expressed doubt that available funds would make operations advisable. When the cold weather set in, the party was to be disbanded, all property disposed of, and the two officers to report in Washington, D.C.[17]

Recognizing that the season's work had been faithfully performed, and that almost $12,000 remained in the road appropriation, Secretary Thompson ordered Sites and Smyth to return to Nebraska in the spring of 1858. Bridges were to be constructed between Dakota City and Niobrara, and the road graded throughout its entire distance.[18] Arriving in the territory in April, the road builders gave first attention to improving the road into and out of Omaha by building several small bridges across sloughs, ditches, and ravines. The bridges erected during the previous season were in good condition until incessant and unprecedented rains came in late May and threatened every structure along the line. The two longest bridges, across branches of Blackbird Creek in the Omaha Reserve, collapsed under the pressure of driftwood and high water. The next disaster occurred at Omaha Creek, lying to the north of the reserve. Subsequent high water took out three more bridges. In each instance, a party of laborers attempted to disassemble the structures to keep the timber from being carried downstream. Reconstruction was made where possible, and the officials apparently judged the events no reflection upon engineering skill: "Considering the great amount of damage done by high water through western Iowa and Missouri, where scarcely a bridge was left standing, it should be a matter of congratulation that the bridges upon this road escaped with comparatively so small a loss."[19] Engineer Smyth and his working party, in the meantime, erected twenty-four bridges in all, which ranged from ten to 78 feet in length across the streams on the western sector of the route. The excavation and grading necessary to make the road passable and permanent were finished by early August.[20]

This federal road construction was a major interest of Nebraska residents. The territorial governor had called upon the council of the Legislative

Assembly to "examine into and report the facts," about the project. A select committee report, unanimously adopted by the council, revealed that the federal government had built for a local need:

The entire length of this road is two hundred and eight miles [?], passing through a country of unsurpassed beauty and fertility; rich in mineral wealth. Its course being almost parallel with the Missouri river, it crossed all the streams putting into said river. Being eligibly [*sic*] located, the road is of incalculable benefit and importance to the Territory, and its advantages can only be properly appreciated by the emigrant and hardy pioneer as he wends his way westward in quest of a home he intends to reclaim from the possession of the red man, and improve, beautify, and adorn as a resting place for himself and those dependent upon his exertion and labor.[21]

The legislators had specific recommendations for additional improvements, including the erection of fifteen bridges that would necessitate a $15,000 federal appropriation. The committee recommended the adoption of a joint memorial and resolution asking Congress for this aid, the bridging of the Niobrara River, and an extension of the road to the military post at Fort Randall.[22]

The superintendent endorsed their request for further support:

... [The road] will be of incalculable benefit to the Territory, and will induce an earlier settlement and development of the country. The pecuniary condition of the people of the Territory, brought on by the exhorbitant prices they were compelled to pay for the necessaries of life, will not warrant an undertaking on their part to construct the bridges required for the accommodation of themselves, and indispensable to the emigrant; besides, some of these bridges are remote from the settlements, and only of benefit to the traveller or emigrant.[23]

During the summer of 1858, seventy families moving their belongings in wagons and driving their stock, had settled in the valley of the Niobrara. Sites was of the opinion that "these hardy pioneers, while they brave the dangers and hardships of a frontier life, extending settlement and cultivation, require and should receive all the benefits that the government can consistently bestow."[24]

Although no additional federal support was forthcoming, general superintendent Campbell, in preparing his report on the Pacific Wagon Roads, could say with confidence, "This road ... appears to have given great satisfaction to the people of Nebraska."[25] The businesslike efficiency surrounding its construction provided a short but bright chapter in the dismal story of federal road building by the Interior Department in the 1850's. The wagon road remained the main highway from the Platte to the Niobrara River until railroad connections were made.[26]

Part III

THE ARMY CONTINUES TO BUILD ROADS WEST

CHAPTER XV : *Beale's Wagon Road along the Thirty-fifth Parallel: Fort Smith to the Colorado River, 1857-1859*

ALTHOUGH THE THIRTY-FOURTH CONGRESS attempted to transfer all federal wagon-road construction to the Interior Department, the War Department continued in the business as a result of a clerical omission in the appropriation bill for the overland roads to the Pacific. The section in this law authorizing $50,000 for a road along the thirty-fifth parallel from Fort Defiance, New Mexico, to the Colorado River failed to specify that construction was to be supervised by the Secretary of the Interior.[1] The Secretary, moreover, was not as pleased to assume this new responsibility for road building as the Congress apparently had been in transferring it to him. Applications pouring into the Interior Department for positions on the project were ignored.[2] The Interior Secretary used the ambiguity in the law to leave the construction under the direction of the Secretary of War. When the result of their oversight in drafting the legislation became evident, Miguel O. Otero, New Mexico delegate, and John S. Phelps, of Missouri, appealed directly to President Buchanan, assuring him that they intended to have the road constructed under the same authority as the others.[3] Their plea was to no avail.

The War Department's experience in exploration, survey, and road building had been continuous. The road reconnaissances of Stansbury to the Salt Lake Basin, of Simpson between the Arkansas frontier and Santa Fe, and of Bryan across the plains of Kansas and Nebraska had proved the ability of the Army Engineers to organize on a grand scale. Wagon-road surveys were, in a sense, a preliminary stage in the evolutionary process, culminating in the location and construction of the transcontinental railroads. The significant interrelationship between wagon-road and railroad construction had been recognized as early as 1849 by men of vision like Thomas Hart

Benton of Missouri, who introduced a bill in the Senate to locate and build a national road to the Pacific. On the central route he proposed, iron railways were to be laid where practicable and advantageous; but along other sections, unsuited for rails, a macadamized road would suffice temporarily. Paralleling any railroad across the center of the continent, a right of way 100 feet wide was to be reserved as a common road for wheel carriages, horse and foot travelers, free from tolls or charges.[4]

The Pacific Railroad Convention meeting in St. Louis this same year not only adopted a resolution urging Congress to build a railroad to the Pacific as a national project, but petitioned that body to authorize an immediate survey of one or more practical roads with facilities for emigrant travel across the continent. Provision for a reconnaissance of the nation's possessions west of the Mississippi by the Topographical Engineers to locate such roads was thought imperative.[5] A month later, in Memphis, another railroad promotion meeting suggested three methods of increasing transportation and communication with the Pacific Coast: a transcontinental railroad, a ship canal or railroad through Mexico and Central America, and a military road along the Mexican frontier.[6] Construction of the wagon road was declared not only feasible but essential for the transportation of troops and munitions of war across the new Mexican cession.

Men of science as well as railroad enthusiasts agreed that a wagon road would be the precursor of a transcontinental railroad. In 1851 when the House Committee on Post Office and Post Roads was reviewing a proposal to construct a mail route from Fort Smith to San Diego, Matthew F. Maury, head of the Naval Observatory and an influential member of national scientific circles, was called on to testify as to the type of western transportation improvements the federal government might attempt. As a member of the St. Louis convention and a correspondent of journals advocating a Pacific railroad, Maury was known to support federal aid to transportation.[7] He reported to the committee:

> Ultimately I look to a railroad across the country to California, but it will be some years at least before that can be completed. In the mean time we want a good wagon road; a road over which the western emigrant and fortune-seeker may haul his stuff, drive his cattle, and find, in a well-established line of military posts, security against Indians and banditti.[8]

In response to pressure generated by the combined desires of ardent nationalists, sectional politicians, and railroad promoters, Congress, in 1853, appropriated $150,000 for Pacific railroad surveys to be directed by the War Department. This legislation resulted in the first attempt of the government to conduct a comprehensive and systematic examination of the vast region lying between the Mississippi River and the Pacific Ocean.[9] Six explorations were made between 1853 and 1855. The assignment marked no radical departure from the tasks customarily given to the Army. These so-called surveys were reconnaissances similar to those conducted earlier by the Topographical Engineers; they were made as rapidly and covered as wide an area.

In the selection of personnel, Jefferson Davis was guided by the advice of Colonel John J. Abert, chief of the Topographical Bureau. Orders to the officers in charge of the railroad surveys, for the most part topographical engineers, contained virtually the same procedural instruction, even the same terminology, of those sent to the earlier wagon-road builders of the Corps. An identical plan of organization and similar list of equipment for each railroad expedition, drawn up by the Bureau chief, was based upon the experience gained by the officers of his Corps on their earlier western reconnaissances. The Pacific railroad surveys gained much greater publicity, and have been thought more important historically, because of the simultaneous activity of several explorations, and because of the political and sectional significance attached to the selection of a Pacific railroad site. But for the Army, the surveys represented not an innovation, but an acceleration of the program of exploration to improve transportation routes. In summarizing the activities of the several railway expeditions, Professor Robert R. Russel wrote: "The main wagon train in each case took a route determined upon from the information furnished by earlier explorations or reports brought in from day to day by reconnoitering parties sent in advance. This route had to be immediately practicable for wagons whether or not it seemed practicable for a railroad."[10]

The chain relationship between topographical exploration, wagon-road surveys, and railroad building was recognized by each Army officer directing a railroad exploration. Captain John Pope, in command of the thirty-second parallel railroad survey from the Red River to the Rio Grande River thought: "The construction of a railroad across the plains necessarily presumes the establishment of a wagon road along the route ..."[11] Lieutenant John G. Parke, who first examined the western sector of the thirty-second parallel route between January and March, 1854, and repeated his survey to seek a better way, in May and June, 1855, pointed out the advantages of this terrain for a mail road. He reported: "In considering the adaptability of this country to the establishment of a post route, extending from the Mississippi to the Pacific, the advantages ... are about three hundred miles of hard and smooth road, resembling a macadamized, and almost equal to a plank road."[12]

After declaring this "the most practicable and economical route for a railroad from the Mississippi river to the Pacific ocean," Secretary of War Davis observed:

> ... it appears practicable to obtain, at a small expense, a good wagon road, supplied with water by common wells, from the Rio Grande down the San Pedro and Gila and across the Colorado desert. Such a road would be of great advantage in military operations, would facilitate the transportation of mail across that country, and relieve emigrants pursuing that route from much of the difficulty and suffering which they now encounter.[13]

In short, the construction of a wagon road at federal expense along the preferred route might serve as a stimulus to rail construction.

To the north along the thirty-fifth parallel, Captain Amiel W. Whipple commanded another railroad survey. When traveling over the Simpson-Marcy route from Fort Smith to Santa Fe, he stated in his journal, as had his predecessors in the Topographical Corps, "by the route that we examined a wagon route could easily be cut."[14] Summarizing the accomplishments of this expedition in November, 1856, Captain Andrew A. Humphreys, in charge of the Pacific Railroad Surveys and Explorations Office, suggested: "The construction of the wagon road from Fort Defiance to the Colorado river will probably solve the question of the railroad practicability of the line..."[15]

When the evidence about the various and sundry routes to the Pacific was all reported, a political deadlock in the Thirty-fourth Congress prevented the authorization of construction anywhere. The Pacific Wagon Road program, approved as a substitute, was accepted by the majority in Congress as a stop-gap measure to placate those on the Pacific Coast and along the Mississippi River, who clamored for better communication facilities. At the same time, the program was a continuation of a national policy whereby the federal government assumed the obligation for improving transportation in the territories of the trans-Mississippi West. The new road program under the Interior Department was not a diversion in federal policy but only in administrative responsibility. By securing control of the construction of the wagon road along the thirty-fifth parallel, the United States Army preserved the continuity of its role as the nation's chief road builder.

The man chosen by John B. Floyd to improve this route was Edward Fitzgerald Beale, who had been among those instrumental in engaging the interest of Jefferson Davis, the previous Secretary of War, in an experiment with camels as beasts of burden in the Southwest.[16] Beale had first sailed to the Pacific Coast on the frigate *Congress* as a young naval officer under the command of Commodore Robert F. Stockton. The vessel arrived in

Monterey after hostilities with Mexico had broken out, and Beale forthwith joined a small detachment to march overland to meet Kearny's "Army of the West."[17] The groups united just before the disastrous battle at San Pasqual. After a display of gallantry on the battlefield, Beale was dispatched with Kit Carson to sneak through the Mexican lines and seek relief for Kearny's men from Commodore Stockton in San Diego.

Still suffering from the wounds of this engagement, Beale went east two months later where he served as a witness for John C. Frémont, at the Pathfinder's trial for insubordination in October, 1847. During the next two years, Beale made six journeys across the continent from ocean to ocean. On the second trip additional fame was gained when he brought the first authentic and official news of gold discovery in California and a bag of the precious metal to Washington, D.C. Commissioned a lieutenant in the Navy in August, 1850, Beale resigned within a few months to take up residence in California as manager of the estates of W. H. Aspinwall and Commodore Stockton. Two years later President Fillmore appointed him Superintendent of Indian Affairs for California and Nevada. On his return to the Pacific Coast, May-August, 1853, he traveled with Gwinn Harris Heap and a party of twelve across southern Colorado and Utah, making a preliminary survey for a railroad. When his appointment as Indian superintendent terminated, Beale was commissioned a brigadier general of the California militia. In the meantime, he had demonstrated enthusiasm for the novel proposal to organize a camel corps for the transportation of provisions to the Army posts in the Southwest. The purchase of camels was authorized by an act of Congress in 1855 and, according to Stephen Bonsal, Beale's biographer, seventy-six camels were brought to Indianola, Texas, in two separate voyages in 1856. Beale's appointment as wagon-road superintendent allowed the merging of the camel experiment with road building.[18]

Leaving San Antonio on June 25, 1857, with his wagon train and camel caravan, Beale traveled along the lower military road through the settlements of Castroville and Blacksburg, and on across the Devil's River. The party ascended the valley of the Pecos River as far as Fort Lancaster at the mouth of Live Oak Creek. Crossing the river opposite the fort, Beale and his men went up the stream for a single day's march before striking westward by way of Comanche Spring, through Wild Rose Pass, and into Fort Davis. The military road from this outpost to El Paso, via Van Horn's Wells and Eagle Springs, was followed to the Rio Grande Valley. The river was then ascended to Albuquerque, where the expedition arrived August 10. On this six-weeks' march the commander had gained a reputation for being a strict disciplinarian, forcing his men to break camp before daybreak, to march until mid-morning, and after a short rest for food, to resume forward progress until late afternoon.[19]

At the outset of the journey, several of the camels had become ill and had to be relieved of their loads. Beale recorded with anxiety: "Thus far the camels have not been able to keep up with the wagons, but I trust they will prove better travelers as they become more accustomed to the road."[20] By the time the trans-Pecos Texas country was reached the camels began to prove their mettle and the superintendent joyously recorded:

> The camels arrived nearly as soon as we did. It is a subject of constant surprise and remark to all of us, how their feet can possibly stand the character of the road we have been travelling over for the last ten days. It is certainly the hardest road on the feet of barefooted animals I have ever known. As for food, they live on anything, and thrive. Yesterday they drank water for the first time in twenty-six hours . . . Mark the difference between them and mules; the same time, in such weather without water, would set the latter wild, and render them nearly useless, if not entirely break them down.[21]

The fame of the camels preceded the wagon-road expedition as it moved up the Rio Grande Valley. In one Indian village near Albuquerque, Beale, traveling in a bright red ambulance wagon and driving camels, was mistaken for the head showman of a circus. As crowds gathered, Beale indulged his humor by assuring the doubtful that a circus had come to town, equipped with a monkey and horse act.[22]

The expedition was detained in the Albuquerque vicinity for several days, awaiting the arrival of a military escort. Beale's impatience knew no bounds, and when news of the detachment's approach was received, he mounted his favorite white dromedary, "Seid," and rode across the country at top speed to greet the commander. The wagon-road party was sent forward to the Zuñi villages on the immediate line of the proposed survey. To comply with the congressional stipulation that the eastern terminus be at Fort Defiance, Beale, accompanied by twenty men, rode northward to that post in the Canyon Bonito of the Navajo country, 180 miles west of Santa Fe. His arrival at the fort to start officially the wagon-road reconnaissance was an occasion for celebration:

> As we stood in the warm sun of August, it was most refreshing to see the captain's servant throw off the folds of a blanket from a tub in the bottom of the wagon, and expose several large and glistening blocks of ice, while at the same time the captain produced a delicate flask of "red eye".[23]

On the afternoon of August 27, everything was in readiness and Beale led his mounted men out of the fort into the wilderness. He recorded his emotions in his journal.

No one who has not commanded an expedition of this kind, where everything ahead is dim, uncertain, and unknown, except the dangers, can imagine the anxiety with which I start upon my journey. . . . Today commences it.[24]

The detachment first returned to the Zuñi villages to join the wagon train and camel corps, and the combined party then moved westward to Jacobs Well, a familiar watering place, encamping there for the night. Traveling along the thirty-fifth parallel to the Rio Puerco of the West, the expedition periodically found the trail of Captain Whipple, and traveled along it when possible. The valley of the Puerco was followed to its junction with the Little Colorado. The latter stream could not be descended because of a narrow defile known as Cañon Diablo (Devil's Canyon) which blocked the passage of wagons. A route paralleling the stream was followed. Near the northern outlet of the canyon, the expedition left the river valley and bore due west, searching for water, encamping when it was found, and marking the trail with the wheels of the wagons. Their route westward by way of Leroux and Lewis Springs to Mount Floyd was just north of the present towns of Flagstaff and Williams, Arizona. West of Mount Floyd, the direction of travel was slightly north of west, to take advantage of the waters of Gabriel Spring. Turning southwest from this encampment, the explorers reached the Colorado River above present-day Needles, California, at the southernmost tip of the present state of Nevada.[25]

Beale's trip to the Colorado River was an exploring expedition similar to that of Whipple. The road superintendent's primary task was to seek out terrain passable for wagons. At the journey's outset, he recorded that "the country looks open and promises a level road."[26] Beyond Leroux Springs he noted: ". . . Today . . . we have made nineteen (19) miles. Could any amount of writing say more for a road?"[27] Only on rare occasions were improvements made, and then the obstacles to direct travel were overcome by temporary expedients. In the valley of the Little Colorado, the party ". . . came to a mesa or table-land, the ascent to which occasioned some delay, as it was necessary to cut down the hill before our wagons could go up."[28] Just before reaching the Colorado River, an entire morning was spent cutting down the grades across a mountain, along which the route passed for three-quarters of a mile. The commander warned: "Emigrants cannot pass here until the hill is worked. I estimate the expense of making this mountain pass a good one, and a good road for emigrants, at five thousand dollars."[29]

While road work was kept to a minimum, Beale was preoccupied with the camel experiment. At the start a large majority of the men did not share their leader's fervor and scouted the idea of the animals going on the exploration. This view prevailed among the rank and file all the way to Fort

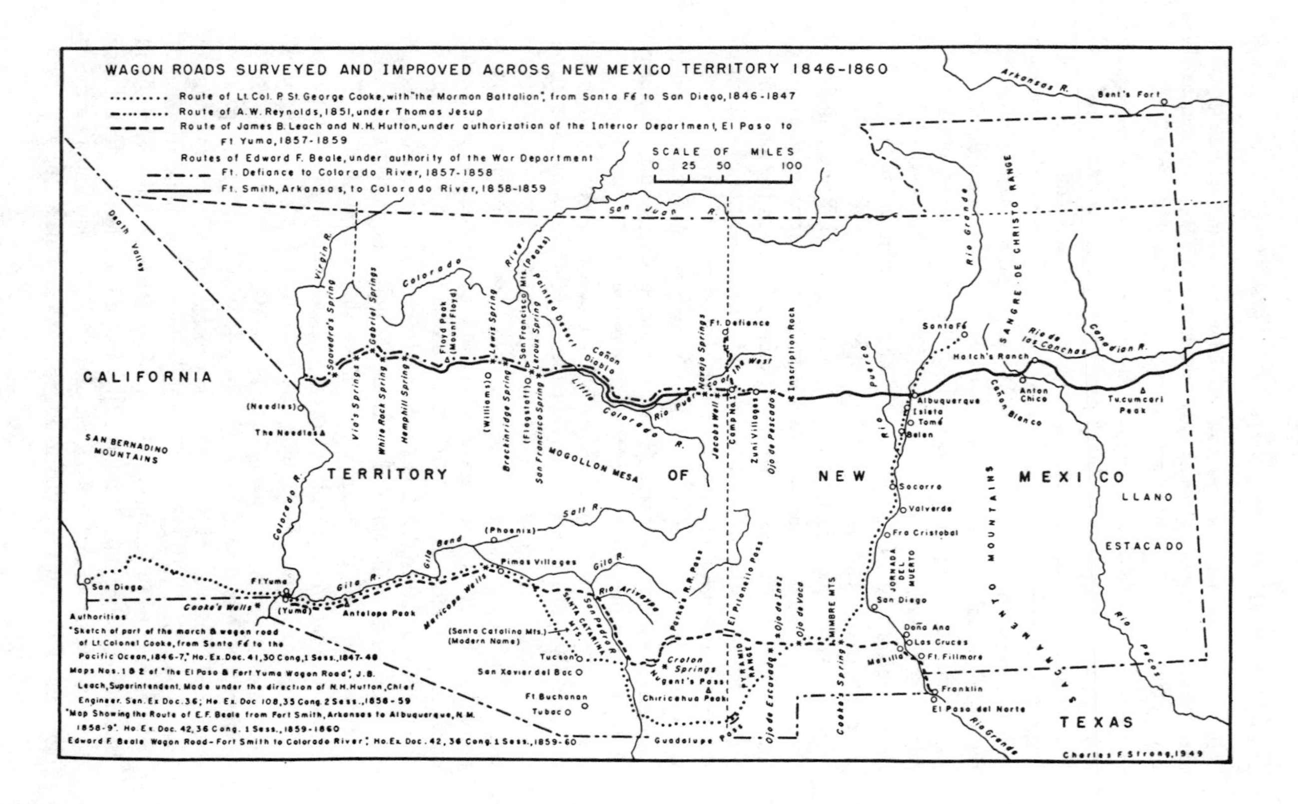

WAGON ROADS SURVEYED AND IMPROVED ACROSS NEW MEXICO TERRITORY 1846-1860
Route of Lt. Col. P. St. George Cooke, with "the Mormon Battalion", from Santa Fé to San Diego, 1846-1847
Route of A. W. Reynolds, 1851, under Thomas Jesup
Route of James B. Leach and N. H. Hutton, under authorization of the Interior Department, El Paso to Ft Yuma, 1857-1859
Routes of Edward F. Beale, under authority of the War Department
Ft. Defiance to Colorado River, 1857-1858
Ft. Smith, Arkansas, to Colorado River, 1858-1859
SCALE OF MILES
0 25 50 100
Arkansas R.
Bent's Fort
Death Valley
CALIFORNIA
SAN BERNADINO MOUNTAINS
(Needles)
The Needles
Virgin R.
Colorado
Colorado R.
San Juan R.
Rio Grande
SANGRE DE CHRISTO RANGE
Rio de las Conchas
Canadian R.
Santa Fé
Hatch's Ranch
Anton Chico
Cañon Blanco
Tucumcari Peak
Albuquerque
Isleta
Tomé
Belen
Rio Puerco
Saavedra's Spring
Vio's Springs
Gabriel Springs
White Rock Spring
Hemphill Spring
Floyd Peak (Mount Floyd)
(Williams)
Lewis Spring
Breckinridge Spring
(Flagstaff)
San Francisco Mts. (Peaks)
Leroux Spring
San Francisco Spring
Painted Desert
Cañon Diablo
Little Colorado R.
Rio Puerco of the West
Navajo Springs
Ft. Defiance
Jacobs Well
Camp Nal
Zuni Villages
Ojo de Pescado
Inscription Rock
River
MOGOLLON MESA
TERRITORY OF NEW MEXICO
LLANO ESTACADO
SACRAMENTO MOUNTAINS
Socorro
Valverde
Fra Cristobal
JORNADA DEL MUERTO
San Diego
Doña Ana
Las Cruces
Ft. Fillmore
Mesilla
Franklin
El Paso del Norte
Rio Grande
Rio Pecos
TEXAS
Salt R.
(Phoenix)
Gila Bend
Pimas Villages
Gila R.
Maricopa Wells
Ft Yuma
(Yuma)
Antelope Peak
Cooke's Wells
San Diego
Rio Arivaypa
San Pedro R.
SANTA CATARINA MTS.
(Santa Catalina Mts.) (Modern Name)
Tucson
San Xavier del Bac
Ft Buchanan
Tubac
Parke's R.R. Pass
El Picacho Pass
Croton Springs
Nugent's Pass
Chiricahua Peak
PYRAMID RANGE
Guadelupe Pass
Ojo de Escavada
Ojo de Inez
Ojo de Vaca
MIMBRE MTS
Cooke's Spring
Authorities
"Sketch of part of the march & wagon road of Lt. Colonel Cooke, from Santa Fé to the Pacific Ocean, 1846-7," Ho. Ex. Doc. 41, 30 Cong. 1 Sess., 1847-48
Maps Nos. 1 & 2 of "the El Paso & Fort Yuma Wagon Road", J. B. Leach, Superintendent. Made under the direction of N. H. Hutton, Chief Engineer. Sen. Ex. Doc. 36; He Ex. Doc 108, 35 Cong. 2 Sess., 1858-59
"Map Showing the Route of E. F. Beale from Fort Smith, Arkansas to Albuquerque, N. M. 1858-9". Ho. Ex. Doc. 42, 36 Cong. 1 Sess., 1859-1860
Edward F. Beale Wagon Road - Fort Smith to Colorado River", Ho. Ex. Doc. 42, 36 Cong. 1 Sess., 1859-60
Charles F. Strong, 1949

Davis. In time their doubt turned to acceptance and finally to enthusiasm. Two weeks after the expedition left Fort Defiance, Beale noticed that the camels were so quiet and gave the party so little trouble that they were forgotten. He recorded that "certainly there never was anything so patient and enduring and so little troublesome as this noble animal.... There is not a man in camp who is not delighted with them."[30] The superintendent's personal zeal mounted to its zenith by the time the westward march was half complete:

My admiration for the camels increases daily with my experience of them. The harder the test they are put to the more fully they seem to justify all that can be said of them. They pack water for others four days under the hot sun and never get a drop; they pack heavy burdens of corn and oats for months and never get a grain; ... No one could do justice to their merits or value in expeditions of this kind, and I look forward to the day when every mail route across the continent will be conducted and worked altogether with this economical and noble brute.[31]

When the expedition approached the Colorado River, Beale was firmly convinced that the experiment was a success. He admitted:

I rarely think of mentioning the camels now. It is so universally acknowledged in camp, even by those who were most opposed to them at first, that they are the salt of the party and the noblest brute alive, that to mention them at all would only be to repeat what I have so often said of them before.[32]

The leader's satisfaction compensated in part for his disappointment and thorough disgust with the guide, Leco. Just west of the Little Colorado River in the San Francisco Mountains, the party became uncertain of its whereabouts and its future course. The guide proved of little assistance and Beale, not a man of patience, exploded:

We unfortunately have no guide, the wretch I employed at the urgent request and advice of every one in Albuquerque, and at enormous wages, being the most ignorant and irresolute old ass extant.

This obliges us to do the double duty of road making and exploring, which is very arduous, besides adding infinitely to my anxiety and responsibility.[33]

The man proved so worthless that the commander was finally obliged to send him to the rear of the expedition, and only regretted that he had not done it sooner. A few days later Leco himself strayed from the main party, lost his mule when alighting to make a fire, and, fearful of further punishment, chose to follow the mule rather than to return to camp on foot. Forty-eight hours later a search party brought him in.[34] Henceforth he was known as the "miserable Leco."

Near the Colorado, the Indians from the Mohave villages came forth to meet the expedition, first in groups of two and three and then by the dozens. Soon the wagon-road party was surrounded, and one Indian who had learned a few words of English from soldiers at the nearest military post, 250 miles distant, became spokesman by saluting Beale with the greeting, "God damn my soul eyes. How de do! How de do!"[35]

After crossing the river, the camel corps and wagon train were sent to the military post at Fort Tejon, 90 miles north of Los Angeles, where a near-by ranch was Beale's personal property. The road superintendent followed the United States surveyor's trail into Los Angeles before returning to the fort to recruit and refit the party. On arrival, he ordered the camels placed in camp a few hundred yards from the summit of the Sierra Nevada to test their ability to withstand cold. They thrived in two and three feet of snow. When six mules were unable to extricate a supply wagon caught in a sudden snow storm, the camels were dispatched to the rescue and brought the load through snow and ice to the expedition's camp.[36]

Beale determined to return over his route in mid-winter to test its feasibility for winter transit. The force was reduced to twenty picked men who could travel rapidly. They crossed the Colorado River aboard a steamboat, *General Jesup,* on January 23, 1858. Within a month, the detachment was at its initial camp site south of Fort Defiance. The wagon wheels of the summer trip clearly defined the road traversed. Old camping places of the outward march and of the Whipple party were periodically passed; Beale recalled the anxious hours spent there when the way ahead had been unknown. At the journey's end, he wrote in his journal:

> A year in the wilderness ended! During this time I have conducted my party from the Gulf of Mexico to the shores of the Pacific ocean, and back again to the eastern terminus of the road, through a country for a great part entirely unknown, and inhabited by hostile Indians, without the loss of a man. I have tested the value of the camels, marked a new road to the Pacific, and travelled 4,000 miles without an accident.[37]

Like other wagon-road surveyors, Beale became an ardent advocate for improvements along his route. The problem of inadequate water was to be solved by the construction of dams across the ravines and canyons at short intervals along the entire road. A $100,000 appropriation was recommended to build bridges, cut off elbows to straighten the road, provide a permanent water supply, and continue exploration during 1859. According to the superintendent, the public lands sold along the road's right of way, once it was opened, would certainly repay the appropriation fourfold within a three-year period.

Beale assured the Secretary of War that no further doubt could exist about the practicability of the country near the thirty-fifth parallel for a wagon road. Although he and Whipple did not travel precisely the same line with their wagons, they went through much the same country. Beale was certain this would "inevitably become the great emigrant road to California," and since he had successfully driven two hundred and fifty head of sheep on his expedition, the route was likewise destined, in his opinion, to be utilized by all the stock drivers of New Mexico pushing droves to the West Coast. The optimistic explorer had no personal doubt that the "whole emigration to the Pacific Coast would pursue this one line" and asked the government to improve it for the emigrant and stock drivers rather than scatter the federal government's endeavors as well as the emigrants over a dozen different routes.[38]

New Mexico's delegate, Otero, introduced a bill into the Thirty-fifth Congress to appropriate the funds Beale had recommended. The legislation, referred to the Military Affairs Committee of the House, received its sanction, but on May 14, 1858, the House of Representatives rejected the measure. Simultaneously an interested group of senators sought action to authorize further improvements along the thirty-fifth parallel. Robert W. Johnson of Arkansas obtained unanimous consent to a resolution calling upon the Committee on Military Affairs and Militia to investigate the advantages of opening a road from Fort Smith, Arkansas, to Albuquerque, New Mexico. On May 26 Senator Trusten Polk of Missouri introduced a bill for completing the Albuquerque–Colorado River road. Although Stephen Douglas' territorial committee reported the latter measure to the floor with a recommendation that it pass, the Senate failed to take action.[39] The War Department succeeded in its objectives, however, by including provisions in the Army Appropriation Act, 1858, for $50,000 to construct bridges and improve the crossings of streams on the road from Fort Smith to Albuquerque, and $100,000 to complete the road west from Albuquerque along the thirty-fifth parallel.[40]

Beale was retained to improve the line explored the previous season. By mid-October, 1858, he was at the rendezvous in Fort Smith, directing preparations for the westward journey. An assistant was dispatched to Fort Arbuckle to secure the services of the Delaware Indian, Black Beaver, and Jesse Chisholm, the half-breed Cherokee, whom Beale and others considered the best qualified guides in the Southwest. At first neither agreed to accompany the expedition because the Comanches were known to be on the warpath, but Beale's personal appeal brought Chisholm into camp. Dick, a well-trained Delaware hunter and guide, substituted for Black Beaver.[41]

Leaving the Arkansas frontier on October 28, Beale's party followed the well-established road paralleling the Canadian River. The route was basically that of Marcy and Simpson, 1849–1850, of Whipple, 1853, and of hundreds of Argonauts, traders, and emigrants traveling from Arkansas to the New Mexico settlements. The country was known to be flat and the soil's surface hard enough to support loaded wagons, so that Beale's only responsibility was to remove obstacles blocking the continuous movement of his train. In the wooded sections, the expedition marked its way by cutting down the trees that obstructed the passage of wagons and blazing others to mark the route.[42] For the use of future travelers, an itinerary was prepared, locating each camp site and recording the time and distance traveled daily. Data was provided on the character of the country at and between the camps, the available water supply, the daily temperatures, prominent landmarks, and the presence or absence of game and Indians.[43]

Since the specific assignment of this company in expending the congressional appropriation east of Albuquerque was to improve stream crossings, N. B. Edwards was placed in charge of a bridge-building detachment on November 15, to make constructions at sites selected by the commander. Wooden structures were erected across the San Bois and Little rivers and at least eight creeks flowing into the Canadian, known locally as Longtown, Gypsum, Elm, Comet, Marcou, Wood, Bear, and Oak.[44] The superintendent believed these wood bridges temporary arrangements because "... all the creeks ... will require permanent bridges of iron so that the Indians cannot burn them, or else the emigrant must follow the divide to the great loss of time and distance."[45] The bridge building was mainly kept within the present boundaries of Oklahoma. The construction team performed its task well in spite of the leader's attempt to deprecate its accomplishments.[46]

The expedition which Beale led across the Plains was a motley crew. His road party, including a military escort, comprised one hundred and thirty men. Two pieces of artillery were carried along for defense, in case of Indian attack. Several mail stages under the direction of R. Frank Green joined the expedition to secure the protection of numbers through the Comanche country. The mail contractor had an agreement to carry the mail from Neosho, Missouri, to Santa Fe, New Mexico, but after the Comanche outbreak he had delayed his departure a month, awaiting Beale's migration. Several emigrant parties also joined the command. Jesse Chisholm continued all the way as guide and Beale's Negro manservant and constant companion, Absolum, was making his third trip from ocean to ocean.[47]

As on his previous explorations, Beale never denied himself the pleasure of his animals, the hunt, and good food and drink. Dick, the Delaware, and Little Axe, the Shawnee, were dispatched daily on hunting excursions to provide fresh meat for the camp. Supplies of deer, buffalo, wild turkey, and small game animals were plentiful and the men seldom came back empty-handed. Beale had brought along a pack of greyhounds—Nero, Fannie, Prince, Buck, and Remus—and periodically they downed a fine deer in a short but gallant race across the hills. The superintendent delighted in the sport.

> My greyhounds this evening brought down and killed a fine fat buck antelope; only two were in the race, which was in full view of the whole camp; the hounds, Buck and Fannie, ran in upon and killed it within a mile of the point from which they started; the pace was tremendous, and, take it altogether, it was one of the prettiest bursts of speed I ever saw.[48]

The hunting proved profitable. On Christmas day, the surplus of twenty-five turkeys, preserved in the commissary wagon by the cold weather, were prepared for a feast.[49] The expedition arrived at Hatch's Ranch on the Gallinas River, a tributary of the Pecos, on December 26, bringing the first lap of their journey to a close.[50] Beale spent the month of January making an official visit to Santa Fe and paying a social call on his old and esteemed friend, Kit Carson.

Returning to Hatch's Ranch, the road builder devoted the second week of February, 1859, to an examination of the country due east of the ranch and the towns of Chaparito and Anton Chico, seeking a connection with the Canadian along its tributary, the Conchas. If successful, future travelers would be able to follow the Canadian for 180 to 200 miles farther before leaving it at the mouth of the Conchas. Finding a satisfactory route for wagons, Beale insisted that the Canadian Valley need not be left until within 14 miles of Anton Chico.[51]

The reconnaissance group departed from the ranch on February 26, traveling north of Anton Chico along a new western route. At the head-waters of the Pecos, they visited the site of Captain Pope's experimental artesian well, which Beale had seen the previous season, and all admired the industry and perseverance with which the work was prosecuted. Arriving in Albuquerque March 3, the crew spent five days preparing for the road work that lay ahead. Beale purchased two hundred and fifty head of sheep to drive to California. A part of these animals were to be used for food, thereby reducing the transportation of provisions; the rest undoubtedly went to the Tejon Ranch of the superintendent.

The first day's march west of Albuquerque was through sand hills which constituted, in Beale's opinion, the worst part of the road to California.

That evening the party encamped on the Puerco River which Beale reported should be bridged at a cost of from $5,000 to $7,000.[52] Using the itinerary of the previous year, the road party found the route of march to Inscription Rock in good condition. Five days were spent, however, in improving the grades across the summit of the dividing ridge of the Rocky Mountains, just west of the Puerco River. A detachment was, in the meantime, sent forward with two wagons to the Zuñi villages to purchase corn for the train. After devoting two days to an exploration of the valley of Inscription Rock, Beale and his men traveled quickly through Zuñi and on to Jacobs Well. Beyond this watering place a party of fifty worked for a day making a side-cut from the summit of an intervening hill to the plain.[53]

No further road repairs were believed necessary until the Little Colorado was reached. There the former route, where the stream was first approached, was relocated, and a troublesome ravine bridged by filling it with loose rock. After following the river for five days, George Beale, an assistant, found an excellent rock ford where this river, usually with muddy banks and bottom, could easily be crossed by emigrants. On the opposite side of the river, the working crew of fifty spent three days removing obstacles from the road as it climbed an imposing mesa. Near by a much-needed spring was discovered, cleaned out, and a basin built in which to store the accumulated water.[54]

Beyond the San Francisco Mountains, the road was graded, large rocks removed from the passage, and several small springs improved by digging larger basins. By carefully relocating his road without lengthening the distance, and by discovering new sources of water, Beale provided watering places not more than 30 miles apart. Day after day was spent in making improvements.

Engaged in road work west of the San Francisco Mountains, Beale was delightfully surprised to see on the western horizon the approach of two dromedaries, one of which was ridden by Ali Hadji, who had accompanied him on the 1857–1858 trip. Beale had previously sent a clerk to California, by way of El Paso, with orders that his camels should be brought to the Colorado River with supplies. The camel party, escorted by a troop movement of seven hundred men, had been forced to engage a thousand Mohave warriors in battle near the banks of the river. Mail carriers, headed east, joined the escort at the river, but when a second battle became necessary they fled the scene, according to messages brought to Beale. The mail had to be strapped to the backs of the trustworthy camels. When the camel party from California joined the road builders, the eastern packet was transmitted to an agent of the mail company traveling with Beale's expedition, and that for California placed on the camels' backs. The road builder recorded: "Thus the first mail of the 35th parallel was brought on my camels both ways."[55]

On May 1, 1859, when Beale and thirty-five men were marching ahead of their column in anticipation of engaging the hostile Mohave Indians, they encountered troops sent out by the Army for their protection. Since the military commander had removed the Indian danger by forcing the tribe to sign a treaty of peace, Beale's men were denied the expected skirmish, to their disappointment and disgust. The expedition moved into camp on the west bank of the Colorado River. The next two months were spent in improving the western end of the road between Saavedra's Spring and the Colorado, a section which had been reported unsatisfactory the previous year. When this final task was completed, the roadway was pronounced excellent for six-mule team wagons carrying 3,500 pounds. Meanwhile, Beale and chosen assistants absented themselves in the California settlements buying supplies for the return march.[56]

The return trip got underway July 2, 1859. With a mounted detachment and two four-mule wagons, Beale retraced his ground in rapid marches. At Floyd Peak he paused long enough to bring out some mint, gathered near a well in this mountain valley, and with one bottle of brandy, brought along for snakebite, prepared the first mint julep ever tasted on that mountain side. By July 29 the expedition was in Albuquerque, having made the entire distance from the Colorado River in 108 hours of marching time. This feat spoke for the excellence of the wagon route.[57] According to their itinerary from Fort Smith to the Colorado River, the men had improved a road 1,422 miles long. The distance was covered in 652 hours of actual marching; but nine months had elapsed between the Fort Smith departure and the return to Albuquerque.[58]

Having exhausted the funds for the road, Beale followed the customary policy of other federal road builders in recommending a new appropriation to continue the project. A sum of $200,000 had been expended by Congress in a two-year period for wagon-road surveys along the thirty-fifth parallel. A new grant of $100,000 for improvements on the road west of Albuquerque was requested in June, 1860; $50,000 additional was needed for bridging streams, particularly the Puerco River and the Rio Grande. Military officers had given the bridging projects a high priority on the list of territorial needs, and a War Department endorsement was procured for the superintendent's request which was forwarded to the Ways and Means Committee of the House of Representatives.[59] No congressional action followed. Beale's career as a wagon-road builder thus came to a close, and he returned to his ranch in southern California. Abraham Lincoln eventually selected him as surveyor general of California and Nevada.[60]

The superintendent's enthusiasm for his route along the thirty-fifth parallel, first aroused during the march of 1857–1858, was increased by the longer journey of 1858–1859. In reporting his exploits to the Secretary of War, he noted the incomparable advantages of the wide and level river bottom of the Canadian for a wagon road all the way from the last settlements of Arkansas to the first settlements of New Mexico. No heavy pull for wagons had been encountered on the entire route, and nature had supplied the country bountifully with the three requisites already mentioned for an overland route—wood, grass, and water.[61] Although the road surveyed directly west of Albuquerque needed improvement, beyond Floyd Peak to the Colorado it was "thoroughly completed, and ready for travel of any kind whatever."[62]

Just as Captains Whipple and Humphreys had recognized the importance of wagon-road construction along the thirty-fifth parallel to prove its practicability for a future railroad, Beale was fully aware that his assignment was mainly that of an explorer and surveyor for the railroad. Moving westward along the Canadian, Beale repeatedly made reference to the valley's advantages for a railroad and on one occasion bluntly recorded, "I am convinced that it [the Canadian valley] offers decidedly the most level line for a railroad to be found for the same distance between the Atlantic and the Pacific."[63] In forwarding his journal of the 1858–1859 expedition to the Secretary of War, he penned an introductory letter elaborating on the superb facilities offered by the thirty-fifth parallel country for a railroad as far west as New Mexico. Detailed construction estimates for the line, conveniently broken into three divisions of proposed building, totaled $21,391,100.[64]

The great promoter's ardent enthusiasm appeared limitless when he wrote to the Secretary of War:

> Without intending to draw invidious comparisons between the various routes from our western border to the Pacific ocean in favor of that by the 35th parallel, I think I can, with safety, say that none other offers the same facilities for either wagon or railroad.
>
> It is the shortest, the best timbered, the best grassed, the best watered, and certainly in point of grade, better than any other line between the two oceans, with which I am acquainted.[65]

Beale's work as a wagon-road surveyor, directing an unusually effective and coöperative group of men, prepared the way for the chartering of the Atlantic and Pacific Railroad in July, 1866, to build along the thirty-fifth parallel, and the final laying of the rails from Albuquerque to Needles, California, along that land grant by the Santa Fe Railroad.

CHAPTER XVI : *Across the Northern Plains: The Mullan Road, 1853-1866*

As late as 1850 the only continuous overland wagon road into the Pacific Northwest was along the Oregon Trail, through the South Pass, and northwest across the Blue Mountains to the Columbia River. Asa Whitney, a New York merchant interested in the China trade, had crusaded during the 1840's in behalf of the construction of an overland route directly west from the Great Lakes to the Oregon country. However, when he outlined a specific proposal for railroad construction before Congress in January, 1845, the majority of the legislators branded him a dreamer; a few insisted that he was a man of vision. In spite of the fact that the first consideration of federal subsidy for a transcontinental route was associated with the northern Plains, this region was the last to receive national aid for emigrant and commercial transit. The War Department, in conducting the Pacific Railroad surveys, had classified the area between the forty-seventh and forty-ninth parallels as a major field for examination; yet Congress, in sanctioning the construction of wagon roads to the Pacific as precursors to the railroad, had overlooked the possibility of a northern route.

Although national attention was focused primarily on the southern and central transcontinental routes, the United States Army, with the aid of settlers in the Pacific Northwest, sought additional outlets to the East. A road was located from Puget Sound to Fort Walla Walla (Wallula) on the Columbia River. This and near-by traces, together with the established emigrant route down the Columbia to the Willamette Valley, were the western links in the northern transcontinental wagon route. The Army next turned its attention to the Rocky Mountains, the most difficult region through which the overland route must pass. The improvement along this segment, so vital to the movement of pack trains and emigrant wagons, was the Mullan road.

John Mullan came to the Pacific Northwest in 1853 as a young West

Point graduate of twenty-three, attached to the railroad surveying party of Isaac I. Stevens. When the expedition reached Fort Benton, Lieutenant Mullan was sent forward to the Bitterroot Valley to greet the Flathead Indians near old St. Mary's Mission and Fort Owen. En route he was to seek out the best path across the Rocky Mountains from the headwaters of the Missouri.[1] This assignment led to the discovery of Mullan's Pass, used today for both road and rail.[2] Mullan's diligence as an explorer commended him to his superior,[3] and when the main party moved west to Washington, the young officer was left in charge of a meteorological detachment to observe and collect data on winter conditions.[4] Exploration was to be continued between the Rocky and Bitterrroot ranges, with examinations northward to Flathead Lake and southward to Fort Hall, there joining the earlier Frémont survey.[5] The first task was the location of a suitable site for winter quarters in the Bitterroot Valley, just south of John Owen's trading center, where the fifteen men of the detachment could find shelter in hastily erected log huts. These quarters, named Cantonment Stevens, became the center for a series of exploring expeditions and for the interrogation of traders, hunters, and Indians familiar with the region.[6]

Within a month Mullan headed south to the source of the Bitterroot River and across the divide into the Big Hole Basin of Idaho, going as far south as Fort Hall on the Snake River.[7] Soon after, another journey to Fort Hall was made by way of the Jefferson Fork of the Missouri. On the return, the party explored all the forks of the Missouri River to the east, and returned to the Bitterroot Valley via the Deer Lodge and Hell Gate valleys.[8] From a half-breed voyageur, Gabriel Prudhomme, who had accompanied the Jesuit fathers on their entry into the Northwest, Mullan learned of a feasible route for wagons between the Bitterroot Valley and Fort Benton on the Missouri River. Accompanied by a small pack train, he traversed this route to the fort, recovered a wagon train left there the preceding fall, and returned to Cantonment Stevens. The return journey had taken only fourteen days.[9] The next trip of April–May, 1854, took him north to Flathead Lake and the Kootenai River.[10] Within a single year he had crossed the Continental Divide at least six times and had traveled an aggregate distance of more than 1,000 miles. All his youthful ardor and enthusiasm had been devoted to this, his first military command.

The final phase of his work under Stevens was to find a suitable westward route from the Bitterroot Valley to Fort Walla Walla. After examining all known or reported passes, Mullan selected a line across the Coeur d'Alene Mountains through a pass named for his guide and interpreter, Sohon. In September, the Mullan crew abandoned winter quarters and followed this survey to The Dalles.[11]

Although the Stevens expedition was engaged in an official railway survey, the commander recognized the immediate need for wagon-road improvements. Instructions to Mullan repeatedly referred to the selection of "routes practicable for a railroad or wagon road." The young officer sensed that his primary mission in the winter of 1853 was to solve the problem of connecting the plains of the Columbia with those of the Missouri through mountain passes practicable for emigrant wagons. "In connection with the proper location and construction of a railroad, one of the most essential aids in advance was a good wagon-road line," he recorded in his official report, "and reduced the spirit of my own work almost to exploring the country for practical wagon-road locations, which, in time, should lend themselves as aids to the construction of our railroad lines."[12]

In January, 1855, Lieutenant Mullan left Puget Sound with reports for the War Department from Governor Stevens and with resolutions from the legislative assembly of Washington Territory recommending the continuation of his labor. On arrival in the national capital, he learned that Congress had appropriated $30,000 for a military road westward from the Great Forks of the Missouri River, in the Territory of Nebraska, to intersect the military road leading from Walla Walla to Puget Sound.[13] This legislation, sponsored by the War Department, was instigated by Stevens' preliminary reports on Mullan's success as an explorer. Although the War Department favored the project, the Secretary was personally adverse to its commencement at the time, because the appropriation was inadequate for the assignment. The outbreak of Indian hostilities in the Northwest during 1855–1856 served as an additional excuse for postponement.[14] In the interim Mullan served a tour of duty in Florida, participating in a campaign against the Seminoles, and earned an A.M. degree at St. John's College, Annapolis, before returning to frontier duty at Fort Leavenworth.[15]

In 1856, when Stevens appeared in Washington, D.C., as the delegate from Washington Territory, he reminded the War Department of the neglected road enterprise and inaugurated a campaign to cut Army red tape so that a start might be made. The subject of overland communication had taken on national importance; it was no longer a matter for speculation but a reality. Moreover, the Mormon War soon compelled the federal government to consider all possible routes for reaching the Great Basin. Majors and Russell, freighters, approached Mullan for his opinion on the advisability of transporting supplies by steamboat up the Yellowstone River to the mouth of its tributary, the Bighorn, and thence overland to Salt Lake. Although the boldness of the scheme fired Mullan's enthusiasm and imagination, he knew it was impracticable, and the idea was abandoned. Simultaneously with the Mormon difficulty, the federal government pursued its Pacific Wagon Road program. Frederick W. Lander was concentrating on shortening and improving the central Oregon Trail for emigrants; John Butterfield was bringing fame to the southern route with his overland mail. Stevens, Mullan, and other advocates of the northern route urged, in fairness, some consideration for their line. Captain Andrew A. Humphreys, in charge of the Office of Explorations and Surveys, secured a special assignment for Mullan to begin work on a military road across the northern Rockies on March 15, 1858. To establish the eastern terminus of the road, the United States Army secured the coöperation of Charles P. Chouteau of the St. Louis trading house, who was to push a steamboat to the farthest possible point up the Missouri.[16]

Mullan traveled by steamer to the Columbia River, arriving at Fort Dalles on May 15. He was organizing a party of road laborers, purchasing supplies, and negotiating for an authorized military escort of sixty men to be furnished at Fort Walla Walla, when news reached The Dalles that a military command under Colonel Edward J. Steptoe had been defeated by the Indians on the immediate line of the proposed survey. Although the earlier Indian disturbances in the Northwest had been quelled temporarily, tribes east of the Cascades were not subdued. The Cœur d'Alene, Spokanes, and Palouse, agitated by miners' activities in the Colville region, had resumed the warpath. Mullan halted his preparations and dispatched a courier to Steptoe, to inquire about the strength of the Indians in the field and the advisability of road work through their domain. Meantime his crew improved the road between The Dalles and the Des Chutes River. Steptoe, quite naturally, believed any attempts at road construction foolhardy. Disbanding his working party, Mullan requested attachment to the retaliatory expedition of Colonel George Wright, as a topographical officer, in the hope of familiarizing himself further with the regional geography and of performing incidental surveys.[17] At the end of Colonel Wright's successful campaign, Mullan "...had seen sufficient of the western approaches to the Bitter Root range..." to be convinced "...that the physical difficulties were such that a longer period of time, more ample means, and a larger force of men than either myself or others had imagined were requisite to accomplish our object."[18]

Mullan repaired once more to the national capital, and with the aid of Stevens, prevailed upon the War Department to include a $100,000 item for the road from Fort Benton to Fort Walla Walla in the annual appropriation for the Army. The measure became law on March 3, 1859.[19] New instructions were immediately issued Mullan emphasizing the more permanent character of the road than was originally contemplated by the Department. As usual first improvements were to be made on those sections presenting the most difficult obstructions to the passage of wagons. In addition to the mechanics and laborers, one hundred enlisted men of the infantry or artillery were to be attached to the command. Mullan was instructed to organize working parties of these soldiers to aid in the road construction, granting them in turn additional compensation allowed by Army regulations.[20]

By the time Mullan took the field he had employed ninety men; the escort, with its employees, numbered one hundred and forty, making a combined expedition of two hundred and thirty. This force was sufficiently strong to repel any Indian attack, most likely to come, if at all, in the hostile Blackfoot country. Sohon and Theodore Kallecki, the topographer, who had been employed by Mullan for the abortive 1858 exploit, were retained for this expedition. A capable engineering corps was organized, including P. M. Engle of the Topographical Engineers, W. W. Johnson, Conway R. Howard, and W. W. DeLacy, civil engineers with eastern railroad experience. DeLacy had just completed a survey of the Bellingham Bay road along Puget Sound under the direction of Captain George H. Mendell of the Topographical Corps. The explorations and surveys of these engineers made possible the scientific examination of much of the area between the road termini.[21]

Leaving Fort Walla Walla on July 1, the expedition moved northward across Dry Creek and the Touchet River, headed for the Snake River.[22] Both streams were bridged. Uncertain as to the route to be followed eastward across the Bitterroots, Mullan dispatched Sohon to ascend the Palouse River to its headwaters, and Engle to go up the Snake to the mouth of the Clearwater and beyond. Both reported the terrain "impossible," the timber virtually impenetrable, and the Indians hostile. The lieutenant decided, therefore, to follow the line he had explored in 1854, leaving the Snake River to cross the Spokane plains northeasterly, skirting the Coeur d'Alene Lake at its southern edge, and arriving at the Coeur d'Alene Mission by the valleys of the Coeur d'Alene and St. Joseph rivers. Mullan expected to reach the hospitable Bitterroot Valley in time to go into winter quarters. As the road work progressed, wooden post markers were periodically sunk along the right of way with the designation "M. R." (military road), recording the measured distance from Walla Walla.

No major construction tasks were necessary on this first sector of the road. As Mullan noted, ". . . the route may be termed a natural wagon road, needing but light improvement." Delay in crossing the Snake River had proved an annoyance, and one soldier was drowned when he jumped from a

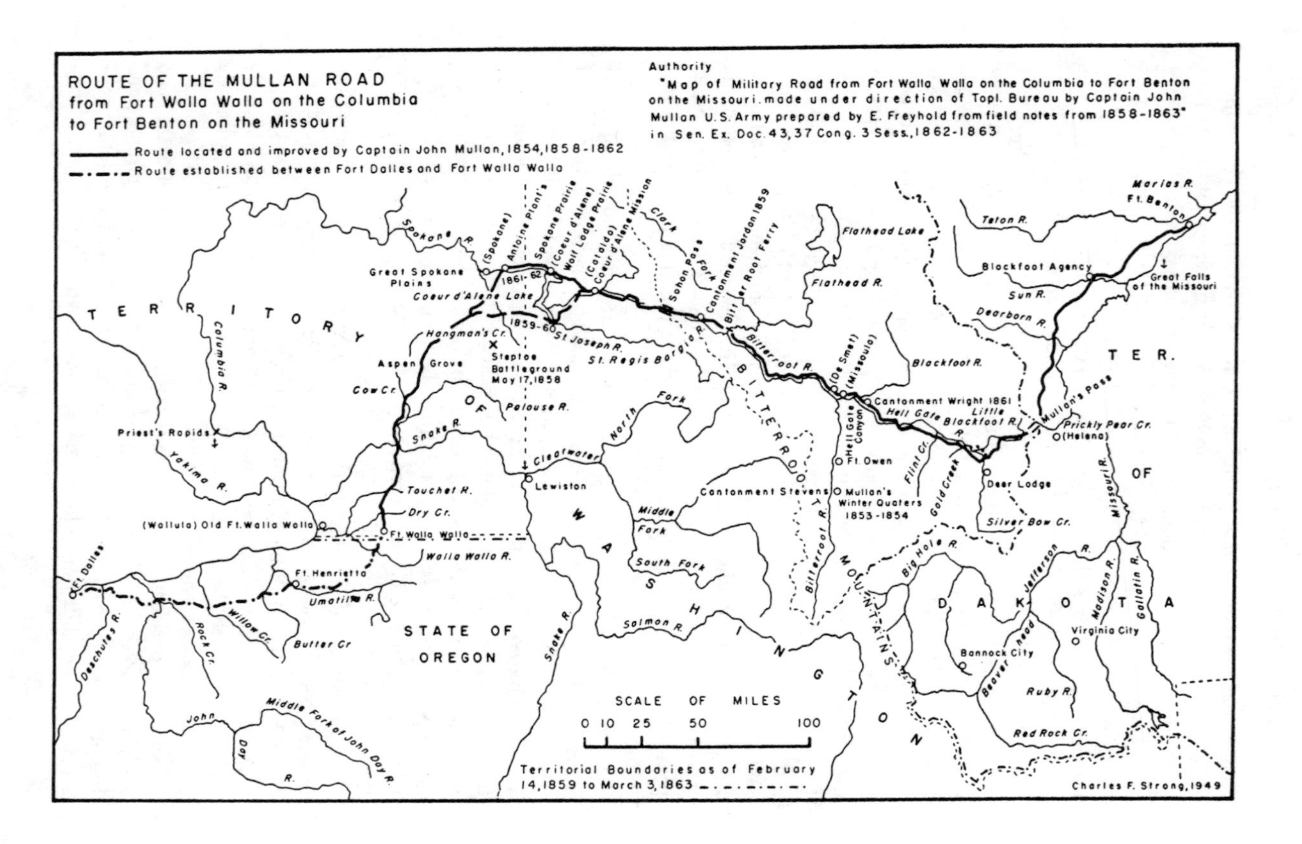
ROUTE OF THE MULLAN ROAD
from Fort Walla Walla on the Columbia
to Fort Benton on the Missouri
Route located and improved by Captain John Mullan, 1854, 1858-1862
Route established between Fort Dalles and Fort Walla Walla
Authority
"Map of Military Road from Fort Walla Walla on the Columbia to Fort Benton on the Missouri, made under direction of Topl. Bureau by Captain John Mullan U.S. Army prepared by E. Freyhold from field notes from 1858-1863"
in Sen. Ex. Doc. 43, 37 Cong. 3 Sess., 1862-1863
TERRITORY OF WASHINGTON
TER. OF DAKOTA
STATE OF OREGON
BITTERROOT MOUNTAINS
Columbia R.
Spokane R.
(Spokane)
Antoine Plant's
Spokane Prairie
(Coeur d'Alene)
Wolf Lodge Prairie
(Cataldo)
Coeur d'Alene Mission
Clark Fork
Sohon Pass
Cantonment Jordan 1859
Bitter Root Ferry
Flathead Lake
Flathead R.
Teton R.
Marias R.
Ft. Benton
Blackfoot Agency
Great Falls of the Missouri
Sun R.
Dearborn R.
Great Spokane Plains
Coeur d'Alene Lake
1861-62
1859-60
Hangman's Cr.
St. Joseph R.
St. Regis Borgia R.
Steptoe Battleground May 17, 1858
Aspen Grove
Cow Cr.
Palouse R.
Bitterroot R.
(De Smet)
(Missoula)
Blackfoot R.
Cantonment Wright 1861
Hell Gate
Little Blackfoot R.
Mullan's Pass
Prickly Pear Cr.
(Helena)
Priest's Rapids
Yakima R.
Snake R.
North Fork
Clearwater
Hell Gate Canyon
Ft. Owen
Flint Cr.
Gold Creek
Deer Lodge
Touchet R.
Lewiston
Cantonment Stevens
Mullan's Winter Quaters 1853-1854
Missouri R.
Dry Cr.
(Wallula) Old Ft. Walla Walla
Ft. Walla Walla
Middle Fork
Silver Bow Cr.
Walla Walla R.
Big Hole R.
Jefferson R.
Madison R.
Gallatin R.
Ft. Dalles
Ft. Henrietta
South Fork
Bitterroot R.
Umatilla R.
Deschutes R.
Rock Cr.
Willow Cr.
Butter Cr
Snake R.
Salmon R.
Virginia City
Bannock City
Beaverhead
Ruby R.
Red Rock Cr.
John Day R.
Middle Fork of John Day R.
SCALE OF MILES
0 10 25 50 100
Territorial Boundaries as of February 14, 1859 to March 3, 1863
Charles F. Strong, 1949

wooden raft carried downstream by the water current. When the party reached the St. Joseph River valley, the ground proved so marshy that trees were felled to corduroy the impassable places. Deciding not to bridge this stream or the Coeur d'Alene, Mullan ordered the construction of two flatboats, 42 feet long and 12 feet wide. One was rowed across the lake and up the Coeur d'Alene to await the party's arrival. During construction the topographers traced the sources of the St. Joseph, mapped its tributaries, and surveyed the Coeur d'Alene Lake. With a road crew of fifty men Mullan established a line through the timber and underbrush across the divide between the two rivers and up the Coeur d'Alene to the mission. Laborers, teamsters, and twenty soldiers worked for a week constructing three bridges, cutting three miles of timber, and excavating sidehills along the survey. The expedition arrived at the mission on August 16. Two hundred miles of road survey and improvement had been completed within six weeks.[23]

Before leaving the mission, Lieutenant Mullan dispatched Engle with instructions to explore up the Clark's Fork as far as the Pend d'Oreille Mission and to return across the divide between that stream and the Spokane to determine the advantages of the line for a wagon or rail route. Sohon went on a scouting party to select the best point of passage over the Coeur d'Alene Mountains and to continue an examination of the valleys of the St. Regis Borgia, Bitterroot, and Hell Gate rivers.[24] Mullan's immediate plan of operation was to send a pack train forward eleven miles to work back through the timber to the mission while he, with forty workers and three wagons, would push eastward to join them. Once this junction was made, the party was divided into four working crews, to improve the road on up the valley of the Coeur d'Alene toward the summit of the mountains or divide, separating its waters from those of the Bitterroot. Those of the military escort, not choosing to engage in road work, remained behind with the wagon train to keep open communications between Fort Walla Walla and the Coeur d'Alene Mission.

Eastward from the mission, the members of the expedition encountered grave difficulties, since a forest 100 miles wide lay between them and the St. Regis Borgia River. Besides the dense standing timber, the fallen trees lay in an intricate tangle on the floor of the forest, appearing at times to be an insurmountable obstacle. Crews of axmen worked from dawn to sunset, hacking a passage for the wagon forge and supply train. To complicate the train's movement further, the valleys they followed were narrow defiles, at times verging toward canyons, and the serpentine streams left alternate flats and spurs along their banks. To avoid the most difficult spurs, the waters were crossed repeatedly. Mullan put teamsters and packers to work alongside the laborers, and secured the services of thirty to forty soldiers in clearing the timber, grading, and bridging. At each day's close the axes, spades, and picks had to be reconditioned for further use. The Bitterroots were finally crossed at Sohon Pass and the valley of the St. Regis Borgia reached on December 4, 1859. Mullan recorded: "Justice cannot be done to the industry and fortitude of the men while mastering this wilderness section." Some had not escaped injury: two men were cut with axes, one accidentally shot in the knee, and another injured by a falling tree.

Improvements on the third 100-mile sector had taken fifteen weeks. Mullan frankly admitted that "the amount of work required was immense, and very much underestimated by myself and others." The party had been challenged, however, by the receipt of a message announcing the arrival of the little steamer *Chippewa* at Fort Benton with rations for the crew and construction equipment. The lieutenant was certain for the first time that his task, if successful, would connect the navigation heads of the Missouri and the Columbia.[25]

On November 1, snow began to fall in the mountains and within five days it was 18 inches deep. On the eighth, the temperature dropped below zero and winter's warning was at hand. The anticipated break in the weather did not come, and all hope of reaching the Bitterroot was finally abandoned. A cluster of log huts in the St. Regis Borgia Valley became winter quarters known as Cantonment Jordan. A storehouse and office were erected on December 5, in forty-two-degree-below-zero weather. Many of the expedition's animals perished from exposure and exhaustion before reaching this station. The cattle that remained were slaughtered for beef, and the horses driven 100 miles farther to the warmer climate of the Bitterroot. Most of them died of exposure or starvation on the way.

These winter months were used by the road builders to compile field notes, complete maps, and write memoirs and reports for the War Department. As the mailman came and went on snowshoes each month he was charged with the responsibility of recording the snow depths from gauges posted along the way. By midwinter, P. E. Toohill, the expressman, was forced to give up travel on the western sector of the Mullan road, but came into camp by way of Clark's Fork to the north. On inquiry, the lieutenant learned that the Indians had never been known to cross the mountains in winter by the Coeur d'Alene route, but continuously used the Clark's Fork. Too late the commander had discovered that climatic advantages made the longer water-grade way to the north preferable, but he was candid in admitting it.

The men constantly suffered from the cold. Many cases of frostbite resulted from open boots torn to pieces by the wear and tear on the road. One hunter, lost in the mountain forest, remained away from camp four days and nights without food or blankets. In time, he crawled into the cantonment with frozen feet and a nearly deranged mind, only to learn that both legs had to be amputated to save his life. By the winter's close, there were twenty-five cases of scurvy, so Mullan made an overland trip to the Pend d'Oreille Mission to procure fresh vegetables from the priests there.

In the winter Mullan had dispatched Johnson as a special messenger to Washington, D.C. with letters to the Secretary of War, the Chief Topographical Engineer, the Quartermaster General, and delegate Stevens. In each he outlined a proposal, first considered by Jefferson Davis during the Pierce administration, to move a military train from the headwaters of the Missouri to those of the Columbia. Mullan was to arrive at Fort Benton in August, 1860, with an empty train that either had to be abandoned, or teamsters hired to return it to Fort Walla Walla. If a detachment of three hundred recruits needed to supply the vacancies in the military companies in Oregon and Washington could meet his party at Fort Benton, their return over the road would prove its practicability for military transit, and by using the equipment of the road party the federal government would be saved $30,000 in transporting them to the Pacific Northwest.[20]

As spring approached, Mullan realized that his party was virtually stranded, due to the loss of livestock. The supplies at Fort Benton had to be procured, and he was compelled to seek the aid of the Flathead Indians near Fort Owen. This friendly tribe furnished one hundred and seventeen horses with pack saddles and the services of twenty men to accompany Sohon on a trek to Fort Benton for eleven hundred essential rations.

Early in March a working party was sent forward to the Bitterroot Ferry, 15 miles distant, to begin the second season's operations. Here the snowfall had been light and the weather was mild compared to the site of Cantonment Jordan. The intervening stretch of dense timber was by-passed until the season was far enough advanced to clear the accumulated snow. By May 10 the road builders had improved the route for 30 miles along the north bank of the Bitterroot. Lieutenant James L. White, with thirty-two soldiers of the escort, and engineer DeLacy, with twenty-five axmen, returned to open the 15 miles of dense timber passed over in March. The main work crew soon encountered a mountain spur, six miles across, that proved unavoidable. This so-called "Big Mountain," the greatest obstacle of the entire route, required the constant labor of one hundred and fifty men for six weeks to establish the proper grades for wagons. Periodic blasting was necessary, and in a premature explosion one worker lost the sight of an eye, and another man was severely stunned.

From this obstruction the remaining 60 miles to the junction of the Bitterroot and Hell Gate was chiefly through open timber. The only improvements necessary were one-half mile of sidehill excavating and the construction of a 150-foot bridge over a slough too deep for the wagons to cross. This occupied forty men for ten days. On July 1 the expedition crossed the Big Blackfoot River by means of a wagon boat and small bateau.[27]

At this point news was received of the arrival of recruits at Fort Benton, who impatiently awaited Mullan's appearance with the horses and wagons for their transportation. The road builders now hurried forward along the Hell Gate to Gold Creek and then ascended the Little Blackfoot to Mullan's Pass across the Continental Divide. Turning northward they marched rapidly to the Dearborn River, pausing only once for a four-day delay to grade the crossing of Medicine Rock Mountain. On the Dearborn the party was overtaken by Johnson, direct from Washington, who reported congressional approval of a second $100,000 allotment for the continuation of the road project.[28] Mullan and his men crossed the prairie between the Dearborn and the Sun which was a distance of 32 miles, without attempting any road work. At the Blackfoot Agency on the latter stream the expedition was divided into four contingents, each exploring a different way into Fort Benton. All arrived August 1, 1860.

At this upper Missouri outpost the road builders were greeted by the members of Captain William F. Raynolds' expedition, who had been exploring the country west of the Yellowstone River mouth between the Missouri and the Platte. Under the auspices of the Office of Explorations and Surveys of the Army, Raynolds, a topographical officer, was sent out from St. Louis in May, 1859, to study the topography and climate of the region, to record its agricultural and mineral resources, and note the numbers, habits, and dispositions of the Indians. Although this reconnaissance was primarily for exploration, in accord with the Army's interest in transportation, the party was to observe the facilities and obstacles of the terrain "to the construction of rail or common roads, either to meet the wants of military operations or those of emigration through, or settlement in, the country."[29] Captain Raynolds was given specific instructions to determine first, the most direct and feasible route between Fort Laramie and the Yellowstone in the direction of Fort Union at the river's mouth; second, the best route from Fort Laramie northwest along the base of the Bighorn Mountains toward Fort Benton, thence westward along the Mullan road; and third, a way from the South Pass to the Yellowstone, ascertaining the practicability of a route from the headwaters of Wind River to those of the Missouri.[30]

The Raynolds party ascended the Missouri River to Fort Pierre, arriving June 28. The remainder of the summer was spent in an overland march westward across the Cheyenne River, around the north slopes of the Black Hills, thence to the Yellowstone River at Fort Sarpy—a trading post of Chouteau and Company, 30 miles below the mouth of the Bighorn. After obtaining supplies, the explorers moved south, examining the headwaters of the Bighorn and Powder rivers before going into winter quarters.[31]

For the exploration of 1860 the expedition was divided into two groups. Lieutenant Henry E. Maynadier, in charge of one detachment, journeyed up the Bighorn to the Greybull River, turned westward to the Shoshone River, and across to the Clark Fork of the Yellowstone. Following this tributary to its mouth, he ascended the Yellowstone in a westerly direction across the mountains into the Gallatin Valley. At the three forks of the Missouri, the party awaited the arrival of Raynolds.[32]

With Jim Bridger as guide, Raynolds' detachment headed westward to the sources of the Wind River. The mountain guide had told many weird tales about the natural phenomena at the headwaters of the Yellowstone, and the Army officer was determined to be the first scientific explorer of the region. Bridger insisted that a military train could not leave the Wind River area and cross the dividing ridge northward to the Yellowstone. Raynolds tried it and failed after repeated reconnaissances, because the mountain gorges and slopes were buried in the deep snow of early summer. He skirted the future Yellowstone Park area to the south and west via Jackson's Hole, Pierre's Hole, and across the Continental Divide to Madison River, thence to the three forks. Here the party divided again. Maynadier descended the Yellowstone to its junction with the Missouri, and Raynolds followed the Missouri to Fort Benton where, before Mullan's arrival, he conferred with Major George Blake, in charge of the troops on the way to Oregon.[33]

By the time Raynolds' expedition returned to St. Louis, the different divisions of his party had traveled, since leaving Fort Pierre, a total distance of 5,000 miles by land over routes heretofore unknown. In addition the explorers had journeyed nearly 3,000 miles by boat.[34] In spite of the great distances covered, the Raynolds' reconnaissance had not gained any clear-cut picture of the topography or resources of the region traversed. Confusion as to their whereabouts and the best route and methods of further travel so delayed both Raynolds and Maynadier that they were unable to observe a total eclipse of the sun in southern Canada, a major assignment of the expedition.

Either from their experience or from the information gained from Bridger and others, Raynolds nevertheless recommended several routes as possible wagon roads leading from the Platte Valley, or Oregon Trail, to the headwaters of the Missouri. From the Platte, at Deer Creek, there was no serious obstruction for wagons along the northeast base of the Bighorn Mountains and down the Bighorn River to the Yellowstone. To go directly to Fort Benton, the Bighorn Valley should be left before the stream's junction with the Yellowstone, in order to cross the latter below the big bend in the present Billings vicinity, thence across the plains to the Judith River and on northward to the fort. A direct route also was reported from the Red Buttes along the Platte, through the northern sector of the Bighorn Mountains to the three forks of the Missouri, thence to the Continental Divide near Mullan's Pass, where a junction could be made with Mullan road leading to Walla Walla. On the basis of Raynolds' report, Captain Humphreys was convinced that the Fort Benton–Walla Walla road could be connected with the Platte route and a new line of communication opened from the Missouri River to the settlements of Washington Territory.[35]

At Fort Benton, Raynolds, who was preparing to descend the Missouri, transferred his pack animals to Mullan. The road builder did not tarry long at the eastern terminus. Discharged civilians and soldiers whose term of service was soon to expire were loaded in a Mackinac boat to descend the Missouri River. By reducing his force to twenty-five men and transporting essential supplies on pack animals, Mullan planned to keep in advance of the recruits under Major Blake's command and make necessary road repairs to facilitate their passage. The remaining personnel, including Sohon, were transferred to Blake's jurisdiction, and the available means of transportation, including the wagons, were to be used by the major's party.

Improvements were systematically made as Mullan returned over the road from east to west. The crossings of many dry gulches and ravines on the eastern sector were graded. West of Mullan's Pass, marshy areas were made passable by corduroying, additional timber was cut from the right of way, and streams were bridged. Just west of the Divide, the lieutenant took time to relocate the road by turning south from the Little Blackfoot River into the Deer Lodge Valley and descending the Hell Gate along its south bank. Although the road was lengthened three miles, several crossings of the Little Blackfoot and Hell Gate were avoided. In the mountains hundreds of stumps of trees, felled during the winter snows, had to be trimmed closer to the ground. Mullan was convinced that the marshy country south of Coeur d'Alene Lake made it impracticable for travel during the spring and early summer and decided to relocate the road to the north of the lake during another season.

On the entire march daily communications were dispatched to Major Blake, advising him of advantageous camping sites and providing information to facilitate travel through difficult terrain. At the Coeur d'Alene Mission a pack train from Fort Colville met the command to receive one hundred and fifty recruits for that station. The remaining soldiers marched into the parade ground at Fort Walla Walla, having traversed more than 600 miles in fifty-seven days. A few were stationed here and others moved on to The Dalles and Fort Vancouver. Mullan recorded in his final report:

> Thus ended this military experiment *via* the upper Missouri and Columbia rivers; and the success that attended it, the good effects that it induced, the economy resulting, and the eulogistic manner in which each officer of the command referred to the trip, all constitute a sufficient commentary upon its feasibility for future military movements toward the north Pacific.[36]

During the autumn in Walla Walla, Mullan drafted details for the next season's work in conformity with War Department instructions. His twofold purpose was to relocate the road north of the Coeur d'Alene Lake by July 1, 1861 and then begin bridging operations, involving eighty constructions, to the east of the Coeur d'Alene Mission.[37] While "comparatively idle," as he described his condition during the winter because an assistant had taken the field notebooks to Washington, D.C. as the basis for cartographic work, the road builder conceived a new plan of operation. Rather than outfit in the Northwest, a wagon-road party would be organized at Fort Leavenworth, march overland to Fort Laramie, and then head to the Deer Lodge Valley where the winter of 1861 could be spent. The practicability of this route could thus be tested before improving his road westward from the Deer Lodge Valley to Walla Walla in the season of 1861–1862. Mullan was obviously seeking an overland connection to the States from the eastern terminus of his road. He hoped to take up the unfinished business of Captain Raynolds and, in doing so, foreshadow the work of John Bozeman in locating the "Bozeman Trail" and the Army's post-Civil War improvement of the Fort Laramie–Montana cutoff known as the "Powder River Road."[38] Going East to present his new scheme, Mullan arrived at the national Army headquarters only to learn that his original proposal had been approved and the authorization forwarded to Oregon. Chaos prevailed in the capital because of the rapidly changing government personnel with the inauguration of Abraham Lincoln and the breakup of the Union. Mullan hastily returned to his post at Walla Walla to intercept his orders.[39]

By May 13, 1861, he had a force of sixty civilians and a detachment of twenty-one soldiers of the Quartermaster Corps ready to take the field. At the crossing of the Spokane River, an additional force from Fort Colville was to join the expedition, bringing the escort to one hundred men under the command of three officers.[40] The first section of the road from Fort Walla Walla to the Snake River was in excellent condition for wagons and, with all bridges completed, no improvements were necessary. Beyond the Snake a new line northward to the crossing of the Spokane at Antoine Plant's ferry was surveyed. Plant, a half-breed Flathead Indian speaking both French and English, had received a charter from the legislative assembly of Washington Territory to operate his flatboat on the river. Here a delegation of Spokane Indians, visiting the expedition's camp, assured Mullan that the tribe was not adverse to the opening of a road through their country. Between the Snake and the Spokane the major improvement had been the construction of a 105-foot bridge across the Little Spokane River. Mullan reported:

This road is now excellent, can be travelled at all seasons, and has, except during the summer months, all the requisites for an emigrant or military line. All the streams which are affected by spring freshets are now bridged, and there is no point along the entire line where wagons will be compelled to double teams.[41]

Beyond the Spokane Crossing the road party encountered a difficult belt of timber, 30 miles wide, far heavier than that south of the lake. Mullan recorded: "We are now in very dense timber, that renders the proper exploration, location and making of the road, a difficult and irksome task, and, what with the myriads of mosquitos to annoy us, are well calculated to try one's patience."[42] The party was divided into two detachments, one of axmen, another of graders, who moved from point to point cutting a road through the wilderness. They continued the work of chopping, clearing, and grading day after day, and if by nightfall the axmen had moved forward two miles, all were satisfied. The daily goal of the graders was always 500 yards of sidehill excavation. Mullan recorded "in all this excavation . . . the trees had to be first felled, and the removing of stumps by grubbing was slow and difficult, and, what with the rocks between the roots, hard on the tools."[43] The expedition reached the Coeur d'Alene Mission on August 4, a month late according to the timetable, and the commander admitted again that he had misjudged the extent of the job. In the end, however, he considered the road opened.

. . . a creditable piece of mountain work, . . . [which] will compare favorably with any turnpike of the same length and through a similar difficult country. All the creeks and streams are bridged, muddy or marshy flats corduroyed, and no grade so difficult but that single teams with laden wagons can pass over with ease.[44]

Reaching the mission, the expedition joined its old road of 1859. The flatboat, stationed on the St. Joseph, was brought across the lake to a new crossing of the Coeur d'Alene River. Mullan, now certain of the task which lay ahead, reorganized his force to establish six small contingents, four to specialize in bridge constructions, one to cut trees, and another to grade improvements. The lieutenant brought up the rear with the supply wagons. Inspired by a spirit of competition, the bridge builders progressed up the Coeur d'Alene River completing twenty structures before crossing the summit of the Bitterroots on September 15. A similar system was initiated along the St. Regis Borgia.[45] The completed road was 25 feet wide, hard and solid and dry, according to the builder.[46]

The first snow of winter in early November drove the road party out of the mountains, down the Bitterroot Valley, and on to the mouth of the Blackfoot River. The commander decided to split his force into five sections—one to bridge the Blackfoot, the others to be distributed in winter camps along Hell Gate River at points where side-cuts were to be made to avoid stream crossings. Each party was instructed to build log huts for winter quarters and then continue to work until the severity of the weather put an end to it. The main encampment, Cantonment Wright, was constructed near the Blackfoot.[47]

Events of the winter of 1861–1862 were a repetition of those of 1859–1860, only the suffering and disaster were greater. During the autumn, the grass along the route had been so scarce that many animals wandered off in search of food. Due to the negligence of the herders, the Indians stampeded the remaining stock, and some forty or fifty head of beef cattle and work oxen were lost in the dense timber.[48] Many of those remaining perished in January for lack of forage. Mullan maintained communication with the men in the scattered winter cabins by sleds, until the cold weather forced a suspension of all travel. One laborer who attempted to pass between two camps was overtaken by night before he reached his destination. Building a fire, he attempted to remove his moccasins that had become wet during the day, only to find them frozen to his feet. Alarmed at his plight, he retraced his steps to the original encampment. Upon placing his feet in a tub of water to thaw them, the flesh fell from the bones. After intense suffering, his life was saved by amputating both legs above the knees. His comrades raised a purse for his temporary support and left him to the kind charity of the fathers at the Pend d'Oreille Mission. In view of his previous season's experience with winter weather in the northern Rockies, Mullan seems to have been guilty of inadequate preparation, faulty planning, and a relaxation of discipline which inevitably resulted in isolation, hardship, and even tragedy for the men. When spring brought an end to this season that "would long be remembered throughout the length of the Pacific Coast," the expedition was demoralized. The bridges along the Bitterroot left half completed at the beginning of the winter and the side-cuts along the Hell Gate were finished, but a large number of the employees, wishing to be discharged, forced Mullan to go directly to Fort Benton. On this trip no work was performed other than that necessary to reach the fort by June 8.

Returning to his party and escort in the mountains, Mullan observed that the high water of spring had destroyed a major bridge across the Hell Gate, thrown the Blackfoot bridge out of shape, and washed away six structures along the St. Regis Borgia and two on the Coeur d'Alene. Not having an adequate labor force of his own, Mullan could make no repairs. In accord with Army practice, he therefore contracted with Samuel Hugo, a resident of Deer Lodge Valley, to repair and replace bridges on the eastern section of the road. Funds were to come from the federal appropriation; supervision was entrusted to the commissioners of Missoula County.[49]

Mullan's road work was now complete. His trace, as finally located, was northward from Fort Walla Walla to the Snake River at the mouth of the Palouse. From here the road ran east of north to the crossing of the Spokane River and around the north shore of Coeur d'Alene Lake to the Coeur d'Alene Mission. The direction of the Mullan road now turned toward the southeast, ascending the Coeur d'Alene River to its source and across the Bitterroot Mountain Divide through Sohon Pass. On the east side of this summit, the route followed the St. Regis Borgia to its junction with the Bitterroot and along the latter stream to the scenic Hell Gate Canyon. Thence the road followed the Hell Gate Valley as far as its Deer Lodge tributary, before turning northeast through the mountains along the Little Blackfoot River to the summit of the Rocky Mountain Divide at Mullan's Pass. East of the divide, the military highway followed, in general, the course of the Missouri River, first north and then northeast to Fort Benton, the eastern terminus.[50]

Mullan summarized the improvements made and the terrain to be traversed by future travelers:

> Our road involved one hundred and twenty miles of difficult timber cutting, twenty-five feet broad, and thirty measured miles of excavation, fifteen to twenty wide. The remainder was either through an open, timbered country, or over open, rolling prairie. From Walla-Walla eastward the country might be described in succinct terms as follows: First one hundred and eighty miles, open, level, or rolling prairie; next one hundred and twenty miles, densely timbered mountain bottoms; next two hundred and twenty-four miles, open timbered plateaus, with long stretches of prairie; and next one hundred miles, level or rolling prairie.[51]

"Thus ended my work in the field," wrote Mullan, "costing seven years of close and arduous attention, exploring and opening a road of six hundred and twenty-four miles from the Columbia to the Missouri river, at a cost of $230,000."[52]

Although the Mullan road was ostensibly constructed for military purposes with the expectation that contingents and supplies for the Northwest posts could be moved with more rapidity and economy, it never attained importance as a through military highway.[53] No evidence points to its extensive use by the United States Army after the initial march of Major Blake's command.[54] Apparently the military urgency for the road had mainly passed by the time of its completion in 1862 due to exceptional quiescence among the Northwest Indian tribes.[55]

More important was its use as an emigrant route for those wishing to settle in the Northwest, or for the army of mining prospectors who followed the gold rush into Idaho and Montana. Rumors about gold discoveries in the northern Rockies had been prevalent since the days of '49 in California. As early as 1853 gold had been found by members of the Stevens surveying party along the Hell Gate River. Although the "color" was so meager that it failed to excite the explorers, they named the stream Gold Creek. Systematic prospecting was started by the Stuart brothers, James and Granville, in 1857–1858. In August, 1860, on the eastern slopes of the Rockies, Mullan's party "found traces of quartz and continual indications of gold." The leader recorded, "one of my men found ten percent prospects in the Big Prickly Pear ... and the Indians gave me to understand that two miles higher up the stream, in another cañon, gold had been found by them."[56] When he tried to organize his second wagon-road expedition in May, 1861, news of major discoveries on the Clearwater branch of the Snake in the lands of the Nez Percés was known, and securing personnel for an exploring party was almost impossible.[57] Mullan reported the following development of June, 1862:

> We halted at the American Fork and visited the Deer Lodge gold mines, where we found Messrs. Blake, McAdam, Higgins, Dr. Atkinson, and Gold Tom at work sluicing, and at that time they were taking out about ten dollars per day to the hand, and with fair prospects of extensive digging. Where-ever parties had prospected in ravines or river bottoms they had found prospects from one to fifteen cents to the pan. Convincing ourselves by indubitable proofs of the existence of gold at this point, we continued our journey....[58]

At Fort Benton, Mullan was met by 364 emigrants, brought up the Missouri in four steamers from St. Louis. Some, bound for Walla Walla, came with saw and grist-mill equipment; others were miners, intending to prospect the country en route. The young lieutenant predicted "the commencement of a long line of emigrant travel, that in years to come must course along these rivers in search of new homes toward the Pacific slope."[59] Some emigrant parties, which failed to reach Fort Benton before Mullan's departure, overtook his expedition before it reached Walla Walla. On arrival the commander furnished supplies to those in need and presented each party with an itinerary recommending suitable camping places. Mullan recorded:

> All these emigrants were a very proper class of persons, mostly from the western States, where the civil troubles had caused a number to look towards the Pacific in quest of new homes. They had made the journey on the steamers in from thirty to forty days from St. Louis, and, purchasing their land outfit at Fort Benton at very moderate figures, were journeying safely and pleasantly toward the setting sun. The safe passage of these emigrants during this season proves the value of the line for emigrant purposes, and will yet cause it to stand in competition with other lines across the continent.[60]

Among those emigrants using the Mullan road in 1862 was a party escorted by Captain James L. Fisk under the auspices of the War Department. During 1861 the representatives from the Pacific states and territories had prevailed upon Congress to appropriate $50,000 for military escorts to emigrant trains headed for California, Oregon, and Washington. The next year the allotment of an additional $25,000 was justified because the withdrawal of troops from the Great Plains for the eastern campaigns of the Civil War had denied emigrants and mail carriers the usual protection against the Indian tribes. These funds for the so-called Emigrant Overland Escort Service of the United States were administered by the Secretary of War.[61]

Many men in Minnesota, Wisconsin, and Iowa in 1862 were eager to reach the newly discovered gold fields in the "inland empire" of Washington Territory. They hesitated, however, to start without an escort, because of the difficulty in finding a direct route across the barren plains to Fort Benton, inhabited by the Sioux. Minnesota citizens prevailed upon the War Department to use $5,000 of its escort fund to guide and protect a Minnesota-Montana migration in 1862 and at the same time to determine the feasibility of a northern overland route. Captain Fisk received orders "to organize, equip, and conduct an escort to an emigrant train from Fort Abercrombie, across the plains to the north, to Fort Benton, Dakota Territory; thence across the mountains, via Captain Mullan's government wagon road, to Fort Walla Walla—there dispose of the expedition's property, and return via Oregon and San Francisco."[62]

Leaving Fort Abercrombie on the Red River of the North, 250 miles northwest of St. Paul, the Fisk expedition of one hundred and twenty emigrants, including thirteen women, traveled northwest to the headwaters of the James River, thence to the Coteau du Missouri, and on to Fort Union. West of this trading post, the party ascended the Missouri and its tributary, the Milk, before turning south to Fort Benton. They were two months on the road, usually averaging 16 to 17 miles each day.[63] At the request of the emigrants, Fisk and his assistants remained in charge of the train as it started west along the Mullan road. When the expedition reached the point on the route where the trail to the new Prickly Pear diggings branched off, the majority of the emigrants decided to remain in the valley and try their luck prospecting. After receiving a written commendation for his escort service, Fisk, with a remnant of the group, continued along the military road to Fort Walla Walla.[64] Thus a northern overland route from the Mississippi River to the Pacific became a reality.

In 1863 Captain Fisk escorted another smaller emigrant train across the northern plains with fifty of its men selected as a guard. Congress appropriated $10,000 for this second venture. From Fort Abercrombie to the Missouri River the route traveled was much the same as that used in the summer of 1862. Beyond the Missouri, the expedition turned northwest, by-passed Fort Union, and approached the Milk River by way of Wood Mountains, Porcupine River, and Frenchman's Creek. After moving along the Milk in a westerly direction, they turned south along the route of the previous year to Fort Benton. The search for gold was now reaching its most exciting stage. The newcomers were greeted at Bannock by a crowd of miners including many Fisk emigrants of the previous year. The bonanza of the Montana gold fields at Virginia City had been discovered, and the captain visited the area to make an official report before catching the Bannock City Express to Salt Lake where stagecoach connections could be made for a return eastward along the central overland route.[65]

Reports of the rich gold discoveries in Montana resulted in the greatest enthusiasm for westward migration to the mines in 1864. Major General John Pope, commanding the Department of the Northwest, somewhat alarmed at the prospects of heavy migration and the indications that the Sioux were organizing to obstruct the passage of pioneers, issued instructions that all trains should select a place of rendezvous, organize for defense, and seek an escort. Captain Fisk was authorized by the Army to escort his third overland expedition. Taking advantage of a cavalry movement across the plains, the emigrant party traveled along a more southerly route than in previous seasons, starting at Fort Ridgely and reaching the Missouri River at Fort Rice. One hundred and fifty miles beyond the Missouri, a band of

Sioux Indians attacked the emigrant party, cutting off and killing seven of the eight men of the rear guard and capturing two overturned wagons loaded with rifles, ammunition, and commissary supplies. Continuous attacks on the train forced the emigrants to form a corral and fortify it. For sixteen days they withstood attack, until an Army relief party could arrive from Fort Rice.

The emigrants had, in the meantime, quietly organized a council for their government, and decided to go no farther. Fisk was eager to move west and urged the commander of the rescue detachment to escort them for a two-day march. When he refused, Fisk made an issue of the situation and challenged the emigrants to take their choice of returning with the Army to Fort Rice or going forward with him. Army officers, already concerned because the increasing number of emigrant expeditions was agitating the Sioux to the point of open hostility, resented Fisk's attitude and action. The result was a curtailment of the Army Escort Service.[66]

The disaster of 1864 did not discourage Fisk. Unable to obtain financial support from the government, the intrepid captain determined to conduct a privately organized expedition to Montana for commercial purposes. Although an elaborate colonization scheme did not materialize in the summer of 1865, during the next year Fisk led the largest of all his expeditions across the Plains. Homeseekers, prospectors, and pioneer merchants made up a party of 325 moving in 160 wagons.[67] The movement of population over the northern overland route was heavier in 1866 than in any earlier year. At least 1,400 persons used Fisk's route and the eastern sector of the Mullan road in reaching the Northwest between 1863 and 1866.[68]

Recognizing the national importance of transportation to and through the Northwest gold fields, the United States Senate, in the darkest hours of the Civil War, passed a resolution authorizing the preparation, printing, and distribution of Captain Mullan's final report on the military road.[69] In contrast with his military communiqués written, for the most part, in the field between 1859 and 1861, this document was prepared with the verdict of history in mind. Mullan's road project became a partial fulfillment of the dreams of Thomas Jefferson and the schemes of Jefferson Davis. The readable publication was full of advice for the immigrant homesteader, including an itinerary and detailed instructions on the outfitting and conduct of an overland expedition. Mullan described the agricultural and grazing potentialities of the area through which his road ran, the known mineral wealth, the physical and human resources destined to make the region flourish. The Northwest's greatest need at the close of his road-building career was the same as at its beginning—a railroad. The young engineer presented all the data, compiled during his nine years' labor, for possible routes and desirable methods of construction and finance. Two years later his *Miners' and Travelers' Guide to Oregon, Washington, Idaho, Montana, Wyoming, and Colorado, via the Missouri and Columbia Rivers* became available to the public.[70] Mullan was the victim or the beneficiary of his own promotional literature for the Northwest, and in 1863 resigned his Army commission to take up residence on a homestead near Walla Walla.[71]

In spite of the sudden burst of migration across the northern plains immediately after the completion of the Mullan road, this passage through the Rockies never became the through emigrant road any more than the through military road that its originators and promoters had anticipated.[72] Traffic was frequently confined to the eastern sector, particularly during the first two or three years after the Montana gold discoveries. By the middle 'sixties the Mullan road had become of greater value as a route of commerce than as an avenue of immigration. A keen competition developed between St. Louis, Missouri, and Portland, Oregon, for the trade with the Montana settlements. Steamboats brought supplies up the Missouri to Fort Benton and up the Columbia to Wallula, and there mule pack trains were organized to carry cargo into the mountains. Both eastern and western ends of the Mullan road were now used extensively. Deliveries of goods converged on the new towns near the road's center between the Bitterroots and the Rockies, but few supplies were carried straight through. In spite of Mullan's interest in making his road usable for wagons, very few actually traveled it.[73] Physical obstacles and neglect made it virtually impossible for emigrants' wagons to get through, and for all practical purposes the Mullan road was nothing more than a pack trail. Some freighters even complained of the difficulty in traveling with heavily loaded mules.[74]

In a memorial of 1866 the Washington territorial assembly reminded Congress that the "Mullan wagon road" was impassable for wagons through the Coeur d'Alene and Bitterroot mountains because of fallen timber and destroyed bridges. A plea was made for $100,000 to be expended in repairs under the direction of a competent engineer of the United States Topographical Bureau. The argument for federal aid was along established lines:

> ... The population of this vast region of country is too new and too poor to be able to take hold of and rapidly complete such a great enterprise as the opening of this military wagon road.
>
> The inhabitants, coming as they have from all parts of the United States, are unacquainted with each other, and admitting that they have all the necessary means within themselves for the opening of this road, a few months' acquaint-

ance with each other is insufficient to establish the necessary confidence to organize a company and put forward to completion so great an undertaking....

...You will not permit so important a work as this road, together with what the Government has already expended, to go to ruin, nor permit individuals to seize upon available portions of it and claim a franchise whereby they will be enabled to use a part of an improvement erected at public expense, and convert the proceeds drawn from the toiling miner or travelworn emigrant to their own use, which in some instances is now being done.[75]

Without the road's improvement Washington settlers were deprived of a valuable market for their agricultural products, and the inhabitants of western Montana would continue to pay exorbitant, if not extortionate, prices for necessities. Government supplies stored in warehouses at Wallula could not be delivered economically to Forts Walla Walla and Colville. All the Northwest soon hoped to see "a mail coach on the route instead of a train of pack horses."[76] Such reasoning in urging federal appropriations for internal improvements had become shopworn and Congress turned a deaf ear.

During the short period of the Montana gold rush, when thousands of men hastened to the diggings from the north central states, the northern overland route temporarily challenged the Oregon Trail's monopoly on migration to the Pacific Northwest. But as a continuous avenue of transportation the northern overland route never rivaled the central route, either as it was known to the early pioneers, or as subsequently relocated in part and improved by the federal engineers. Yet the mountain sector of the northern route, the Mullan road, was recognized by residents of the Northwest as the federal government's major contribution in improving the region's transportation system before the railroad era.[77] This wagon-road project, conceived as a part of the government's topographical surveys for the Pacific railways, provided the information necessary at a later date for the proper location of the northern transcontinental lines. The Northern Pacific Railroad tracks west of Helena follow the Mullan trace over the Divide of the Rockies and on across western Montana. The Bitterroots are crossed by the rail line along the Clark's Fork, recognized by the Army engineer as a route superior to his Coeur d'Alene crossing. The Chicago, Milwaukee, St. Paul, and Pacific joins the Mullan survey in the Deer Lodge Valley and parallels the Northern Pacific tracks across the state, almost to the western border, before following one of the lines south of the Coeur d'Alene Lake established by Mullan. The Great Northern has extended a branch line from Fort Benton to Helena along the eastern part of Mullan road in the upper Missouri Valley. Today a broad paved federal highway, U. S. 10, winds its way across western Montana and Idaho along much of the roadway located and improved by Mullan's men.

Part IV

AN ATTEMPT AT COÖPERATION IN WAGON ROAD BUILDING BY THE INTERIOR AND WAR DEPARTMENTS

CHAPTER XVII : *The Sawyers Wagon Road from the Mouth of the Niobrara River to Virginia City*

THE UNITED STATES ARMY's task of policing the water route up the Missouri River and the central overland route across the Great Plains to California and the Pacific Northwest was a comparatively simple task in the period before the Civil War. At the close of the war, however, many routes across the Plains were being used by a steady flow of emigrants to the rapidly developing Montana and Idaho regions. The encroachment of these travelers upon the land and food supply of the northern Plains Indians was forcing them, in desperation, to acts of depredation. Although in 1865 the Army did not lack the men or equipment to maintain peace, an agreement with the Interior Department on a policy protecting the white emigrant, yet equally fair to the Indian, was not easily obtained.[1]

General John Pope, commander of the Division of the Northwest, surveyed the entire problem in 1866. In the first place, travel across the Plains could be restricted to one or two routes. This he believed impracticable, even though it might be expedient, since such a policy would necessitate a greater military force to require the emigrants to pursue designated routes than to protect all the byways they might wish to explore. A second solution was to establish Indian reservations away from the preferred overland routes and force the tribesmen to remain there. The only other alternative to these proposals was a continuation of the joint military and civilian supervision of the Red Men with its confusion, conflict, and inefficiency.[2]

Congress, meanwhile, had taken steps to reactivate the federal wagon-road program that had been dormant during the war years, except for the activities of John Mullan and men like James L. Fisk with the Emigrant Overland Escort Service. The impact of the Civil War, with its spirit of retribution, removed the southern plains routes from the favorable attention of Congress. Therefore, after 1865, road construction in the territories was to be concentrated north of the Platte River. As a result of gold discoveries in the Northwest, that area, more than any other, appeared to be a land of opportunity in the postwar world, and each ambitious town in Minnesota, Iowa, and Dakota hoped to become the eastern terminus of the key route westward. Merchants and land promoters were eager for the profits of migration and commerce. Representative Asahel W. Hubbard of Iowa introduced a bill in the House of Representatives on January 5, 1865, authorizing the construction of a wagon road from the Missouri River to Virginia City, Montana. The legislation was referred to the House Committee on Roads and Canals of the Thirty-eighth Congress. Within the month, William H. Wallace, delegate of Idaho Territory, proposed a measure for road construction from Lewiston, Idaho, across the mountains, to Virginia City, Montana. In the Senate, Minnesota's Alexander Ramsey was concurrently advocating the construction of federal roads from the western boundary of his state across the plains to the Northwest. After a series of legislative conferences to consolidate forces, a single act was framed, providing for the construction of wagon roads in the territories of Idaho, Montana, Dakota, and Nebraska. Under the guidance of Senator Morton S. Wilkinson of Minnesota, the measure speedily and satisfactorily jumped the hurdles in the legislative process and received President Lincoln's signature on March 3, 1865.[3]

This law provided for four wagon roads: first, from Niobrara, at the mouth of that river to Virginia City, with an eastern branch leading to Omaha; second, from the mouth of the Big Sioux, near Sioux City via Yankton, northwestward to the mouth of the Big Cheyenne River, thence up this river to intersect with the Niobrara road; third, from the western boundary of Minnesota to a point near the mouth of the Big Cheyenne River; and finally, from Virginia City to Lewiston. Congress had approved the construction of a series of wagon roads located in relation to geographic features which, when completed, would form a continuous line from Lewiston, Idaho, via Virginia City, to the Powder River, then divide into two branches across the plains—one by way of the Niobrara River, the other by the Big Cheyenne River. These two branches would provide connections with Omaha, Sioux City, and the western Minnesota boundary, there joining the federal and state roads into St. Paul. Undoubtedly, the eastern terminus at the mouth of the Niobrara was chosen because the federal government had surveyed and improved a road in 1858, under the direction of George L. Sites, along the banks of the Missouri between Omaha and Niobrara. A total of $140,000 was allotted for the project with

$50,000 designated for the Omaha–Niobrara–Virginia City road and a like sum for the Lewiston–Virginia City route. Of the remaining funds $20,000 was to be applied to construction from the mouth of the Big Cheyenne to its intersection with the Niobrara road, and $10,000 used to build a bridge across the Big Sioux River. Any funds remaining were to be used to improve the road from Sioux City, Iowa, to the mouth of the Big Cheyenne.

The hastily drawn legislation did not reveal the exact intentions of the Congress. No funds were designated for the road from the western Minnesota boundary to the mouth of the Big Cheyenne; no specification was made as to the percentage of funds to be used on the Niobrara–Virginia City road and its branch to Omaha; and the route from the Big Sioux to the Big Cheyenne seemed to be of less importance than a bridge across the river opposite Sioux City. On one point, however, the congressional desire was explicit—the road program was to be directed by the Department of the Interior. Not only was the Secretary empowered to survey, locate, and construct the wagon roads, but the unexpended funds for an earlier authorized military road from Sioux City to Fort Randall, Dakota Territory, were to be transferred from the War Department to his jurisdiction.[4] No evidence points to congressional dissatisfaction with the road-building methods of the Army, nor to an attempt to transfer this phase of federal activity to the Interior on a permanent basis, as in the Pacific Wagon Road program of 1857–1860. Congressional leaders envisioned a coöperative program. President Lincoln, in turn, made clear to E. M. Stanton, Secretary of War, and John P. Usher, Secretary of the Interior, his wish that the civil and military branches of federal service should work together in solving the policing and the migration-transportation problems of the plains.[5]

The key road of the new federal project, from Niobrara to Virginia City, was another attempt to solve the basic problem of locating the most feasible and the safest passage for emigrant trains from the central Plains to the Idaho and Montana region. The outbreak of Civil War precluded serious consideration of Raynolds' and Mullan's project for such a road, and soon the United States Army was so preoccupied with the conflict that all western road building was neglected.

During the war years the need for a cutoff from the Oregon Trail east of the mountains to the Montana mining region became the concern of private traders. With the news of gold discovery, John M. Bozeman, a Georgian, came to Montana in 1862 after an unsuccessful period of prospecting in Colorado. Convinced that a more desirable route to the Bannock and Virginia City diggings could be located than along the devious way of the Oregon Trail, dipping south to Fort Bridger, and then north through

Idaho by way of Fort Hall, Bozeman and his partner, John M. Jacobs, blazed a trail along the eastern slopes of the Bighorn Mountains to the Platte Valley, mapping as they traveled during the summer of 1863. After organizing a wagon train in Nebraska, they returned to Montana over the new route. Leaving the Oregon Trail near present-day Casper, Wyoming, the trace ran northward across the plains, just east of the Bighorns to the vicinity of Buffalo, Wyoming. Here it turned northwest, following the base of the mountains across the Bighorn River into the Yellowstone Valley. Beyond the Bighorn the preferable route was by the Clark Fork of the Yellowstone, but several alternative passages were available in case of Indian hostilities. A large band of Sioux Indians intercepted the 1863 party on the plains south of Buffalo, accused Bozeman of violating their hunting grounds, and warned him against proceeding further. The train turned back to the Oregon Trail. Jacobs escorted the wagons along this longer route to Montana, whereas Bozeman, with nine companions, crossed over to the headwaters of the Bighorn River west of the Bighorn Mountains and continued down that valley to Montana.

Bozeman organized another train in 1864 to cross the forbidden domain of the Sioux east of the Bighorns. Jim Bridger, the famous fur trader and trapper of the Rockies who had gained equal renown as a guide and scout for government expeditions, was also serving as a guide to emigrant parties this season. Bridger preferred the route west of the Bighorns, passing through the Wind River Canyon and down the Bighorn River valley into Montana. Although the grass supply was less than on the plains to the east, the Sioux lands were not violated and there was no danger of attack. Bozeman had examined the passages on both sides of the mountains, however, and preferred the eastern route. As he planned to move sizable trains of heavily loaded commercial wagons along the route, an abundance of grass for animals was essential. The Indians, he reasoned, would not dare to attack these large expeditions as they crossed the country. Although this route was referred to as the Jacobs cutoff during the war, it soon became known as the Bozeman trail.[6]

The Emigrant Overland Escort Service, instigated by Congress in 1861, was a partial answer to the pioneer's need for protection across the Great Plains en route to California, Oregon, and Washington. Annual appropriations of $25,000 to $50,000 continued throughout the war, and allotments were made for escorts along specific routes. Captain Medoram Crawford performed this service along the Oregon Trail from Omaha to Walla Walla, but the Fisk expeditions of 1862 and 1863 along the northern overland route were the most widely advertised. In 1864 a $10,000 stipend was designated to protect emigrants on still a third route from Niobrara on the Missouri River to the Gallatin Valley of Idaho.[7] Major Henry E. Maynadier, who had served with the Raynolds' expedition, was ordered to Sioux City in the summer of 1864 to await the arrival of emigrants wishing to follow the trail. Unable to raise a party of twenty-five, Maynadier declined to make the trip, and the expedition disbanded.[8]

With the end of the war the Army expanded its program for the protection of emigrant trains. General William T. Sherman, commander of the Division of the Missouri, concurred with General Pope about the need for expansion of the reservation policy for the Plains Indians, but did not consider it inadvisable to confine plains travel to one or two routes. As a result of the unfavorable reports of Gouverneur K. Warren after his expedition of 1857 along the Niobrara, the Army judged this route impracticable for wagons. The general came to the conclusion that travel to Montana in 1866 would naturally follow three lines: along the Missouri River; by land along the Platte to Fort Laramie, thence along the eastern base of the Bighorns to the headwaters of the Yellowstone and to Virginia City; and from Fort Pierre to the valley of the Big Cheyenne, thence to the Black Hills, and on west to form a junction with the route near the Bighorns. The Army committed itself to police these routes to protect travelers from Indian attack.[9] Emigrants along the Minnesota-Montana or north overland route were expected to be few in number, and they would continue to receive the services of armed escorts; no attempt would be made at continuous policing. General Sherman recognized Pope's wisdom in proposing an understanding with the Interior Department on Indian policy. Soon after he presented his proposal that the Sioux should be limited to the area north of the Platte, west of the Missouri, and east of the Bighorns, and the Arapaho and Cheyenne should be restricted to the country below the Arkansas. Thus a wide belt of territory running from east to west between the Platte and Arkansas rivers, through which the bulk of California-Oregon migration had passed, would become the exclusive domain of the emigrant. The United States Army was still determined to construct a military road from Fort Laramie to Virginia City along the Powder River.[10]

The Army had surveyed this Powder River road along the Bozeman trail during the summer of 1865 by marching a column of troops through the country east of the Bighorn Mountains. This goaded the Sioux into action because the road would cut through the most desirable hunting grounds still available to the tribe. Red Cloud, their chief, first protested and then threatened resistance. The soldiers returned to the disputed region in 1866 with orders to secure the proposed route for military and emigrant travel by constructing forts along the trail. Colonel Henry B. Carrington, in charge of the "mountain district," moved to the headwaters of the Powder River to enlarge an outpost built in 1865, known as Cantonment, or Fort, Connor. Relocated and strengthened, the fortification became known as Fort Reno. Three hundred miles to the northwest Fort Philip Kearny was next constructed, and the chain was completed in August by the establishment of Fort C. F. Smith where the road crossed the Bighorn River. Throughout the summer Red Cloud's warriors maintained a sniping resistance. Troops working on the forts were periodically attacked, any straggler venturing beyond the walls of the fort was likely to be cut down, and wagon supply trains were raided. The skirmishing reached a climax in December, 1866, when a wood train near Fort Philip Kearny was assaulted by the Sioux. After relieving the train, the rescue party under Captain William J. Fetterman, foolishly and against orders, pursued the attackers, was ambushed, and all eighty-two of its members slaughtered. The Army henceforth spoke of the incident as the Fetterman Massacre.

In these same years, 1865–1866, the Interior Department was engaged in its road-building assignment to connect the Middle West with Montana and Idaho. Before field work could begin, however, Secretary Usher was forced to consider the patronage claims of Nebraska's delegate, Phineas W. Hitchcock, and of Representative A. W. Hubbard of the Sioux City district in Iowa. The latter proposed an allotment of $5,000 to $8,000 for the Omaha branch and urged the Secretary to permit Hitchcock to name its superintendent. This coöperation, the congressman thought, would lessen the rivalry between Omaha and Sioux City and would assure some assistance from Omaha residents in making the Niobrara route the most important to the West. The Nebraskan, in turn, asked that at least $12,000 be set aside for the branch. No allocation of funds was agreed upon, nor did the Interior Department employ an engineer to build the Omaha road.[11] However, Secretary Usher urged Hubbard to select the superintendent of the Niobrara–Virginia City road and to seek the coöperation of General Pope in obtaining military protection and supplies.[12] James A. Sawyers of Sioux City, Iowa, a man with military experience and some knowledge of the Plains Indians, was appointed to lead the expedition. As a young Tennessean he had served in the cavalry during the Mexican War. Soon after he moved to Iowa and during 1861 enlisted in the Sioux City volunteer cavalry used by the Indian service in defending the frontier. After the Sioux outbreak in Minnesota in 1862, Sawyers accepted a commission as lieutenant colonel of the Northern Border Brigade of the Iowa militia which had been created as a policing unit.[13]

The Secretary of the Interior displayed little interest in the wagon-road project, once the appointments were made. Sawyers' instructions for his expedition were anything but restrictive: he was to determine the precise location of the road and the amount of work requisite, such as bridging the streams and improving fords. No stipulation was made for the number of mechanics and laborers to be hired. The problem of subsistence was to be solved in consultation with the military.[14] Hubbard procured through General Pope the promise of an escort of at least two hundred cavalry equipped with two howitzers. In May, when Sawyers returned to Sioux City from Chicago where he had gone to purchase supplies, he learned that two infantry companies, with one hundred and eighteen men, awaited his arrival at the Niobrara River mouth. They had rations only for three months and no wagons to transport civilian supplies. He protested about the inadequate escort to General Pope who, in turn, put pressure upon General Alfred Sully to detail an additional detachment of twenty-five cavalrymen, authorize rations for the six-month period Sawyers expected to be in the field, and furnish rifles and ammunition to the civilian employees of the train.[15]

Sioux City residents were keenly interested in promoting the Niobrara route, and the *Sioux City Journal* had spearheaded the movement to get federal approval for the wagon road. Hubbard's activities in procuring the necessary appropriation and escort were reported in the press, and soon Sawyers' preparations were followed with similar interest. Extensive advice was given to prospective emigrants about essential supplies needed on the Montana frontier. The anticipation of profits from new gold discoveries or trade with those who had already found wealth in the diggings was a recurrent theme.[16]

The road-building party, assembled at Niobrara, consisted of fifty-three men, including Lewis H. Smith, engineer; D. W. Tingley, physician, a clerk, guides, scouts, pioneers, herders, and drivers. Fifteen wagons were to haul the chains, tools, tents, camp equipage, and subsistence. To pull the wagons Sawyers purchased forty-five yoke of oxen. The escort train had twenty-five wagons, each drawn by six mules. A private freighting company of Sioux City, C. E. Hedges and Company, added thirty-six wagons loaded with goods destined for Virginia City. Five emigrant families also attached themselves to the expedition. In all, eighty wagons were to roll across the Plains in a single column to mark a new trail.[17]

Leaving Niobrara on June 13, 1865, the expedition moved westward along the southern bank of the Niobrara River, paralleling the meanderings of that stream. Sawyers paused briefly at the crossings of its tributaries to

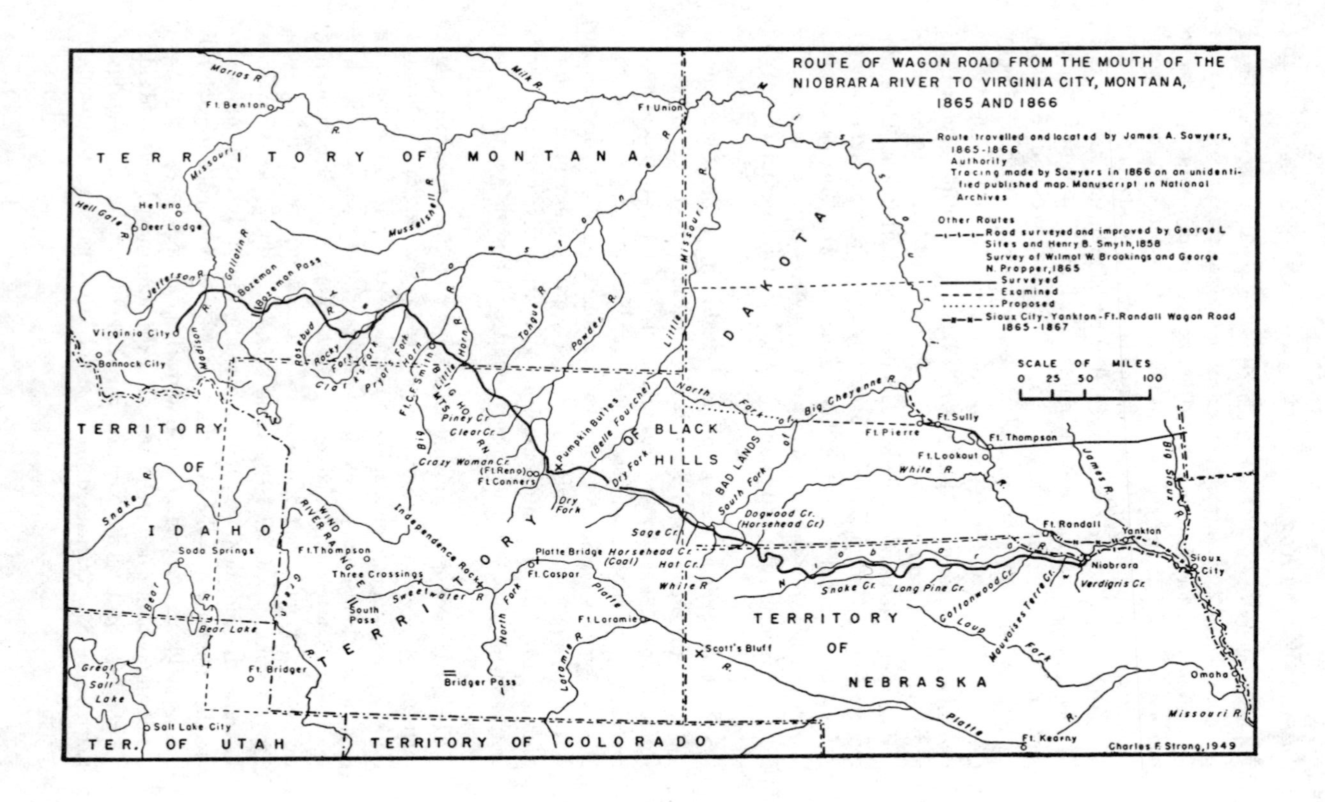

ROUTE OF WAGON ROAD FROM THE MOUTH OF THE NIOBRARA RIVER TO VIRGINIA CITY, MONTANA, 1865 AND 1866
Route travelled and located by James A. Sawyers, 1865-1866
Authority
Tracing made by Sawyers in 1866 on an unidentified published map. Manuscript in National Archives
Other Routes
Road surveyed and improved by George L Sites and Henry B. Smyth, 1858
Survey of Wilmot W. Brookings and George N. Propper, 1865
Surveyed
Examined
Proposed
Sioux City-Yankton-Ft. Randall Wagon Road 1865-1867
SCALE OF MILES
0 25 50 100
TERRITORY OF MONTANA
TERRITORY OF IDAHO
DAKOTA TERRITORY
TERRITORY OF NEBRASKA
TER. OF UTAH
TERRITORY OF COLORADO
BLACK HILLS
BAD LANDS
BIG HORN MTS.
WIND RIVER RANGE
Marias R.
Milk R.
Ft. Benton
Ft. Union
Missouri R.
Musselshell R.
Helena
Deer Lodge
Hell Gate R.
Gallatin R.
Jefferson R.
Bozeman
Bozeman Pass
Virginia City
Madison R.
Bannock City
Rosebud R.
Rocky Fork
Clark's Fork
Pryor Fork
Ft. C.F. Smith
Little Horn R.
Tongue R.
Powder R.
Little Missouri R.
Piney Cr.
Clear Cr.
Crazy Woman Cr.
(Ft. Reno)
Ft. Conners
Pumpkin Buttes
(Belle Fourche)
North Fork of Big Cheyenne R.
Dry Fork
Ft. Sully
Ft. Pierre
Ft. Thompson
Ft. Lookout
White R.
James R.
Big Sioux R.
South Fork
Dogwood Cr. (Horsehead Cr)
Sage Cr.
Horsehead Cr. (Coal)
Hat Cr.
Ft. Randall
Yankton
Sioux City
Niobrara R.
Niobrara
Verdigris Cr.
Snake Cr.
Long Pine Cr.
Cottonwood Cr.
Mauvaises Terre Cr.
Loup Fork
Platte R.
Omaha
Ft. Kearny
Scott's Bluff
Ft. Laramie
Laramie R.
North Fork
Platte Bridge
Ft. Caspar
Independence Rock
Three Crossings
Ft. Thompson
Sweetwater R.
South Pass
Bridger Pass
Green R.
Ft. Bridger
Soda Springs
Snake R.
Bear R.
Bear Lake
Great Salt Lake
Salt Lake City
Charles F. Strong, 1949

improve the fords by grading river embankments and, where necessary, erecting bridges. No improvements other than those essential to the movement of the train were attempted. The expedition started off in an optimistic spirit. Periodic accounts of progress were reported to the *Sioux City Journal.* One reporter described the train's organization:

. . . A scouting party with guides go in advance; then comes a section of Artillery with a division of Infantry; the train follows, divided into three sections, with a division of Infantry between each section, while a division of Infantry and a section of Artillery close up the road.[18]

After traveling across two-thirds of Nebraska along the Niobrara River, the party decided to leave the river and strike out northwest for the White River. Beyond this stream they encountered the Bad Lands of present western South Dakota and were obliged to clear the roadway by cutting down a few trees and grading several bluffs. After traversing the divide between the White River and the south fork of the Big Cheyenne, Captain George W. Williford, in charge of the escort, insisted that his supplies were dangerously low and his nearly barefoot men needed shoes. New gear was to be obtained from Fort Laramie. On July 21 his quartermaster left the road party with an escort of fifteen cavalrymen to ride 75 miles to the fort for clothing and subsistence.[19]

From the outset Sawyers had trouble with his escort. The superintendent insisted that 143 men were insufficient to protect the train when crossing hostile Indian country. Some emigrants and several men whom he had hired, turned back because of the escort's size. Nor was he convinced that the soldiers were brave! On July 3 engineer Smith was ordered on a reconnaissance to the mouth of Snake Creek to determine if a better road could be found more closely paralleling the Niobrara. Although he started out with four cavalrymen, according to Sawyers, "three of the men straggled and came back to camp being afraid of Indians, but he made the reconnaissance accompanied by one man." A similar incident occurred on a survey beyond the White River when a cavalry escort abandoned the guide, Baptiste Dufond. The road superintendent was further aggrieved over the delay necessitated by the side trip to Fort Laramie. The detachment rejoined the party on the Dry Fork of the Cheyenne on August 1, but with no supplies. Fearful of an Indian attack upon the supply wagon, the lieutenant in charge had sent it with a wagon train scheduled to join the military forces constructing Fort Connor on the Powder River.[20]

The expedition's affairs went from bad to worse. Sawyers became increasingly enraged over the two weeks' delay caused by the escort's unpreparedness. To retaliate Williford reminded the superintendent that the military had awaited the road-building crew for a month at the Niobrara's mouth, meanwhile exhausting its subsistence. Williford insisted upon receiving his supplies before moving into the Powder River Valley. Sawyers dispatched a searching party toward Fort Laramie with the hope of intercepting the train so that the road crew could move forward, but they returned empty-handed. Further ill-will was engendered when the military commander suggested that the expedition's guides, Ben F. Estes and the Yankton half-breed Baptiste, had proved their incompetence and urged that a replacement be secured at Fort Laramie. Sawyers refused the advice and irritably charged that the soldiers were too cowardly to follow the guides on reconnaissance.[21] Chaos prevailed because of the divided command.

The train finally moved across the north folk of the Cheyenne within view of the Pumpkin Buttes and headed northwest to the Powder River around the north side of these two red circular mounds rising abruptly on the plains. This route led the party into the Bad Lands of the Powder River valley where the terrain was broken and water supplies inadequate. The passage proved impracticable for wagons and the decision was made to retrace the route to the Pumpkin Buttes. The Army officers insisted that the guides were lost.

The expedition had now reached the Indian hunting grounds where Red Cloud was determined to prevent road constructions. When engaged in a scouting assignment August 13, south of the Pumpkin Buttes, Nat D. Hedges, a lad of nineteen in charge of the freight train, was surprised by a party of Cheyennes, killed, scalped, and his body mutilated. The Indians continued to follow and harass the expedition by stealing and stampeding stock. At sunrise August 15, the bluffs surrounding the party's corral were covered with five or six hundred warriors, who charged onto the plains, repeatedly shooting into the encampment. After a three-day siege, the majority of the warriors agreed to withdraw the attack in exchange for supplies of tobacco, coffee, and sugar, but the young men of the Sioux wanted a war of annihilation. Two members of the cavalry escort, venturing into the Indian camp, were shot. Williford considered this warning sufficient, and urged the abandonment of the expedition and a return to Fort Laramie. Sawyers preferred to send a scouting party to the Powder River in search of General Patrick E. Connor who was constructing a fort on the river and traversing the Bozeman trail with a military party as a demonstration of the Army's power and resolve. The superintendent called for volunteers from the cavalry to escort the guides, but none were forthcoming and the officers refused to detail them. Instead, soldiers joined employees in discussing the advantages of burning the train and each man making his own way back to Fort Laramie.

A civilian reconnoitering party set out for the Powder River, 50 miles distant, on August 17. They returned in three days, reporting a good road to the river and evidence that a large military train had recently moved down the valley. The expedition hesitatingly made a few short marches, but when the Dry Fork of the Powder was reached, Williford refused to go farther. The guides reported this impasse to officials at Fort Connor, 13 miles from camp. Williford's infantry command was relieved of duty and a new cavalry unit ordered to accompany the Sawyers' party as far as the Bighorn River. Since the route of General Connor was followed northwest to the Tongue River, the pioneers and workmen no longer had to grade ravines to keep the wagons from overturning. At the Tongue crossing a cavalry officer on a scouting mission was surrounded and killed by Indians. The following day, while the wagon train was fording the stream, the Indians attacked the rear guard, stampeded the loose cattle, and drove off thirty head. When the train attempted to proceed, the surrounding hills were again covered with well-armed and threatening Indians. With only a small escort of thirty-five cavalrymen, the party was obliged to corral. A stock driver and an emigrant were mortally wounded in the mélee. Firing was continuous on both sides throughout the day, cattle were slain and wagons damaged. A tentative peace agreement was signed the next noon, and Sawyers was able to move forward with his wagons across the divide between the Tongue and the Bighorn. Two days later, however, a mail detail riding between Connor's camp and Fort Laramie was assaulted by Indians and had to seek refuge with the wagon-road party. The expedition was again forced to corral, and for several days the men were besieged by the enemy. The bodies of three comrades were wrapped in a blanket and placed in a common grave while several men played fiddles and danced jigs in a near-by tent to divert the Indians who had arrived in camp under a flag of truce to discuss peace terms. In frontier fashion, cattle were driven across the grave so the Indians could not locate and dig up the corpses after the expedition moved forward.

Because the second escort was ordered to go no farther than the Bighorn under any circumstances, the majority of civilian employees refused to move beyond that point without military protection. In the midst of this exasperating predicament, Sawyers uncovered a plot among the employees to seize the lighter wagons, leave the emigrants stranded, and make a dash for Virginia City. Angry when rebuked, these men threatened revenge and ultimately formed an agreement to force the expedition's retreat to Fort

Connor. After a single day's return march, a company of cavalry with Indian scouts, under the command of Captain A. E. Brown, arrived as a relief escort. The cavalrymen of the earlier detachment had completed their enlistment period and were to return to Fort Laramie to be mustered out of service. Brown aided Sawyers in suppressing the insurrection among his men and assigned eight cavalrymen to accompany the road builders who were now able to move quickly beyond the Bighorn, over the Bozeman trail, and to their destination at Virginia City. The expedition's odometer recorded 1,039 miles of travel.[22]

Sawyers had originally planned for an immediate return over the route to make improvements and correct any unnecessary winding of the trail. Then, during the autumn, the Omaha branch was to be located. Delay because of bickering with the Army and the later Indian attacks had forced a cancellation of this scheme, so the employees were paid their wages and discharged. Rather than experience additional delay in disposing of his equipment in a market glutted by the wagons and camp equipage brought in by gold seekers, Sawyers transferred his outfit to a commission merchant for sale. He returned to Sioux City by the Salt Lake City stage.

The superintendent was satisfied with the location of his overland route and pleaded that it be kept open for travel by protecting emigrants and freighters from Indian attack. According to Sawyers, the Niobrara road shortened the distance between the Missouri River and the Montana settlements by 600 miles; stages took sixteen days to make the journey over the Platte route by Fort Bridger and Fort Hall, but only eight would be required on the new route. No mountain range would have to be crossed, and the superintendent testified to the presence of sufficient quantities of fuel, grass, and water at camping places.

Although the route was reported as satisfactorily located, Sawyers' work as a road builder was severely criticized by his military and civilian colleagues. The Army insisted that the road was not practical because of the roughness of the terrain and refuted his claims about water and grass. Improvements were reported to have consisted of only three or four bridges, not of a permanent nature, and the filling of gulches and ravines with dirt. The Army officials were sure that the superintendent was more intent on getting the freight wagons through than in building an adequate wagon road. Civilian employees added to these criticisms by presenting evidence to support charges of negligence of duty and misapplication of federal funds, accusing the superintendent of using the appropriation to pay the drivers of freight wagons transporting goods in which he shared a financial interest. According to others, the expedition equipment was used to repair settlers' wagons, and all privately owned cattle lost by emigrants and freighters during Indian raids was replaced with government stock.[23]

Soon after the organization of the wagon-road expeditions to Montana and Idaho, John P. Usher left the Interior Department. The new Secretary, James Harlan, prevailed upon President Andrew Johnson to assign an experienced military engineer to his Department as superintendent of railroad and wagon-road constructions. Lieutenant Colonel James H. Simpson was selected for the assignment by Secretary of War Stanton.[24] Simpson assured Generals Pope and Sully that the expedition's affairs would be investigated, and requested Representative Hubbard to call so that they "might talk about matters of interest to Sawyers and the government." Sawyers was notified to report for a conference with Secretary Harlan.[25]

The road superintendent denied the validity of the charges about his work. Instead, he accused the majority of his critics of negligence and cowardice. These men had failed, he contended, to supply the coöperation essential to an expedition of hazardous nature. He defended the presence of the emigrant and freighting parties in the wagon train; since the route was to be used by such groups, it was advantageous that they test its practicability. Moreover, the latter plan had been endorsed by Secretary Usher. No comment was made about his personal investment.[26] Sawyers' conduct was approved by the Interior Department, but Lieutenant Colonel Simpson continued to struggle with the confused accounts. Finally he sent Sawyers a sharp message: "Your inattention to the instructions of the office relative to the completion and timely rendition of your accounts has been a source of greatest annoyance. Please supply deficiencies and prevent any recurrence."[27]

Plans for a second expedition from Sioux City to Virginia City were initiated upon Sawyers' return from the first. The superintendent recommended an additional $20,000 appropriation for the Niobrara road. On January 8, 1866, Hubbard of Iowa introduced a bill into the House of Representatives authorizing the sum, but the Committee on Roads and Canals presented an adverse report.[28] The first expedition had cost only $20,000, and the remainder of the initial allotment, including that credited to the Omaha branch, was thought sufficient for the second trip.

In view of the Indian difficulties experienced in 1865, an escort for the new expedition seemed imperative. Hubbard presented requests to the War and Interior Departments. General U. S. Grant decided that General Sherman now had so few troops for service in the trans-Mississippi West that an escort could not be furnished. The Army should make every effort to scatter troops along the main routes across the Indian country to aid emigrants, but unless Congress granted an increase in the size of the Army, according to Grant, all escort service elsewhere would have to cease.[29] The Army thus planned a boycott of the Niobrara road. In March, Simpson ordered Sawyers to make no improvements on the Omaha branch of the road. The trace of the previous season was to be abandoned and the Platte Valley route followed to Fort Connor. This order was later countermanded by Secretary Harlan and instructions issued to move along the Niobrara.[30]

Field officers, unaware of Grant's decision, had promised Hubbard an escort and two howitzers for the expedition. General Pope was forced to withdraw the authorization. A mass meeting in Sioux City adopted a resolution emphasizing the road's importance and pleading for the Army escort.[31] Sawyers went to St. Louis to confer with Generals Sherman and Pope, only to learn that the Army was temporarily opposed to all wagon-road building across the Plains north of the Platte River.[32] Rifles and revolvers were secured for each member of the road party, and Sawyers headed west with no escort on June 12. Confusion over the escort was not the only factor forcing a late start. Supplies coming up the Missouri from St. Louis were delayed when a steamer ran aground. In the interim, twenty-five employees, already hired, were sent forward to improve the first sector of the road along the Niobrara.

Sawyers planned this second expedition to be more than a reconnaissance. Mules were secured for heavy plowing and scraping, and heavy grading was anticipated.[33] The 1866 expedition, when finally under way, was composed of fifty-seven employees and seven scouts and guards.[34] The route followed was practically the same as that of the preceding year. Many of the fords, previously improved, and several bridges were in good condition, and the trail was still marked. Where necessary, bridges were repaired or new ones constructed. Rapid progress was made to the crossing of the south Cheyenne, but beyond there to the Yellowstone Valley the party was again hindered in its efforts by relentless Indians. On July 8 a half-dozen braves, returning from peace talks at Fort Laramie, attempted to steal the expedition's mules. One brave was killed by the camp guard as the others fled. As a result the road party was under the constant surveillance of Indian scouts as it moved up the Dry Fork of the Cheyenne. The inevitable retaliatory attack came at dusk on July 13. The next morning an ambushed scouting party was saved from annihilation by the timely arrival of work crews. Three days later the road builders reached Fort Reno, five miles from the old Cantonment Connor. Here fourteen emigrants, with four wagons, joined the train for the protection gained by large numbers. In the dangerous Powder River country a corral was formed each evening and the sentry doubled. Night after night the Indians assailed the herd in the hope of starting a stampede and thus stranding the men. On July 21 Sawyers' crew came upon Colonel Carrington's command of seven infantry companies constructing Fort Philip Kearny on the Piney Fork of the Powder River. Stories of Indian depredations were exchanged, and Carrington assured Sawyers that Red Cloud was determined to stop migration along the Powder River. The superintendent requested the colonel to provide an escort as far as the Tongue River, but Carrington refused to do so on the grounds that it was inadvisable to divide his command. The road-building party was further reinforced by an emigrant train of thirty-two wagons and sixty-one men who had encamped near the new fort.

Indian attempts to scatter the oxen were made at the crossings of the Tongue and the Little Bighorn. As Sawyers described it, "the boys were too quick . . . and they 'skedaddled' into the brush with bullets flying like hail among them." At the Bighorn crossing the wagons were placed on a flat-boat and ferried across the river.

The Indian danger was now left behind. Sawyers procured local guides who led him over the shortest known routes of the Bozeman trail up the Yellowstone, across the divide, and into Virginia City. From the outset Sawyers had used his previous season's experience to shorten the road, and the total distance was cut more than 100 miles. The traveling time had been reduced from four months to two and a half. In Virginia City, Sawyers once more encountered difficulty in selling his outfit and the expedition's funds were so nearly exhausted that he had to borrow money to pay the crew. Once the accounts were balanced, Sawyers again transferred the remaining equipment to commission merchants in Virginia City.[35]

The *Sioux City Journal,* unofficial sponsor of the Sawyers' expeditions, received and published progress reports of this journey as of the first. When the superintendent arrived in Iowa, the newspaper announced:

> In a brief conversation with him since his return, he states that he found the route from Niobrara to Virginia City all that was claimed for it last year; and in fact, upon making the trip the second time, he is more than ever convinced of its superiority over any route heretofore travelled from the Missouri river. Mr. Sawyers is satisfied that the greater portion of the overland travel to the mines is destined to take the Niobrara route as shorter, safer, and better in every way than all others. Mr. Sawyers is a man of no idle words, and his statements can be safely relied upon. He could certainly have no object in trying to deceive the people by misrepresentation, as his work is now done, and his connections with the route at an end, and any effort to induce the people to travel by this route against their interests must react upon himself.[36]

In spite of the enthusiasm of Sawyers and the optimism of Sioux City residents, plans by private enterprisers to outfit an emigrant and freighting train to traverse the route in 1867 did not materialize. The expedition of 1866 was the last. Construction of the Union Pacific Railroad had earlier made unnecessary the construction of the Omaha branch of the road, and now the Niobrara wagon road, so closely paralleling the rail line along the Platte, could not meet the competition.

Even in view of the minimum standards expected by the emigrant and required by the government for an adequate road for covered wagons, the Sawyers road did not meet the specifications. The two trips west were less road-building expeditions than reconnaissances to explore the country for a new commercial outlet, primarily for Sioux City merchants. The pattern of activity served as a sharp reminder of the difficulties involved in civilian construction of the Pacific wagon roads between 1858 and 1860. The events of 1865–1866 also indicated that the contest between the War and Interior Departments over the administration of Indian affairs had extended to the federal road-building program. Both agencies were trying to do the same job, but there existed a basic conflict in organization and procedure between the civilian and military arms of the government that had made difficult, if not impossible, a coöperative project.

CHAPTER XVIII : *The Big Cheyenne River Road The Dakota Territory Route to the Northwest*

By an act of March 3, 1865, Congress had authorized a federal wagon road to run from the mouth of the Big Cheyenne River westward through the Black Hills to join the Powder River road. From the Big Cheyenne's outlet a northern branch was to extend eastward to the Minnesota boundary and a southern route was to follow the banks of the Missouri as far as Sioux City. Interior Secretary Usher considered the project primarily a patronage problem. Dakota's territorial delegate, Dr. Walter A. Burleigh, elected by local Republicans in 1864, was eagerly seeking appointments for the leaders of his political machine in the territory. As an agent for the Yankton Indians, with a small salary, Burleigh had amassed a fortune in a four-year term. The genial doctor made little effort to hide his graft and machinations from Dakota residents, but he had great facility in diverting the investigation of federal inspectors. With the profits of the agency business, he purchased sheep and flour to distribute among the destitute residents during the political campaign of 1864. By the time the scandalous proportions of his corruption became known to the Indian Office, the delegate had attained such influence in the Andrew Johnson administration that prosecution was never instituted. Few men in the early days of the territory were "so popular, so able, so big-hearted, so unscrupulous."[1]

For superintendent of the road from Minnesota to the Big Cheyenne toward the Black Hills, Burleigh nominated Wilmot W. Brookings, who had come to Dakota in 1857 and was among the first to settle in Sioux Falls. Brookings, a fortune seeker, was manager of the Western Town Company, and periodically a member of the territorial legislature. In 1864 he was elected speaker of the assembly.[2] Gideon C. Moody was made superintendent of the southern division, which included a bridge to be constructed across the Big Sioux and the road along the Missouri River to the Cheyenne. His political training had been received in Indiana as a leader of the "young

Republicans" supporting Oliver P. Morton in the 1850's. Moody's scathing attacks upon one who had challenged the position of the party leader precipitated a duel which the authorities stopped short of bloodshed. After a military career with an Indiana infantry company during the Civil War, where the sobriquet "colonel" was earned, he headed west to Yankton with the promise of a federal appointment.[3]

Moody was instructed to begin construction on the Big Sioux bridge, for which $10,000 had been designated. In the meantime, the Secretary of the Interior was to inquire at the War Department about the unexpended funds for the Sioux City–Fort Randall military road to determine the total amount available for the southern branch of the Big Cheyenne route.[4] Brookings was notified that his first job was to make a survey eastward from the mouth of the Big Cheyenne to the western boundary of Minnesota along a way which would serve emigrants and military trains and also would establish the best connections for St. Paul. For road surveyor and engineer, Brookings, in conference with Burleigh, selected George N. Propper, secretary of the 1864 legislature.[5] John B. S. Todd, former Dakota delegate who had aided Representative Hubbard of Iowa in securing the road program, only to be defeated for reëlection by Burleigh, protested the selection of politicians, rather than professional engineers, for the projects. In vain he sought the patronage.[6]

Brookings assured the Interior Department that he knew the wishes of the people of Minnesota and Dakota. The northern road should meet the Minnesota boundary near the forty-fourth parallel, or between it and Lake Benton. Travel converged on this point and connections could be made with the principal railroads of the state—the Southern Minnesota Railroad from St. Paul through St. Peter, thence southwest to the Big Sioux River, and the Transit Railroad located by charter from Winona via St. Peter to a point on the Big Sioux below the forty-fifth parallel. With departmental approval of this line for the road, he planned an early start on May 1, to prepare for the spring migration.[7]

Interest in the eastern sector of the road was short lived after Brookings was informed by General Alfred Sully, to whom he had addressed a request for an escort, that the Army would make an expedition westward to the Black Hills along the Big Cheyenne with 1,000 cavalry. Many gold seekers had arrived in Yankton and were urging the Army to give them protection. Brookings enthusiastically expected a train of 500 emigrants to be organized for the Black Hills and Montana, and urged the Interior Department to request from the Secretary of War an order to Sully emphasizing the desirability of coöperation with the wagon-road party.[8] Stanton was asked for an escort and authorization for the issuance of 50 Sharp's rifles, 50 Colt pistols, and 10,000 rounds of ammunition from Fort Randall.[9] Brookings was refused cash advances, however, and Washington authorities questioned his certainty that this single survey would establish the most satisfactory junction with the Niobrara road which, in fact, had not yet been located. The line of Brookings' survey was to be approved by the Interior Department before any permanent improvements were authorized.[10]

Leaving Yankton on June 12 the road party traveled 230 miles northwest to Fort Sully on the Missouri River. On arrival General Sully notified Brookings that his instructions from General Pope had been changed, and that he was positively ordered not to cross the Missouri River.[11] Brookings then appealed for an escort which was denied, and Sully urged him not to conceal from his men the danger beyond the Missouri:

> If you chose to go and run the risk you do so on your own responsibility. I can not assist you even if you are in danger. I would not send you a company of 60 men. It would invite an attack and you would be caught in the same trap that Capt. Fisk was last year with his large party of emigrants and 50 or 60 soldiers.[12]

The road superintendent was determined to explore a part of the route to the forks of the Cheyenne. He read the general's letter to his men and called for volunteers. Some stepped forward but a few individuals were frightened and left the expedition. After awaiting the Army's departure up the Missouri on July 10, and certain that the hostile Indians would follow Sully's command at a respectful distance to check its activities and destination, Brookings, with sixteen mounted men and one friendly Indian, headed west with three two-horse wagons and six additional saddle horses. The party traveled north of west to the old military road between Fort Pierre and Fort Laramie which was followed to the Cheyenne just below the junction of the river forks. On the return march the reconnaissance was due east striking the Missouri River at Fort Pierre. The distance measured by odometer was 86 miles.[13]

By July 20 Brookings had decided to transfer his work crew, stranded 230 miles from the Dakota villages, to the road toward the Minnesota boundary without awaiting a preliminary survey. Engineer Propper first traversed the country northward from Fort Sully to the mouth of the Big Cheyenne, the place designated for crossing the Missouri River. Directly across the river, bluffs were seen rising 250 feet above the bottom lands. The discovery of the bluffs was not a keen disappointment. The country lying due east was known to have a limited water supply and was destitute of wood for stretches as long as 80 miles. A line was, therefore, surveyed

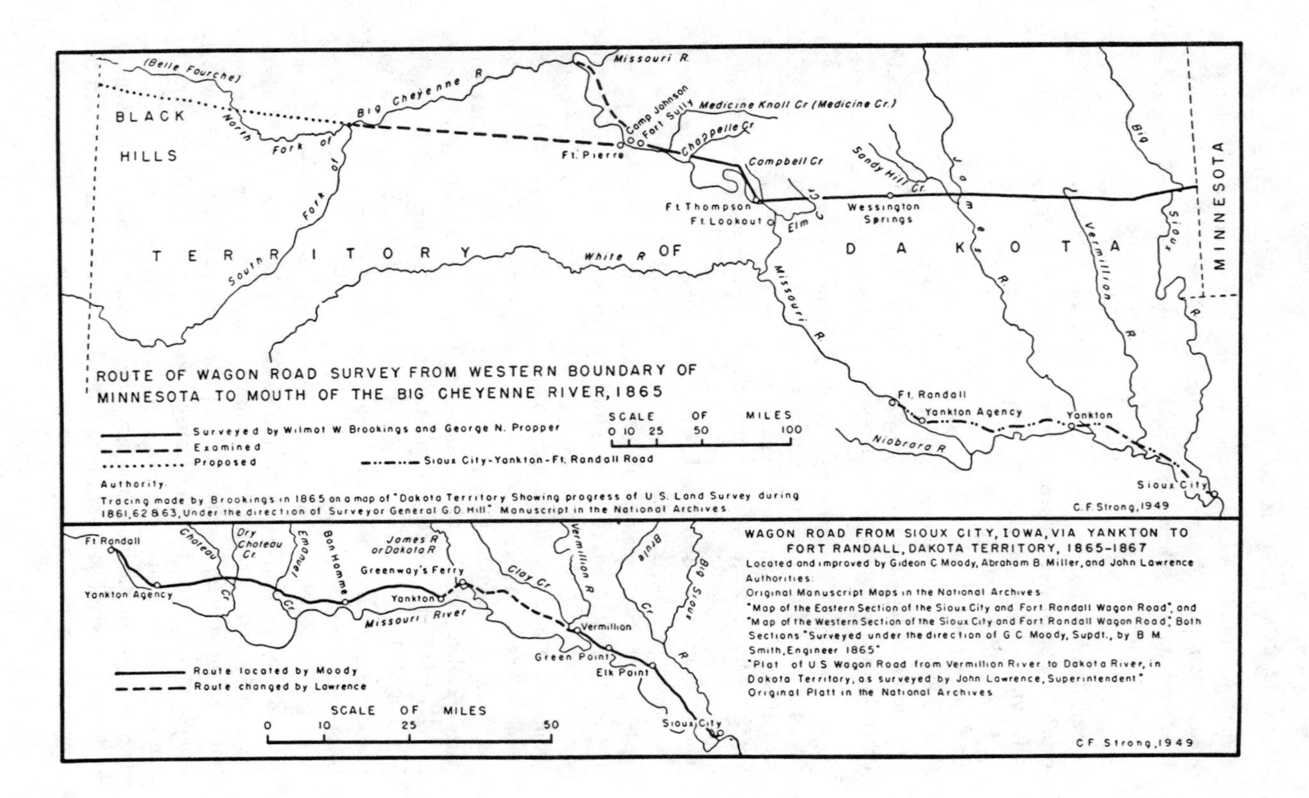

ROUTE OF WAGON ROAD SURVEY FROM WESTERN BOUNDARY OF MINNESOTA TO MOUTH OF THE BIG CHEYENNE RIVER, 1865
Surveyed by Wilmot W. Brookings and George N. Propper
Examined
Proposed
Sioux City-Yankton-Ft. Randall Road
SCALE OF MILES
0 10 25 50 100
Authority.
Tracing made by Brookings in 1865 on a map of "Dakota Territory Showing progress of U.S. Land Survey during 1861, 62 & 63, Under the direction of Surveyor General G.D. Hill." Manuscript in the National Archives.
C.F. Strong, 1949
(Belle Fourche)
BLACK
HILLS
North Fork of
Big Cheyenne R
South Fork of
Missouri R.
Camp Johnson
Fort Sully
Medicine Knoll Cr (Medicine Cr.)
Chappelle Cr
Ft. Pierre
Campbell Cr
Ft. Thompson
Ft. Lookout
Elm Cr
Sandy Hill Cr.
Wessington Springs
James R.
Vermillion R.
Big Sioux R.
MINNESOTA
TERRITORY OF DAKOTA
White R
Missouri R.
Ft. Randall
Yankton Agency
Yankton
Niobrara R
Sioux City
WAGON ROAD FROM SIOUX CITY, IOWA, VIA YANKTON TO FORT RANDALL, DAKOTA TERRITORY, 1865-1867
Located and improved by Gideon C Moody, Abraham B. Miller, and John Lawrence
Authorities:
Original Manuscript Maps in the National Archives
"Map of the Eastern Section of the Sioux City and Fort Randall Wagon Road", and "Map of the Western Section of the Sioux City and Fort Randall Wagon Road", Both Sections "Surveyed under the direction of G.C. Moody, Supdt., by B.M. Smith, Engineer 1865"
"Plat of U.S. Wagon Road from Vermillion River to Dakota River, in Dakota Territory, as surveyed by John Lawrence, Superintendent"
Original Platt in the National Archives
Route located by Moody
Route changed by Lawrence
SCALE OF MILES
0 10 25 50
C.F. Strong, 1949
Ft Randall
Choteau Cr
Dry Choteau Cr.
Emanuel Cr
Bon Homme
James R or Dakota R
Greenway's Ferry
Yankton Agency
Yankton
Missouri River
Clay Cr
Vermillion R
Brule Cr
Big Sioux R
Vermillion
Green Point
Elk Point
Sioux City

29 miles south to the road headquarters, Camp Johnson, opposite Fort Pierre. Here Brookings proposed the federal wagon road should cross the Missouri River, and rationalized his decision in reporting that the Cheyenne's forks were directly west, and a lengthy detour north to that river's mouth could be avoided. Leaving Camp Johnson, the road builders moved south of east to the Crow Creek Indian Agency, thence east of south to old Fort Thompson before passing over the plains in a due easterly direction to the Minnesota boundary.[14]

At the crossings of main streams, Elm and Sandy Hill creeks, the James, Vermillion, and Big Sioux rivers, fords were built by placing two rows of large boulders 24 feet apart. The road bed between these rows was filled with smaller stones, then covered with earth and gravel to make a smooth pavement. The banks of the streams were ploughed, scraped, and graded to facilitate the approach of wagons to these fords. The exact route was not fixed until the return trip. To guide future travelers, a monument of earth or stone was erected each half-mile with a four-foot stake inserted in the center, and the letters "M.W.R." burned into a flat surface at its top. In the level country these markers could be seen for several miles.[15]

The superintendent had completed the season's work with less than $7,000 and reported that additional expenditure east of the Missouri River was useless. Wagon trains would encounter no difficult terrain to the Missouri Valley where an old fur traders' and military road ran from Fort Thompson north to Camp Johnson. This road was packed as solid as stone by the passage of three thousand cavalry on Indian expeditions between 1862 and 1865. Like other federal road builders, Brookings announced that his survey would become the great thoroughfare to Montana, Idaho, and the Pacific and proposed that the remaining funds be used for a continuation along the north fork of the Cheyenne, skirting the Black Hills to the north, and joining the Niobrara road at the Tongue River crossing. Mountaineers and Indians reported that a good road could be located as far as the Little Missouri, though no credible information about the country beyond to the Powder River was available.[16]

Dakota residents hoped for a renewal of federal aid in establishing the Cheyenne road in 1866. The territorial assembly adopted a memorial requesting the establishment of a mail route between Fort Sully and Virginia City, Montana, with a weekly or semimonthly service. The War Department was asked to erect a military post at the north base of the Black Hills halfway between Forts Sully and Connor. The Cheyenne road was reported to be 500 to 600 miles shorter than the Omaha–Fort Laramie route to Montana. The Army could thereby save tremendous sums in transporting supplies to forts on the Powder River road. Opening the new route would facilitate exploration of the Black Hills where reliable informers reported the presence of precious metals. Moreover, timber on the hillsides, desperately needed by settlers on the Dakota plains, would become accessible.[17]

General Sully assured Dakota's governor, Newton Edmunds, that the United States Army favored the Big Cheyenne route, protected by a fort in the Black Hills, and urged him to secure unity among the territorial politicians in seeking congressional aid. According to the general, Dakota and Iowa insisted upon both the Niobrara and Cheyenne routes; Wisconsin and Minnesota wanted a line by way of Lewis Lake and Fort Pierre; those to the south championed the Platte and the Arkansas. In spite of Indian concessions, the Army could not guarantee the maintenance of any route without the construction and garrisoning of forts near by. Since emigrants to the south had roads from Omaha and Fort Leavenworth guarded by troops, those in the north should agree on the Cheyenne road, which was shorter than the Niobrara road to Montana. People from Wisconsin and Minnesota as well as northern Iowa and Illinois could benefit by this road across Dakota.[18]

General Sully was attempting to heal the breach between the military and civil authorities, growing out of Indian treaty negotiations the previous year. Governor Edmunds visited President Lincoln in Washington in January, 1865, just as Congress was drawing to a close, and suggested the appointment of a commission to enter the Indian country west of the Missouri and negotiate a peace with all the inimical tribes. The President, impressed with Edmunds' plan, introduced him to the congressional committees handling Indian affairs. A treaty commission was authorized and $20,000 appropriated for expenses. The President named a panel of six: two civilian officials, Governor Edmunds and Edward B. Taylor, superintendent of Indian affairs; two military men, Major General S. R. Curtis and Brigadier General Henry H. Sibley; and two prominent citizens, Henry W. Reed and Orrin Guernsey. To the annoyance and consternation of Edmunds, General Pope would not permit the commission to enter the Indian country or engage in peace negotiations. The ensuing conflict was terminated by the intervention of Secretary of the Interior Harlan, who secured a revocation of the military order against the commission.[19] At Fort Sully, during October, the Teton bands of Sioux came to terms with the white men. The Indians were to withdraw from the overland routes already established or to be established in their domain, and agreed not to interfere with the persons or property of citizens traversing these routes. The United States was to pay $6,000 annually for twenty years to hold this concession. A reservation for the Lower Brules was established along the east bank of the Missouri River for a 20-mile stretch between the White River and old Fort Lookout, running ten miles back from the river. The United States reserved the right to construct roads through the reservation. The Senate ratified this agreement, and on March 17, 1866, President Johnson proclaimed the treaty in force.[20]

Brookings, though not a participant in these affairs, knew what had taken place and in January, 1866, wrote to the Interior Department for support in securing an escort of two hundred infantrymen, twenty-five cavalrymen, and two pieces of artillery for an 1866 road expedition. Since this party would be the first to cross the Indian country after the Fort Sully treaties, he argued, it should be so well protected that the Indians would not be tempted to break their compact. Men of science, prospectors, and Montana emigrants all hoped for military protection.[21] As the ratification of the Indian treaty was still pending, Secretary Harlan directed Brookings to suspend, for a time, all operations on the Big Cheyenne River road and close the accounts.[22]

Dakota residents were dismayed by this turn of events. Moses K. Armstrong, clerk of the Territorial Supreme Court and secretary of the Dakota Historical Society who was also on the opposite side of the political fence from Burleigh and Brookings, tried to capitalize on the discontent. Rumors were circulated that the road work was suspended because Brookings refused to go without a military escort and had caused a rift in relations between the War and Interior departments. Armstrong proposed to organize two hundred mounted men from the Missouri Valley, interested in opening up the Black Hills, to ride to Fort Connor and back before November, 1866.[23] In the meantime, Lieutenant Colonel Simpson, in charge of the Wagon Road Office in the Interior Department, instructed Brookings to transfer all property belonging to the Big Cheyenne road to James A. Sawyers, superintendent of the Niobrara road.[24] Armstrong now withdrew his proposal, but suggested that Sawyers be instructed to open up a route from the head of the Niobrara northward to the Black Hills, and offered his services as journalist and topographer.[25]

The Army insisted on opening the Big Cheyenne road; the Interior Department was equally determined to champion the Niobrara road. Brookings, now promised an Army escort, approached Edmunds, who had guaranteed the right of transit for wagon roads in the peace treaty of 1866, and together they appealed to Burleigh, always an ardent supporter of his constituents' interests.[26] The upshot of the affair was a congressional inquiry of Secretary Harlan's action and motivation. The Interior Department produced evidence to show that the majority of the treaty commissioners, including Governor Edmunds, recommended an end to road building. They wrote the Secretary:

... it would be highly impolitic and dangerous to make any further survey of routes through the country inhabited by the Teton band of Sioux Indians before the negotiations contemplated with the bands and tribes ... have been fully completed, and the annuities under the latter fully paid as stipulated. There is a serious doubt on the part of all the savages heretofore hostile as to the good faith of the government towards them, and it is therefore highly desirable that the treaties already made be ratified, and the annuities provided paid to the several bands with the least practicable delay.[27]

Moreover, agreed Harlan, a military guard of two hundred men was thought necessary by Brookings. The maintenance of troops on the plains, including pay, subsistence, and transportation, cost the federal government $1,000 yearly for each soldier, and the Secretary of the Interior considered this expense for an escort excessive. More important, the best qualified engineers and topographers doubted that a route could be opened through the Bad Lands, and it was probable that the proposed road would have to diverge toward the north and follow the Missouri River, or to the south in the vicinity of the Niobrara route, then in construction. Either eventuality would render the expenditure useless. The Secretary concluded: "The department did not, therefore, deem it wise to attempt to disburse so trivial a balance on a work of doubtful utility, at so great a cost of treasure for military guard, and considerable peril to life, in face of the remonstrance of the commissioners ..."[28] Secretary Harlan stood by the decision until the close of his administration in December, 1866.

Immediately upon the installation of Orville H. Browning as Secretary of the Interior, the campaign was resumed to locate the Big Cheyenne road which would connect with the Niobrara–Virginia City road near Fort Philip Kearny, 300 miles west from the initial point on the Missouri. Former superintendent Brookings reviewed the project's history and urged that the unexpended balance of the appropriation be used to aid the spring migration of 1867.[29] Burleigh enlisted the support of his old friend Oliver P. Morton of Indiana in prevailing upon the Interior Department for a renewal of the survey.[30] General U. S. Grant produced an order to the military commander of Dakota authorizing protection for the Cheyenne road if it were consistent with the demands for troops elsewhere.[31] In an attempt to present a united front, political feuds in Dakota were forgotten, and Burleigh nominated his former opponent, J. B. S. Todd, as superintendent of the road.[32] Brookings, Governor Edmunds, Asa Bartlett, chief justice of the territory, and Lucian O'Brien, editor of the *Dakota Republican,* organized a pressure group to petition the Interior Department for a change in policy.[33] Apparently all was to no avail, for Browning announced in the annual report of 1867:

I have declined ordering a resumption of work on the projected road from the mouth of the Big Cheyenne to a point on the Niobrara road, in consequence of the hostile attitude of the Indians. The unexpended appropriation applicable to this road is twelve thousand, one hundred and fifty-seven dollars and seventy cents ($12,157.70).[34]

The treaty with the Sioux at Fort Laramie, April 29, 1868, ended all hope for the construction of the Big Cheyenne wagon road. All lands west of the Missouri River to the one hundred and fourth meridian and north of Nebraska to the forty-sixth parallel, encompassing the western half of present-day South Dakota, were set aside for an Indian reservation. The United States, as usual, reserved the right to transport the mail and build wagon roads through the Indian country. However, Red Cloud had forced the federal government to accede to his demand that the Powder River road be closed and troops be withdrawn from Forts Philip Kearny and C. F. Smith.[35] With the closing of this route, there was little point in the construction of a military road from the Missouri River to the Black Hills with no western outlet.

Colonel Moody had meanwhile assumed the role of dispenser of federal patronage in the towns of southeastern Dakota Territory. The delay in securing the transfer of $15,000 for the Sioux City–Fort Randall road from the War Department permitted concentration on the Big Sioux bridge. By May 31, 1865, the site had been chosen and the superintendent was busily engaged in organizing working parties. The summer was mainly taken up in procuring timber, ten to 60 miles distant, and hauling it to the site.[36] Due to administrative changes in the Interior Department with the retirement of Secretary Usher, there was little supervision. When Lieutenant Colonel Simpson assumed responsibility for the road program, he immediately instructed Moody to make a survey of the road from the Big Sioux mouth via Yankton to the Big Cheyenne. Once the survey was approved, the superintendent was to build a road 30 feet wide and to cut any timber along the right of way that would obscure the sun. Major streams were to be bridged. Wet sections were to be covered with a causeway of logs at least 18 feet wide, topped with brush and earth.[37]

In October Moody submitted plans for a 626-foot bridge across the river opposite Sioux City and indicated that preliminary construction was under way. Great delay and unusual expense had been incurred in locating suitable material for building such a substantial structure. He reported that a bridge of cottonwood logs, available along the Big Sioux, would have been cheaper, but of no permanence. A shorter bridge would have been swept away by floods. The initial $10,000 presumably was expended on gathering supplies, and Moody proposed to Representative Hubbard and Governor Edmunds that funds from the road appropriation be transferred to the bridge construction. The three informally agreed that a better plan would be to ask Congress for an additional $10,000 earmarked for the bridge.

During this first season no road work had been done. In October, quite late in the working season, the superintendent admitted planning the departure of a surveying party to Fort Randall along a road which he described as already well established and well traveled.[38] In reply to Simpson's request for a full report on his year's activities, Moody could only forward a detailed estimate of $12,972 for completing the bridge.[39]

Rumors were rife throughout the territory that Moody had squandered, or misappropriated, the federal funds. One persistent story suggested that the bridge money had been used by the superintendent to purchase a flock of sheep. Enos Stutsman, a political opponent, secured the passage of a resolution by the territorial assembly, requesting a statement of disbursements. When this was not immediately produced, a second resolution was passed condemning Moody's conduct. Doane Robinson, South Dakota historian, has concluded:

> ... if Colonel Moody erred in the disbursement of this large bridge and road fund it was in the interests of the half-starving, drouth and grasshopper-stricken pioneers of Dakota.... From the standpoint of strict economy the money may have been improvidently used, but no evidence has been found that it was used corruptly, dishonestly, or for the pecuniary profit of Colonel Moody.
>
> ... [Moody] employed to a very large extent the Scandinavian farmers numerously populating the southeastern counties of the territory, and so arranged the work of construction that they were able to give their farm duties proper attention and build the road during seasons when farm work was slack ... Moreover, having learned by careful calculation that the road could be built for much less than the appropriation, he voluntarily paid the workmen almost double the ruling price for men and teams.... But it endeared him to the people of the southeastern counties, and made the Scandinavian farmers ... always with him to a man in politics and business.[40]

The "colonel" was preparing the way for a territorial judgeship and a seat in the United States Senate.

Representative Hubbard filed a protest with the Interior Department in

August, 1865, because of Moody's failure to construct the road in conjunction with the bridge building.[41] When news of the vote of censure of the territorial assembly reached the Department, Colonel Simpson directed his assistant, John R. Gillis, to go to Sioux City for a general inspection and report.[42] The inspector examined the Big Sioux bridge, one-third completed, and found the structure faulty in design, insecure, and unnecessarily expensive because of the superintendent's lack of technical knowledge. Gillis later met Moody in Yankton and together they traveled the survey from Sioux City to Yankton, which generally followed the Missouri River bottom. Although a harder surfaced road, more likely to be free of bogs, could be located on the bluffs, it was longer, and wood and water were scarce. The inspector certified the river route. Beyond Yankton to Bon Homme the federal road would, on the contrary, follow a prairie road away from the river. Between Bon Homme and the Yankton Agency, two alternative routes could be followed: Brown's route via his trading post, or the older Cooper route along the river. The stage lines were moving to the new route and the inspector preferred it. From the Yankton Indian Agency to Fort Randall, the federal road would run along the Missouri River.[43] The entire distance was 126 miles, with Yankton a half-way point between termini. Apparently everyone had forgotten the original congressional plan to extend the road beyond Fort Randall to the mouth of the Big Cheyenne.

As soon as the superintendent's plan was announced, various local pressure groups protested the location of some segment that adversely affected their economic holdings. Committees in the territorial assembly made some relocations, then secured legislative approval of a favored route by designating it a territorial road and increasing the width to 80 feet.[44] Byron M. Smith, the road engineer, reported these developments to the Interior Department and requested the Secretary to resolve the conflicting representations. The majority of Dakotans agreed that the controversy should be settled permanently by some disinterested individual, so that fields along the road could be fenced and plans made for its maintenance. Smith agreed that the road should follow the river except between Yankton and Bon Homme, to connect the farms and towns. The back routes ran through lonely desolate country, and some settlers, losing their way, had been frozen to death in the winter months.[45]

Back in Washington, inspector Gillis prepared an elaborate report concluding that more than anything else a good engineer was needed because of the inexperience of both Moody and Smith.[46] Simpson, seeking a disinterested party, recommended that Secretary Harlan appoint A. B. Miller of New York, "a prudent, economical, honest, and capable engineer."[47]

Simpson notified Moody to make no further expenditures except to safeguard the timber at the bridge site.[48]

Iowa and Dakota politicians continued the effort to secure another congressional grant. Through the influence of Burleigh, the appropriation committee called upon Secretary Harlan to present estimates for the project's completion. The Interior Department proposed $15,000 to finish the Big Sioux bridge, $6,000 each for those over the Vermillion and Dakota rivers, a total expenditure of $27,000.[49] On April 7, 1866, Congress appropriated $10,000.[50]

Miller, Gillis, and Simpson had conferred in Washington in January, 1866, and agreed upon construction plans. The new engineer headed for Dakota in February, and immediately upon arrival inspected the Big Sioux and Vermillion bridge sites. He then traversed the route surveyed to Fort Randall together with the alternate proposals. In his enthusiasm, Miller attempted to work during the winter months, was caught in a Dakota blizzard with his wagon team on the road, and was forced to abandon it and to proceed on foot. Lost when darkness came, he fortunately found a fence leading to a farmer's barn where he passed the night in violent exercise to keep from freezing in fifteen-below-zero weather. With hands and feet "frosted" by this experience and unable to wear gloves or boots, Miller took the warning to await spring.[51] At that time the Moody-Gillis survey was approved except for the section between Yankton and Bon Homme, where Miller, on examination, preferred Cooper's river road to Brown's route across the prairie. The road beyond the Yankton Indian Agency to Fort Randall was considered dangerous for loaded wagons, some having overturned and injured the drivers.[52] By June 1 the road was cleared from Sioux City to Yankton, and construction crews were busily engaged on the Big Sioux and Vermillion bridges.[53]

The location of the James, or Dakota, River bridge had become a source of bickering among the local economic factions and the federal engineers. Before 1862 the Vermillion-Yankton road had crossed the river at Greenway's Ferry. One mile below the ferry a settler, LeBlanc, had in that year built a bridge across the James on property he hoped to sell to the federal government. A mile nearer the river's mouth, one Van Osdel operated a ferry which he proposed the government might use free from tolls if the bridge was located on his site.[54] Moody and Smith accepted his offer.[55] LeBlanc immediately appealed to General Sully, whose army had used his bridge, for help against the "political hacks, Moody, Edmunds, and Burleigh" whose favor he could not gain. Sully notified the Interior Department that "the road now travelled to Yankton is a good and direct route as any and moreover has the advantage of a good bridge over the James River, lately built by citizens which obviates the necessity of the government going to the expense of building another bridge."[56] LeBlanc's original request for his bridge was $2,000, but he soon was prepared to take any offer. Inspector Gillis had approved the Van Osdel site and Miller preferred it.[57]

Brookings instigated a campaign in May, 1866, to change the crossing of the James to Greenway's Ferry farther north, for the benefit of his Sioux Falls friends. A petition from Yankton County requesting the change went to delegate Burleigh who presented it in person to Secretary Harlan.[58] Simpson prepared a brief attempting to justify the earlier decision, which only evoked from Brookings a bitter denunciation of the procedures used by the department inspector in gaining information, and further denounced him for misrepresentation.[59] In defense Gillis suggested that all personal testimony in such matters was of little value: the question was simply one of engineering, and his judgment concurred with that of Moody and Smith. The decision, possibly erroneous, had been confirmed by Miller who had no personal stake involved. Gillis informed the Secretary that Brookings and his land company virtually owned Sioux Falls.[60] Simultaneously he warned Miller that the politicians were gaining control of policy. Iowa's Hubbard joined Burleigh in forcing the Interior Secretary to order the crossing of the James at Greenway's.[61] The Department's engineer protested that he had already constructed the road to Van Osdel's site, but would erect the James River bridge as instructed when the other structures were completed.[62]

By November, 1866, a road along the 60-mile stretch from Sioux City to Yankton was opened 40 feet wide. Thirty bridges, varying in length from ten to 150 feet, had been constructed to aid wagon travel. The bluffs at Fort Randall were graded. Miller considered the bridge and road work two-thirds completed, though beyond Yankton the road was no more than passable for wagons.[63]

Lieutenant Colonel Simpson and his assistant, John Gillis, were removed from the administrative responsibility when Secretary Harlan left the Interior Department. Dakota's Burleigh urged Secretary Browning to replace Miller with a local engineer.[64] Miller in the meantime concentrated work on the Big Sioux and Vermillion rivers. Burleigh, sensing a stall, appealed to the Secretary of the Interior to press for construction across the James at Greenway's, the site now sanctioned by the Dakota territorial legislature.[65] The engineer had decided to capitulate to the local interests, and insisted that only shortages of funds made impossible the immediate construction of a bridge at the new site to the north. When the Vermillion bridge was finished, he reported the remaining funds insufficient to launch the new project and requested permission to dispose of all road equipment. Excited by this predicament the Yankton County residents petitioned their commissioners not to lose any part of the federal money but to guarantee the possible deficit.[66] The commissioners therefore protested to the Interior Department that thirty-five months had passed since the initial appropriation and still no bridge was completed over the James River. They would now assume the responsibility and finance the improvement by voluntary subscriptions.[67] At this juncture, Miller died in Sioux City.[68] On his person $6,000 in government bonds and more than $4,000 in cash were found. No one knew if these funds were public or private, and the road work was further delayed pending investigation.

Burleigh was determined to secure more federal money for Dakota roads.[69] Since the Fort Laramie treaty (1868) contemplated the occupation by Sioux Indians of that part of the territory where the Big Cheyenne road was to have passed, he prepared to ask Congress for the remaining funds to complete the James River bridge and the road leading west from Sioux City. Secretary Browning provided the figures on the unexpended balance, $12,157, for incorporation in the legislation.[70] During February, 1868, Burleigh succeeded in getting the approval of the House of Representatives. In April the Senate Committee on Territories endorsed the House measure. Senator George F. Edmunds of Vermont challenged the committee report, insisting that the Secretary of the Interior had estimated $1,500 sufficient for the bridge. Edmunds suggested an amendment limiting the appropriation to $2,500, but John Conness of California secured an agreement on the $1,500 allotment. On June 17 the territorial committee submitted a resolution requesting the House to return the bill to the Senate for correction because of an error by the Interior Department. Once more the Senate committee recommended the original $12,157 appropriation. Senator Edmunds again raised an objection, but within a week a compromise agreement to authorize $6,500 was reached. The bill was signed by President Grant on July 13, 1868.[71]

Delegate Burleigh meanwhile secured the appointment of his Dakota colleague, John Lawrence of Yankton, as the federal engineer.[72] Permission was once again requested to build the James River bridge at Greenway's, but Browning hesitated because his $1,500 estimate to Congress had been based on the assumption that Van Osdel's site would be used. However, the change was approved and Burleigh advised that $6,500 was required; hence the necessity for the speedy reconsideration by Congress.[73] During September and November, Lawrence erected the bridge but found he had expended

$1,000 in excess of the appropriation.[74] Browning appealed to Burleigh, who secured another federal grant.[75] By April 1, 1869, the engineer inquired if he might resume work. The new Secretary of the Interior, Jacob D. Cox, was informed by the Treasury that the money could not be expended until the new fiscal year, July 1. Burleigh was dissatisfied and protested in person to both cabinet members. Secretary Cox finally wrote Lawrence: "Go ahead and complete work. *Complete said bridge.*"[76]

When the bridge was finished, Dakota's new delegate, S. L. Spink, who had come to the territory in 1865 as secretary, carried on the Burleigh tradition in requesting permission to use unexpended funds on the road from Yankton to Vermillion.[77] The following year the delegate forwarded a petition from his constituents urging additional improvements on the road west of the James River to Fort Randall. Stage drivers insisted that the route on the bluffs near the fort imperiled life and property. Had the funds given out before the federal agents got this far? Since this wagon road was located in the Yankton Indian reservation only the government was empowered to make improvements.[78] Spink failed to get another appropriation, so the Dakota legislative assembly in 1873 placed the responsibility for the federal wagon road under the various counties through which it ran.[79]

Between 1865 and 1869 Congress had awarded $52,000 to Dakota Territory for the southern branch of the Big Cheyenne road. No consideration was given to possible improvements north of Fort Randall. On the southern third of the road between the fort and Sioux City, the expenditure of funds was concentrated east of Yankton. Thus, three times the amount of money was spent on a federal road one-sixth the length of that envisioned by the Congress which launched the project in 1865.

The Dakota and Iowa politicians were doomed to disappointment in their plans for two overland routes to the Northwest. The Big Cheyenne construction was a fiasco. Throughout, the Interior Department had found itself in the unenviable position of having been assigned responsibility for a road project which would have undermined its Indian policy. Moreover, the events associated with the construction of the southern branch of the Big Cheyenne road provided perfect examples of maladministration, the continuous pressure for federal patronage in the territories, and the procedures used in obtaining it in the years after the close of the Civil War. Five Secretaries of the Interior had participated in the struggle to locate and complete the road.

CHAPTER XIX : *Route from Lewiston, Idaho Across the Bitterroots, to Virginia City Montana*

THE SELECTION of a superintendent for the wagon road from Lewiston, Idaho, to Virginia City, Montana, was the most difficult personnel problem faced by the Usher administration in launching the 1865 road program. The position carried a stipend of $2,000 a year, and no competent engineer desired to work for that amount on this western sector running through the gold fields where prices were inflated. Although Dakota's John B. S. Todd and Minnesota's Ignatius Donnelly proposed candidates for the superintendency, the Secretary of the Interior early determined this appointment would be made by James Harlan, at the time an Iowa senator.[1] John Connell of Toledo, Iowa, who had lost an arm on a Civil War battlefield, received the assignment but refused to accept on the advice of his physician.[2] In late May, 1865, Wellington Bird of Mount Pleasant, Iowa, was offered the post, but declined because the season was too late to cross the Plains and inaugurate the road program before winter. Bird suggested a mutual friend to Harlan. Nothing came of this recommendation, and in January, 1866, Bird reconsidered and accepted the appointment for the next season.[3]

Idaho residents were sorely disappointed that the first working season after their delegate had secured the $50,000 appropriation had been allowed to pass without construction. The council of the territorial assembly addressed a memorial to the Secretary of the Interior, requesting a beginning on the road from Lewiston by way of the Lolo Fork of the Clearwater River to the summit of the Bitterroot Mountains, thence by way of the Lolo Fork of the Bitterroot to intersect the Mullan military road at Hell Gate Canyon, near present-day Missoula. Such a road would shorten the distance from Walla Walla to Virginia City by 160 miles. According to the petitioners, the Mullan road, notwithstanding the expenditure of $230,000, could not be traveled in the spring, even by pack trains, because of the marshy ground in some sections. The road had not been used by loaded wagons for several years.[4]

Although Bird originally planned to outfit his expedition in Iowa and carry a year's supplies across the Plains in six or seven wagons pulled by mules, the Secretary of the Interior ordered him to postpone preparations pending a conference in the national capital.[5] George B. Nicholson of Ohio was named engineering assistant.[6] The superintendent and engineer, in conference with Lieutenant Colonel Simpson, agreed to go by steamer to Portland and procure supplies as well as laborers in the Northwest. A scientist was thought desirable. For a time, A. Winchell, professor of geology, zoölogy, and botany at the University of Michigan considered the post, but upon the recommendation of John Evans of Colorado, Oliver Marcy of Northwestern University in Evanston finally was appointed.[7] The Washington conference also discussed the possibility of using Nez Percés Indians as laborers in preference to Chinese, and the Superintendent of Indian Affairs asked the local Indian agent to consider any proposals made by superintendent Bird.[8]

Leaving New York on March 10, the superintendent, engineer, and scientist arrived in Lewiston on May 1, much too early to commence road building. Bird spent several weeks studying local geography and currently traveled routes from the Columbia River to Montana, and also evaluating the conflicting opinions of Idaho citizens about transportation needs. The only known route used by the stage and wagons ran from Walla Walla southward to Fort Boise, on to Fort Hall where travel divided, some going to Salt Lake City, others to Virginia City. To the north, the Mullan road ran from Walla Walla to the Coeur d'Alene Lake and along the Bitterroot River to Missoula. Even farther north was the circuitous route by way of Clark's Fork of the Columbia and Lake Pend d'Oreille. On these roads pack trains were used. Directly across the mountains east of Lewiston were two passages: the Lolo route striking the Bitterroot Valley halfway between Fort Owen and Missoula, and the south Nez Percés trail to the headwaters of the Bitterroot by way of Elk City. Mining parties had crossed the mountains by both routes but no pack trains had come through. Although the Idaho legislature had endorsed the Lolo route, the superintendent planned to examine both for comparative advantages.[9]

Lewiston citizens, interested in the road, provided compensation for a local guide, one Colonel Craig, who had been a member of the Stevens party. Bird and Nicholson agreed to hire a former government surveyor in Oregon, Major Sewell Truax, as assistant. Truax had also commanded the

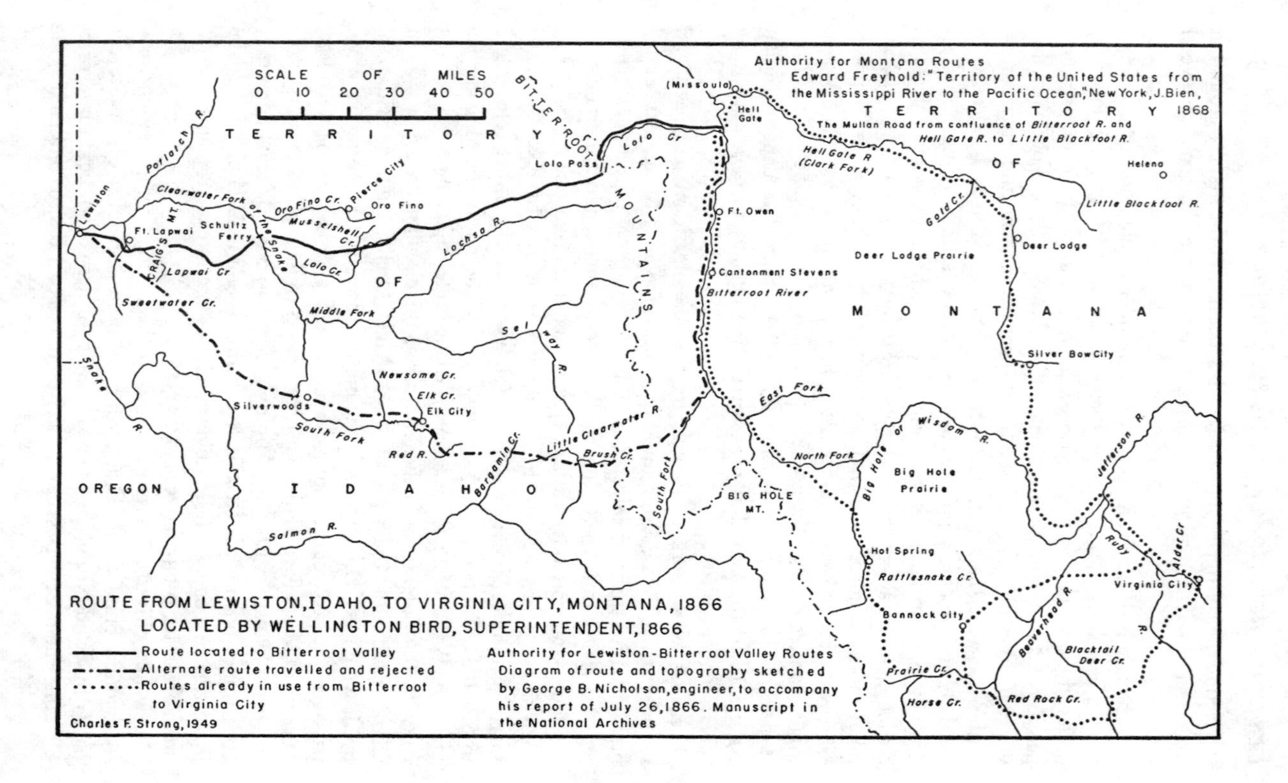

ROUTE FROM LEWISTON, IDAHO, TO VIRGINIA CITY, MONTANA, 1866
LOCATED BY WELLINGTON BIRD, SUPERINTENDENT, 1866

Route located to Bitterroot Valley
Alternate route travelled and rejected
Routes already in use from Bitterroot to Virginia City

Authority for Lewiston-Bitterroot Valley Routes
Diagram of route and topography sketched by George B. Nicholson, engineer, to accompany his report of July 26, 1866. Manuscript in the National Archives

Charles F. Strong, 1949

military post at Fort Lapwai and acquired many properties in Idaho. A Nez Percé, Tah-tu-tash, was to go along as interpreter. Bird described the party as consisting of "... one engineer, one surveyor, one physician and geologist, one interpreter, one wagon master, one assistant wagon master, one carpenter, one blacksmith, two night herders, and fifty men for cooks, drivers, and laborers."[10] Large scale preparations were made for the party. One spring wagon pulled by mules was to travel in the forefront, with twelve covered wagons hauling equipment behind the crew. Forty yoke of oxen were to pull supplies including blacksmith's equipment, medicine chests, tents, five mess chests, five sheet-iron cook stoves, buffalo robes, blankets, and six months' subsistence for sixty men. Large quantities of working tools were procured: "one grading plow, twenty-four mining spades, forty-eight railroad spades, twenty-four picks, twenty-four chopping axes, and twelve heavy hatchets." By the time surveyor's instruments, field glasses, and odometers were purchased, $20,000 of the appropriation had been expended.[11]

Leaving Lewiston on May 24 the party followed a military road as far as Fort Lapwai. Beyond the fort and Indian reservation, the surveyors crossed Craig Mountain to the Clearwater Fork of the Snake at Schultz's ferry, 62 miles east of Lewiston. They then pushed forward to Musselshell Creek, 30 miles distant, by June 6, and were forced to go into camp to await the melting of winter's snow. Reconnoitering parties ascending the Bitterroot Range reported it covered with dense forests or heavy underbrush, and in the canyons at the divide, 6,000 feet high, the snow was six feet deep. No better way was found, however, than the Lolo trail used by Lewis and Clark in passing from the Missouri to the Columbia in 1805. Superintendent Bird notified the Interior Department that once the survey was complete, he would cut a narrow path through the forested mountains for pack mules. Businessmen in Idaho would be satisfied, and the trail could become the basis for a permanent wagon road. Fifty thousand dollars was a mere pittance to inaugurate the construction of a road through these mountains.[12]

Five miles beyond Musselshell Creek the surveyors crossed the Clearwater Lolo, followed that stream to its headwaters, and then began to ascend the mountains. Vegetation was so thick the surveying instruments were useless. Parties of workmen blazed the way by hacking through the forest, finding each others' direction periodically by shouts. The instrumental survey was made after the trail was cut. Where possible the route was kept on the backbone of the ridges to avoid descending into canyons, dropping from 1,000 to 4,000 feet below the mountain crests. The divide was crossed at Lolo Pass and the Lolo Fork of the Bitterroot followed to its mouth, where the party arrived July 7 in a state of exhaustion.[13]

From the Bitterroot Valley many local roads led to the Montana towns, and the superintendent saw no need to explore farther. At John Owen's fort, just below the Lolo mouth, the proprietor convinced Bird that the mining resources of Virginia City were of secondary importance to the business activities of Deer Lodge, Helena, and Fort Benton—all reached by a junction with the Mullan road at Hell Gate Canyon. Professor Marcy, who had busily accumulated specimens of natural phenomena and taken copious notes on scientific and economic observations, was ready to return East by way of the Missouri River. To comply with the technical requirements of his assignment, the road superintendent ordered the professor to prepare an official report on his journey to Virginia City en route to Fort Benton. From Missoula to Deer Lodge, Montana, a distance of 91 miles, the professor's party traveled the Mullan road. By now the original bridges across the Big Blackfoot and the Hell Gate had been swept away; private citizens had erected a toll bridge at the first site and a ferry was used at the second. Marcy noted that heavily loaded pack animals were constantly passing. A local road from Deer Lodge to Silver Bow was used more by emigrant wagons than by freighters. South of the Silver Bow, a new road had been made across the Divide between the waters of the Columbia and the Missouri to join the thoroughfare connecting Virginia City, Helena, and Fort Benton. Stage coaches and freight wagons traversed this route daily. The scientist notified Bird that there was no need for government aid on the 200-mile stretch from Missoula to Virginia City because local businessmen would make necessary repairs.[14]

On July 12 Nicholson, Truax, and Tah-tu-tash ascended the Bitterroot River to its headwaters and traveled westward across the mountains along the southern Nez Percés trail. Bird and the rest of the party retraced the Lolo route between Fort Owen and Fort Lapwai. The Nicholson detachment hurriedly climbed the 6,000-foot summits and periodically descended 2,000 to 4,000 feet into canyons such as those at Brush Creek and the Little Clearwater. Within eight days they arrived at the mining community, Elk City. They had established a new speed record for the 177-mile journey from Fort Owen. Nicholson left his companions here and headed westward to Silverwoods Mountain House where a good stage road led into Lewiston. He reported the distance by this route 233 miles as compared with 200 by the Lolo forks, with mountainous terrain extending 35 miles farther.[15]

The superintendent therefore planned to improve the Lolo route as the most direct passage to the Hell Gate, thence to the commercial centers of

Montana. The Interior Department was informed, however, that this route, in common with all others, would be obstructed by snow from six to eight months of the year. From road headquarters, on Musselshell Creek, a party of sixty laborers under Major Truax attempted to cut a trail ten to twelve feet wide across the mountains during August and September. Colonel Craig was employed in advance of the workmen reëxamining mountain passes to keep the heavy grading at a minimum. Nicholson followed the work crews, making a detailed survey. Subsistence stores and camp fixtures were moved by pack animals bringing up the rear. By the last week of September road work stopped along the Lolo Fork of the Bitterroot, 25 miles from its mouth.[16]

On examining the financial accounts, Bird discovered that $42,000 had been expended. In his opinion the remaining balance did not justify keeping the stock during the winter so he disposed of all expedition equipment except road tools and camp fixtures. After making a personal arrangement with Major Truax to continue the road work in the spring of 1867, Bird returned to New York by steamer.[17] Because of communication difficulties, the superintendent had received no instructions or supervision from his friend, Secretary Harlan. When he reported in Washington, Lieutenant Colonel Simpson rebuked him for assuming the responsibility for shutting down operations, naming a successor, and leaving his post of duty without departmental authorization. Moreover, citizens of the superintendent's home, Mount Pleasant, had charged Bird with swindling government funds.[18]

The Browning administration launched an investigation. Bird argued that if he had remained in the mountains and drawn his salary from the small balance of the appropriation he would have been liable to censure. The written agreement with Truax and the latter's bond were produced as evidence of good faith. No plans for road work were contemplated until the arrangement had the sanction of the Interior Department. On learning the displeasure of officials, the road superintendent notified Truax to make no improvements or expenditures. Although Bird continued to insist his arrangement with the major was both reasonable and logical, and members of the expedition in conjunction with northwestern politicians certified to Truax's ability, the Interior Department refused to consider his confirmation as superintendent.[19] The charges of dishonesty against Bird had been raised by three Iowa laborers who accompanied him on the ocean trip to the Northwest. Denied daily wages for the period of travel and the time required to outfit the party, these men had demanded a salary increase once road work began. When Bird refused, they quit the expedition, threatening vengeance. The superintendent termed them the "embodiment of insubordination," Iowa wood choppers unaccustomed to handling mules in the mountains, and "about as much use as an elephant would be to a farmer in a cornfield."[20] The Interior Department dismissed the charges.[21]

The Idaho territorial assembly had, meanwhile, addressed a memorial to Congress complaining about the unfinished condition of the Lewiston-Virginia City road. These lawmakers commended Bird's faithful and judicious application of the original appropriation toward the construction of the road, but "... owing to the distance of the superintendent's residence from its location, the expense of his fit-out to engage in the enterprise, and the cost of viewing out the route, more than one-half of the appropriation was consumed before the work upon the road was actually commenced."[22] The unexpended balance of $8,000 was inadequate to prosecute the work, so a new appropriation of $60,000 was requested. In April, 1867, Bird made a final effort to secure Browning's authorization for Truax to continue improvements on the basis that every dollar spent before the trail was overgrown would accomplish what five would do later. The superintendent also pleaded in vain with the Secretary to endorse the Idaho legislature's request before the Congress.[23]

No more federal funds were forthcoming. In April, 1870, D. M. Sells, Indian agent at the Lapwai reservation, joined the post commander at Fort Lapwai in asking the new Secretary of the Interior, Jacob D. Cox, to expend the remaining balance, $8,000, in improving the road through the Lapwai Valley.[24] Although the fort and agency would obviously benefit, the Secretary demanded evidence that the proposal was of more than local importance.[25] The Indian agent reported that the road was a sector of the thoroughfare between Lewiston, Oro Fino, and Pierce City. In summer, stages were on the road to all the mining camps, but in winter only pack trains could get through. If the grades could be improved along the way so heavily loaded wagons could be used throughout the year, freight rates would be reduced for the benefit of the government and residents. Major Truax, who had prepared the construction estimates filed with the Department, still hoped to direct the road repairs.[26] Secretary Cox refused to authorize the expenditure. Apparently the construction of a wagon road across the Bitterroots from Lewiston to Virginia City had been given up as a lost cause. Certainly the federal government had no more to show for the expenditure on this Idaho-Montana segment of the federal road program of 1865 than on the overland routes through the Nebraska and Dakota territories.

CHAPTER XX : *Wagon Roads West*

THE AMERICAN FRONTIERSMAN expressed his individualism by seeking an untrod path into the wilderness for a new home. Yet the pioneers' individualism and adaptability did not preclude their willingness to call upon the government for practical help in solving problems of migration and transportation. When projects, because of size or financial outlay, were beyond the means of private enterprise or the collective action of a western community, the resources and sponsorship of the national government were unhesitatingly demanded. Local groups constantly besieged Congress with requests for roads and other internal improvements. In the process localism was broken down, and a great desire to expand national power soon permeated most western communities. The pioneer became a nationalist as well as an individualist.

Although the desirability of wagon-road construction by the federal government was recognized by presidents from Thomas Jefferson to Martin Van Buren, no agreement was reached among the politicians on the constitutionality of internal improvements. After repeated failures to secure an amendment authorizing federal participation, James Monroe concluded that although the national government might not construct or hold jurisdiction over post roads, military roads, or canals in the states, it was not beyond the power of Congress to appropriate money for improvements. He insisted that federal financial aid could be given only to projects for common defense and of national interest, but not to works for state or local benefit. Monroe's successor, John Quincy Adams, did not adhere to such rigid requirements. Andrew Jackson, on the other hand, attempted to return to Monroe's basic requirement of the national importance for projects receiving federal funds. An obvious uncertainty of opinion existed as to where the line should be drawn between federal works that were constitutional and should receive federal appropriations and those that were unconstitutional.

Over the protest of a militant minority, the Democratic party platform of 1840 incorporated some fundamental concepts on the question of internal improvements. First, the party proclaimed its belief in a federal government of limited and expressed powers, to be strictly construed, and warned against the inexpediency of doubtful constitutional action. Moreover, it was felt that the Constitution did not confer on the national government the power to carry out a general system of internal improvements. In view of this policy of the major political party, only those road projects which were the exception to the general rule were likely to receive federal support. Throughout the 1840's a process was inaugurated to wear down gradually the reservations of congressmen by pressure of interested groups, chiefly from the West. Those improvements declared and proved necessary to the execution of a specific power of Congress, such as the maintenance of post roads, were most certain to gain approval. Even so, constructions within state boundaries were closely scrutinized and usually rejected by the strict constructionists attempting to preserve State rights in the federal system. Westerners successfully placed emphasis upon the exclusive power of Congress to make regulations for the territories of the United States. Military roads were also justified on the basis of providing for the common defense, a criterion early approved by President Monroe. The recognition that Congress possessed power in the territories not admissable in the states, plus the responsibility for the common defense, usually proved an unbeatable combination in securing a majority vote in both houses of Congress. Territorial delegates, introducing road legislation, never failed to describe the improvement as a *military* road. Every effort was made in the 1840's and 1850's to avoid a discussion of constitutionality. Once the project was launched, no serious question about its legality was likely to arise, and funds for its continuation or expansion were more easily obtained.

With emphasis upon national defense as a constitutional justification of the federal road program, these projects were assigned to the Secretary of War. The United States Army thus became the government's road builder. The year 1846 signalized the establishment of the "Great West." The Oregon country below the forty-ninth parallel was incorporated into the United States, and a state of war was declared to gain the northern Mexican provinces for an expanding democracy. The responsibilities of the United States Army thereby became manifestly greater; the purposes, scope, and methods of the military-road program of the federal government were revolutionized.

Although the Thirtieth Congress, in March, 1849, had allotted $50,000 to the Army for surveying routes from the Mississippi Valley to the Pacific Coast, the next Congress was so preoccupied with the debate over the political organization of the Mexican cession that scant attention was given to the establishment of a realistic policy to meet the transportation needs of the expanded national boundaries. The customary procedure of aiding the new territories with a federal subsidy for the road system was approved for Minnesota.

When the Thirty-second Congress assembled in 1851, the role played by the federal government in improving transportation through the new territories organized in the Mexican cession had become a subject for public discussion. The legislators were obliged to consider, if not to resolve, the nature and extent of national aid. Improvements in Minnesota were continued, and Joseph Lane of Oregon Territory prevailed upon his colleagues to extend national support for military roads to the Pacific Northwest, since Oregon and Minnesota, both in the initial stages of territorial status, had repeatedly been granted identical concessions by Congress.

An inordinate amount of debate in the Thirty-second Congress related to the question of federal aid in the construction of a Pacific railroad. Sectional interests and constitutional scruples combined to block the passage of any legislation. In the midst of the struggle Senator Richard Brodhead of Pennsylvania suggested the absurdity of authorizing a railroad before surveys proved its practicability and proposed an appropriation for a reconnaissance of all possible routes by the Corps of Topographical Engineers. Upon congressional authorization of a $150,000 expenditure, the administration of these Pacific railroad surveys was entrusted to Jefferson Davis, Secretary of War in the cabinet of Franklin Pierce. In the selection of personnel and the organization of these reconnaissances, Davis accepted the guidance of Colonel John J. Abert, chief topographical engineer. However, the program was not assigned to Abert's bureau, but to the Office of Explorations and Surveys established under the direction of Captain A. A. Humphreys, another member of the Corps. Thus a new War Department agency entered the field of transportation improvement in the trans-Mississippi West. With larger appropriations available, the War Department made the administrative adjustments to continue the topographical exploration of the West on a more thoroughgoing and systematic basis.

The interrelationship between topographical exploration, wagon-road reconnaissance, and railroad survey was generally recognized. The railroad reconnaissances were but the initial activities of the Office of Explorations and Surveys. The systematic examination of western terrain by topographical engineers, under the Agency, continued until the Civil War. The work of exploration by the Army soon gained unusual attention in the scientific world. Geologists, botanists, and zoölogists vied with each other to secure attachment to a western command as compilers of data on natural history.

Far more generous than its two predecessors, the Thirty-third Congress, 1853–1855, provided for military roads in every organized territory in the trans-Mississippi West. So many constructions were started in the Pacific Northwest that Colonel Abert was obliged to establish a Pacific Coast Office for Military Road Constructions in San Francisco. Military roads were launched for the first time in the territories of Utah, New Mexico, Kansas, and Nebraska. Financial support for federal wagon roads granted at this session totaled $564,000.

The tremendously expanded military road systems in the territories were approved by a Congress with an overwhelming Democratic majority, the largest since the Mexican War. Yet this party had been, throughout its history, notoriously unsympathetic to internal improvements at federal expense. In his inaugural address President Pierce cautiously suggested that the government could assist the territories by all constitutional means but expressed doubt of the legality of Pacific railroad construction and some concern about its propriety. In August, 1854, the President vetoed an omnibus River and Harbors' Bill, again proclaiming that opinion in the country was not divided about the value and importance of internal improvements, but was separated into many factions on the question of their constitutionality. In Pierce's judgment, a *general* system of internal improvements by the federal government was unconstitutional, and he insisted on holding the party to its position taken in 1840 and reaffirmed before each presidential election. According to his veto message, works of national importance in the territories necessary to the execution of an enumerated power of the federal government were, however, to be sanctioned. The President illustrated the principle by stating that when any road was manifestly required by the military service, it seemed undeniable that it was constitutional. Moreover, he specified that the appropriation for each project should be made in a separate bill so that it might stand on its independent merits. Western Democrats, accepting these stipulations, discovered a means of increasing federal aid to transportation and communication without encountering a rebuff from the administration and a further disruption of the party. Their votes combined with the Whigs', pushed many road bills through the Congress, and all received the chief executive's approval.

Jefferson Davis became the spokesman for the administration on questions of western roads. Although during his first months in office he turned to Colonel Abert for advice in supervising the railroad surveys, the Secretary of War generally assumed personal responsibility for the administration of the military road programs throughout the West. Field officers were annoyed by his rigid adherence to administrative procedures, his apparent lack of understanding of the physical obstacles to be overcome, and the necessity for larger appropriations. Although work was sometimes delayed thereby, no evidence is available in his direction of the territorial road programs to reinforce the accusations of unbending preference for the southern route, as revealed in the railroad survey report. Knowing that the War Department was under constant scrutiny because of his personal sympathies for southern interests, Davis consciously sought to avoid criticism of his administration of northern roads.

After the election of Franklin Pierce as President, the two-party political system experienced an upheaval. The Whigs disappeared as a potent organization. The Democrats learned that sectional differences of opinion on major political issues of the time might create a schism within the national organization. Many northern and western men of all parties joined forces in the new Republican party and won surprising victories in the by-elections of 1854. When the Thirty-fourth Congress assembled in December, 1855, the senators from California were vociferous in their demands for an immediate improvement of communications with the Mississippi Valley. If a railroad was to be delayed, wagon roads were immediately practicable. Senator Weller urged the construction of two wagon roads to the California state boundary: one from the Missouri by way of South Pass, Salt Lake, and Carson Valley; and a second from El Paso to Fort Yuma across New Mexico Territory.

Although the Democratic party maintained its control of the Senate, the House of Representatives was dominated by members of the young Republican party. To placate them, the Senate endorsed an appropriation for a wagon road from Fort Ridgely, Minnesota, to the South Pass. This internal improvement was to be supervised by the Interior Department, apparently because the House Republicans wanted to avoid the influence of Jefferson Davis. A constitutional justification was found in the central government's enumerated power to regulate commerce with the Indian tribes. The legislation for the central and southern wagon roads to California died in House committees.

Simultaneously with these congressional discussions, the major political parties were preparing for the presidential campaign of 1856. Proponents of governmental sanction and assistance for the Pacific railroads attempted to put both parties on record in favor of the project. The Democrats in Cincinnati recognized "the great importance, in a political and commercial point of view, of the safe and speedy communication, by military and postal roads, through our own territory, between the Atlantic and Pacific coasts of this Union, and that it is the duty of the Federal Government to exercise promptly all its proper constitutional power to the attainment of that object." This carefully worded statement represented the first departure of policy on internal improvements since 1840 and precipitated a row before adoption. The Republicans were more emphatic. In their eyes the railroad to the Pacific along the most central and practicable route was imperatively demanded, and the party thought the federal government should render immediate aid to the construction of an emigrant road along the line of the railroad. In the month following the Democratic convention, July, 1856, President Pierce signed the Fort Ridgely–South Pass Wagon Road bill, authorizing the Interior Department rather than the War Department to supervise construction. There was no pretense that it was a military road. Either in conformity to the shifting policy of his political party as expressed in the Cincinnati platform, or in acceptance of the principle that roads could be built to regulate commerce with the Indians as well as to provide for the common defense, the President believed the legislation constitutional.

When Congress reconvened after the election of James Buchanan, the House of Representatives resumed consideration of the Senate's "Military Roads" to the Pacific. The Republican majority reported a new bill starting the central route at Fort Kearny and changing the western terminus from Carson Valley to Honey Lake, farther north on the California border. An omnibus bill approving this route, the El Paso–Fort Yuma road, and a construction along the thirty-fifth parallel from Fort Defiance, New Mexico Territory, to the Colorado River cleared all the legislative hurdles. The first two projects were to be supervised by the Interior Department, but in the excitement no specification was made as to the executive department responsible for the Fort Defiance–Colorado River road. Apparently reconciled, Pierce approved this appropriation bill on the last day of his term.

In contrast with its generosity toward the emigrant roads to the Pacific, the Thirty-fourth Congress demonstrated little interest in the territorial military road system. The legislators refused to inaugurate any new works in Minnesota because statehood was pending. Debate over Oregon appropriations precipitated a bitter attack on federal internal improvements by the strict constructionists, highlighting the intense party and sectional feeling. Some Southerners publicly repudiated the Cincinnati platform of their party. Others insisted that if the military power was a justification for road building, then the western territories had no more right to a federal subsidy than the eastern states. A third group protested the use of a military justification as a ruse for federal patronage. The divergent lines of the opposition's thinking proved their undoing. It also represented their swan song. In all, this Congress approved almost $800,000 for wagon roads in the trans-Mississippi West.

Jacob Thompson, Secretary of the Interior, established a separate administrative agency, the Pacific Wagon Roads Office, to direct the public works assigned to his department. The record of this civilian agency was the gloomiest in the history of federal aid to road building in the trans-Mississippi West. The disaster accompanying the program was chiefly administrative. The civilian engineers completed their survey and improvement assignments with reasonable satisfaction, but the superintendents, as frontiersmen with political influence, proved incapable managers. The confusion, moreover, tempted the dishonest.

Although national legislators conceived of these Pacific roads as a comprehensive transportation program to compensate for the failure to agree upon a railroad route, residents of each territory were unwilling to sacrifice federal support for their own military road system. The Thirty-fifth Congress, 1857–1859, was bombarded with requests for new roads or appropriations to continue construction, all carrying the endorsement of the Topographical Engineers. Statehood had now eliminated Minnesota from serious consideration. Scant attention was given to Utah's pleas because of the niggardly attitude toward the Mormon community. Territorial assemblies in Washington and New Mexico memorialized Congress for expanded road networks, involving the expenditure of several hundred thousand dollars. Their delegates pressed as never before for favorable consideration, only to be denied. The Thirty-sixth Congress approved minor appropriations, totaling $50,000 for Washington and New Mexico military roads in 1860 and early 1861, but the concession came too late for practical use because of the outbreak of war.

After the congressional rebuke of the Topographical Engineers in 1856, the United States Army continued to locate wagon roads west, with limited appropriations, always in hopes of regaining the legislators' confidence. Military road systems in the territories remained under the Bureau of Topographical Engineers. The larger reconnaissance and constructions across country were placed under the Office of Explorations and Surveys, and trained engineers were transferred from the Topographical Bureau for these purposes. Because of the congressional failure to specify that the Fort Defiance–Colorado River road was to be directed by the Interior Department in conjunction with the other overland routes to the Pacific, the unenthusiastic Secretary transferred its administration to the War Department. For many years the United States Army favored a road to connect the headwaters of the Missouri with those of the Columbia. The plan culminated in the construction of the Mullan road, a five-year project, that was the federal government's greatest contribution to transportation development in the Pacific Northwest. In 1859–1860, another topographical engineer of the Office of Explorations and Surveys, W. F. Raynolds, explored a new route to connect the central emigrant road, along the Platte, with the Pacific Northwest, but the Civil War postponed further development of the route by the Army.

In 1865 Congress resumed the wagon-road program by appropriating $140,000 for a series of constructions leading to the Pacific Northwest. The direction of these roads was placed under the Secretary of the Interior, and President Lincoln urged the War and Interior departments to coöperate in coördinating policy relative to the transportation and policing problems of the Great Plains and Rocky Mountain area. Events of 1865–1866 only served to emphasize the futility of the attempt. The conflict between the two departments over Indian policy was extended to the road program. Both agencies were attempting to do the same job.

The Interior Department's postwar road-building endeavors were no more successful than the attempt to improve the routes to the Pacific Coast in 1856–1860. Less emphasis was placed upon the failure because transportation enthusiasts in the Far West and elsewhere in the nation, were absorbed in the prospects of extensive federal land subsidies to transcontinental railroads. Future wagon roads could only serve as feeder lines after the east-west connection of the Union Pacific and the Central Pacific railroads in 1869.

One of the most important activities in the trans-Mississippi West, 1846–1869, was topographical exploration for trails and roads that were to become avenues of migration, communication, and commerce. The dominant roles were played by agents of the national government, either as officers of the United States Army or civilian employees of the Interior Department. As a result of federal appropriations, the old established emigrant and trade routes known as the Santa Fe, Oregon, and California trails were shortened and improved. New ways of migration were opened. The emigrants' interest was always of paramount importance to the military road builders as well as to the civilian contractors. Wood for the camp fires, grass for the animals, and water for both men and beast were essential. Where they were known to be inadequate, adjustments were made, as in the digging of artesian wells and the construction of water tanks in New Mexico Territory.

The Argonaut of 1849–1850 was often indebted to an Army surveyor who had examined and reported upon a route of travel to California. One of the most popular avenues was the Marcy-Simpson trail along the banks of the Canadian River. Between El Paso and San Diego, many traveled along much of Cooke's wagon road. In the 1859 rush to the Colorado Rockies, the routes of Stansbury and Bryan were sometimes followed across the Kansas and Nebraska plains, either going to or returning from the Pike's Peak region. When the mining frontier moved to Idaho and Montana in the early 1860's, prospecting parties and freighters from both east and west used the Army's Mullan road to reach the sites of the latest discoveries. The continued search for gold in the Northwest was a dominant factor behind the government's sponsorship of the Sawyers expeditions of 1865–1866.

The road program did not facilitate the demands of a mobile population to the exclusion of the settler. Just as the military roads had provided an access to the agricultural lands of Iowa and Arkansas, so the pioneers in Nebraska and Kansas pushed westward along the pathway of similar surveys to take up farm land. The rich agricultural resources of the Rogue and Umpqua river valleys in southern Oregon were made easily accessible to the homeseeker. Lumbermen in the upper Mississippi Valley of Minnesota, in Oregon between the Willamette and Columbia rivers, and along the Puget Sound of Washington were indebted to federal road surveyors who penetrated the evergreen forests. Improved transportation facilities accelerated the volume of land sales and in some areas were a contributing factor to the pattern of settlement. Many frontier communities used these wagon roads to move products of the farm and forest to market and thereby laid the foundations for the earliest commerce of the region.

Before the Civil War, at least one, if not both, of the termini of most federal roads was at a military installation. Although the Indian-fighting Army was never so fortunate as to conduct its campaigns along military road surveys in the trans-Mississippi West, these improvements did provide a way for new recruits and supply trains dispatched to western outposts, and a continuous connection between the forts or Indian agencies and the population centers.

Many entrepreneurs engaged in mail deliveries under the subsidy of a government contract or carrying freight to the western forts for War Department compensation, received additional, though indirect, aid in using roads already explored and surveyed at federal expense. The freighting companies, such as Russell, Majors & Waddell whose headquarters were at Leavenworth, made extensive use of the roads laid out from the Kansas forts into the Great Basin and to the New Mexico settlements on the upper Rio Grande. The Pony Express, and later, the transcontinental telegraph followed, for the most part, the Simpson survey across western Utah Territory. The Pacific wagon road, located and partly improved by the Interior Department west of El Paso across New Mexico Territory, was traveled extensively by the stage and mail coaches on the San Antonio–

San Diego route and by the Butterfield Overland Service. The government of the United States thus provided continuous and sustaining support for many businessmen whose financial success and promotional publicity brought them fame as builders of the West.

The federal engineers made a direct contribution to the location of the highways and railroads of the late nineteenth and twentieth centuries. As trained explorers and topographers, government surveyors succeeded in finding the natural passages for transportation routes along the river valleys, across the plains, or through the mountain passes. These were surveyed and mapped, and recommended for wagon travel. When the railroad reconnaissances were authorized in 1853, they represented not an innovation but a continuation of federal aid to transportation and communication along the established policy. It was inevitable that modern communication lines should follow, to a large extent, the recommendations of those who first scientifically examined the terrain of the trans-Mississippi West. The Army engineers established the basic pattern of the modern highway systems of the states of Minnesota and New Mexico during the territorial period. The Southern Pacific Railroad followed much of Cooke's first wagon road to California; the Santa Fe ran along the general route of Whipple's railroad survey and Beale's wagon road between Albuquerque, New Mexico, and Needles, California; the tracks of the Northern Pacific, the Great Northern, and the Milwaukee system all utilizd parts of Mullan road.

The federal government's contribution in building wagon roads west is a vital chapter in the nation's development. The significance of this "Engineer's Frontier" has been aptly appraised by Professor Samuel Flagg Bemis:

> This mapping and measuring of the far west, the scientific foundation of our present day civilization, was the task of a numerous group of hard-working, conscientious, and adventurous young engineers mostly selected from the United States Army officers and working under the command of the Secretary of War.... With a few notable exceptions, the patient mappers of the West have passed out of memory. Their voluminous reports, replete with scientific data, have gone into the gloomy repositories of forbidding government documents. Yet, like many another pathfinder of science the work of the army "topographer", to use a once familiar word, lives with us today in our railroads, our reclamation works,... in short in the whole material foundation of the west. The "Engineers Frontier" of the fifties and sixties has long since passed by, after that of the fur trader and the prospector, but into a deeper oblivion. The American people are all too ignorant of the great services of these engineers who surveyed the west for no profit to themselves, while at their side other men were making fortunes and contributing nothing to posterity.[1]

Notes

NOTES TO CHAPTER I

[1] Seymour Dunbar, *A History of Travel in America,* IV, 1.

[2] Charles B. Quattlebaum, "Military Highways," *Military Affairs,* VIII (Fall, 1944), 225–238.

[3] W. Stull Holt, *The Office of the Chief of Engineers of the Army: Its Non-Military History, Activities, and Organization,* pp. 1–11; "Annual Report, Bureau of Topographical Engineers, December 30, 1839," *Senate Document 58,* 26 Cong., 1 sess. (1839–1840), pp. 10–12.

[4] Jeremiah Simeon Young, *A Political and Constitutional Study of the Cumberland Road,* introduction.

[5] Robert Bruce, *The National Road,* pp. 10–12. An excellent biographical note listing governmental publications on the National Road may be found in Philip D. Jordan, *The National Road.*

[6] See the annual reports of the Chief Engineer in the Reports of the Secretary of War, 1829–1838. For example, "Report of the Chief Engineer, November 23, 1833," *Senate Document 1,* 23 Cong., 1 sess. (1834–1835), pp. 81–84.

[7] Young, *op. cit.,* p. 106.

[8] Lee Burns, "The National Road in Indiana," *Indiana Historical Society Publications,* VII (1923), 237.

[9] "Annual Report, Bureau of Topographical Engineers, December 30, 1839," *Senate Document 58,* 26 Cong., 1 sess. (1839–1840), pp. 21–24; Carl E. Pray, "An Historic Michigan Road," *Michigan History Magazine,* XI (January, 1927), 325–341; George B. Catlin, "Michigan's Early Military Roads," *Michigan History Magazine,* XIII (Spring, 1929), 196–207.

[10] "Annual Report, Bureau of Topographical Engineers, December 30, 1839," *Senate Document 58,* 26 Cong., 1 sess. (1839–1840), pp. 24–30; H. E. Cole, "The Old Military Road," *Wisconsin Magazine of History,* IX (September, 1925), 47–62.

[11] Henry Putney Beers, *The Western Military Frontier, 1815–1846,* pp. 44–46, 50, 103.

[12] *Ibid.,* pp. 100–101. Carolyn T. Foreman, ed., "Report of Captain John Stuart on the Construction of a Road from Fort Smith to Horse Prairie on Red River," *Chronicles of Oklahoma,* V (September, 1927), 333–347.

[13] Beers, *op. cit.,* p. 111.

[14] *Ibid.,* p. 113.

[15] *Ibid.,* pp. 118–119, 121, 127, 131, 135–136.

[16] "Petition of Samuel Dickens *et al.,* Memphis to Little Rock," *House Executive Document 151,* 22 Cong., 1 sess. (1831–1832), pp. 1–2.

[17] *U.S. Statutes at Large,* IV, 5.

[18] *Ibid.,* pp. 244, 557.

[19] *Ibid.,* p. 650.

[20] "Report of the Chief Engineer, November 23, 1833," *Senate Document 1,* 23 Cong., 1 sess. (1833–1834), pp. 80–81; "Report of the Chief Engineer, November 1, 1834," *Senate Document 1,* 23 Cong., 2 sess. (1834–1835), pp. 111–112; "Report from the Engineer Department, November 15, 1835," *Senate Document 1,* 24 Cong., 1 sess. (1835–1836), p. 112.

[21] "Report from the Engineer Department, November 30, 1836," *House Executive Document 2,* 24 Cong., 2 sess. (1836–1837), pp. 207, 285–289; "Report from the Engineer Department, November 30, 1837," *Senate Document 1,* 25 Cong., 2 sess. (1837–1838), pp. 378–381.

[22] *U. S. Statutes at Large,* IV, 462–463.

[23] *Ibid.,* p. 557.

[24] *Ibid.,* p. 753.

[25] "Road—Chicot County to Little Rock," *House Report 374,* 22 Cong., 1 sess. (1831–1832).

[26] *U. S. Statutes at Large,* IV, 712, 724.

[27] "Documents Relating to Bill Making Appropriations for Certain Roads in Arkansas," *Senate Document 300,* 24 Cong., 1 sess. (1835–1836); "Documents Relating to Arkansas Roads," *Senate Document 40,* 24 Cong., 2 sess. (1836–1837).

[28] W. Turrentine Jackson, "The Army Engineers as Road Builders in Territorial Iowa," *Iowa Journal of History,* XLVII (January, 1949), 15–33.

NOTES TO CHAPTER II

[1] Marcy to Kearny, June 3, 1846, "Report of the Department of War, Including Military Correspondence Relative to California and New Mexico, 1846–1850," *House Executive Document 17,* 31 Cong., 1 sess. (1849–1850), p. 237.

[2] Ralph P. Bieber and Averam B. Bender (eds.), *Exploring Southwestern Trails, 1846–1854,* Vol. VII of *The Southwest Historical Series,* pp. 22–25. For general accounts of the Mormon Battalion, see also Hubert Howe Bancroft, *History of California, 1542–1890,* V, pp. 469–472; Frank A. Golder, Thomas A. Bailey, and J. Lyman Smith (eds.), *The March of the Mormon Battalion from Council Bluffs to California,* pp. 28–80; and Leland H. Creer, *Utah and the Nation,* pp. 20–22, 31–36.

[3] William H. Emory, "Notes of a Military Reconnaissance, from Fort Leavenworth, in Missouri, to San Diego, in California, including part of the Arkansas, Del Norte, and Gila Rivers," *House Executive Document 41,* 30 Cong., 1 sess. (1847–1848), pp. 7–11, 43. Hereafter cited: Emory, "Notes of a Military Reconnaissance..."

[4] *Ibid.,* p. 45, p. 551.

[5] *Ibid.,* p. 53; see also "Journal of Captain A. R. Johnston, First Dragoons," *House Executive Document 41,* 30 Cong., 1 sess. (1847–1848), p. 572.

[6] Bieber and Bender, *op. cit.,* pp. 25–26.

[7] Emory, "Notes of a Military Reconnaissance . . . ," p. 56. See also Thomas Kearny, *General Philip Kearny, Battle Soldier of Five Wars, Including the Conquest of the West by General Stephen Watts Kearny,* pp. 113–135.

[8] *Ibid.,* p. 62.

[9] "Cooke's Journal of the March of the Mormon Battalion, 1846–1847," in Bieber and Bender, *op. cit.,* pp. 65–67. Professors Bieber and Bender have edited and published the manuscript copy of Cooke's Journal. An inaccurate reproduction of the original is available in *Senate Executive Document 2,* 30 Cong., special sess. (March, 1849), pp. 1–85, under the title of "Journal of the march of the Mormon battalion of infantry volunteers, under the command of Lieutenant Colonel P. St. George Cook, (also captain of dragoons) from Santa Fe, New Mexico to San Diego, California, kept by himself by direction of the commanding general of the army of the west." A brief summary of this journal also appears in the federal documents as a report from Cooke to Kearny, February 5, 1847, *House Executive Document 41,* 30 Cong., 1 sess. (1847–1848), pp. 551–562. In this study, all citations and quotations are from the manuscript edited by Professors Bieber and Bender and hereafter cited as *Cooke's Journal.*

[10] *Cooke's Journal,* p. 69. In the Senate Executive Document publication the word "determined" has been italicized for emphasis. This was probably done by an editor, as was often the practice.

[11] *House Executive Document 41,* 30 Cong., 1 sess. (1847–1848), p. 560.

[12] *Cooke's Journal,* pp. 85–86.

[13] *Ibid.,* p. 106.

[14] For diagram of the route, see map p. 248.

[15] *Cooke's Journal,* p. 111.

[16] *Ibid.,* p. 106.

[17] *Ibid.,* p. 199.

[18] *Ibid.,* p. 93.

[19] *Ibid.,* p. 166.

[20] *Ibid.,* pp. 222–223.

[21] *Ibid.,* pp. 239–240.

[22] *Ibid.*, p. 29.

[23] *Fort Smith Herald*, December 22, 1847.

[24] *Ibid.*, March 1, 1848.

[25] *Ibid.*, March 31, 1848.

[26] *Ibid.*, August 23, 1848.

[27] *Ibid.*, September 27, 1848.

[28] *Ibid.*, October 11, November 1, and November 3, 1848. W. Sheridan Warrick, "James Hervey Simpson, Military Wagon Road Engineer in the Trans-Mississippi West, 1849–1867," unpublished M.A. thesis, University of Chicago, 1949. After examining the files of the *Fort Smith Herald*, Warrick concludes that the initial agitation for a wagon road west from Fort Smith was stimulated by the ratification of the Treaty of Guadalupe Hidalgo. The nature of the project had become clear and the decision to seek federal government aid had been made before the news of gold discovery in California. An additional and undoubtedly the most important incentive was then provided. Grant Foreman in his study *Marcy and the Gold Seekers*, and Ralph P. Bieber in *Southern Trails to California in 1849*, Vol. V of The Southwest Historical Series, have left the impression that the news of gold discovery initiated the move for this national wagon road project.

[29] Ralph P. Bieber has summarized his extensive research on the gold rush period in three publications: "The Southwestern Trails to California in 1849," *Mississippi Valley Historical Review*, XII (December, 1925), 342–375; "California Gold Mania," *Mississippi Valley Historical Review*, XXXV (June, 1948), 1–28; and *Southern Trails to California in 1849*.

[30] *Fort Smith Herald*, November 22, 1848, February 7, 1849.

[31] Senator Solon Borland to the editor of the *Arkansas Democrat*, January 24, 1849. Reprinted in the *Fort Smith Herald*, February 21, 1849. Bieber, *Southern Trails to California in 1849*, p. 42.

[32] Borland to William L. Marcy, January 10, 1849. Reprinted from *Arkansas Democrat* in *Fort Smith Herald*, February 21, 1849.

[33] Borland to the editor of the *Arkansas Democrat*, January 24, 1849. Reprinted in *Fort Smith Herald*, February 21, 1849.

[34] *Ibid.*

[35] *Fort Smith Herald*, February 14, 1849.

[36] James H. Simpson, "Report of Exploration and Survey of Route from Fort Smith, Arkansas, to Santa Fe, New Mexico . . . ," *Senate Executive Document 12*, 31 Cong., 1 sess. (1849–1850), p. 2. Hereafter cited as Simpson, *Report on Route from Fort Smith to Santa Fe*. This journal together with Marcy's report is published as *House Executive Document 45* of the same Congress. In writing the study, *Marcy and the Gold Seekers*, Grant Foreman has interrelated the Simpson and Marcy accounts with other materials to present a continuous study of the venture.

[37] *U. S. Statutes at Large*, IV, p. 372.

[38] *Fort Smith Herald*, March 14, 21, 1849.

[39] "Report of Captain R. B. Marcy," *House Executive Document 45*, 31 Cong., 1 sess. (1849–1850), p. 23. Hereafter cited as Marcy, *1849 Report*.

[40] *Fort Smith Herald*, April 11, 1849.

[41] Simpson, *Report on Route from Fort Smith to Santa Fe*, p. 2.

[42] Marcy, *1849 Report*, p. 26.

[43] For a diagram of the route, see map p. 224.

[44] Bieber, *Southern Trails to California in 1849*, pp. 44–45. James W. Abert, "Journal of Lieutenant J. W. Abert from Bent's Fort to St. Louis, 1845," *Senate Executive Document 438*, 29 Cong., 1 sess. (1845–1846). Simpson used Abert's map on his 1849 survey.

[45] Simpson, *Report on Route from Fort Smith to Santa Fe*, p. 24.

[46] Marcy, *1849 Report*, p. 31.

[47] Simpson, *Report on Route from Fort Smith to Santa Fe*, p. 19.

[48] *Ibid.*, pp. 4–5.

[49] Marcy, *1849 Report*, p. 29.
[50] *Ibid.*, pp. 32–33.
[51] *Fort Smith Herald*, July 15, 1849.
[52] Marcy, *1849 Report*, p. 48.
[53] Simpson, *Report on Route from Fort Smith to Santa Fe*, pp. 22–23.
[54] *Ibid.*, p. 2. Averam B. Bender, "Governmental Explorations in the Territory of New Mexico, 1846–1859," *New Mexico Historical Review*, IX (January, 1934), 14–15.
[55] Simpson, *Report on Route from Fort Smith to Santa Fe*, p. 2. The Bureau had referred him to the maps of the second expedition of Lt. J. W. Abert and of the second expedition of John C. Frémont. Simpson did not associate the "caravan route" with the Old Spanish Trail but his description seems to establish the fact. The Old Spanish Trail was annually traveled by trading caravans in the 1830's and early 1840's. See LeRoy R. Hafen, "The Old Spanish Trail, Santa Fe to Los Angeles," *Huntington Library Quarterly*, XI (February, 1948), 149–160.
[56] James H. Simpson, "Journal of a Military Reconnaissance from Santa Fe, New Mexico to the Navajo Country . . .", *Senate Executive Document 64*, 31 Cong., 1 sess. (1849–1850), p. 64. Hereafter cited as Simpson, *Navajo Journal*. A map of the expedition may be found in an unofficial publication of Simpson's journal: *Journal of a Military Reconnaissance from Santa Fe, New Mexico to the Navajo Country.*
[57] Richard Campbell's visit to California in 1827 has often been regarded as fiction. Alice B. Maloney attempted to prove otherwise in "The Richard Campbell Party of 1827," *California Historical Society Quarterly*, XVIII (December, 1939), 347–354.
[58] Simpson, *Navajo Journal*, p. 137.
[59] Lorenzo Sitgreaves, "Report of an Expedition Down the Zuni and Colorado Rivers," *Senate Executive Document 59*, 32 Cong., 2 sess. (1852–1853). Simpson was later convinced that the Sitgreaves exploration was prompted by his recommendation. See "Annual Address delivered before the Minnesota Historical Society, January 19, 1852," *Annals of the Minnesota Historical Society*, III (1852), 16.
[60] Marcy, *1849 Report*, pp. 196–199.
[61] Howard Stansbury, "Exploration and Survey of the Valley of the Great Salt Lake of Utah," *Senate Executive Document 3*, 32 Cong., special session, March, 1851. Hereafter cited as *Stansbury Report*. An interesting summary and contemporary review of the Stansbury report can be read in the *Princeton Review*, XXIV (October, 1852), 687–696. Brief accounts of Stansbury's activities are found in the histories of Utah: Hubert Howe Bancroft, *History of Utah, 1540–1886*, pp. 463–467; Orson F. Whitney, *History of Utah*, I, pp. 412–415; J. Cecil Alter, *Utah, the Storied Domain: A Documentary History of Utah's Eventful Career*, I, pp. 108–112; Andrew Love Neff, *History of Utah, 1847–1869*, ed. by Leland H. Creer, pp. 152, 231–232, 253–254; and Leland H. Creer, *The Founding of an Empire: the Exploration and Colonization of Utah, 1776–1856*, pp. 129–139. Professor Creer in his study of early Utah has presented a scholarly summary of the contributions of Army men as surveyors of new routes for wagons and rails through the Great Basin to the Pacific.
[62] *Stansbury Report*, pp. 13–76.
[63] *Stansbury Report*, p. 78; Creer, *op. cit.*, pp. 130–131.
[64] *Stansbury Report*, pp. 79–80.
[65] *Ibid.*, p. 84; J. Cecil Alter, *James Bridger*, pp. 214–220. Alter carefully explains the role of Bridger as guide to Stansbury. For each of the Stansbury surveys, see map p. 30.
[66] *Stansbury Report*, p. 88.
[67] *Ibid.*, p. 93.
[68] J. Cecil Alter notes that this road down the Blacksmith's fork was never constructed. *Utah, the Storied Domain*, I, p. 110.
[69] *Stansbury Report*, pp. 97–119; Creer, *op. cit.*, pp. 133–134.
[70] Stansbury's account has become a classic source of information on the Mormons and their way of life. John W. Gunnison also was responsible for the preparation of a favorable appraisal of the Mormon community: *The Mormons or Latter-Day Saints in the Valley of the Great Salt Lake.*

[71] *Stansbury Report,* pp. 225–227. Of great value in tracing Captain Stansbury's surveys is his "Map of a Reconnaissance between Fort Leavenworth on the Missouri River and the Great Salt Lake in the Territory of Utah made in 1849 and 1850." This map is published with the *Report* in the congressional document series. It is seldom found with the separately published volumes of the report. The actual drawing of the map was entrusted to Gunnison and Charles Preuss. The original may be seen in the Division of Cartographic Records, The National Archives.

[72] *Stansbury Report,* p. 229.

[73] *Ibid.,* p. 232; Alter, *James Bridger,* pp. 221–227.

[74] The waters of this tributary of the Yampah eventually flowed into Green River in Utah.

[75] *Stansbury Report,* pp. 235–260.

[76] *Ibid.,* p. 261.

[77] In 1862 the Overland Stage moved its route from the Sweetwater River southward. Until 1868 the stages crossed the Laramie Plains and Bridger's Pass through the Divide. The Union Pacific decided to use a pass through the mountains slightly to the north of Bridger's but followed much of the Stansbury survey route. Bridger's Pass is approximately 20 miles southwest of Rawlins, Wyoming.

NOTES TO CHAPTER III

[1] Averam B. Bender, "Opening Routes Across West Texas, 1848–1850," *Southwestern Historical Quarterly,* XXXVII (October, 1933), 117–119; Ralph P. Bieber, *Southern Trails to California in 1849,* pp. 36–37; Bieber and Bender, *Exploring Southwestern Trails,* pp. 31–33. Basic research on southwestern exploration and travel has been completed by Professors Bieber and Bender and the writer is indebted to them for much of the material in this chapter. For a diagram of this and other Texas routes, see map p. 38.

[2] Bender, "Opening Routes Across West Texas, 1848–1850," *loc. cit.,* pp. 119–120; Bieber, *Southern Trails to California in 1849,* p. 38.

[3] Bender, "Opening Routes Across West Texas, 1848–1850," *loc. cit.,* pp. 120–121; Bieber and Bender, *Exploring Southwestern Trails,* pp. 34–35.

[4] "Journal of William Henry Chase Whiting, 1849," in Bieber and Bender (eds.), *Exploring Southwestern Trails,* p. 256. Hereafter cited as *Whiting Journal.* The manuscript of this journal is in the records of the Chief of Engineers, War Department Records, The National Archives. The journal was largely copied from a diary kept on the journey and published in part in the *Publications of the Southern History Association,* VI (1902), 283–294, 389–397, and X (1906), 1–18, 78–95, 127–140. A brief summary statement of the account may be read in Whiting's report to General J. G. Totten, Chief Engineer of the United States, June 10, 1849, *House Executive Document 5,* 31 Cong., 1 sess. (1849–1850), pp. 281–293. The report of Lieutenant William F. Smith to Lieutenant Colonel J. E. Johnston of the Topographical Engineers is located in *Senate Executive Document 64,* 31 Cong., 1 sess. (1849–1850), pp. 4–7. This federal document also contains reports of Texas reconnaissances by J. E. Johnston, F. T. Bryan, N. H. Michler, and S. G. French.

[5] *Whiting Journal,* pp. 269–277.

[6] *Ibid.,* pp. 297–298.

[7] *Ibid.,* pp. 301–302.

[8] Bieber and Bender, *Exploring Southwestern Trails, 1846–1854,* pp. 35–36.

[9] Smith to Johnson, May 25, 1849, *Senate Executive Document 5,* 31 Cong., 1 sess. (1849–1850), pp. 6–7.

[10] *Whiting Journal,* p. 349.

[11] Report of Captain S. G. French, Quartermaster's Corps, December 21, 1849, *Senate Executive Document 64,* 31 Cong., 1 sess. (1849–1850), pp. 40–41.

[12] J. E. Johnston to Major General Brooke, December 28, 1849, *Senate Executive Document 64,* 31 Cong., 1 sess. (1849–1850), pp. 26–27.

[13] *Senate Executive Document 64,* 31 Cong., 1 sess. (1849–1850), pp. 26–27, 41–52.

[14] *Ibid.*, p. 50.

[15] *Ibid.*, pp. 26–27.

[16] Francis T. Bryan to J. E. Johnston, December 1, 1849, *Senate Executive Document 64,* 31 Cong., 1 sess. (1849–1850), p. 17.

[17] *Ibid.*, pp. 14–25.

[18] Bieber, *Exploring Southwestern Trails,* p. 37.

[19] Nathaniel Michler, "Routes from the Western Boundary of Arkansas to Santa Fe and the Valley of the Rio Grande," *House Executive Document 67,* 31 Cong., 1 sess. (1849–1850), pp. 1–3. This document is also printed in *Senate Executive Document 64,* 31 Cong., 1 sess. (1849–1850), pp. 29–39.

[20] *Ibid.*, p. 11.

[21] *Ibid.*, pp. 1–12; Bender, "Opening Routes Across West Texas, 1848–1850," *loc. cit.*, pp. 131–134.

[22] Michler to Major George Deas, July 31, 1849, *Senate Executive Document 64,* 31 Cong., 1 sess. (1849–1850), pp. 7–13.

[23] Bender, "Opening Routes Across West Texas, 1848–1850," *loc. cit.*, pp. 134–135; Gouverneur K. Warren, "Memoir to Accompany Map of the Territory of the United States from the Mississippi River to the Pacific Ocean," *Senate Executive Document 78,* 33 Cong., 2 sess. (1854–1855), p. 62. Hereafter cited as *Warren's Memoir.*

[24] Averam B. Bender, "The Texas Frontier, 1848–1861," *Southwestern Historical Quarterly,* XXXVIII (October, 1934), 137; *Warren's Memoir,* p. 62.

[25] Bender, "The Texas Frontier, 1848–1861," *loc. cit.*, pp. 139–140; *Warren's Memoir,* p. 88.

[26] The camels used in west Texas were probably a remnant of the herd used by Edward F. Beale in his wagon-road survey across New Mexico along the thirty-fifth parallel. For a detailed discussion, see pp. 245–250; 254–255. See also, Lewis B. Lesley (ed.), *Uncle Sam's Camels.*

[27] Bender, "The Texas Frontier, 1848–1861," *loc. cit.*, pp. 142–146. See also the diary report of Hartz in *Senate Executive Document 2,* 36 Cong., 1 sess. (1859–1860), pp. 424–441.

[28] Bender, "The Texas Frontier, 1848–1861," *loc. cit.*, pp. 146–148; William H. Echols, "Report of a Reconnaissance West of Camp Hudson," *Senate Executive Document 1,* 36 Cong., 2 sess. (1860–1861), pp. 36–50.

NOTES TO CHAPTER IV

[1] Return I. Holcombe (ed.), *Minnesota in Three Centuries,* II, pp. 419, 421–422, 430–431, 436.

[2] James H. Baker, "History of Transportation in Minnesota," *Collections of the Minnesota Historical Society,* IX (April, 1901), 9, 18–20; Edward Van Dyke Robinson, *Early Economic Conditions and the Development of Agriculture in Minnesota,* Studies in the Social Sciences, Bulletin No. 3, p. 34.

[3] John Pope, "Report of an Exploration of the Territory of Minnesota," *Senate Executive Document 42,* 31 Cong., 1 sess. (1849–1850), p. 11.

[4] *Ibid.*, pp. 8–9, 15–16, 18.

[5] Arthur J. Larsen, "Roads and Trails in the Minnesota Triangle, 1849–1860," *Minnesota History,* XI (December, 1930), 387–388.

[6] Arthur J. Larsen, "The Development of the Minnesota Road System," unpublished doctoral dissertation, University of Minnesota, 1938. Hereafter cited as Larsen Dissertation. Mr. Larsen has made an exhaustive investigation and the writer is indebted to him for the citations in this work and in his two articles in *Minnesota History* that have pointed the way to many source materials in The National Archives and the Minnesota Historical Society Library.

[7] Arthur J. Larsen, "Roads and the Settlement of Minnesota," *Minnesota History,* XXI (September, 1940), 231–232.

[8] James H. Simpson to John J. Abert, September 15, 1851, *House Executive Document 2*, Part II, 32 Cong., 1 sess. (1851–1852), pp. 441–442. Hereafter cited as Simpson, *1851 Report*.

[9] "Address of Henry H. Sibley, of Minnesota, to the People of the Minnesota Territory, June, 1849." Printed document, Sibley Papers, Minnesota Historical Society; *Cong. Globe*, XVIII (1848–1849), 407, 599, 615.

[10] William W. Folwell, *A History of Minnesota*, I, pp. 246–255; *Minnesota Council Journal, 1849*, pp. 13, 15, 16; *Laws of the Territory of Minnesota, 1849*, pp. 165, 169, 172–173.

[11] Numerous letters from Minnesotans urging Sibley to secure funds for roads are filed in the Sibley Papers, Minnesota Historical Society.

[12] *House Report 172*, 31 Cong., 1 sess. (1849–1850), p. 2.

[13] *Cong. Globe*, XIX (1849–1850), 1074–1075, 1089.

[14] Sibley to Ramsey, May 30, 1850, Ramsey Papers, Minnesota Historical Society. Sibley also confided, "I think I shall allow myself a *leetle* frolic by way of relaxation, for thus far I have had but a dog's life of it."

[15] *Cong. Globe*, XIX (1849–1850), 1348–1349, 1356.

[16] *U. S. Statutes at Large*, IX, p. 439. For a diagram of the Minnesota roads, see map p. 50.

[17] Larsen Dissertation, p. 51.

[18] "Address of Henry H. Sibley to the People of Minnesota Territory, July 29, 1850." Printed document in the Sibley Papers.

[19] Abert, Report of the Colonel of the Corps of Topographical Engineers, November 14, 1850, *Senate Executive Document 1*, 31 Cong., 2 sess. (1850–1851), p. 390.

[20] *Ibid.*, pp. 390–393; Abert to Charles M. Conrad, September 2, 1850, and December 22, 1851, The National Archives, War Department Records, Letter Book of the Chief of Topographical Engineers to the Secretary of War, 1850–1854. All letters from Colonel Abert to the Secretary used in this chapter may be found in this collection unless otherwise indicated.

[21] Abert, Report . . . November 14, 1850, *loc. cit.*, p. 390; Abert to Potter, October 5, 1850, The National Archives, War Department Records, Bureau of Topographical Engineers, Letters Sent. Copies of letters written by the Bureau are available in bound letter books arranged chronologically. Letters Received are located by using the Bound Registers of Letters which summarize each communication. An alphabetical list of addressers accompanies each volume. All outgoing letters cited in this chapter may be found in this collection unless otherwise indicated.

[22] Larsen Dissertation, pp. 52–53.

[23] Ramsey to Conrad, February 15, 1851. The Minnesota legislature's resolution was enclosed in this letter; Sibley to Abert, March 8, 1851, The National Archives, War Department Records, Bureau of Topographical Engineers, Letters Received. All incoming correspondence used in this chapter may be found in this collection unless otherwise indicated.

[24] Abert to Conrad, December 22, 1851, *House Executive Document 12*, 32 Cong., 1 sess. (1851–1852), p. 5; Simpson, *1851 Report*, p. 438.

[25] Simpson, *1851 Report*, pp. 438–439; W. Sheridan Warrick, "James Hervey Simpson, Military Wagon Road Engineer in the Trans-Mississippi West, 1849–1867," unpublished M.A. thesis, University of Chicago, 1949. Warrick has carefully summarized Simpson's road-building activities as presented in that officer's reports.

[26] Quoted in Larsen Dissertation, pp. 58–59.

[27] Simpson, *1851 Report*, p. 441.

[28] *Cong. Globe*, XXI (1851–1852), 21, 100.

[29] Abert to Conrad, December 22, 1851, *loc. cit.*, p. 6. In replying to this House Resolution, the Chief Topographical Engineer followed the customary practice of submitting extracts from his earlier reports and those of Lieutenant Simpson containing the information requested.

[30] *Cong. Globe*, XXI (1851–1852), 1451–1452.

[31] *Ibid.*, pp. 1451–1455, 1535–1536.

[32] Simpson to Abert, March 4, 1852.

[33] *Cong. Globe*, XXI (1851–1852), 1174.

[34] Abert to Conrad, April 14, 1852.

[35] Simpson to Sibley, April 3, 1852, Sibley Papers.

[36] Larsen Dissertation, p. 61.

[37] Abert, "Report of the Colonel of Topographical Engineers, November 18, 1852," *House Executive Document 1*, 32 Cong., 2 sess. (1852–1853), pp. 217–218.

[38] Larsen Dissertation, pp. 58–59.

[39] *U. S. Statutes at Large*, X, p. 150.

[40] Simpson to Abert, September 17, 1853, *House Executive Document 1*, Part III, 33 Cong., 1 sess. (1853–1854), pp. 28–29. Hereafter cited as Simpson, *1853 Report*. Simpson to Abert, June 1, 1853.

[41] Henry M. Rice to Abert, April 27, 1853; Edmund Rice and W. H. Rice to Abert, April 27, 1853; Larsen Dissertation, pp. 79–80.

[42] Larsen Dissertation, pp. 64–65; J[esse] L. Reno, "Survey, etc., of Road from Mendota to Big Sioux River, April 1, 1854," *House Executive Document 97*, 33 Cong., 1 sess. (1853–1854).

[43] Reno, "Survey . . . from Mendota to Big Sioux . . . ," *loc. cit.*, pp. 2–12. For a diagram of the route, see map p. 50.

[44] Rice to Abert, May 6, 1854.

[45] *Cong. Globe*, XXIII (1853–1854), 1031–1032, 1051–1052, 1389; *U. S. Statutes at Large*, X, pp. 306–307.

[46] *U. S. Statutes at Large*, p. 581.

[47] Simpson to Abert, September 15, 1854, *Senate Executive Document 1*, Part II, 33 Cong., 2 sess. (1854–1855), p. 346.

[48] Folwell, *A History of Minnesota*, I, pp. 329–350; Larsen Dissertation, pp. 82–83.

[49] Rice to Jefferson Davis, October 20, 1854; Simpson to Abert, November 8, 1854.

[50] *U. S. Statutes at Large*, X, pp. 1172–1176; Larsen Dissertation, pp. 89–91.

[51] *Cong. Globe*, XXIV (1854–1855), 697, 767, 773; *U. S. Statutes at Large*, X, p. 610.

[52] *Cong. Globe*, XXIV (1854–1855), 522, 1017, 1144; *U. S. Statutes at Large*, X, p. 635, 638.

[53] Simpson to Abert, September 20, 1855, *Senate Executive Document 1*, Part II, 34 Cong., 1 and 2 sess., pp. 469–486, Appendix D, pp. 498–501. Hereafter cited as Simpson, *1855 Report*.

[54] *Ibid.*, pp. 475, 480–481, 487–492.

[55] *Ibid.*, pp. 483–485.

[56] *Daily Minnesota Pioneer*, August 17, 1855; Larsen Dissertation, pp. 85–86.

[57] Larsen Dissertation, pp. 86–87; Fletcher to Sibley, August 10, 1855, Sibley Papers; Rice to Davis, August 22, 1855; Simpson to Abert, August 11 and September 10, 1855.

[58] *Minnesota Democrat*, September 5, 1855, quoted in Larsen Dissertation, p. 88.

[59] The legislature of 1854 adopted eight memorials pertaining to roads; in 1855, five additional. *Laws of the Territory of Minnesota, 1854*, pp. 152, 154, 159–161, 163, 166, 170; *Laws . . . , 1855*, pp. 177–182, 185.

[60] *Cong. Globe*, XXIV (1854–1855), pp. 414, 486, 492, 697, 767, 773.

[61] *Laws of the Territory of Minnesota, 1856*, p. 343; *House Committee Report 191*, 34 Cong., 1 sess. (1855–1856); Abert to Davis, April 2, 1856.

[62] *Laws of the Territory of Minnesota, 1856*, pp. 351, 355, 358, 360, 367, 371, 373; *Laws . . . , 1857*, pp. 292, 294.

[63] The particular measure under discussion provided $10,000 for bridging streams on the territorial road from St. Paul to Elliota; $5,000 to improve a road from Kasota to the Winnebago agency; $5,000 for a Shakopee–Le Seur road; $5,000 for a Faribault to Traverse des Sioux road—all to be built by the Commissioner of Indian Affairs.

[64] *Cong. Globe*, XXV (1855–1856), 1493.

[65] *Ibid.*, p. 1493.

[66] *Ibid.*, p. 1494.

[67] *Cong. Globe*, XXVI (1856–1857), 391.

[68] *Ibid.*, p. 1111.

[69] *Ibid.*, pp. 1111–1112.

[70] *U. S. Statutes at Large*, XI, p. 203.

[71] Manypenny to McAboy, April 19, 1855, June 26 and August 6, 1856; Manypenny to Rice, February 19, 1856, The National Archives, Interior Department Records, Office of Indian Affairs, Letter Books.

[72] Thom to Abert, September 5, 1857, *Senate Executive Document 11*, 35 Cong., 1 sess. (1857–1858), pp. 348–355; Larsen Dissertation, pp. 105–110; *U. S. Statutes at Large*, XI, p. 204.

[73] Stansbury to Abert, October 15, 1858, *Senate Executive Document 1*, 35 Cong., 2 sess. (1858–1859), pp. 1193–1202.

[74] Stansbury to Abert, September 30, 1859, *Senate Executive Document 2*, 36 Cong., 1 sess. (1859–1860), pp. 857–866; Stansbury to Abert, November 5, 1860, *Senate Executive Document 1*, 36 Cong., 2 sess. (1860–1861), pp. 532–540; Stansbury to Lieutenant Colonel Hartman Bache, October 22, 1861, *Senate Executive Document 1*, 37 Cong., 2 sess. (1862–1863), p. 546.

[75] Larsen Dissertation, p. 117. The quotation is from Simpson, *1855 Report.*

NOTES TO CHAPTER V

[1] Parts of this chapter have been published: "Federal Road Building Grants for Early Oregon," *Oregon Historical Quarterly*, L (March, 1949), 3–29.

[2] Charles H. Carey, *A General History of Oregon Prior to 1861*, II, pp. 467–468; Hiram F. White, "Samuel Royal Thurston," *Iowa Journal of History and Politics*, XIV (April, 1916), pp. 239–264; Cornelius H. Hanford, "Pioneers of Iowa and the Pacific Northwest," *Annals of Iowa*, X (January–April, 1912), 331–342. Many early Iowa politicians transferred their residence to the Pacific Northwest and continued their political activity. In these last two articles, the lives of three men mentioned in this study, Samuel R. Thurston, W. W. Chapman, and George H. Williams, have been described in some detail.

[3] *House Report 348*, 31 Cong., 1 sess. (1849–1850). This report on "Roads in Oregon" was made on May 24, 1850. Hubert Howe Bancroft, *History of Oregon, 1848–1888*, II, p. 152 n.

[4] Oscar Osburn Winther, "The Place of Transportation in the Early History of the Pacific Northwest," *The Pacific Historical Review*, XI (December, 1942), 384.

[5] Oscar Osburn Winther, *The Old Oregon Country: A History of Frontier Trade, Transportation and Travel*, pp. 122–125.

[6] Oscar Osburn Winther, "The Roads and Transportation of Territorial Oregon," *Oregon Historical Quarterly*, XLI (March, 1940), 42. Bancroft, *op. cit.*, p. 305 n.

[7] Winther, "The Place of Transportation in the Early History of the Pacific Northwest," *loc. cit.*, p. 383. This interpretive study emphasizes the political, economic, and social impact of transportation in the development of the Northwest. Professor Winther's scholarly investigations on this subject have been summarized in his volume, *The Old Oregon Country.* The broad pattern of northwest transportation has been presented recently by Randall V. Mills, "A History of Transportation in the Pacific Northwest," *Oregon Historical Quarterly*, XLVII (September, 1946), 281–312.

[8] *Senate Miscellaneous Document 5*, 31 Cong., 2 sess. (1850–1851), p. 5. This memorial of the legislature of Oregon, referred to the Committee on territories on January 10, 1851, stated in part: "It is important, both for civil and military purposes, that a road be opened from Puget Sound to some point on the Columbia below the Cascades; one from Puget Sound to a point on the Columbia near Walla-Walla; one from the Dalles of the Columbia to the Willamette Valley; and one from Astoria, or some other point at the mouth of the Columbia, to Willamette falls; one up each side of the Willamette, from the Columbia to the upper end of the valley, and thence one to some point in Sacramento valley, in California. For these roads we ask your liberal consideration." Bancroft, *op. cit.*, II, p. 75.

[9] Bancroft, *op. cit.*, p. 306; Carey, *op. cit.*, p. 483.

[10] For a discussion of Minnesota developments, see chap. iv.

[11] *Cong. Globe,* XXII (1852–1853), 165, 179, 233.

[12] *U. S. Statutes at Large,* X, p. 151. For a discussion of the Steilacoom-Fort Walla Walla road, see chap. vi.

[13] *Cong. Globe,* XXIII (1853–1854), 46, 140, 1031, 1115, 1621; *U. S. Statutes at Large,* X, p. 303. Brief references to the Oregon "military roads" constructed during the territorial period are to be found in contributions to the *Quarterly of the Oregon Historical Society.* As examples, see Thomas W. Prosch, "Notes from a Governmental Document on Oregon Conditions in the Fifties," VIII (June, 1907), 195; Lester B. Shippee, "The Federal Relations of Oregon—VII," XX (December, 1919), 365; Robert C. Clark, "Military History of Oregon, 1849–59," XXXVI (March, 1935), 53–54.

[14] Bancroft, *op. cit.,* pp. 305–306 n; Carey, *op. cit.,* p. 730. A detailed map of this survey transmitted to the Bureau of Topographical Engineers of the War Department may be seen in the Division of Cartographic Records, The National Archives. The map is entitled "Sketch of the Military Road from Myrtle Creek, Umpqua Valley to Camp Stuart, Rogue River Valley, Oregon, located in the autumn of the year 1853 by Major B. Alvord, 4th Infantry, U. S. Army," with an additional notation "surveyed and drawn by Jesse Applegate, of Yoncalla, Oregon, with Burt's Solar Compass." A table of distances has been included and someone in the Bureau has written "no field notes of this survey furnished the Bureau."

[15] August 2, 1854, The National Archives, War Department Records, Bureau of Topographical Engineers, Letters Sent. All correspondence used in this chapter is located in the War Department Records unless otherwise designated.

[16] Withers to Davis, December 6, 1854. Accompanying this correspondence was a map of the road, copies of proposals for bids, handbills, abstract of bids received, and a copy of each contract. Clippings from the *Umpqua Weekly Gazette,* Scottsburg, for October 21, 1854, containing the advertisement for bids, and for November 18, 1854, including a defense of Withers relative to the granting of contracts, were also forwarded as inclosures to this report.

[17] Chapman to Davis, November 7, 1854. See note 1.

[18] Withers to Davis, February 6, 1855, *Senate Executive Document 1,* Part II, 34 Cong., 1 sess. (1855–1856), pp. 507–508.

[19] Withers to Davis, June 15, 1855, *Ibid.,* pp. 508–509. Fort Lane, formerly known as Camp Stuart, was located at the beginning of the road in the Rogue River Valley.

[20] This commendation of the Secretary of War is on the original letter in The National Archives. It does not appear in the printed version of his official report. For a diagram of the Myrtle Creek–Scottsburg road, see map p. 74.

[21] *Cong. Globe,* XXIV (1854–1855), 3, 367, 483, 492, 509, 697; *U. S. Statutes at Large,* X, p. 608.

[22] Davis to Bache, May 14, 1855. In addition to the Astoria-Salem project, Congress had approved $25,000 for a road from Fort Vancouver (Columbia City Barracks) to The Dalles and $30,000 from Vancouver to Fort Steilacoom on Puget Sound, both in Washington Territory. For a discussion of these roads, see chap. vi.

[23] Lane to Davis, June 8, 1855.

[24] Bache to Derby, July 9, 1855; Bache to Abert, July 10, 1855. Bache thought the road from Fort Vancouver to The Dalles should have precedence over that to Fort Steilacoom, but authorized Derby to use his judgment as to which of the two should receive first attention following the Oregon survey.

[25] Bache to Abert, September 11, 1855. A map of the town of Astoria accompanied this report showing the site of the villages, mountainous terrain, and proposed surveys. Extensive correspondence may be found in the records of the Bureau.

[26] Derby to Bache, July 25, 1855, *Senate Executive Document 1,* 34 Cong., 1 sess. (1855–1856), pp. 502–503. This suggestion of a 100-foot roadway had undoubtedly originated with Simpson in Minnesota.

[27] Derby to Bache, July 30, 1855, *Ibid.,* pp. 503–504. A party of men from Astoria had been through the woods the previous season and one of these accompanied the Derby party as guide.

Their earlier path, although somewhat filled with timber, and the trails of the elk were used to get through the underbrush. Six able axemen, who labored from daylight to sunset cutting this path, seldom made more than five miles a day. After twelve days, the party emerged upon a cart path along the middle branch of the Tualatin River that led to Salem. For a diagram of the route, see map p. 78.

[28] Derby to Bache, January 15, 1856. The report was accompanied by a detailed map of the road, advertisements for contractors, and table of distances. These are available in The National Archives.

[29] Bache to Abert, October 19, 1855.

[30] Bache to Abert, January 18, 1856.

[31] Davis to Abert, March 19, 1856; Derby to Bache, February 8, 1856; Bache to Abert, February 19, 1856; Abert to Davis, March 19, 1856.

[32] Abert, "Report of the Chief Topographical Engineer, 1856," *House Executive Document 1*, Part II, 34 Cong., 3 sess. (1856–1857), pp. 371–372; Derby to Bache, May 13, 1856; Bache to Derby, August 1, 1856. The official reports and letters of Lieutenant Derby, full of interesting descriptive detail, total hundreds of pages. Derby was a recognized writer. Carey states he was "well known as a facetious writer over the signature of 'Phoenix,' alias 'Squibob,' alias 'Butterfield,' and sundry other nommes de plume." p. 483.

[33] Bache to Abert, September 15, 1856.

[34] Abert to Davis, March 19, 1856; Davis to Quitman, March 29, 1856.

[35] *Cong. Globe*, XXV (1855–1856), 1473.

[36] *Ibid.*, pp. 1473–1474.

[37] *Ibid.*, p. 1491.

[38] *Ibid.*, p. 1492.

[39] *Ibid.*

[40] *Ibid.*

[41] *U. S. Statutes at Large*, XI, p. 168.

[42] Mendell to Bache, April 23, 1857. Mendell wrote, "I have examined a large portion of the constructed part of the Astoria road. The road bed is much the same state in which it was left last fall as there has been but little travel over it during the past winter. The only effect the severe storms of the past season have had was to fell a great many trees across the road."

[43] Mendell to Bache, September 1, 1857, *House Executive Document 2*, 35 Cong., 1 sess. (1857–1858), pp. 521–524. The Mendell-Bache-Abert correspondence is extensive, including approximately one hundred unpublished letters.

[44] Hardy C. and Thomas Elliff had a contract for $8,000 to build thirteen miles of road from Jacksonville to Cow Creek; David W. Ranson and Jep Roberts were contractors for the bridging.

[45] Mendell to Captain George Thom, September 16, 1858, *House Executive Document 2*, 35 Cong., 2 sess. (1858–1859), pp. 1213–1214.

[46] Robert M. Hutchinson contracted for a bridge over Deer Creek at Roseburg; A. B. and L. L. Kellogg and Hill and Company also were granted bridge-building contracts.

[47] Mendell to Bache, April 2, 1858.

[48] *Cong. Globe*, XXVII (1857–1858), 2115–2116, 2133.

[49] *Ibid.*, p. 2995.

[50] *U. S. Statutes at Large*, XI, p. 337.

[51] Thom to Abert, July 6, 1859, August 5, 1859, September 17, 1859, October 6, 1859. The last letter is published in *Senate Executive Document 2*, Part III, 36 Cong., 1 sess. (1859–1860), pp. 875–876.

[52] Thom to Abert, September 17, 1859.

[53] Floyd to Abert, November 14, 1859.

[54] Abert to Floyd, March 24, 1860; Floyd to Abert, March 27, 1860.

[55] In the post-Civil War period, federal aid to Oregon road building was in the form of land allotments from the public domain. The state was authorized to dispose of the land through legislative action; the governor was charged with the responsibility of supervising survey and

construction. For more details, see my "Federal Road Building Grants for Early Oregon," *loc. cit.*, pp. 23–29.

[56] Louis Scholl to Ingalls, December 27, 1857; Robert Newell to Ingalls, December 31, 1857; Joel Palmer to Ingalls, January 3, 1858; Ingalls to Captain Alfred A. Pleasonton, November 22, 1858; Harney to Scott, November 29, 1858 in "Affairs in Oregon," *House Executive Document 65*, 36 Cong., 1 sess. (1859–1860), pp. 96–105.

[57] Harney to the Adjutant General, January 17, 1860; Pleasonton to Wallen, April 28, 1859, *Senate Executive Document 34*, 36 Cong., 1 sess. (1859–1860), pp. 1–2, 3–4.

[58] Thom to Dixon, May 16, 1859, *Ibid.*, p. 31.

[59] Special Orders No. 40, April 27, 1859; Wallen to Pleasonton, n.d., *Ibid.*, pp. 4–5.

[60] Wallen to Pleasonton, n.d.; Dixon to Pleasonton, n.d., *Ibid.*, pp. 6–9, 31–37. For a diagram of this route, see "Map of a Reconnaissance for a Military Road from 'the Dalles' of the Columbia River to the Great Salt Lake" by Lieutenant Joseph Dixon included in this federal document.

[61] Wallen to Bonnycastle, June 29, 1850; Bonnycastle to Johnston, July 1, 1850; Johnston to Bonnycastle, July 6, 1859; Bonnycastle to Pleasonton, n.d., *Ibid.*, pp. 18–22; Harney to the General-in-Chief, August 1, 1859, *Senate Executive Document 2*, 36 Cong., 1 sess. (1859–1860), p. 113.

[62] Wallen to Pleasonton, n.d.; Dixon to Pleasonton, n.d., *Senate Executive Document 34*, 36 Cong., 1 sess. (1859–1860), pp. 10–12, 37–40.

[63] Scholl to Pleasonton, December 3, 1859, *Ibid.*, pp. 22–28.

[64] Wallen to Pleasonton, October 1, 1859, Harney to General-in-Chief, October 6, 1859, *Senate Executive Document 2*, 36 Cong., 1 sess. (1859–1860), pp. 119–120; Wallen to Pleasonton, n.d., *Senate Executive Document 34*, 36 Cong., 1 sess. (1859–1860), pp. 14–16.

[65] Wallen to Pleasonton, n.d., *Senate Executive Document 34*, 36 Cong., 1 sess. (1859–1860), pp. 44–45.

[66] Harney to the Adjutant General, January 17, 1860, *Ibid.*, p. 1.

NOTES TO CHAPTER VI

[1] Winther, *The Old Oregon Country*, pp. 127–129.

[2] *U. S. Statutes at Large*, X, p. 151.

[3] Thomas W. Prosch, "The Military Roads of Washington Territory," *The Washington Historical Quarterly*, II (January, 1908), 118–119.

[4] Philip H. Overmeyer, "George B. McClellan and the Pacific Northwest," *The Pacific Northwest Quarterly*, XXXII (January, 1941), 3–60. Overmeyer has examined the McClellan papers in the Library of Congress and, with the use of the correspondence and journal of this Army officer, has presented much additional evidence on what transpired between Stevens and McClellan in the Pacific Northwest, 1853–1854.

[5] Davis to McClellan, May 9, 1853, *Senate Executive Document 1*, Part II, 33 Cong., 1 sess. (1853–1854), p. 67.

[6] Stevens to McClellan, May 9, 1853, *Senate Executive Document 78*, Part I, 33 Cong., 2 sess. (1854–1855), p. 203.

[7] "General Reports of the Survey of the Cascades: General Report of Captain George B. McClellan, Corps of Engineers, U.S.A., in command of the Western Division, February 25, 1853," *Ibid.*, pp. 188–202. This published report is a summary outline of a longer manuscript journal available in the McClellan Papers, Library of Congress. Overmeyer has used the unpublished source extensively in "George B. McClellan and the Pacific Northwest," *loc. cit.*, pp. 3–60. My account is based upon both. McClellan corresponded with Stevens in the autumn of 1854 requesting that the reports of junior officers in the western command be omitted from Stevens' report to the War Department or that the latter make no comment on them or upon Stevens' own report. Stevens refused. He demanded the official journal which McClellan had been ordered to keep, but gave him the alternative of submitting the original or a copy. The 180 pages of manuscript were reduced to 15 printed pages.

[8] Overmeyer, "George B. McClellan and the Pacific Northwest," *loc. cit.*, pp. 14–15, 37. Thomas W. Prosch is in error in stating that no contracts were let.

[9] McClellan Journal, p. 161. Quoted in Overmeyer, "George B. McClellan and the Pacific Northwest," *loc. cit.*, pp. 49–50.

[10] *Ibid.*, pp. 59–60.

[11] Winther, *op. cit.*, p. 130. Winther quotes from the newspaper.

[12] Prosch, "The Military Roads of Washington Territory," *loc. cit.*, p. 121.

[13] For a diagram of Arnold's route, see map p. 94.

[14] Arnold to Jefferson Davis, January 26, 1855, *Senate Executive Document 1*, 34 Cong., 1 sess. (1855–1856), pp. 532–538.

[15] Winther, *op. cit.*, p. 131.

[16] *U. S. Statutes at Large*, X, pp. 603–604.

[17] Davis to Bache, May 14, 1855, The National Archives, War Department Records, Bureau of Topographical Engineers, Letters Sent. All outgoing correspondence used in this chapter is in this collection unless otherwise designated.

[18] For further discussion, see chap. v.

[19] Derby to Bache, September 15, 1855; Bache to Abert, October 1, 1855, The National Archives, War Department Records, Bureau of Topographical Engineers, Letters Received. All incoming correspondence used in this chapter is in this collection unless otherwise designated.

[20] Derby to Bache, October 1, 1855. For a diagram of the route from Fort Vancouver to The Dalles, see map p. 94.

[21] Abert to Davis, December 20, 1855.

[22] Derby to Bache, December, 1855; Bache to Abert, December 29, 1855. Derby forwarded a detailed map of the road survey from The Dalles to Vancouver, together with profiles of the Portage road on February 13, 1856.

[23] Derby to Bache, April 1, April 4, May 3, and June 30, 1856. The lieutenant prepared extensive lists of provisions purchased and through cost accounting tried to prove how economical his operations were.

[24] Derby to Bache, July 12, 1856; Bache to Abert, July 31, 1856.

[25] Derby to Bache, May 13, June 30, August 1, and September 1, 1856. For a summary of the season's work on this road, see "Report of the Chief Topographical Engineer, November 22, 1856," *House Executive Document 1*, 34 Cong., 3 sess. (1856–1857), pp. 372–373. Derby was a prolific writer and appears to have spent more time preparing reports than directing field operations.

[26] For a diagram of the route from Fort Vancouver to Steilacoom, see map p. 94.

[27] Derby to Bache, March 1, 1856; Bache to Abert, March 4, 1856; Abert to Davis, March 16, 1856.

[28] Bache to Derby, May 6, 1856; Davis to Bache, June 21, 1856.

[29] Bache to Derby, August 1, 1856.

[30] *Ibid.*

[31] Derby to Bache, August 10, 1856.

[32] Bache to Abert, August 30, 1856.

[33] Several copies of this printed notice are filed with the Derby reports in the War Department Records.

[34] Derby to Bache, September 26, 1856. This petition was forwarded with several dozens of letters from residents supporting various policies and procedures.

[35] Stevens to Davis, June 8, 1856; Abert to Bache, July 27, 1856; Bache to Abert, August 28, 1856.

[36] Derby to Bache, October 6 and 10, 1856; Bache to Abert, October 31, 1856.

[37] Mendell to Bache, December 1, 1856, April 1, April 23, May 1, and September 1, 1857. Letter of September 1, 1857 is printed in *Senate Executive Document 11*, Part I, 35 Cong., 1 sess. (1857–1858), pp. 522–523.

[38] Mendell to Bache, October 26, 1856.

[39] Mendell to Bache, November 3, 1856.

[40] Bache to Abert, November 3, 1856. A long note was penned to this report and signed by Jefferson Davis, December 19, 1856.

[41] Bache to Abert, January 25, 1857; Bache to Mendell, March 2, 1857; Mendell to Bache, March 25, 1857; Bache to Mendell, April 1, 1857; Mendell to Bache, April 25, 1857. Lieutenant Mendell noted that if Tower failed to meet requirements, then Carter would be the lowest bidder and the other route would be the one constructed. He feared this contingency because the Tower bid was so low. Major Bache assured him Tower would keep his promise because farmers along the route would do most of the work.

[42] Mendell to Bache, August 20, 1857; Mendell to Bache, September 1, 1857, *Senate Executive Document 11*, Part I, 35 Cong., 1 sess. (1857–1858), p. 522.

[43] *Statutes of the Territory of Washington, 1855–1856*, p. 63.

[44] *Cong. Globe*, XXV (1855–1856), 1451, 1492, 1496–1497; XXVI (1856–1857), 391–392, 399, 498, 1115.

[45] *U. S. Statutes at Large*, XI, p. 252.

[46] Bache to Mendell, May 16, 1857; Mendell to Bache, May 24, 1857. Lieutenant Mendell unable to locate an engineer immediately, prevailed upon Bache to seek one in San Francisco. Several men approached by Bache refused the assignment.

[47] Mendell to Bache, August 20, 1857, and September 1, 1857, *Senate Executive Document 11*, Part I, 35 Cong., 1 sess. (1857–1858), pp. 523–524.

[48] DeLacy to Mendell, n.d., *Senate Document 1*, 35 Cong., 2 sess. (1858–1859), pp. 1216–1230.

[49] Mendell to Bache, January 16, 1858; Bache to Abert, January 18, 1858; Mendell to Bache, May 1, 1858; Bache to Abert, July 15, 1858; and Mendell to George Thom, September 16, 1858, *Ibid.*, pp. 1215–1216.

[50] Mendell to Bache, December 1, 1857; April 19, May 1, and July 18, 1858; Bache to Abert, July 19, 1858; Mendell to George Thom, September 16, 1858, *Ibid.*, p. 1215.

[51] These memorials asked for federal aid for the following roads: from the military post at Port Townsend to intersect the military road from Fort Vancouver to Fort Steilacoom; from Fort Vancouver to The Dalles; from a new military post at New Dungeness to Port Townsend; from Fort Steilacoom to Bellingham. They also memorialized Congress for remuneration to citizens who built the military road over Naches Pass.

[52] *Cong. Globe*, XXVII (1857–1858), 1148, 2032, 2057, 2116, 2117, 2133, 2399. The quotation is from p. 2117. Abert's letter to Floyd was dated February 13, 1858; that of Mendell, April 2, 1858.

[53] Thom to Abert, November 4, and December 2, 1858; April 1, June 7, June 27, July 6, August 5, and September 1, 1859.

[54] Thom to Abert, October 6, 1859, *Senate Executive Document 2*, 36 Cong., 1 sess. (1859–1860), pp. 694–696, 875–879.

[55] *U. S. Statutes at Large*, XII, p. 19.

[56] Thom to Abert, September 19, 1860, *Senate Executive Document 1*, Part II, 36 Cong., 2 sess. (1860–1861), pp. 543–545; Thom to Lieutenant Colonel Hartman Bache, October 19, 1861, *Senate Executive Document 1*, Part II, 37 Cong., 2 sess. (1861–1862), pp. 547–548.

NOTES TO CHAPTER VII

[1] *Laws of the Territory of New Mexico, 1851–1852*, p. 224.

[2] James W. Abert, "Report of Lieutenant J. W. Abert of His Examination of New Mexico in the Years 1846–1847," *House Executive Document 41*, 30 Cong., 1 sess. (1847–1848), p. 448. See the map following page 584 in this report.

[3] Whittlesey to Captain Thomas L. Brent, September 10, 1849, *House Executive Document 1*, 31 Cong., 2 sess. (1850–1851), pp. 297–300.

[4] *Laws of the Territory of New Mexico, 1851–1861.*

[5] Jack L. Cross, "Federal Wagon Road Construction in New Mexico Territory, 1846–1860," unpublished M. A. thesis, University of Chicago, 1949. Cross has carefully analyzed the contents of the New Mexico memorials to Congress for transportation aid.

[6] *Laws of the Territory of New Mexico, 1851–1852,* pp. 224–225.

[7] *Laws of the Territory of New Mexico, 1852–1853,* pp. 135, 137, 139.

[8] *Laws of the Territory of New Mexico, 1853–1854,* pp. 162.

[9] *Ibid.*, p. 155.

[10] *U. S. Statutes at Large,* X, p. 303.

[11] *Cong. Globe,* XXIII (1853–1854), 562–564.

[12] Ralph Emerson Twitchell, *The Leading Facts of New Mexican History,* II, p. 309.

[13] *Cong. Globe,* XXIII (1853–1854), 1030–1031, 1621.

[14] Scammon to Abert, October 28, 1854, The National Archives, War Department Records, Bureau of Topographical Engineers, Letters Received. All correspondence used in this chapter is in the War Department Records unless otherwise designated.

[15] Requisition by Scammon, November 21, 1854.

[16] Scammon to Abert, October 4, 1854.

[17] Davis to Scammon, November 28, 1854, *House Executive Document 1,* Part II, 33 Cong., 2 sess. (1854–1855), pp. 42–43. This letter is quoted in F. T. Cheetham, "El Camino Militar," *New Mexico Historical Review,* XV (January, 1940), 5–6.

[18] Abert to Davis, December 4, 1855.

[19] Scammon to Abert, March 1, 1856.

[20] Jesup, "Report of the Quartermaster General, November 26, 1856," *House Executive Document 1,* Part II, 34 Cong., 3 sess. (1856–1857), p. 252. Scammon's shortages were excessive. According to Army custom, only one defalcation, usually the most flagrant, was singled out as a basis for removal.

[21] Davis, "Annual Report of the Secretary of War, December 3, 1855," *Senate Executive Document 1,* Part II, 34 Cong., 1 sess. (1855–1856), p. 19.

[22] Abert, "Annual Report of the Chief Topographical Engineer, November 22, 1856," *House Executive Document 1,* 34 Cong., 3 sess. (1856–1857), p. 361.

[23] *U. S. Statutes at Large,* X, p. 638.

[24] Macomb to Abert, May 14, 1856.

[25] Macomb to Abert, May 27, 1857. The winter months were spent in Washington, D.C. obtaining information about Scammon's accounts and securing data on the details of financial administration in an attempt to avoid more confusion. From Texas, Macomb traveled west with Brigadier General John Garland, who was to be commander of the New Mexico Military Department.

[26] Macomb to Abert, June 30, 1857.

[27] Macomb to Abert, July 13, 1857.

[28] Macomb to Abert, July 28, 1857.

[29] Macomb to Abert, July 13 and November 30, 1857.

[30] Macomb to Abert, January 31, 1858.

[31] Macomb to Abert, April 5, 1858.

[32] Macomb to Abert, April 5, May 5, June 5, July 5, August 5, September 6, 1858; and September 29, 1858, *Senate Executive Document 1,* 35 Cong., 2 sess. (1858–1859), pp. 1206–1207, 1210. For a diagram of this route, see map p. 114.

[33] Macomb to Abert, June 5, July 5, August 5, September 6, 1858; and September 29, 1858, *Ibid.*, pp. 1207, 1210–1211.

[34] Macomb to Abert, August 5 and September 6, 1858; and September 29, 1858, *Ibid.*, pp. 1207–1208.

[35] Macomb to Abert, April 5, 1858; and September 29, 1858, *Ibid.*, pp. 1208–1209; Cheetham, "El Camino Militar," *loc. cit.*, pp. 6–9.

[36] Macomb to Abert, September 29, 1858, *Senate Executive Document 1,* 35 Cong., 2 sess.

(1858–1859), p. 1209; Macomb to Abert, June 6, 1859, *Senate Executive Document 2,* 36 Cong., 1 sess. (1859–1860), pp. 871–874.

[37] Macomb to Abert, May 6, 1859.

[38] Macomb to Abert, March 26, 1859.

[39] Macomb to Abert, October 5, 1858.

[40] Otero to Garland, November 19, 1857; Macomb to Garland, March 9, 1858; and Macomb to Abert, March 11, 1858.

[41] *Cong. Globe,* XXVII (1857–1858), 1517, 1956, 2032, 2056, 2116, 2118, 2133, 2259, 2399, 2664; XXVIII (1858–1859), 10, 200.

[42] Macomb to Abert, July 29, 1857.

[43] Macomb to Abert, August 28, 1858.

[44] Macomb to Abert, June 3, 1859.

[45] Averam B. Bender, "Government Explorations in the Territory of New Mexico, 1846–1859," *New Mexico Historical Review,* IX (January, 1934), 1–32.

[46] Macomb to Abert, June 3, 1859. A letter to Macomb from Captain A. A. Humphreys, in charge of the Office of Explorations and Surveys, April 6, 1859, was included in this communication.

[47] Macomb to Humphreys, December 1, 1860, *Senate Executive Document 1,* Part II, 36 Cong., 2 sess. (1860–1861), pp. 149–152. This report was separately published by the Government Printing Office in 1876 as *Report of the Exploring Expedition from Santa Fé, New Mexico to the Junction of the Grand and Green Rivers of the Great Colorado of the West, 1859.*

[48] *Ibid.,* pp. 149–150. "Ojo Verde" was reported to be 340 miles northwest of Santa Fe.

[49] *Ibid.,* pp. 150–151.

[50] Humphreys to John B. Floyd, November 12, 1860, *Senate Executive Document 1,* Part II, 36 Cong., 2 sess. (1860–1861), p. 146. For a discussion of Beale's wagon road along the thirty-fifth parallel, see chap. xv.

[51] Macomb to Abert, June 30, 1860, *Senate Executive Document 1,* Part II, 36 Cong., 2 sess. (1860–1861), pp. 540–541.

[52] *Ibid.*

[53] Bender, "Government Explorations in the Territory of New Mexico, 1846–1859," *loc. cit.,* pp. 28–29; Steen to J. D. Wilkins, August 10, 1859, Bonneville to the General in chief, April 30, August 31, 1859, *Senate Executive Document 2,* 36 Cong., 1 sess. (1859–1860), pp. 295, 312–314, 330–331.

[54] Bender, "Government Explorations in the Territory of New Mexico, 1846–1859," *loc. cit.,* p. 29; Claiborne to J. D. Wilkins, August 9, 1859, Bonneville to the General in chief, August 31, 1859, *Senate Executive Document 2,* 36 Cong., 1 sess. (1859–1860), pp. 312–314, 328–330.

[55] Bender, "Government Explorations in the Territory of New Mexico, 1846–1859," *loc. cit.,* pp. 29–30. Lazelle to J. D. Wilkins, July 10, 1859, Bonneville to the General in chief, August 31, 1859, *Senate Executive Document 2,* 36 Cong., 1 sess. (1859–1860), pp. 312–316.

[56] Humphreys, "Report of the Topographical Bureau, November 14, 1860," *Senate Executive Document 1,* Part II, 36 Cong., 2 sess. (1860–1861), p. 296.

[57] *U. S. Statutes at Large,* XII, p. 208.

[58] Bache, "Report of the Topographical Bureau, November 14, 1861," *Senate Executive Document 1,* 37 Cong., 2 sess. (1861–1862), p. 124.

[59] *U. S. Statutes at Large,* XVII, p. 621.

NOTES TO CHAPTER VIII

[1] Parts of this chapter have been published: "The Army Engineers as Road Surveyors and Builders in Kansas and Nebraska, 1854–1858," *The Kansas Historical Quarterly,* XVII (February, 1949), 37–59.

[2] *U. S. Statutes at Large,* X, p. 608.

[3] "Military Roads in Kansas," *House Report 36,* 33 Cong., 2 sess. (1854–1855), p. 3.

[4] *U. S. Statutes at Large,* X, p. 64.

[5] "Military Roads in Kansas," *loc. cit.,* p. 3.

[6] Bryan to John J. Abert, June 14, 1855. Bryan had been assigned the duty in Kansas and Nebraska on April 28, 1855. Within two weeks he was on his way to St. Louis. The National Archives, War Department Records, Bureau of Topographical Engineers, Letters Received. All correspondence and manuscript reports used in the preparation of this chapter are in the War Department Records unless otherwise designated.

[7] For a diagram of this route, see map p. 122.

[8] Fort Atkinson was just west of present Dodge City and Bent's Fort was near present Prowers, Colorado. For a history of Bent's Fort, see George Bird Grinnell, "Bent's Old Fort and Its Builders," *Kansas Historical Collections,* XV (1923–1925), 28–91.

[9] Bryan to Abert, December 15, 1855. This annual report of Bryan contains many interesting details of the survey too extensive to be included in this account.

[10] Abert, "Report of the Chief Topographical Engineer, November 22, 1856," *House Executive Document 1,* Part II, 34 Cong., 3 sess. (1856–1857), p. 370.

[11] Bryan to Abert, December 15, 1855.

[12] Bryan to Abert, October 30, 1855.

[13] Bryan to Abert, February 8, 1856.

[14] June 26, 1856.

[15] Lombard to Bryan, November 22, 1856.

[16] *Ibid.* The Secretary of War had agreed to modifications of the contract if the total payment was not in excess of the contract figure of $38,400. By omitting the icebreakers at the Saline the contractor had saved the time needed for the water to go down. On the Smoky Hill it would have been necessary to haul piles for 52 miles.

[17] Bryan to Abert, February 10, 1857. Bryan deducted $50 from Sawyers' payment to complete the grading of the approach to one of the bridges. The contractor produced evidence required by law that he had paid his laborers, with the exception of four men. In time, Bryan discovered each of these four had wages coming, one for as much as $143.75. The administration of contracts was one of the greatest problems confronted by the Topographical Engineers.

[18] Lombard to Bryan, November 22, 1856.

[19] Bryan to Abert, February 10, 1857. Bryan reported that trains traveling over the route could be saved detention and much labor if the small streams and sloughs could be bridged and their approaches graded. The remaining $910.95 of the appropriation of January 1, 1857 was not enough, however, to commence operations. The engineer also renewed his request that a large train be sent over the road to New Mexico so that its wagon wheels would make a trace that could not be effaced before the emigrants followed and permanently marked the route. The road, as far as the Smoky Hill, was already thus marked.

[20] Bryan to Abert, October 30, 1855.

[21] Bryan to Abert, November 12, 1855. In reading the correspondence between Bryan and Abert, the historian will discover what appears to be a growing friction between the officers. Bryan felt his chief was unsympathetic with his problems and overly critical. Abert seems to have lacked confidence in the young officer and considered him at times disrespectful, if not insubordinate.

[22] Bryan to Abert, April 14, 1856. On April 29, Bryan wrote again, "The appointment of this agent is necessary if these two roads are to be surveyed in the same summer as it is impossible for one person to attend to both at the same time on account of the distance between them and the difficulty of moving about from one point to another in such a wild and unsettled country.... Early action is requested as the season is fast approaching when parties destined for the plains should take the field."

[23] Bryan to Abert, May 28, 1856. Bryan also notified the Bureau of the equipment which he might provide for Dickerson's work.

[24] The "Laramie Crossing" of the Platte was the established ford for emigrants on the Oregon Trail traveling to Fort Laramie.

[25] For a diagram of this route, see map p. 122.

[26] Bryan to Abert, February 19, 1857, *House Executive Document* 2, 35 Cong., 1 sess. (1857–1858), pp. 455–464.

[27] *Ibid.* For a continuation of this discussion, see chap. ix.

[28] Bryan to Abert, December 1, 1856, and January 1, February 25, May 14, 1857. These reports were all published by the Secretary of War in his annual report for 1857. See *House Executive Document* 2, 35 Cong., 1 sess. (1857–1858), pp. 455–520. Two maps were forwarded during the winter: "Military Road from Fort Leavenworth to Fort Riley, Kansas; profiles Rock, Vermillion, Grasshopper & Stranger creeks, & Blue and Republican Rivers" and "Reconnaissance of a Road from Fort Riley, Kansas to Bridger's Pass made in obedience to instructions from the War Department in June, July, August, September, and October, 1856." On the latter map Bryan listed J. Lambert, C. T. Larned and S. M. Cooper as assistants. These maps may be seen in The National Archives, Division of Cartographic Records.

[29] Bryan to Abert, April 24, 1857.

[30] Bryan to Abert, December 10, 1857.

[31] *Ibid.*

[32] Bryan to Abert, March 29, 1858. The bridges were at the following creeks: Madison, Miry, Middleton, Loup, Parson's Uphill, Rocky Ford, Crooked, Goodale's Branch, and Bryan's Fork.

[33] Beckwith to Lieutenant Colonel J. H. Long, February 12, 1859. This report includes extensive specifications for each of ten bridges which are of interest primarily to the engineer.

[34] Beckwith to Long, September 27, 1858. This report written at "Camp on the Wagon Road from Fort Riley to Bridger's Pass of the Rocky Mountains on Parson's creek of the Republican Fork of the Kansas River," was published in "Report of the Chief Topographical Engineer, November 11, 1858," *House Executive Document* 2, 35 Cong., 2 sess. (1858–1859), pp. 1097–1098.

[35] Beckwith to Long, February 12, 1859. Beckwith also prepared a map showing the location of bridges constructed in the valley of the Republican fork which is available in The National Archives, Division of Cartographic Records.

[36] Bird B. Chapman to Abert, March 28, 1855.

[37] Izard to Robert McClelland, September 18, 1855. McClelland was Secretary of the Interior. The governor obviously did not know where the blame for the delay should be placed.

[38] Dickerson to Abert, July 20, 1856.

[39] For a diagram of this route, see map p. 122.

[40] Dickerson to Abert, December 15, 1856, *House Executive Document* 2, 35 Cong., 1 sess. (1857–1858), p. 530.

[41] The information for this account of Dickerson's work as a road surveyor has been obtained from his reports to the Bureau dated July 20, August 13, and December 15, 1856. Only the last of these has been published in the "Report of the Chief Topographical Engineer, November 23, 1857," *House Executive Document* 2, 35 Cong., 1 sess. (1857–1858), pp. 525–532. Two maps were forwarded to the Bureau with the following titles: "Map showing survey made for a Territorial Road from a point on the Missouri River opposite Council Bluffs, Iowa (Omaha, Nebraska) showing located road and line of reconnaissance" and "Map and Profile of a survey made for a Territorial Road from a point on the Missouri River (Omaha), opposite Council Bluffs to New Fort Kearney [*sic*], Nebraska Territory." Both are available in The National Archives, Division of Cartographic Records.

[42] *House Report 180*, 34 Cong., 3 sess. (1856–1857); *Cong. Globe*, XXVII (1857–1858), pp. 2056–2057, 2118.

[43] Beckwith to Abert, October 1, 1857, *House Executive Document* 2, 35 Cong., 1 sess. (1857–1858), p. 533.

[44] Twenty-five hundred troops under Colonel Albert S. Johnston, engaged in the so-called "Mormon War" to force Mormon recognition of the authority of the federal government, were

stationed at Fort Bridger during the winter of 1857–1858 and the following summer were in the Salt Lake Basin. A large percentage of the $15,000,000 spent on this military expedition went for the transportation of supplies.

[45] The information relative to Beckwith's work on the road is obtained from his reports to the Bureau on October 1, November 1, December 1, 1857, and September 26, 1858. The first and last of these have been published in the "Report of the Chief Topographical Engineer" for 1857 and 1858.

[46] Gouverneur K. Warren, "Explorations in the Dacota Country, in the Year 1855," *Senate Executive Document* 76, 34 Cong., 1 sess. (1855–1856). A summary of the Warren explorations is to be found in Doane Robinson, *History of South Dakota,* I, pp. 156–160. See also Emerson Gifford Taylor, *Gouverneur Kemble Warren: the Life and Letters of an American Soldier, 1830–1882,* pp. 18–44.

[47] Warren, "Preliminary Report . . . to Captain A. A. Humphreys, Topographical Engineers, in Charge of Office of Explorations and Surveys, War Department, November 24, 1858," *House Executive Document* 2, 35 Cong., 2 sess. (1858–1859), pp. 625–627. Hereafter cited Warren, *Preliminary Report*. The complete and final report was never printed.

[48] *Ibid.*, pp. 627–628.

[49] For a more detailed discussion, see chap. iv.

[50] Floyd to Warren, May 6, 1857, in Warren, *Preliminary Report,* pp. 623–624.

[51] Warren, *Preliminary Report,* pp. 628–634.

[52] For a more detailed discussion, see chap. xi.

[53] Warren, *Preliminary Report,* pp. 649–662; Humphreys to Floyd, November 20, 1858, *House Executive Document* 2, 35 Cong., 2 sess. (1858–1859), pp. 585–588.

[54] Warren, *Preliminary Report,* p. 661.

NOTES TO CHAPTER IX

[1] Bernhisel, formerly of Pennsylvania, had been a member of the Whig party and was fifty-three at the time of his election to Congress in 1851. Orson F. Whitney, *History of Utah,* I, p. 458.

[2] Robert R. Russel, *Improvement of Communication with the Pacific Coast as an Issue in American Politics, 1783–1864,* pp. 187–201. Professor Russel has written a comprehensive discussion, "Congressional Deadlock, 1853–1856."

[3] *U. S. Statutes at Large,* X, p. 304; *Cong. Globe,* XXIII (1853–1854), 1432, 1472, 1621, 1641, 1701.

[4] Andrew Love Neff, *History of Utah, 1847–1869,* ed. by Leland H. Creer, pp. 179–186; Hubert Howe Bancroft, *History of Utah, 1850–1886,* pp. 492–494; and C. V. Waite, *The Mormon Prophet,* pp. 26–29, contain three differing interpretations of the Steptoe period in Utah. The letter to President Pierce, signed by Steptoe and others, is printed in Edward W. Tullidge, *Life of Brigham Young, or Utah and Her Founders,* p. 239.

[5] Davis to Steptoe, September 19, 1854, *Senate Executive Document 1,* Part II, 34 Cong., 1 sess. (1855–1856), pp. 504–505.

[6] Steptoe to Davis, February 1, 1855, *Ibid.*, pp. 505–506.

[7] "Statement of Contracts Made During the Year 1855 for work under the Charge of the Bureau of Topographical Engineers," *House Executive Document 17,* 34 Cong., 1 sess. (1855–1856), pp. 45–47.

[8] Steptoe to Davis, March 28, 1855, *Senate Document 1,* Part II, 34 Cong., 1 sess. (1855–1856), pp. 506–507.

[9] Steptoe to Davis, February 1, 1855, *Ibid.*, pp. 505–506. For a diagram of this southern route to California, see map p. 142.

[10] Steptoe left the territory before Leach had completed his work and John F. Kinney, territorial chief justice, was apparently given the authorization to make the final payment. A

receipt was forwarded to Steptoe who, in turn, mailed it to Abert, topographical chief, from Fort Monroe, Virginia, November 27, 1855, The National Archives, War Department Records, Bureau of Topographical Engineers, Letters Received. All correspondence and manuscript reports used in this chapter are in the War Department Records unless otherwise designated.

[11] Captain E. G. Beckwith, assistant to Captain John W. Gunnison at the time of his death, succeeded to his command and continued exploration for a railroad route west of the Great Basin to the California boundary.

[12] Rufus Ingalls to Major General Thomas S. Jesup, August 25 and November 22, 1855, *Senate Executive Document 1*, Part II, 34 Cong., 1 sess. (1855–1856), pp. 152–168. Captain Ingalls, unlike Captains Stansbury and Gunnison, was very critical of the Mormons and their methods. He was convinced that the elders had interfered in the trial of Gunnison's murderers and that they encouraged the Indians in their hostility toward non-Mormons. His brief report is a classic account of the problems of the Quartermaster Corps in the West.

Photostatic copies of several unpublished reports relative to the Steptoe Expedition, deposited in The National Archives, are available in the Bancroft Library, University of California. More valuable documents include Steptoe's "Report of the March from Fort Leavenworth to Benicia, August 1, 1855," and "List of Camps and Distances from Great Salt Lake City, Utah Territory, to Fort Tejon, California via . . . by Lieut. J. G. Chandler, 3d Arty. on a March under Lieut. S. Mowry, in Spring, 1855."

[13] Bernhisel to Douglas, January 29 and February 2, 1855. Stephen A. Douglas Papers, The University of Chicago Library.

[14] *Cong. Globe,* XXV (1855–1856), 1495.

[15] This Congress apparently thought the occasion one for great levity. Letcher announced, "I am very much oblidged to the gentleman from Pennsylvania [Galusha A. Grow] for stopping this discussion, for, at the rate it was going on, the Virginia character for abstractions was in a fair way to be taken from her and run away with. (Laughter)." Robert T. Paine of North Carolina in closing discussion on the bill said, "I am surprised that there is any objection to it. The circumstances under which it is situated has silenced the tongue of the gentleman from Tennessee [George W. Jones, a glib debator and ardent opponent of federal roadbuilding]. I ascribe his silence to his high-toned chivalry; for are we not told—we do not know whether it is fabulous or not—that there are in Utah from seventeen to forty women to one man. (Laughter) Are these women to be turned out to make the road? (Renewed laughter)." *Ibid.,* p. 1496.

[16] *Ibid.,* XXVIII (1858–1859), 10.

[17] For further discussion, see pp. 147–149.

[18] John M. Bernhisel to John B. Floyd, January 20, 1859; Abert to Floyd, January 24, 1859; Floyd to Bernhisel, February 1, 1859. See also Andrews to Colonel S. Cooper, August 5, 1858; Bryan to Andrews, July 19, 1858; and Bryan to Andrews, July 22, 1858 in "Affairs in Utah," *House Executive Document 2,* 35 Cong., 2 sess. (1858–1859), pp. 104–105, 207–220.

[19] Fitz-John Porter, Assistant Adjutant General, to Simpson, August 24, 1858 in Simpson, "Wagon Routes in Utah Territory," *Senate Executive Document 40,* 35 Cong., 2 sess. (1858–1859), p. 34. Hereafter cited as Simpson, *Utah Wagon Roads.* The writer's account of the season's activity is based upon this official report.

[20] Simpson described the irrigation experiments of the Mormons north of Lake Utah and told of their success in agriculture. Lehi was reported to be a town of one hundred log and adobe houses surrounded by a wall, with one thousand population. American Fork and Pleasant Grove were only half that size and Simpson thought the inhabitants lived in squalor and penury. Provo, five miles south of the Timpanogos canyon, was pictured as a town laid out in "regular squares" with the central building a tabernacle. Simpson, *Utah Wagon Roads,* pp. 6–7.

[21] For a diagram of this route, see map p. 30.

[22] Simpson copied the rates from a sign posted on the toll gate, including a statement that permission was necessary to use the road. He observed, however, that no permission was asked

by his party and no objection was raised. Creer has quoted this notice from another source. See Leland H. Creer, *The Founding of an Empire: The Exploration and Colonization of Utah, 1776–1856*, pp. 162–163.

[23] The Parley's Park road was a segment of the highway used by the Mormon residents between Salt Lake City and Fort Bridger.

[24] Simpson, *Utah Wagon Roads*, pp. 8–13. Simpson to Abert, September 3, 1858.

[25] Simpson, *Utah Wagon Roads*, pp. 14–23. Accompanying this report was a preliminary statement on the geology between Fort Bridger and Camp Floyd, prepared by Henry Engelmann. Simpson also appended an itinerary of his wagon route, a table of temperatures and weather reports at each encampment, and a vocabulary of English words translated into Indian language. It was customary for the officers of the Topographical Corps to include materials on geology, astronomy, and meteorology in their reports and Simpson was overly conscientious.

The route can best be traced on a "Preliminary Map of Routes Reconnoitered and Opened in the Territory of Utah by Captain James H. Simpson, Corps of Topographical Engineers, in the Fall of 1858" that is attached to this report. The original is in The National Archives, Division of Cartographic Records.

[26] Porter to Simpson, October 15, 1858, *Utah Wagon Roads*, pp. 23–24.

[27] This pass was named in honor of Major John F. Reynold, who had encamped here in the spring of 1855 and had examined the pass to determine its practicability for wagons.

[28] Simpson, *Utah Wagon Roads*, pp. 24–25. For a diagram of this route see map p. 30.

[29] LeRoy R. Hafen, *The Overland Mail, 1849–1869*, pp. 63–67.

[30] Simpson, *Utah Wagon Roads*, pp. 34–41.

[31] James H. Simpson, *Report of Explorations Across the Great Basin of the Territory of Utah for a Direct Wagon-Route From Camp Floyd to Genoa, in Carson Valley in 1859*, pp. 41–44. Hereafter cited as Simpson, *Explorations Across the Great Basin*. This document is the primary source for the writer's account. Simpson to A. A. Humphreys, Chief of Engineers, United States Army, March 1, 1875. This correspondence relative to the official publication of the Simpson reports and journals includes a summary of all his explorations.

[32] Locally this route used by Chorpenning and Simpson was referred to as Egan's Route. Howard Egan, a major in the Nauvoo Legion, was a noted guide who had driven stock to California before becoming the superintendent of Chorpenning's road. Bancroft, *op. cit.*, p. 752 n. Howard R. Egan, *Pioneering the West 1846 to 1878*, pp. 194–195.

[33] Simpson established friendly relations with the traveling agents of Chorpenning. Besides the superintendent, he met Ball Robert, agent of the Salt Lake–Pleasant Valley district and Lott Huntington of the Pleasant Valley–Humboldt district. He recorded in his journal: "Then they have an agent called a station agent, from three to seven persons at each station, one being the mail carrier. The number of mules varies at these stations from 8 to 15. The mail during this winter was carried on a pack-mule which was sometimes led and sometimes driven. The required rate of travel (which was accomplished) was 60 miles for every twenty-four hours, changing every 20 to 30 miles. The superintending agent is said to get from $200 to $250 per month, the district agent $100, the station agent from $50 to $75, and the hands from $25 to $50, according to worth. One of the mail company informs me that along the route from this station to the Humboldt they had last winter to subsist themselves on mule and coyote meat. Their stock was transferred from the old road so late last fall as to have caused the death of one man, who died from cold on his last trip over the Goose Creek Mountains." Simpson, *Explorations Across the Great Basin*, p. 61.

[34] *Ibid.*, pp. 64–65. Creer, *The Founding of an Empire*, p. 152. For a diagram of this route, see map p. 142.

[35] Attached to Simpson's 1859 report is a "Map of Wagon Routes in Utah Territory explored and opened by Captain J. H. Simpson, Topographical Engineers," on which his route has been traced through this new and complicated terrain. The original is in The National Archives.

[36] The expedition's cook was troubled by dirty Indians, who hovered over his food in preparation with "their uncombed and *lively* hair," until he jokingly pointed to his revolver one

day. They scattered in alarm but Simpson felt it necessary to issue orders against such indiscretions for fear of losing the Indians' good will. Simpson, *Explorations Across the Great Basin,* pp. 67–68, 72.

[37] Simpson, *Explorations Across the Great Basin,* pp. 80–81, 83–84.

[38] *Ibid.,* pp. 87–92.

[39] Since 1855, when the town was started, two hotels, two general stores, one printing shop and an electric telegraph office had been established. In the vicinity were two grist and four saw mills.

[40] Frederic A. Bee and his brother Albert W. Bee headed the Placerville, Humboldt and Salt Lake Telegraph Company, and later indicated their ambition by changing the company name to the Placerville and St. Joseph Telegraph Company. Under the incentive of a California law pledging a subsidy to the first and second companies to make connections with an eastern line, the Bee organization had constructed a line from San Francisco to Genoa by the time of Simpson's arrival. This line was far from satisfactory. Robert Luther Thompson, *Wiring a Continent, The History of the Telegraph Industry in the United States, 1832–1866,* pp. 348–349.

[41] Simpson writes: "Driver a famous whip, but who, unfortunately had all night long been carousing with some others at the station, and was quite drunk when he started.... Had scarcely left, before, on account of the darkness of the night, the mules got out of the road, and came near breaking the stage by passing between two stumps.... The next obstacle was a bridge, from the farther half of which the puncheon flooring had been removed by some mischievous persons during the night, and piled up on the bank. I got off, and, with the assistance of one of the passengers, who was, like the driver, a little boosy, replaced the flooring, a space of about 2 feet being left on the farther side, on account of deficiency of material. Nothing daunted, however, the driver rushed over, and fortunately gained the other bank without accident.... At last just before reaching the summit, the stage upset and broke the tongue. Luckily, at my suggestion, all were out at the time." *Explorations Across the Great Basin,* p. 102.

[42] *Ibid.,* p. 113.

[43] *Ibid.,* pp. 114–132.

[44] *Ibid.,* pp. 121, 124–130.

[45] Simpson to Smith, August 5, 1859, *Ibid.,* p. 133.

[46] Porter to Simpson, August 5, 1859, *Ibid.*

[47] *Ibid.,* p. 139. For a diagram of this route, see map p. 30.

[48] Simpson to Porter, August 20, 1857 in Simpson, *Explorations Across the Great Basin,* pp. 141–143.

[49] *Ibid.,* pp. 144–147. The report and journal of Simpson is but a single part of an elaborate publication of the War Department relative to his activities in Utah. In the nineteen appendices to the report are itineraries of the three routes explored, a table of astronomical observations and geological positions, barometrical profiles of the routes, estimates for appropriations to improve the roads, and comments on the possible railroad routes to the West. Associates of Simpson also published their reports on geology, paleontology, botany, and icthyology. There were included memoirs on the population and resources of Utah as well as on the Indians of the territory with a comparative vocabulary of various Indian languages. Twenty-five maps, plates, and diagrams completed this publication, one of the most comprehensive released by the Army.

[50] For a short time after leaving Genoa, Lowry had rallied sufficiently to travel on horseback for a few hours daily, but by the time the group reached Camp Floyd he was confined to the carriage. Slowly declining in health, he was hospitalized in Laramie just before his death. Simpson, *Explorations Across the Great Basin,* p. 148.

[51] Arthur Chapman, *The Pony Express,* pp. 128–129.

[52] James Gamble, "Wiring a Continent," *The Californian, A Western Monthly Magazine,* III (June, 1881), 556–563.

[53] Johnston to Samuel Cooper, Adjutant General, August 26, 1859 in "Estimate of Appro-

priations Needed from Congress to Properly Improve the Routes in the Territory of Utah," in Simpson, *Explorations Across the Great Basin*, pp. 217–218.

[54] Simpson to Hooper, December 6, 1859, *Ibid.*, p. 218.

[55] Hooper to Simpson, March 2, 1860, *Ibid.*

[56] This estimate was $10,000 lower than that of the previous year when he had just returned to Camp Floyd.

[57] Williams to Simpson, January 18, 1860, in Simpson, *Explorations Across the Great Basin*, pp. 218–220.

NOTES TO CHAPTER X

[1] *Dictionary of American Biography*, XIX, pp. 628–629.

[2] *Cong. Globe*, XXV (1855–1856), 1252.

[3] *Ibid.*, p. 1297.

[4] *Ibid.*

[5] *Ibid.*, p. 1298.

[6] *Ibid.*, pp. 1298–1299.

[7] Chester Lee White, "Surmounting the Sierras: The Campaign for a Wagon Road," *Quarterly of the California Historical Society*, VII (March, 1928), 3–6. This is a detailed and careful study of the newspapers and state documents and presents in outline form the steps by which Californians located and built a road to the eastern boundary of the state.

[8] "Message of the Governor," *California Senate Journal, 1855*, pp. 45–47. The governor devoted a separate section of his message to a discussion of the overland route to California.

[9] White, "Surmounting the Sierras," *loc. cit.*, p. 5.

[10] *California Statutes*, VI, pp. 180–181. For a record of the legislative action, see *California Senate Journal, 1855*, pp. 663–669, and *California Assembly Journal, 1855*, pp. 14–24. The sectional interests relative to the location of the road are apparent. The seventh session of the California legislature, to raise money for this road, passed a law levying a road tax of $4 on every able bodied male between twenty-one and fifty and a property tax not to be more than five cents on a one hundred dollar evaluation. Incorporated communities were exempt from the property tax, and the road tax could be paid by two days' labor. *California Statutes*, VII, pp. 144–145.

[11] "Annual Report of the surveyor-general of the state of California," *Assembly Document 5*, 7 sess. (1856), pp. 5–9. Marlette had protested to the state legislature that $5,000 was insufficient for any satisfactory survey. After the bill's passage, he asked the attorney general if a part of the $100,000 appropriation for the road might be used for the reconnaissance, but the attorney general insisted that it was designated for construction and not survey.

[12] *Ibid.*, pp. 9–26.

[13] White, "Surmounting the Sierras," *loc. cit.*, p. 10.

[14] Constitution of the State of California, *California Statutes*, 1 sess. (1850), pp. 31–32. Section VIII stated in part that any appropriation "law shall provide ways and means, exclusive of loans, for the payment of the interest on said debt, . . . but no law shall take effect until, at the general election, it shall have been submitted to the people."

[15] The People *v.* Johnson, *California Supreme Court Reports*, VI, pp. 499–506.

[16] White, "Surmounting the Sierras," *loc. cit.*, pp. 11–15. The surveys were made by O. B. Powers, John A. Brewster, William Gamble, and Job Taylor, and Thomas Young.

[17] *California Statutes*, VIII, pp. 272–273.

[18] Quoted in the *Sacramento Daily Union*, October 1, 1855.

[19] *Ibid.*

[20] "Joint Resolution Relative to the Construction and Establishment of Military and Post Roads Across the Plains, April 19, 1855," *California Statutes*, VI, p. 308.

[21] "Joint Resolution Relative to Wagon Road Across the Plains," *Ibid.*, VII, p. 241.

[22] White, "Surmounting the Sierras," *loc. cit.*, p. 15.

[23] Arthur J. Larsen, "The Development of the Minnesota Road System," unpublished doctoral dissertation, University of Minnesota, 1938, pp. 89–92. Hereafter cited as the Larsen Dissertation.

[24] *Minnesota Democrat*, October 19, 1853. In this issue are reprinted many of Nobles' letters that had previously appeared in California and New York papers as a part of his educational campaign.

[25] *Ibid.*, January 11, 1854.

[26] *Ibid.*, February 12, 22, 1854; "Speech of the Hon. Wm. H. Nobles, together with other Documents Relative to An Emigrant Route to California and Oregon through Minnesota Territory, February 10, 1854." This is a summary of Nobles' remarks at the public meeting, telling of a survey of the Shasta Route from the Humboldt River through Nobles' Pass. The Minnesota Territorial Assembly ordered one hundred copies printed and fifty of these were used by Nobles in Washington for promotional work. Extensive quotations from the Shasta *Courier* were included.

[27] *Minnesota Territorial Laws*, 1854, pp. 45, 163–164. Nobles was one of the commissioners to locate the road to the Missouri. The expenses of construction were to be paid by the counties through which it passed. His arguments in support of the northern road to the Pacific were also incorporated in the Congressional memorial requesting $15,000.

[28] At least 60 per cent of the memorials to Congress from Minnesota dealt with road subsidies. The 1854 session passed six such memorials, the 1855 and 1856 sessions approved five each. The legislature of 1857, realizing that statehood was impending, adopted only two petitions for roads.

[29] *Minnesota Territorial Laws*, 1856, pp. 346–348.

[30] *Cong. Globe*, XXV (1855–1856), 1616.

[31] *Ibid.*, p. 1631. The Minnesota memorial also urged the Congress of the United States to appropriate money or grant land for William H. Nobles in recognition of his expenses, labors, and discoveries in opening routes across the Plains and Rocky Mountains into California.

[32] *Ibid.*

[33] *Ibid.*, p. 1632.

[34] *U. S. Statutes at Large*, XI, p. 27.

[35] *Cong. Globe*, XXV (1855–1856), 1485.

[36] *Ibid.*, p. 1964.

[37] *Ibid.*, pp. 2187–2188.

[38] Robert R. Russel, *Improvement of Communication with the Pacific Coast as an Issue in American Politics, 1783–1864*, pp. 200–201. Professor Russel presents a different interpretation of the political developments, suggesting that Weller's diligent work for the Minnesota-Nebraska road, to be constructed by the Secretary of the Interior, a northerner, was an attempt to placate those in the House opposed to southerner Davis. To the writer it is obvious that sectional and political bargaining were present, but chiefly over the guarantee of a southern as well as a central, overland route. The Californian sincerely preferred civilian engineers; he also referred to the Secretary of the Interior as a "western" and not a "northern" man although Robert McClelland's home was in Michigan. From his remarks in the Senate, these views appear clear-cut.

Another student of these legislative developments in the Thirty-fourth Congress has suggested that the southern bloc in the House rejected the central route and therefore the Senate took no action on the southern wagon road. E. Douglas Branch, "Frederick West Lander, Road-Builder," *Mississippi Valley Historical Review*, XVI (September, 1929), 176. The Senate had already passed the appropriation for the southern route. It is difficult, however, to determine whether the Republican majority or the southern bloc checked the legislation for no recorded vote was taken in the House. *Cong. Globe*, XXV (1855–1856), 2187–2188.

The most revealing information comes from a statement by William H. Nobles who claimed that the Republicans in the House, members of his own party, defeated the central route because of their hostility toward the War Department and Secretary Davis. He said, "It met with such opposition from the Republican members that it could not be reached." Nobles to Thompson,

April 16, 1857, The National Archives, Interior Department Records, Letters Received relative to the Fort Kearny, South Pass, and Honey Lake Road. All subsequent letters and reports addressed to the Department of the Interior which are cited in this chapter may be found in this record group, unless otherwise designated.

[39] Russel, *op. cit.*, p. 220.

[40] *Cong. Globe*, XXVI (1856–1857), 611.

[41] *Ibid.*

[42] *Ibid.*, p. 612.

[43] *Ibid.*

[44] *U. S. Statutes at Large*, XI, p. 162. The third section of the law providing for the Fort Defiance–Colorado River road did not state how or by whom it was to be built.

[45] *Ibid.*, p. 252.

[46] McClelland, "Report of the Secretary of the Interior, 1850," *Executive Document 1*, 31 Cong., 2 sess. (1850–1851), p. 30. The Secretary had urged federal surveys to find the best route for "a highway, commencing at some point in the valley of the Mississippi and terminating on the coast of the Pacific." He was not certain if this construction should be a railway, a turnpike, a plank road, or a combination of the three.

[47] *Dictionary of American Biography*, XI, pp. 586–587.

[48] *Ibid.*, XVIII, pp. 459–460.

[49] Petition for Wozencraft and Nobles signed by John B. Weller, Henry M. Rice, and fifty others.

[50] Nobles to Thompson, March 26, 1857. Theodore D. Judah was also an applicant who received serious consideration for the engineering assignment. Judah to Buchanan, March 6, 1857.

[51] Persifer F. Smith to Buchanan, February 14, 1857; James S. Green to Buchanan, March 12, 1857; Magraw to Thompson, March 26, 1857; I. W. Whitfield, A. J. Isaacs and George W. Clarke to Thompson, March 31, 1857; Missouri legislative petition to Buchanan, March, 1857. Included in the forwarded Post Office records were two petitions to James Campbell, Postmaster General, from the General Assembly of Virginia and the Democrats in the Pennsylvania Assembly, as well as a statement to Franklin Pierce from party members in the Maryland Senate and House. In a batch of twenty-five to thirty endorsements was the personal letter from Buchanan, March 17, 1853.

[52] Nobles to Thompson, March 26, 1857. From Nobles' letters, the writer has come to the conclusion that he was an egotist unjustifiably afraid of criticism. He wrote to Thompson of the great significance of his explorations and the discovery of the pass in the Sierra Nevada, of his great privations and of the failure of Congress to recognize his worth. Nobles resented the fact that he had no paid newspapers, no party to sound his praise "trumpet-tongued over the broad union." According to his story, he drafted the Fort Ridgely bill, and the idea of transferring the roads from the War to the Interior Department was exclusively his own as a means of getting a job to which he was entitled. Nobles to Thompson, April 16, 1857.

[53] Thompson to Magraw, May 1, 1857, The National Archives, Interior Department Records, Letters Sent relative to Pacific Wagon Roads. The outgoing correspondence of the Interior Department on wagon-road construction before 1871 is in separate bound volumes, arranged chronologically. All subsequent letters written by the Secretary or his administrators that are cited will be found in this record group.

[54] Thompson to Nobles, April 25, 1857.

[55] John A. Quitman to Thompson, March 5, 1857; John B. Weller to Thompson, March 31, 1857; S. E. Holmes to Thompson, April 16, 1857; and W. B. Weller to Thompson, April 19, 1857.

[56] Petition of California citizens in Washington to Thompson, March 6, 1857; P. T. Herbert to Thompson, March 11, 1857; S. W. Inge to Thompson, March 19, 1857; Charles S. Scott to Thompson, March 30, 1857; and John Bigler, William Bigler and Charles H. Hemstead to Thompson, March –, 1857.

[57] Thompson to Kirk, May 1, 1857.

[58] Leach to Thompson, May 2, 1857. The biographical data was in an enclosed clipping from the *New York Herald*. Leach was the author of the article.

[59] Chorpenning to Thompson, March 18, 1857; Phelps to Thompson, March 19, 1857.

[60] Thompson to Sites, May 19, 1857.

[61] Branch, "Frederick West Lander, Road-Builder," *loc. cit.*, pp. 172–176.

[62] Lander, *Report of the Reconnaissance of a Railroad Route, from Puget Sound via the South Pass to the Mississippi River.* This report may be seen in The National Archives, War Department Records. A photostat copy is available at the Library of Congress. A synopsis of the report is published in "Reports of Explorations and Surveys to Ascertain the Most Practicable and Economical Route for a Railroad from the Mississippi to the Pacific Ocean," *House Executive Document 91*, Vol. II, 33 Cong., 2 sess. (1854–1855), pp. 29–45.

[63] Thompson to Bishop, May 1, 1857.

[64] "Epitome of G. K. Warren's Memoir, 1800–1857, Giving a Brief Account of the English Explorations since A. D. 1800," in *Report Upon United States Geographical Surveys West of the One Hundredth Meridian*, I, pp. 580–581, 583, 594. Hereafter cited as *Warren's Memoir.*

[65] Thompson to Smith, May 15, 1857; to Sites, May 19, 1857.

[66] Thompson to Medary, May 4, 1857.

[67] Thompson to J. G. Cooper, April 22, 1857; to J. R. McCay, May 1, 1857; and to J. D. Goodrich, June 3, 1857.

[68] Thompson to M. A. McKinnon, May 10, 1857; to Jerome R. Gorin, May 23, 1857; and to Kirk, May 1, 1857.

[69] William G. Byers to Buchanan, April 24, 1857; W. H. Welch to Thompson, April 18, 1857; Asa Bell to Thompson, April 20, 1857; William H. Hope to Thompson, April 29, 1857.

[70] Floyd to Thompson, April 27, May 30, 1857; Howell Cobb to Thompson, May 26, 28, 30, 1857.

[71] Abert to Floyd, May 11, 1857.

[72] Thompson to Nobles, April 25 and June 2, 1857; to Magraw, June 4, 1857.

[73] Thompson to J. W. Denver, Commissioner of Indian Affairs, May 1, and May 9, 1857.

[74] Thompson to Kirk, May 1, 1857.

[75] Thompson to Floyd, April 27, 1857.

[76] *Warren's Memoirs*, pp. 580–581, 583. Davis to Thompson, March 13, 1857; Campbell to Rusk, March 13, 1857.

[77] Campbell to McCay, May 23, 1857.

NOTES TO CHAPTER XI

[1] Manypenny to Nobles, September 18, 1856, The National Archives, Interior Department Records, Indian Office Letter Books, LV, pp. 105–107; Manypenny, "Report of the Commissioner of Indian Affairs, 1856," *Senate Executive Document 5*, 34 Cong., 3 sess. (1856–1857), p. 570.

[2] Nobles to Manypenny, December 7, 1856, The National Archives, Interior Department Records, Letters Received Relative to the Fort Ridgely–South Pass Wagon Road. All letters or reports addressed to the Interior Department or to its officers and cited in this chapter may be found in this record group. Thompson to Nobles, April 25, 1857, The National Archives, Interior Department Records, Letters Sent Relative to Wagon Roads. All instructions and correspondence written by the Secretary or his administrators are in this record group.

[3] For a discussion of the political complications leading to this appointment, see chap. x.

[4] Thompson to Nobles, April 25, 1857.

[5] Nobles to Thompson, August 9, 1857.

[6] Albert H. Campbell, "Report upon the Pacific Wagon Roads," *House Executive Document 108*, 35 Cong., 2 sess. (1858–1859), pp. 3–4. This route may be traced on a map prepared by Medary and published with the House document. For a diagram, see map p. 180.

[7] Medary to Nobles, December, 1858; Nobles to Thompson, January 18, 1858; *Ibid.*, pp. 13–29. The report of engineer Medary furnishes the details; he also included an elaborate set of field notes. Nobles left the party on the Big Sioux, and the last part of the survey between that river and the Missouri was supervised by Medary.

[8] Nobles to Thompson, May 15, 1857.

[9] Philo P. Hubbell to Thompson, May 11, 1857; William S. Hubbell to Thompson, May 23, 1857.

[10] Thompson to Nobles, June 2, 1857. The complaint originated with the Treasury, was lodged with Campbell, and proclaimed by Thompson. Raisins, sardines, and codfish were the prohibited items.

[11] Thompson to Nobles, June 3, 1857. Thompson also wrote Gorin to try to straighten out Nobles' accounts so he could depart. The superintendent was paying the unheard of price of $200 a head for oxen.

[12] Nobles to Thompson, June 11, 1857.

[13] Doane Robinson, *History of South Dakota,* I, pp. 166–167; William Divier to Thompson, June 12, 1857.

[14] Rice to Thompson, June 11, 1857.

[15] June 17, 1857.

[16] June 15, 1857.

[17] Rice to Thompson, June 17, 1857. An account of this incident, printed in the *Pioneer and Democrat* of June 16, was inclosed. Gorman also forwarded statements from "the Black Republican organ and from the pet paper of your Republican appointee Nobles." Gorman to Thompson, June 17, 1857.

[18] Thompson to Nobles, June 25, 1857.

[19] *Ibid.*

[20] Gorin to Thompson, June 6 and 8, 1857.

[21] Nobles to Thompson, June 18, 1857.

[22] Gorin to Campbell, June 22, 1857; Campbell to Thompson, June 26, 1857.

[23] Nobles to Thompson, July 14, 1857.

[24] Kintzing Pritchette to Thompson, July 28, 1857.

[25] Nobles to Thompson, July 30, 1857.

[26] Thompson to Nobles, August 8, 1857.

[27] Pritchette to Nobles, August 16, 1857; Campbell to Gorin, August 27, September 1, October 17, 1857. Gorin issued a diatribe against Secretary Thompson for publication in the St. Paul newspapers before departure. Later he claimed Nobles gave him permission to leave Minnesota, though Nobles reported that he had dispatched a messenger to bring Gorin to the headquarters of the expedition and was surprised to learn of his departure.

[28] Gorin to Campbell, November 27, 1857.

[29] Thompson to Nobles, October 14, 1857; Nobles to Thompson, November 29, 1857.

[30] *The Deseret News,* November 1, 1857, Vol. VII, p. 288. This story reprinted from the *New York Daily Times* was dated Washington, August 6, 1857 and entitled, "How Things Are Done at Washington." Both the content and language of the article correspond to those printed in the Minnesota *Pioneer and Democrat.* Much of this newspaper propaganda must have originated with Nobles' friends in Congress who were associated with him in land promotions.

[31] *Ibid.* The newspaper told only part of the story. Gorin had cashed a draft for $15,000 when Thompson, in an attempt to put the brake on rapid expenditures, had placed $12,000 to his credit. As was the custom with the Treasury every draft was rejected unless there were sufficient funds to cover it. On learning of the situation, Thompson forwarded another $12,000 immediately. Much confusion was due to delay in communications. Gorin to Thompson, June 6, 1857; Campbell to Thompson, June 8, 1857; Gorin to Thompson, June 8 and 22, 1857; Campbell to Thompson, June 26, 1857.

[32] *The Deseret News,* November 1, 1857.

[33] July 25, 1857. Clipping in Letters Received Relative to the Fort Ridgely–South Pass Wagon Road.

[34] July 29, August 3, 1857. Clipping in Letters Received Relative to the Fort Ridgely–South Pass Wagon Road. Nobles wrote the editor of the *Star,* February 1, 1859, asking his source of information. The editor replied that, as usual, his paper went to the source of authority, the wagon-road headquarters in the Interior Department.

[35] Pritchette to Thompson, August 26, October 10 and 27, 1857.

[36] McAboy to Thompson, October 24, 1857; Thompson to McAboy, October 30, 1857; Thompson to Nobles, October 30, 1857; McAboy to Campbell, December 24, 1857. *Minnesota Democrat,* July 4, 1855. Medary was to be retained as engineer.

[37] McAboy to Campbell, January 12, 13, 28, 1858; Thompson to McAboy, November 30, 1857.

[38] August 6, 1857. Quoted in *The Deseret News,* November 11, 1857.

[39] Thompson to McAboy, February 3, 1858; Thompson to Nobles, February 4, 1858. McAboy was assured that the change did not result from want of confidence in his ability. He was paid $752 for "travelling expense."

[40] The Treasury insisted that his action in placing the money in charge of a St. Paul civilian who was unbonded was both ill-advised and illegal.

[41] Campbell to Gorin, December 4, 1857.

[42] Memorial of the Minnesota Legislature, June 10, 1858. Records of the United States Senate.

[43] Thompson to Charles H. Mix, June 17, 1858; to Nobles, June 17, 1858. Indian Office funds for these purchases were to be credited to the account of the Fort Ridgely road.

[44] Nobles to Thompson, August 17, 1858; Thompson to Nobles, August 25, 1858.

[45] Cullen to Thompson, August 20, 1858; Thompson to Solicitor of the Treasury, September 22, 1858. Cullen and Mix reported that Nobles refused to transfer the property unless the funds for it came to him directly. He started selling the federal supplies privately, and the two commissioners went to Minneapolis to secure the support of the United States district attorney in stopping him. Tracing these events in "The Development of the Minnesota Road System" Arthur J. Larsen is unusually sympathetic to Nobles and occasionally appears to plead his cause. See p. 102. This is largely due to Larsen's reliance on the biased territorial newspapers.

[46] Nobles to Thompson, February 8, 10, 26, 1859.

[47] Junius Hillyer to Thompson, September 23, October 19, 1858; E. M. Wilson to Hillyer, June 21, 1860.

[48] S. A. Pough to Campbell, July 19, August 29, 1860; Campbell to Pough, September 5, 1860.

[49] Memorial of William H. Nobles, February 15, 1861.

[50] Kelly to Hillyer, January 21, 1861; Hillyer to Kelly, January 22, 1861; Kelly to T. J. D. Fuller, January 26, 1861; E. M. Wilson to Kelly, February 6, 1861.

[51] M. S. Wilkinson to Kelly, February 12, 1861; Kelly to Wilkinson, February 15, 18, 20, 1861.

[52] *U. S. Statutes at Large,* XII, p. 204.

[53] W. C. Dodge to Nobles, February 26, 1861; Alexander Ramsey to Caleb Smith, April 18, 1861; W. A. Gorman to Smith, July 15, 1861; Pritchette to Smith, October 3, 1861. Nobles secured access to inspector Pritchette's reports in the Interior Department files and challenged the statements. Pritchette notified the Secretary that he had anticipated his reports would be treated as confidential official papers, otherwise he would never have transmitted them. However, the leading men in Minnesota had now convinced him the statements were untrue and he would be pained to think of doing an injustice to Nobles.

[54] *Senate Journal,* 37 Cong., 2 sess. (1861–1862), pp. 134, 186; *U. S. Statutes at Large,* XII, p. 911.

NOTES TO CHAPTER XII

[1] Magraw to Thompson, March 26, 1857; April 13, 1857, The National Archives, Interior Department Records, Pacific Wagon Road Office, Fort Kearny–South Pass–Honey Lake Road,

Letters Received. All incoming correspondence, reports, field journals, and diaries cited hereafter may be found in this collection unless otherwise designated.

[2] Lander to Thompson, April 7, 1857. The writer has examined the Lander Papers in The Library of Congress. With the exception of a few personal and family mementos, however, there are no papers in this collection not duplicated in the official records of the expedition in The National Archives. Early drafts or abbreviated copies of Lander's official reports were preserved by him in his personal letter book.

[3] Magraw's early reports make frequent references to Lander's understanding with the Secretary of the Interior by which his separate and advanced command had been authorized. Lander's position was made explicit in Magraw's instructions of May 1, 1857.

[4] Thompson to Magraw, May 1, 1857, The National Archives, Interior Department Records, Pacific Wagon Road Office, Letters Sent. All letters about wagon-road construction sent by the Secretary or his administrators to the field representatives will be found in this record group.

[5] Journal from Independence, Missouri, to Fort Thompson, on the eastern slope of Wind River, by M. M. Long, June 4–August 25, 1857 and by O. H. O'Neill, September 5–29, 1857. Long's journal records the events in Independence. Arriving on June 4, he described the village of two thousand as having "more stores than private houses." The following day, the men, mostly of German, French, and Irish stock, went into camp. Magraw entertained on June 8, with an evening devoted to "ladies, wine and song." Lander was host next evening to a dinner party. The ladies of Independence sponsored a dance lasting until the early morning hours of June 11.

[6] Lander to Magraw, June 8, 15, 17, 1857. See also E. Douglas Branch, "Frederick West Lander, Road-Builder," *Mississippi Valley Historical Review,* XVI (September, 1929), 172–187. Branch has quoted extensively from these letters which he obtained in Lander's letter book among the Papers of General Frederick West Lander in The Library of Congress. Many of the engineer's reports to the superintendent and the Secretary of the Interior were copied into this book. Remaining documents in the Lander Papers are of a personal nature.

[7] Lander to Magraw, June 15, 1857.

[8] Lander to Magraw, June 21, July 6, 1857.

[9] The proposed route had been suggested by William H. Nobles in a letter to Thompson on March 26, 1857. This letter contained detailed recommendations for the central overland route as a part of Nobles' application. For the location of these geographical features, see map p. 194.

[10] Lander to Mullowney, July 16, 1857.

[11] Lander to Ficklin, July 16, 1857.

[12] Lander to Ingle, August 26, 1857.

[13] Lander to Magraw, October 7, 1857; "Preliminary Report of F. W. Lander, chief engineer, upon his explorations west of the South Pass. . . . November 30, 1857," *House Executive Document 108,* 35 Cong., 2 sess. (1858–1859), pp. 30–35. These two major reports provide the summary of Lander's activities for 1857. The published report is also available in *Senate Executive Document 36,* 35 Cong., 2 sess. (1858–1859). The numerous and complex reconnaissances of the season can be located most satisfactorily on William H. Wagner's "Preliminary Map of the Central Division, Fort Kearny, South Pass, Honey Lake Road, 1857–1858," printed to accompany the congressional document. For a diagram of the two proposed routes of Lander, see map p. 194.

[14] Campbell to Magraw, July 4, 1857.

[15] Campbell to Annan, June 15, 1857.

[16] Magraw to Thompson, July 6, 1857.

[17] Thompson to Magraw, July 13, 1857.

[18] Magraw to Thompson, July 31, 1857. Magraw never prepared an extensive report of his activities for the Department. The members of the engineering corps provided most of the information through journals and field books. The Long-O'Neill journal has been cited. Magraw copied these two journals, with modifications, particularly in the early entries, and added a few additional notes for the period from September 30 to October 7. This copy "with considerable

corrections," to use his own terminology, was forwarded to the Department. M. M. Long also kept an "Itinerary of the Fort Kearny, South Pass, Honey Lake Wagon Road, Eastern Division from Independence City to Fort Laramie, July 2, to August 27, 1857." An "Odometer Book from Independence to Fort Thompson, July–September, 1857" started by Long, was completed by another member of the expedition. Richard L. Poor, a young engineer, kept a "compass book" between Fort Kearny and Fort Laramie, August 3–27, 1857. As a companion piece, a "Field Book for Latitudes and Departures" was also prepared for the same inclusive dates.

[19] Campbell to Annan, July 22, 1857; Thompson to Magraw, July 25, August 22, 1857. Annan received an additional $10,000, making the total expenditures $90,000, with a notice that no more funds would be sent without a report of "operations and expectations." The Department refused to pay for flannel shirts purchased by Magraw; he was not authorized to clothe his men.

[20] Thompson to Magraw, September 5, 1857.

[21] Magraw to Thompson, August 22, 1857.

[22] The bill of indictment against Magraw contained innumerable petty and some amusing complaints. His critics claimed that the superintendent was well known on the plains and apparently had an old quarrel to settle with every trader on the road. Moreover, he had breathed anathemas and bitter denunciations against the Mormons, but as he approached Brigham Young's domain his courage failed him. The disgruntled branded him as a coward. The crowning absurdity of his tyrannical egotism, they reported, was his urging each traveler headed east to take a personal letter introducing him to Magraw's crony, the President of the United States, and advising the emigrant that the best possible introduction to Buchanan was to appear with the smell of good liquor on the breath.

[23] Nichols to Lander, September 4, 1857; Nichols to Campbell, September 5, 1857; J. C. Cooper to Campbell, September 4, 1857; Richard Poor to Charles Poor, September 3, 1857. Nichols also recorded his version of events in a "Private Journal kept . . . during the Construction of Fort Kearny, South Pass and Honey Lake Overland Wagon Road, May 12–Sept. 18, 1857." The original is in the William Robertson Coe Collection, Yale University Library.

[24] Mullowney kept an "Odometer and Compass Book and Rough Field Notes, recording sketches of the route from Camp 29 to Wind River, July 17–October 17, 1857" and "A Compass Book from Fort Laramie to the Forks of Rocky Ridge Road, September 16–October 3, 1857." The first field book is the best detailed source of information about the nature of Lander's surveys. Illustrations of terrain, the course of travel, and locations of important geographical features are included in this professional account. The second book, more technical, records the direction and distance of travel under orders from Magraw.

[25] Lander to Nichols, September 21, 1857, quoted in Branch, "Frederick West Lander, Road-Builder," *loc. cit.*, pp. 180–181.

[26] Magraw to Lander, October 7, 1857; Lander to Thompson, November 18, 1857.

[27] Lander to Magraw, October 7, 1857.

[28] Lander to Thompson, December 1, 1857.

[29] Magraw to Ficklin, September 26, 1857; Ficklin to Magraw, October 7, 1857. Magraw's notes appended to the Long-O'Neill journal tell of his personal role.

[30] C. F. Smith to A. S. Johnston, October 15, 1857; Magraw to Thompson, October 20, 1857. Magraw listed the property transferred, the names of the volunteers, and the wages due them by the Interior Department. He apparently considered this report a final accounting.

[31] Journal kept in winter quarters at Fort Thompson, October 14, 1857, to March 9, 1857, by O. H. O'Neill. This journal records the day-by-day events and the attitude of the men toward their confinement in the mountains.

[32] November 21, 1857.

[33] Lander to Thompson, November 30 and December 1, 1857. Only the first two-thirds of the former report has been published. Lander submitted the letters written to him by the engineering corps, his replies, and the official bill of indictment against Magraw. He also forwarded all his correspondence with the superintendent relative to the investigation and his

orders to Mullowney, and summarized his own actions since arrival in Washington. He reminded Thompson that his choice for first assistant engineer had been John Lambert; that Nichols was the choice of Campbell. However, he was convinced that Nichols would not have left the expedition for trivial reasons. If he had conducted all relations with the superintendent in writing, as instructed, written proof of the controversy would be available.

Annan had started home with Lander but could not keep up so he decided to remain behind and travel with Mullowney. Annan to Thompson, October 13, 1857. Lander wrote Thompson on December 14, 1857, that Annan and Long were in Washington.

[34] Thompson to Annan, December 26, 1857; Annan to Thompson, December 28, 1857. The disbursing officer answered Magraw's accusations that he had failed to keep adequate records by submitting written evidence to prove he was prohibited from communicating with the men of the train and was forced to eat alone because he once questioned expenditures. He also forwarded written evidence of the agreement between Magraw and Goodale, the award of the officers at Fort Laramie, the acceptance by Magraw, and Goodale's receipt for $3,617.28 given Magraw. Several men had certified that to please Magraw, they signed vouchers though receiving no goods. Long testified that Magraw admitted to him that he drank too much and charged him with the responsibility of rationing the personal liquor supply.

Mullowney returned from the mountains with commendation from Magraw and the recommendation to the effect that as a good Democrat he was qualified for the vacated post of first assistant engineer. Magraw to Thompson, October 14, 1857. As a defense, the Department received dozens of depositions from the clerk, blacksmith, guide, interpreter, and laborers of the party denying Magraw's drunkenness and tyranny. Lander to Campbell, December 28, 1857. Annan and his cohorts retaliated with numerous letters from observers along the route sustaining their position.

[35] Thompson to Lander, January 13, 1858.

[36] Thompson to Magraw, January 20, 1858.

[37] Lander to Thompson, January 23, 1858. The engineer stated further, "I take the opportunity of expressing my full sense of the obligation under which I am placed by your offer and acknowledge with gratitude the compliments by which it was accompanied."

[38] Lander to Thompson, January 27, 1858.

[39] Magraw to Thompson, January 22, 1858; Magraw to Thompson, January 24, 1858; Magraw to Campbell, February 15, 1858; Magraw to Thompson, February 28, 1858.

[40] Magraw to Thompson, March 25, 1858.

[41] Magraw to Buchanan, April 17, 1858.

[42] Wozencraft to Thompson, April 13, 1857.

[43] Thompson to Kirk, May 1, 1857. For the location of these geographic features, see map p. 202.

[44] Kirk to Thompson, June 30, 1857. This report was written at Silver Creek Range, Sierra Nevada Mountains, El Dorado County.

[45] Kirk to Thompson, July 1, 1857.

[46] Campbell to Kirk, August 17, 1857.

[47] For a diagram of this route, see map p. 202.

[48] The story of the Wood party can serve as an example. This family migration was composed of Wood, his wife and three-year-old son, two brothers, a sister and her husband. At 9 A.M. on August 12, seventy-five Indians, half of whom were mounted, closed in on their wagons. Wood succeeded in cutting a mule from his team and mounting his wife and child. As they were about to start, the Indians fired and wounded the mule which threw Mrs. Wood to the ground. When assisting her to mount again, Wood was struck in the elbow and disabled. His wife became frantic with fear and ran around the wagon several times bearing the child in her arms. At last she was shot, her child seized, and its brains dashed out with the butt of a gun. Wood succeeded in running the gantlet of Indians, reached the front wagons and was dragged inside in a state of exhaustion. Both brothers were wounded and their wagon's sides and covers were pierced with bullet holes. The group was on the verge of annihilation when relief came.

[49] Itinerary of the route prepared by F. A. Bishop, engineer, January, 1858. This detailed document, compiled jointly by Kirk and Bishop, accompanied their annual reports but was never published. The reports may be found in *House Executive Document 108,* 35 Cong., 2 sess. (1858–1859), pp. 36–45. "A Map of the Western Division of the Fort Kearny, South Pass and Honey Lake Road, 1857" was printed in connection with the annual reports.

[50] *Ibid.*

[51] "Report of Superintendent Kirk upon the Western Division of the Fort Kearny, South Pass, and Honey Lake Wagon Road, January 4, 1858," *House Executive Document 108,* 35 Cong., 2 sess. (1858–1859), pp. 36–38. For a diagram of the proposed route, see map p. 202.

[52] Kirk to Thompson, October 18, 1857.

[53] Wozencraft to Thompson, September 17, 1857.

[54] Collection of undated newspaper clippings from California papers forwarded by Wozencraft with his letter of September 17, 1857.

[55] Kirk to Thompson, January 19, April 18, August 3, and October 2, 1858; Campbell to Kirk, October 3, 1858; Thompson to Kirk, November 19, 1858.

[56] Bishop to Kirk, January 4, 1858 in *House Executive Document 108,* 35 Cong., 2 sess. (1858–1859), pp. 38–45.

[57] Thompson to Kirk, June 17, 1858. Branch in his study of "Frederick West Lander, Road-Builder" describes Kirk's annual report as "pessimistic," and his season's work as "fruitless explorations," *loc. cit.,* p. 182. The official sources do not support this evaluation, nor is there evidence that Lander was in any way responsible for Kirk's "removal," as suggested by Branch. Kirk was not removed, but resigned with commendation from the Department when his work was completed.

[58] Lander to Thompson, December 1, 1857.

[59] Thompson to Lander, February 2, 1858.

[60] Lander to Thompson, February 3, 1858. Lander, who had been requested by Congress to prepare a statement about the most satisfactory railroad route across the Rockies informed the Secretary of the Interior that he had never thought of his wagon-road work as having any connection with railroad legislation. He confided to Thompson that, in his opinion, Congress should cease the costly mapping and engraving of elaborated rough field notes until some accurate surveys could be made.

[61] Thompson to Lander, March 25, 1858.

[62] Lander to Thompson, May 15, 1858. Lander was pleased with his progress and his party. He submitted a list of employees and their respective salaries. The officers were paid on a yearly basis; thirty-two laborers were to receive from $30 to $75 a month each.

[63] Lander to Thompson, May 31, 1858.

[64] Lander to Thompson, August 2, 1858. The engineer was greatly worried about the patriotism of his Mormon workers from foreign lands and informed Secretary Thompson that he had tried to explain the nature of the United States government to them. He suggested that hard work and with native Americans on public projects along the frontier was an excellent means of assimilation.

[65] Lander to Thompson, January 20, 1859, *House Executive Document 108,* 35 Cong., 2 sess. (1858–1859), pp. 47–49. The printed report is for the most part identical with Lander's manuscript. The sections dealing with the Mormons and Indian affairs have been transposed and a lengthy discussion of the Pacific railroad is omitted from the published account.

[66] Campbell to Thompson, February 19, 1859, *House Executive Document 108,* 35 Cong., 2 sess. (1858–1859), p. 7. The most accurate description and construction sketches of this road were prepared by William H. Wagner, assisted by M. M. Long, in a Field Book of the South Pass and Honey Lake Wagon Road, Central Division from Gilbert's Post to City of Rocks, June 10–September 15, 1858. Twice each day these engineers drew the route and made sketches of every section of travel. The infinite detail was used by Wagner in drafting the published map. The present-day traveler would have no difficulty in locating the road, once known as Lander's cutoff, with this field book in hand.

[67] Thompson to Lander, September 17, 1858. Before submitting this request to Secretary Thompson, general superintendent Campbell noted that Lander had received $60,000 during the season, disbursed $40,000, and that an estimated $93,200 remained in the account of the Honey Lake Road. No definite balance could be ascertained because of the confusion of Magraw's accounts.

[68] Lander to Thompson, September 29, 1858.

[69] Observations of the Weather at South Pass, November 2, 1858–June 22, 1859, Fort Kearny, South Pass and Honey Lake Road by Charles H. Miller. Miller's records were continued by Gilbert. The trader reported the altercation leading to Miller's death to the Interior Department and asked assistance from the federal government in exterminating "evil characters" in the mountains.

[70] Lander to Campbell, November 26, 1858; Campbell to Lander, November 27, 1858. Specially designed rooms were set aside in the south wing of the Interior Department headquarters for the preparation of these reports.

[71] Lander to Thompson, January 20, 1859, *House Executive Document 108,* 35 Cong., 2 sess. (1858–1859), pp. 61–73.

[72] *Ibid.,* p. 58.

[73] Lander to Thompson, September 29, 1858.

[74] Undated newspaper clipping in the records of the Fort Kearny, South Pass, Honey Lake Road, The National Archives. A similar clipping, in the Lander Papers, is quoted by Branch, "Frederick West Lander, Road-Builder," *loc. cit.,* p. 181 n.

[75] Thompson to Lander, June 8, 1859.

[76] Thompson to Floyd, June 9, 1859.

[77] The sutlership was explained by the assertion that concessions to Goodale were essential to procure his services as guide, that he purchased such extensive supplies that the superintendent was forced to loan him the use of personal mules to pull the wagons and therefore expected a percentage of the profits. However, Goodale was paid off, the goods belonging to him sold at cost, and Magraw claimed to have made no profit. Magraw to Thompson, July 22, 1859.

[78] Thompson to Magraw, October 12, 1857.

[79] Newspaper clipping in records of the Fort Kearny, South Pass, Honey Lake Road, The National Archives. An inked inscription identifies the clipping from the *Washington Star,* March, 1860.

[80] William R. Drinkard, Acting Secretary of War, to Thompson, July 12, August 24, 31, 1859.

[81] Second Auditor to Thompson, February 11, 1860.

[82] Thompson to Magraw, September 13, 1860.

[83] April 11, 1861; Edward Jordan to Smith, April 20, 1861. The results of this case are not available as the records for the Pacific Wagon Roads terminated abruptly on May 1, 1861, due to the preoccupation with the Civil War. It was probably dropped or compromised out of court.

[84] Lander to Thompson, January 20, 1859, *House Executive Document 108,* 35 Cong., 2 sess. (1858–1859), pp. 64–65.

[85] Lander to Thompson, March 16, 1859.

[86] Thompson to Charles Mix, March 25, 1859; Thompson to Lander, March 25, 1859.

[87] Journal or Notes of Travel taken in the Field by the Advance Party under W. H. Wagner, chief engineer, from Bellmont, Kansas Territory, to Oroville, California, April 29–October 8, 1859.

[88] Lander to Thompson, May 30, 1859.

[89] Wagner to Lander, January 29, 1860, *House Executive Document 64,* 36 Cong., 2 sess. (1860–1861), pp. 20–30.

[90] Lander to Thompson, July 21, 1859.

[91] Lander to Thompson, October 12, 1859.

[92] Lander to Thompson, March 1, 1860, *House Executive Document 64,* 36 Cong., 2 sess. (1860–1861), pp. 2–5.

[93] *Ibid.,* p. 7.

[94] *Ibid.,* pp. 9–10.

[95] *Ibid.,* pp. 10–18.

[96] *Ibid.,* pp. 18–20.

[97] Thompson to Lander, April 16, 1860.

[98] Lander to Thompson, May 18, 1860.

[99] Lander to Thompson, July 22, 24, 1860.

[100] Lander to Thompson, October 31, 1860, *House Executive Document 64,* 36 Cong., 2 sess. (1860–1861), pp. 31–35. Branch has mislocated Rabbit Hole Spring and is confused as to the season in which the work was performed.

[101] Lander to Thompson, September 27, 1860. Lander conscientiously collected and forwarded each account to Washington. The entire conversations between the road builder and the chief were recorded. Lander was convinced that the Paiutes who had suffered from the aggression of the white land-grabbers at Honey Lake in Long and Steamboat valleys could be reconciled by the allotment of sufficient supplies. The western Shoshones and Bannocks, he felt, needed chastisement. Winnemucca soon reported to Fort Churchill from his haunts at Pyramid Lake to discuss the cause of his people.

[102] Lander to Thompson, January 5, 1861, *House Executive Document 63,* 36 Cong., 2 sess. (1860–1861), pp. 2–4.

[103] Moses Kelly to Lander, February 4, 1861. Lander was commissioned with the rank of colonel when the Civil War broke out. He served as William H. Seward's secret agent in Virginia and was later sent by Abraham Lincoln to Texas to offer Sam Houston the support of the United States troops if the governor decided to try to keep his state from seceding or was willing to organize southern unionist support. David M. Potter, *Lincoln and his Party in the Secession Crisis,* p. 350. Promoted to a brigadier generalship and the command of the Eastern Division of the Army of the Potomac, Lander had a distinguished record before his death from a wound received in battle. Branch, "Frederick West Lander, Road-Builder," *loc. cit.,* p. 187.

NOTES TO CHAPTER XIII

[1] Pope to Rusk, March 9, 1857, The National Archives, Interior Department Records, Pacific Wagon Road Office, El Paso-Fort Yuma Road, Letters Received. All incoming correspondence and reports, field journals, and diaries cited hereafter are from this collection unless otherwise indicated.

[2] Campbell to Rusk, March 13, 1857. Pope had suggested to Rusk that Campbell was the best qualified man he knew to supervise the establishment of the southern route. He suggested that the senator meet with Campbell. The similarity in the proposals of the two men reveal a spirit of coöperation resulting from their conferences.

[3] Anderson to Rusk, March 10, 1857.

[4] Leach to Thompson, March 5, 1857. Leach was unable to write clearly or spell correctly so he was forced to employ an amanuensis, A. T. Hawley, who eventually accompanied him on the western trip.

[5] *Ibid.*

[6] Davis to Thompson, March 13, 1857; Campbell to Thompson, March 17, 1857.

[7] For a discussion of the background and political support of these men, see chap. x.

[8] Thompson to J. R. McCay, May 1, 1857; Thompson to M. A. McKinnon, May 10, 1857. The National Archives, Interior Department Records, Pacific Wagon Road Office, Letters Sent. All letters relative to wagon-road construction sent by the Secretary or his administrators to the field representatives may be found in this record group. W. Drayton Cress to Campbell, May 12, 1857.

[9] Petition to Thompson signed by Rusk, Douglas, John A. Quitman of Mississippi, James C. Jones of Tennessee, and others, March 7, 1857; James W. Denver of California to Thompson, April 17, 1857; Leach to Thompson, April 20, 1857.

[10] Thompson to Leach, May 9, 1857. For a diagram of this route, see map p. 248.

[11] *Ibid.*

[12] *Ibid.*

[13] Leach to Campbell, April 27, 1857; to Thompson, May 2, 6, 11, 1857.

[14] Undated memoranda prepared by Leach itemizing supplies. Ordnance Department report to Campbell, August 17, 1857.

[15] Leach to Campbell, June 27, 1857.

[16] The itinerary of Leach's mule-train movements toward El Paso comprises a manuscript of 134 pages; hereafter cited as Leach Itinerary. It was prepared for publication purposes but its content was a perfect testimony to maladministration; therefore, the document was not certified or forwarded to Congress.

[17] Leach to Thompson, July 26, 1857.

[18] Leach Itinerary. For a diagram of the route followed, see map p. 224.

[19] *Ibid.*

[20] Thompson, "Annual Report of the Secretary of the Interior," *House Executive Document 1*, 35 Cong., 2 sess. (1858–1859), p. 11.

[21] McKinnon to Thompson, November 18, 1857.

[22] Welcome B. Sayles to Thompson, June 20, 1858; Leach to Thompson, July 6, 1858.

[23] Leach to Thompson, October 22, 1857.

[24] Hutton to Thompson, January 8, 1858.

[25] Itinerary of the El Paso and Fort Yuma Wagon Road Expedition under the superintendence of James B. Leach, 1858; a forty-three page manuscript. Like the earlier "itinerary" this dealt, in part, with financial troubles and was filed but not published.

[26] Hutton prepared two maps of the El Paso and Fort Yuma road, one of the eastern sector, another of the western. These were published with his annual report in the congressional documents. Originals are in The National Archives, Cartographic Division.

[27] Hutton to Leach, January 29, 1857, *House Executive Document 108*, 35 Cong., 2 sess. (1858–1859), pp. 77–80. Leach forwarded periodic reports to the Interior Department from the field: October 22, December 9, December 20, 1857; January 8, February 8, March 10, April 10, April 30, May 15, June 4, June 20, 1858. Brief reports from assistants were usually inclosed. Hutton's printed annual report summarizes their contents. For a diagram of the route, see map p. 248.

[28] *Ibid.*, p. 85.

[29] Hutton described in detail the location and nature of each water site with the type of improvement made. A summary was written by Campbell to Thompson, February 19, 1859, *Ibid.*, pp. 9–11.

[30] Hutton to Leach, January 29, 1857, *Ibid.*, p. 85.

[31] *Ibid.*, p. 92. Leach to Thompson, January 29, 1857, *Ibid.*, p. 75.

[32] Thompson to Leach, April 22, May 9, 1857.

[33] McKinnon to Thompson, June 1 and 4, 1857.

[34] Thompson to Leach, May 11, 1857.

[35] Campbell to Leach, July 23, 1857.

[36] Campbell to Leach, August 17, 1857.

[37] McKinnon's Statement of Difference, June 11, 1860; Leach's Statement of Difference, May 22, 1860. Funds had been placed to the road builders' credit with the Assistant Treasurer of the United States in New York. They wrote drafts on the balances. Against this credit, their certified vouchers and drafts were compared. In the double entry bookkeeping, the vouchers and drafts had to coincide and both cancel out with the credits. Any difference was the responsibility of the federal employee.

[38] Leach to Campbell, June 27, 1857; Leach to Thompson, October 22, 1857; McKinnon to Thompson, November 18 and 25, 1857.

[39] Leach to Thompson, February 22, 1858.

[40] Sayles to Thompson, June 17, 1858.

[41] Thompson to Sayles, May 4, 1858.

[42] Sayles to Thompson, June 6 and 7, 1858.

[43] Sayles to Thompson, June 20, 1858.

[44] Sayles to Campbell, July 22, 1858; Sayles to Thompson, August 5, 1858; McKinnon to Thompson, July 6 and 22, 1858.

[45] Thompson to Sedgewick, October 11, 1858.

[46] Sayles to Thompson, November 20, 1858. Woods was charged with forgery and fraud under the authority of the act of March 3, 1823, *U. S. Statutes at Large,* III, pp. 771–772.

[47] Sayles to Thompson, December 21, 1858.

[48] Silas H. Handy to Thompson, May 3, 1858; Charles B. Smith to Thompson, May 9, 1858.

[49] Sayles to Thompson, December 20, 1858.

[50] Sayles to Campbell, January 3, 1859.

[51] Copies of the indictment and true bills against James B. Leach, May 16, 1859.

[52] Campbell to Charles H. Hunt, United States District Attorney, New York, May 14, 1859; Campbell to Sayles, May 23, 1859; Sayles to R. B. Hubbard, United States District Attorney, Austin, Texas, May 26, 1859; Thompson to Hubbard, September 6, 1859; Campbell to Robert Ould, March 7, 1861.

[53] Thompson to Hubbard, November 7, 1859; Campbell to James P. Warren, November 25, December 27, 1859; Thompson to James Henry Nash, December 14, 1859.

[54] Campbell to Ould, March 7, 1861. *Nolle Prosequi* is an entry made on the court record whereby the prosecutor or plaintiff declares he will proceed no further.

[55] Campbell to Leach, March 12, 1861.

[56] T. J. D. Fuller to Thompson, September 19, 1860.

[57] Leach to Thompson, October 25, 1860.

[58] Thompson Memorandum, April 18, 1861.

[59] Thompson to Charles P. Stone, April 4, 1860; Thompson to John B. Floyd, April 4, 1860.

[60] *Senate Executive Document 13,* 37 Cong., 2 sess. (1861–1862), p. 13; "Annual Report of the Secretary of the Interior," *Senate Executive Document 1,* 36 Cong., 2 sess. (1860–1861), p. 47.

NOTES TO CHAPTER XIV

[1] *U. S. Statutes at Large,* XI, p. 252.

[2] J. Sterling Morton and Albert Watkins, *History of Nebraska,* I, pp. 219–220.

[3] Chapman to Thompson, April 4, 1857, The National Archives, Interior Department Records, Pacific Wagon Road Office, Platte River and the Running Water River (Nebraska) Wagon Road, Letters Received. All incoming correspondence and reports hereafter cited may be found in this collection unless otherwise designated.

[4] W. A. Gorman to Thompson, March 22, 1857; F. McMullin to Thompson, March 24, 1857; J. A. Biggs to Buchanan, March 27, 1857; James B. Steadman to Buchanan, March 27, 1857.

[5] Fitch and Hendricks to Thompson, May 8, 1857; Chapman and Sites to Thompson, telegram, May 12, 1857; Stephen A. Douglas to Thompson, May 13, 1857.

[6] Chapman to Thompson, April 10, 1857.

[7] Floyd to Thompson, May 8, 1857.

[8] Thompson to Sites, May 8 and 15, 1857; Thompson to Smyth, May 15, 1857, The National Archives, Interior Department Records, Pacific Wagon Road Office, Letters Sent. All correspondence or instructions sent by Secretary Thompson or his administrators may be found in this collection unless otherwise designated.

[9] Sites to Thompson, June 26, 1857.

[10] Peter G. Washington, Assistant Secretary of the Treasury, to Sites, September 3, 1855; Campbell to Sites, July 11, 1857.

[11] Campbell to Sites, July 29, 1857.

[12] Sites to Thompson, July 10, 1857, *House Executive Document 108,* 35 Cong., 2 sess. (1858–1859), pp. 101–107. The details of the route can best be traced on "Map of the Wagon-Road from the Platte River via Omaha Reserve and Dakota City to Running Water River," published with the congressional documents. J. Sterling Morton, *Illustrated History of Nebraska,* pp. 113–116. In an extensive footnote on these pages, Morton has quoted from this published document.

[13] For a diagram of this route, see map p. 236.

[14] Sites to Thompson, August 10, 1857, *House Executive Document 108,* 35 Cong., 2 sess. (1858–1859), p. 112.

[15] *Ibid.,* p. 115.

[16] Sites to Thompson, October 22, November 18, 1857.

[17] Thompson to Sites, November 17, 1857.

[18] Sites to Thompson, January 14 and 22, 1858; Thompson to Sites, March 19, 1858.

[19] Sites to Thompson, January 20, 1859, *House Executive Document 108,* 35 Cong., 2 sess. (1858–1859), p. 120.

[20] Smyth to Campbell, August 12, 1858.

[21] Quoted in Sites to Thompson, January 20, 1859, *loc. cit.,* p. 123. The fifth session of the Territorial Legislative Assembly of Nebraska had printed this "Report of Select Committee Adopted by the Council, October 19, 1858" in circular form. Sites forwarded a copy to the Interior Department, December 13, 1858.

[22] Fort Randall was located on the right bank of the Missouri, approximately 34 miles above the mouth of the Niobrara.

[23] Sites to Thompson, January 20, 1859, *loc. cit.,* p. 124.

[24] *Ibid.,* p. 125.

[25] Campbell to Thompson, February 19, 1859, *House Executive Document 108,* 35 Cong., 2 sess. (1858–1859), p. 12.

[26] Morton, *Illustrated History of Nebraska,* I, p. 357.

NOTES TO CHAPTER XV

[1] *U. S. Statutes at Large,* XI, p. 162.

[2] John B. Weller, M. A. Otero, John S. Phelps, and others to Jacob Thompson, March 16, 1857; N. Henry Hutton to Thompson, March 25, 1857; A. W. Reynolds to Thompson, March 26, 1857; James M. Phillips and R. H. Leonard to Thompson, March 30, 1857, The National Archives, Interior Department Records, Pacific Wagon Road Office, Letters Received.

[3] Phelps to Buchanan, March 30, 1857, *Ibid.*

[4] "Railroad to the Pacific—Senator Benton's Bill," *The Western Journal,* II (January, 1849), 189–195.

[5] "Pacific Railway Convention at St. Louis," *The Western Journal,* III (November, 1849), 71–75.

[6] "The Memphis Convention," *The Commercial Review of the South and West,* ed. by J. D. B. DeBow, VIII (March, 1850), 217–232.

[7] *The Dictionary of American Biography,* XII, p. 428.

[8] *House Report* 95, 31 Cong., 2 sess. (1850–1851), p. 1.

[9] George Leslie Albright, *Official Explorations for Pacific Railroads, 1853–1855.* This study provides a convenient summary of the governmental documents on the railway surveys.

[10] Robert R. Russel, *Improvement of Communication with the Pacific Coast as an Issue in American Politics, 1783–1864,* p. 173.

[11] John Pope, "Report of Exploration of a Route for the Pacific Railroad near the Thirty-

Second Parallel of North Latitude, from Red River to the Rio Grande," *House Executive Document 91,* 33 Cong., 2 sess. (1854–1855), II, p. 39.

[12] John G. Parke, "Report of Explorations for Railroad Routes from San Francisco Bay to Los Angeles, California, West of the Coast Range and from the Pimas Villages on the Gila to the Rio Grande, near the Thirty-Second Parallel of North Latitude," *House Executive Document 91,* 33 Cong., 2 sess. (1854–1855), VII, p. 33.

[13] Andrew A. Humphreys and Jefferson Davis, "Conclusion of the Official Review of Reports Upon the Explorations and Surveys for Railroad Routes from the Mississippi River to the Pacific Ocean," *House Executive Document 91,* 33 Cong., 2 sess. (1854–1855), VII, p. 8.

[14] "Route Near the Thirty-Fifth Parallel under the Command of Lieutenant A. W. Whipple, Topographical Engineers, in 1853 and 1854," *House Executive Document 91,* 33 Cong., 2 sess. (1854–1855), III, p. 17.

[15] Humphreys and Davis, "Conclusion of the Official Review . . . ," *loc. cit.,* p. 36.

[16] Major Henry C. Wayne had proposed to the War Department in 1848 that camels be brought from the Near East to serve as beasts of burden on the military frontier. Jefferson Davis, then Senator from Mississippi, was interested in this scheme. Charles C. Carroll, *The Government's Importation of Camels: A Historical Sketch,* p. 392.

[17] "The Statement from Gen. E. F. Beale for Mr. Hubert Howe Bancroft." In this biographical sketch, available in the Bancroft Library at the University of California, Beale elaborated on his relations with Stockton and Kearny. His praise for Stockton was as boundless as his antipathy for Kearny. Many newspaper clippings relative to Beale's California activities are available there.

[18] Stephen Bonsal, *Edward Fitzgerald Beale: A Pioneer in the Path of Empire 1822–1903.* A more recent biographical account by W. J. Ghent is in *The Dictionary of American Biography,* II, p. 88. The story of the railroad surveying party was published by Gwinn Harris Heap, *Central Route to the Pacific.* The camel corps experiment has been ably presented by Lewis Burt Lesley, *Uncle Sam's Camels.* Beale's account of his Fort Defiance–Colorado River survey, published in the government documents, is reprinted in this volume. A recent book dealing with the subject is Harlan D. Fowler, *Camels to California.*

[19] Beale, "Wagon Road from Fort Defiance to the Colorado River," *House Executive Document 124,* 35 Cong., 1 sess. (1857–1858), pp. 15–32.

[20] *Ibid.,* p. 15.

[21] *Ibid.,* p. 25. Cited by Lesley, *op. cit.,* p. 64 n.

[22] *Ibid.,* p. 32.

[23] *Ibid.,* p. 36. Also cited by Lesley, *op. cit.,* p. 84 n.

[24] *Ibid.,* p. 37.

[25] For a diagram of this route, see map p. 248.

[26] Beale, "Wagon Road from Fort Defiance to the Colorado River," *loc. cit.,* p. 42.

[27] *Ibid.,* pp. 52–53.

[28] *Ibid.,* p. 45.

[29] *Ibid.,* p. 75.

[30] *Ibid.,* pp. 43–44.

[31] *Ibid.,* p. 61. Cited by Lesley, *op. cit.,* p. 104 n.

[32] *Ibid.,* p. 67. Cited by Lesley, *op. cit.,* p. 108 n.

[33] *Ibid.,* p. 51.

[34] *Ibid.,* pp. 52, 56.

[35] *Ibid.,* p. 75. Cited by Lesley, *op. cit.,* p. 113 n.

[36] *Ibid.,* pp. 75–77.

[37] *Ibid.,* p. 87. This closing quotation of Beale's journal has been used by Bonsal, *op. cit.,* p. 229, and by Lesley, *op. cit.,* pp. 123–124. In forwarding his report to the Secretary of War, John B. Floyd, Beale wrote: "The journal which I send you is a faithful history of each day's work, written at the camp fire at the close of every day. I have not altered or changed it in any respect whatever, as I desired to speak of the country as it impressed me on the spot, so as to be as

faithful in my description of it as possible. You will therefore find it very rough, but I hope those who may follow in my foot steps over the road may find it correct in every particular. I have written it for the use of emigrants more than for show, and if it answers the purpose of assisting them I shall be well satisfied." *Ibid.*, p. 2. Among the reports of the federal road builders published in the government documents, this journal of Beale's is unique. The author's keen sense of humor, picturesque language, and storytelling ability combine to produce many paragraphs that only the strong-willed can refrain from quoting.

[38] *Ibid.*, pp. 2–3.

[39] *Cong. Globe*, XXVII (1858), 1517, 2032, 2133, 2259, 2399, 2664.

[40] *U. S. Statutes at Large*, XI, p. 336.

[41] Beale, "Wagon Road—Fort Smith to Colorado River," *House Executive Document 42*, 36 Cong., 1 sess. (1859–1860), pp. 8–9.

[42] *Ibid.*, p. 12.

[43] *Ibid.*, pp. 76–82.

[44] *Ibid.*, pp. 9–10, 12–13, 18–20.

[45] *Ibid.*, p. 19.

[46] The curiosity of Grant Foreman, Oklahoma historian, concerning the two large bridges across the San Bois and Little rivers, destroyed during the Civil War, led to an investigation of their origin. That part of Beale's journal dealing with the transit of Oklahoma was edited by Foreman and published as "Survey of a Wagon Road from Fort Smith to the Colorado River," *Chronicles of Oklahoma*, XII (March, 1934), 74–96.

[47] Beale, "Wagon Road—Fort Smith to Colorado River," *loc. cit.*, pp. 12, 16. Beale named a prominent landmark along the road "Ab's Head."

[48] *Ibid.*, p. 28.

[49] *Ibid.*, p. 29.

[50] Hatch's Ranch was three miles from the village of Chaparito. Beale located it "eighty miles from Santa Fe, fourteen from Anton Chico, twenty-five from Los Vegas, forty from Fort Union, and one hundred and thirty from Albuquerque." *Ibid.*, p. 30.

[51] *Ibid.*, pp. 30–33. This exploration can best be traced on "Map showing the route of E. F. Beale from Fort Smith, Ark. to Albuquerque, N. M. 1858–9," accompanying the federal document.

[52] *Ibid.*, pp. 33–34.

[53] *Ibid.*, p. 40. For a diagram of this route, see map p. 248.

[54] Beale, "Wagon Road—Fort Smith to Colorado River," *loc. cit.*, p. 43.

[55] *Ibid.*, p. 45.

[56] *Ibid.*, pp. 49–50.

[57] *Ibid.*, pp. 52–54.

[58] Calculations compiled from the itinerary, *Ibid.*, pp. 76–91.

[59] "Estimate—Completion of Wagon Road from Fort Smith to the Colorado," *House Miscellaneous Document 98*, 36 Cong., 1 sess. (1859–1860), pp. 1–2.

[60] *The Dictionary of American Biography*, II, p. 88.

[61] Beale to Floyd, December 15, 1859, "Wagon Road—Fort Smith to Colorado River," *loc. cit.*, p. 2.

[62] *Ibid.*, p. 7.

[63] *Ibid.*, p. 22.

[64] *Ibid.*, pp. 1–8.

[65] *Ibid.*, p. 7.

NOTES TO CHAPTER XVI

[1] Isaac I. Stevens, "Report of Explorations for a Route for the Pacific Railroad near the Forty-Seventh and Forty-Ninth Parallels of North Latitude from St. Paul to Puget Sound," *House Executive Document 91*, 33 Cong., 2 sess. (1854–1855), pp. 34–35. Hereafter cited as Stevens, *Pacific Railroad Reports*, I.

[2] Addison Howard, "Captain John Mullan," *The Washington Historical Quarterly,* XXV (July, 1934), 185.

[3] Mullan to Stevens, October 2, 1853 and January 20, 1854, Stevens, *Pacific Railroad Reports,* I, pp. 59–61, 301–319.

[4] *Ibid.*, p. 79.

[5] Stevens to Mullan, October 3, 1853, *Ibid.*, pp. 61–62.

[6] John Mullan, "Report on the Construction of a Military Road from Fort Walla Walla to Fort Benton," *Senate Executive Document 43,* 37 Cong., 3 sess. (1862–1863), p. 2. Hereafter cited as Mullan, *1863 Report.* See also Lieutenant James H. Bradley, "Account of the Building of Mullen's [*sic*] Military Road," *Contributions to the Historical Society of Montana,* VIII (1917), 162–163.

[7] Mullan to Stevens, November 19, 1853, in Stevens, *Pacific Railroad Reports,* I, pp. 319–322.

[8] Mullan to Stevens, January 21, 1854, *Ibid.*, pp. 322–349.

[9] Mullan to Stevens, April 2, 1854, *Ibid.*, pp. 349–352; Mullan, *1863 Report,* p. 4.

[10] Mullan to Stevens, May 8, 1854, in Stevens, *Pacific Railroad Reports,* I, pp. 516–527, 633–635.

[11] Mullan to Stevens, November 12, 1854, *Ibid.*, pp. 527–537.

[12] Mullan, *1863 Report,* p. 3.

[13] *U. S. Statutes at Large,* X, p. 603.

[14] "Reports from the Department of the Pacific," *House Executive Document 1,* 34 Cong., 3 sess. (1856–1857), pp. 147–203. This is an excellent collection of letters and reports from officers of the Army about Indian difficulties in the Northwest.

[15] Howard, "Captain John Mullan," *loc. cit.*, p. 189.

[16] Mullan, *1863 Report,* pp. 7–8.

[17] Pal Clark (ed.), "Journal from Fort Dalles, O. T., to Fort Wallah Wallah, W. T., July, 1858. Lieut. John Mullan U. S. Army," *The Frontier: A Magazine of the Northwest,* XII (May, 1932), 368–375. This is a journal written by Mullan en route to report for duty with Colonel Clark.

[18] Mullan, *1863 Report,* p. 9.

[19] *U. S. Statutes at Large,* XI, p. 434.

[20] Humphreys to Mullan, March 15, 1859, in Mullan, "Military Road from Fort Benton to Fort Walla Walla," *House Executive Document 44,* 36 Cong., 2 sess. (1860–1861), pp. 74–75. Hereafter cited as Mullan, *First 1861 Report.*

[21] *Ibid.*, pp. 3–4; Mullan, *1863 Report,* p. 10.

[22] For the location of Mullan's route, see map p. 262.

[23] Mullan, *First 1861 Report,* pp. 5–15; Mullan, *1863 Report,* pp. 13–17.

[24] Mullan's assistants were all required to prepare written reports of their explorations and surveys. Dozens of these have been included as appendices to the *First 1861 Report* and the *1863 Report.*

[25] Mullan, *First 1861 Report,* pp. 15–28; Mullan, *1863 Report,* pp. 18–19.

[26] Mullan, *First 1861 Report,* pp. 28–36; Mullan, *1863 Report,* pp. 20–21.

[27] Mullan, *First 1861 Report,* pp. 41–54; Mullan, *1863 Report,* pp. 22–23.

[28] *U. S. Statutes at Large,* XII, p. 19.

[29] William F. Raynolds, "Report . . . on the Exploration of the Yellowstone and Missouri Rivers in 1859–1860," *Senate Executive Document 77,* 40 Cong., 2 sess. (1867–1868), p. 4. Hereafter cited as Raynolds, *Final Report.*

[30] *Ibid.;* Merrill G. Burlingame, "The Influence of the Military in the Building of Montana," *Pacific Northwest Quarterly,* XXIX (April, 1938), 138–139; Burlingame, *The Montana Frontier,* pp. 108–109.

[31] Raynolds, "Preliminary Report on the Yellowstone Expedition," *Senate Executive Document 1,* 36 Cong., 2 sess. (1860–1861), pp. 152–153. Hereafter cited as Raynolds, *Preliminary Report.*

[32] Raynolds, *Final Report,* pp. 127–154; Burlingame, *The Montana Frontier,* p. 109.

[33] Raynolds, *Final Report*, pp. 153–154; Burlingame, *The Montana Frontier*, p. 109.

[34] Raynolds, *Preliminary Report*, p. 155.

[35] Humphreys, "Report of the Office of Explorations and Surveys," *Senate Executive Document 1*, 36 Cong., 2 sess. (1860–1861), pp. 146–147.

[36] Mullan, *First 1861 Report*, pp. 54–70; Mullan, *1863 Report*, pp. 27–28; Bradley, "Account of the Building of Mullen's [*sic*] Military Road," *loc. cit.*, pp. 162–169. This is a good summary of the 1863 Mullan report, but shows no evidence of the writer's familiarity with the earlier dispatches sent from the field. Highlights of the 1863 report also appear in Howard, "Captain John Mullan," *loc. cit.*, pp. 185–202. No evidence is available that Blake shared Mullan's enthusiasm of the route for military transit, and local historians insist that the Quartermaster Corps' failure to use the road was due to the difficulties Blake encountered with his wagons.

[37] Mullan, *First 1861 Report*, pp. 70–74. Detailed estimates of personnel and equipment required were catalogued. Wages averaged $4,000 a month, and with a fifteen-month tour of duty, $60,000 would be so used. Rations were estimated at $10,500, and the remaining balance of $85,000 was to go for equipment.

[38] Burlingame, "John M. Bozeman, Montana Trailmaker," *Mississippi Valley Historical Review*, XXVII (March, 1941), 541–568.

[39] Mullan, *1863 Report*, pp. 28–29.

[40] Mullan, "United States Military Road Expedition from Fort Walla-Walla to Fort Benton, W. T.," *Senate Executive Document 1*, 37 Cong., 2 sess. (1861–1862), pp. 549–551. Hereafter cited as Mullan, *Second 1861 Report*.

[41] *Ibid.*, p. 554.

[42] *Ibid.*, p. 557.

[43] *Ibid.*, pp. 555–561.

[44] *Ibid.*, p. 563.

[45] *Ibid.*, pp. 564–569; Mullan, *1863 Report*, p. 31.

[46] Mullan, *Second 1861 Report*, p. 565.

[47] Mullan, *1863 Report*, p. 32.

[48] Mullan, *Second 1861 Report*, p. 569.

[49] Mullan, *1863 Report*, pp. 33–35.

[50] This route can easily be traced on map p. 262.

[51] Mullan, *1863 Report*, p. 37.

[52] *Ibid.*, p. 36.

[53] Samuel F. Bemis, "Captain John Mullan and the Engineers' Frontier," *The Washington Historical Quarterly*, XIV (July, 1923), 203; Henry L. Talkington, "Mullan Road," *The Washington Historical Quarterly*, VII (October, 1916), 305. Bemis mentions the anticipated military use; Talkington suggests its realization. Others emphasize its failure to materialize. Bradley, "Account of the Building of Mullen's [*sic*] Military Road," *loc. cit.*, p. 169.

[54] Howard, "Captain John Mullan," *loc. cit.*, p. 195.

[55] Oscar O. Winther, "Early Commercial Importance of the Mullan Road," *Oregon Historical Quarterly*, XLVI (March, 1945), 280.

[56] Mullan, *1863 Report*, p. 26. This evidence foreshadowed the important discoveries at Helena in 1864.

[57] Mullan, *Second 1861 Report*, p. 549.

[58] Mullan, *1863 Report*, p. 34.

[59] *Ibid.*, p. 35.

[60] *Ibid.*, p. 36.

[61] W. Turrentine Jackson, "The Fisk Expeditions to the Montana Gold Fields," *The Pacific Northwest Quarterly*, XXXII (July, 1942), 265.

[62] James L. Fisk, "Expedition from Fort Abercrombie to Fort Benton," *House Executive Document 80*, 37 Cong., 3 sess. (1862–1863), p. 1.

[63] W. M. Underhill, "North Overland Route to Montana," *Washington Historical Quarterly,* XXIII (July, 1932), 177–195.

[64] "The Expeditions of Jas. L. Fisk to the Gold Fields of Montana and Idaho in 1862 and 1863," *Collections of the State Historical Society of North Dakota,* II (1908), Part II, p. 34.

[65] Fisk, "Expedition of Captain Fisk to the Rocky Mountains," *House Executive Document 45,* 38 Cong., 1 sess. (1863–1864).

[66] "Official Correspondence pertaining to the War of the Outbreak, 1862–1865," *South Dakota Historical Collections,* VIII (1916), 100–588. This is a compilation of source materials relative to the fourth Fisk expedition reprinted from the *War of the Rebellion: Official Records of the Union and Confederate Armies.*

[67] "The Expeditions of Capt. Jas. L. Fisk, 1864–1866," *Collections of the State Historical Society of North Dakota,* II (1908), Part I, pp. 421–461.

[68] Underhill, "North Overland Route to Montana," *loc. cit.,* p. 177.

[69] February 6, 1863.

[70] New York, 1865.

[71] Howard, "Captain John Mullan," *loc. cit.,* p. 197.

[72] Bradley, "Account of the Building of Mullen's [*sic*] Military Road," *loc. cit.,* p. 169.

[73] For example, between January 1 and November 15, 1866, the heaviest year of travel on the Mullan road, an estimated 6,000 mules loaded with freight and 5,000 head of cattle were driven from Walla Walla to Montana, whereas only 52 light wagons, carrying family belongings traversed the road. In the same year only 31 emigrant wagons came from the States via the Mullan road. Memorial in *Statutes of the Territory of Washington, 1866–1867,* p. 237.

[74] Winther, "Early Commercial Importance of the Mullan Road," *loc. cit.,* pp. 30–34. This monograph, with editorial revisions, has been incorporated as a chapter, "The Mullan Road," in Winther's *The Old Oregon Country.*

[75] *Statutes of the Territory of Washington, 1866–1867,* pp. 233–236.

[76] *Ibid.,* pp. 236–239.

[77] T. C. Elliott, "The Mullan Road: Its Local History and Significance," *The Washington Historical Quarterly,* XIV (July, 1923), 206–209; "Mullan Road Markers," *The Washington Historical Quarterly,* XVII (January, 1926), 76–77.

NOTES TO CHAPTER XVII

[1] Merrill G. Burlingame, *The Montana Frontier,* pp. 101–126.

[2] General John Pope, "Report . . . of the Condition and Necessities of the Department of the Missouri," *House Executive Document 76,* 39 Cong., 1 sess. (1865–1866), p. 9.

[3] *Cong. Globe,* XXXV (1864–1865), 116, 694, 850, 1006, 1045, 1085–1086, 1118, 1413.

[4] *U. S. Statutes at Large,* XIII, pp. 516–517.

[5] John P. Usher to A. W. Hubbard, March 11, 1865, The National Archives, Interior Department Records, Pacific Wagon Road Office, 1865–1871, Letters Sent. All outgoing correspondence used in this chapter may be found in this record group unless otherwise indicated.

[6] Grace R. Hebard and E. A. Brininstool, *The Bozeman Trail,* I, p. 219; Burlingame, "John M. Bozeman, Montana Trailmaker," *Mississippi Valley Historical Review,* XXVII (March, 1941), 541–568.

[7] *U. S. Statutes at Large,* XII, pp. 204, 333, 642; XIII, p. 14.

[8] Alice V. Myers, "Wagon Roads West, The Sawyers Expeditions of 1865, 1866," *Annals of Iowa,* XXIII (January, 1942), 218.

[9] "Protection Across the Continent," *House Executive Document 23,* 39 Cong., 2 sess. (1866–1867), pp. 20–21.

[10] "Report of the Secretary of War, 1866," *House Executive Document 1,* 39 Cong., 2 sess. (1866–1867), p. 27.

[11] Hubbard to Usher, March 11, 1865; Hitchcock to Usher, April 18, 1865, The National

Archives, Interior Department Records, Pacific Wagon Road Office, 1865–1871, Letters Received. All incoming correspondence concerning the post-war road program may be found in this record group unless otherwise designated.

[12] Usher to Hubbard, March 11, 1865.

[13] William H. Ingham, "The Iowa Northern Border Brigade of 1862–3," *Annals of Iowa*, V (October, 1902), 491.

[14] Usher to Sawyers, March 14, 1865.

[15] James A. Sawyers, "Wagon Road from Niobrara to Virginia City," *House Executive Document 58*, 39 Cong., 1 sess. (1865–1866), pp. 10–11. This printed report of Sawyers is incomplete. Several pages of the original manuscript, available in The National Archives, containing unfavorable comment about the United States Army were deleted by a judicious editor. Lewis H. Smith, "Diary of the Sawyers Expedition, 1865," in Harvey Ingham, *Northern Border Brigade* is virtually a duplicate of Sawyers' official report. The account may have been prepared jointly, or either may have copied the other. A few items of fact vary and a paraphrase is occasionally noted. A third account is available in Albert H. Holman, "Niobrara-Virginia City Wagon Road," *Pioneering in the Northwest*. A short sketch is given in Arthur Jerome Dickson, *Covered Wagon Days*, pp. 234–239. Dickson admittedly obtained his information from Holman who was a teamster on the expedition.

[16] Myers, "Wagon Roads West," *loc. cit.*, pp. 221–222.

[17] Sawyers, "Wagon Road from Niobrara to Virginia City," *loc. cit.*, p. 11.

[18] *Sioux City Journal*, July 1, 1865, quoted in Myers, "Wagon Roads West," *loc. cit.*, p. 223.

[19] The route may be found on map p. 288.

[20] Sawyers, "Wagon Road from Niobrara to Virginia City," *loc. cit.*, pp. 11, 15, 17, 18.

[21] Sawyers' account of the conflict is summarized in five pages of his final report. Before publication, this part of the manuscript was deleted by an editor in the Interior Department. Williford to J. Q. Lewis, February 1, 1866; Pope to James Harlan, March 15, 1866.

[22] Sawyers, "Wagon Road from Niobrara to Virginia City," *loc. cit.*, pp. 22–28; Sawyers to Harlan, July 20, August 23, and October 14, 1865; Dickson, *op. cit.*, pp. 238–239.

[23] Frank B. Morgan to Harlan, November 29, 1865; Williford to J. Q. Lewis, February 1, 1866; Pope to Harlan, March 15, 1866.

[24] James Harlan to the President, May 19, 1865; Andrew Johnson to the Secretary of War, May 31, 1865, cited in James H. Simpson, *Report of Lieut. Col. James H. Simpson . . . to Honorable James Harlan, November 23, 1865.*

[25] Simpson to Sully, November 8, 1865; Simpson to Sawyers, November 23, 1865; Simpson to Hubbard, January 22, 1866.

[26] Sawyers to Simpson, January 23 and April 7, 1866.

[27] Simpson to Sawyers, May 28, 1866.

[28] *Cong. Globe*, XXXVI (1861), 3351.

[29] Grant to Thomas Eckert, Assistant Secretary of War, March 11, 1866; Eckert to Harlan, March 16, 1866.

[30] Sawyers to Simpson, March 29, 1866.

[31] Myers, "Wagon Roads West," *loc. cit.*, pp. 231–232.

[32] Sawyers to Simpson, March 29, 1866. Colonel Henry B. Carrington's preparations to occupy the Powder River road were simultaneous with those for the Sawyers' expedition. The colonel wrote the Superintendent of Indian Affairs to send all available maps to Fort Laramie in time for a council with the Indian tribes in May, 1866. Simpson forwarded a photograph of Raynolds' map, Sawyers' report, and tracings of the Niobrara-Virginia City wagon route. Carrington to the Indian Office, March 31, 1866; Simpson to Carrington, May 11, 1866.

[33] Sawyers to Simpson, June 6, 1866.

[34] Sawyers to Simpson, August 31, 1866.

[35] Sawyers, "Annual Report on Niobrara–Virginia City Wagon Road, November 26, 1866." This annual report was never printed by the government. Much of its content appeared periodically in the *Sioux City Journal* during the western trip. The chronicler of these Sawyer's expedi-

tions is not known, though his connection with the Sioux City press is obvious. Sawyers and at least two others used this same material in preparing official and unofficial reports. The author was probably a trained journalist who went along as secretary or clerk.

[36] *Sioux City Journal*, December 1, 1866, quoted in Myers, "Wagon Roads West," *loc. cit.*, pp. 234–235.

NOTES TO CHAPTER XVIII

[1] Doane Robinson, *History of South Dakota*, I, pp. 220–221.

[2] *Ibid.*, p. 923.

[3] *Ibid.*, pp. 605–606.

[4] Usher to Burleigh, March 11, 13, 1865; Usher to Moody, March 14, 1865, The National Archives, Interior Department Records, Pacific Wagon Road Office, 1865–1871, Letters Sent. All outgoing correspondence used in this chapter may be found in this collection unless otherwise specified.

[5] Usher to Brookings, March 13, 1865.

[6] Todd to Usher, March 17, 1865; Todd to James Harlan, June 3, 1865, The National Archives, Interior Department Records, Pacific Wagon Road Office, 1865–1871, Letters Received. All incoming correspondence used in this chapter may be found in this collection unless otherwise specified.

[7] Brookings to Usher, April 3, 1865.

[8] Brookings to Usher, April 20, 1865.

[9] Usher to Stanton, April 20, 1865.

[10] Usher to Brookings, April 20, 1865.

[11] Brookings to Simpson, November 27, 1865, "Report of the Secretary of the Interior, 1866," *House Executive Document 1*, 39 Cong., 2 sess. (1866–1867), p. 126.

[12] Sully to Brookings, July 3, 1865.

[13] Brookings to Simpson, November 27, 1865, "Report of the Secretary of the Interior, 1866," *loc. cit.*, pp. 126–127.

[14] For a diagram of this route, see map p. 300.

[15] Brookings to Harlan, August 18, 1865; Propper to Brookings, October 31, 1865, in "Wagon Road from Niobrara to Virginia City," *House Executive Document 58*, 39 Cong., 1 sess. (1865–1866), pp. 2–3.

[16] Brookings to Harlan, November 10, 1865; Brookings to Simpson, November 27, 1865, "Report of the Secretary of the Interior, 1866," *loc. cit.*, pp. 124–127.

[17] *Laws, Memorials and Resolutions of the Territory of Dakota, 1865–1866*, pp. 560–561, 566–569. Copies of these memorials were forwarded by Brookings to Simpson, December 25, 1865.

[18] Sully to Edmunds, December 8, 1865.

[19] Robinson, *op. cit.*, pp. 227–228.

[20] *U. S. Statutes at Large*, XIV, pp. 699–702.

[21] Brookings to Simpson, January 31, 1866.

[22] Simpson to Brookings, February 17, 1866, "Wagon Roads in Western Territories," *House Executive Document 105*, 39 Cong., 1 sess. (1865–1866), p. 3.

[23] Armstrong to the Secretary of the Interior, March 7, 1866.

[24] Simpson to Brookings, February 23, 1866, "Wagon Roads in Western Territories," *loc. cit.*, p. 3.

[25] Armstrong to the Secretary of the Interior, March 16, 1866.

[26] Brookings to Simpson, March 12, 1866.

[27] "Wagon Roads in Western Territories," *loc. cit.*, pp. 2–3.

[28] *Ibid.*, p. 2.

[29] Brookings to Browning, February 2, 1867.

[30] Morton to the Secretary of the Interior, March 28, 1867.

[31] Copy of orders inclosed in Burleigh to Browning, May 1, 1867.

[32] Todd to Simpson, February 2, 1867; Burleigh to Browning, March 5, and May 1, 1867.

[33] Brookings and others to Browning, May 27, 1867.

[34] Orville H. Browning, "Report of the Secretary of the Interior, 1867," *House Executive Document 1*, 40 Cong., 2 sess. (1867–1868), p. 18.

[35] *U. S. Statutes at Large*, XV, pp. 635–640.

[36] Moody to Usher, May 10 and 31, 1865; Moody to Harlan, July 1, 1865. Excerpts from the last two letters are published in "Report of the Secretary of the Interior, 1866," *loc. cit.*, p. 128.

[37] Simpson to Moody, August 25, 1865.

[38] Moody to Harlan, October 12, 1865, "Report of the Secretary of the Interior, 1866," *loc. cit.*, pp. 129–130.

[39] Moody to Simpson, November 10, 1865, *Ibid.*

[40] Robinson, *op. cit.*, pp. 225, 606.

[41] Hubbard to Harlan, August 10, 1865.

[42] Simpson to Gillis, November 8, 1865.

[43] Gillis to Simpson, December 4, 1865. The proposed route may be followed on map p. 300.

[44] *Laws, Memorials and Resolutions of the Territory of Dakota, 1865–1866*, pp. 507–510.

[45] Smith to Gillis, January 11, 1866.

[46] Gillis to Simpson, December 4, 1865.

[47] Simpson to Harlan, December 14, 1865; Miller to Simpson, December 14, 1865.

[48] Simpson to Moody, December 26, 1865.

[49] Simpson to Harlan, January 22, 1866.

[50] *U. S. Statutes at Large*, XIV, p. 21.

[51] Miller to Simpson, February 19, 1866.

[52] Miller to Simpson, March 1, 1866.

[53] Miller to Simpson, June 1, 1866.

[54] Gillis to Simpson, December 4, 1865.

[55] Moody to Simpson, November 23, 1865.

[56] LeBlanc to Sully, October 20, 1865; LeBlanc to the Secretary of the Interior, October 26, 1865.

[57] Gillis to Simpson, December 4, 1865.

[58] Brookings to Burleigh, May 4, 1866.

[59] Brookings to Harlan, May 30, 1866.

[60] Gillis to Harlan, June 9, 1866.

[61] Gillis to Miller, June 18, 1866; Harlan to Miller, August 10, 1866.

[62] Miller to Simpson, August 25, 1866.

[63] Miller to Simpson, November 30, 1866.

[64] Burleigh to Browning, March 5, 1867.

[65] Burleigh to Browning, July 20, 1867; *General Laws, Memorials and Resolutions of the Territory of Dakota, 1866–1867*, p. 134.

[66] James River Bridge Petition, October 9, 1867.

[67] Yankton County Commissioners to the Secretary of the Interior, October 12, 1867.

[68] Burleigh to Browning, November 11, 1867; John P. Allison and T. J. Kinkaid to Browning, November 11, 1867.

[69] The Dakota territorial assembly had adopted a memorial requesting the transfer of funds. *General Laws, Memorials and Resolutions of the Territory of Dakota, 1866–1867*, pp. 132–133.

[70] Burleigh to Browning, February 24, 1868.

[71] Orris S. Ferry to Browning, April 2, 1868; *Cong. Globe*, XXIX (1867–1868), 935, 1470–1471, 2623–2624, 3213, 3242, 3463, 3505; *U. S. Statutes at Large*, XV, pp. 89, 90.

[72] Burleigh to Browning, November 11, 1867; February 3, 1868; Browning to Burleigh, February 4, 1868; Browning to Lawrence, February 4, 1868; Lawrence to Browning, April 18, 1868.

[73] Browning to Burleigh, June 19, 20, 1868; Browning to Lawrence, August 5, 1868.

[74] Lawrence to Browning, September 30, November 16 and 30, 1868.
[75] Browning to Burleigh, February 20, 1869.
[76] Cox to Lawrence, April 15, May 15, 1869.
[77] Spink to Cox, July 26, 1869.
[78] Spink to Cox, February 23, 1870.
[79] Moses K. Armstrong to Columbus Delano, June 18, 1873.

NOTES TO CHAPTER XIX

[1] Todd to Usher, March 17, 1865; Harlan to Usher, March 22, 1865; Donnelly to Usher, March 23, 1865, The National Archives, Interior Department Records, Pacific Wagon Road Office 1865–1871, Letters Received and Sent. All correspondence used in this chapter is found in this collection.
[2] Usher to Connell, March 20, 1865; Connell to Usher, April 28, 1865.
[3] Bird to Harlan, May 29, 1865 and January 12, 1866.
[4] *Laws of the Territory of Idaho, 1865–1866,* pp. 292–293; Caleb Lyon of Lyonsdale to Harlan, January 18, 1866.
[5] Bird to Harlan, January 12, 1866; Bird to Simpson, February 13, 1866.
[6] Simpson to Bird, February 19, 1866; Nicholson to Simpson, February 22, 1866.
[7] Winchell to Harlan, February 12, 1866; Harlan to Winchell, February 16, 1866; Marcy to Evans, March 1, 1866.
[8] Bird to Simpson, February 28, 1866.
[9] Bird to Harlan, May 8 and 28, 1866; Bird to Simpson, February 9, 1867.
[10] Memorandum of February 9, 1867.
[11] *Ibid.;* Bird to Harlan, May 28, 1866.
[12] Bird to Simpson, June 18, 1866.
[13] Nicholson to Bird, January 31, 1867. For a diagram of this route, see map p. 314.
[14] Marcy to Bird, July 18, 1866, and January 31, 1867.
[15] Nicholson to Bird, July 26, 1866.
[16] Bird to Simpson, July 31 and September 15, 1866.
[17] Bird to Simpson, November 7, 1866.
[18] O. B. Porter to Browning, November 21, 1866; John R. Ferris to Browning, December 12, 1866.
[19] Bird to Browning, December 31, 1866 and February 19 and March 11, 1867.
[20] LeRoy G. Palmer to Browning, January 4, 1867; Bird to Browning, March 20, 1867.
[21] Asa Thompson to Browning, April 2, 1867.
[22] *Laws, Memorials and Resolutions, Legislative Assembly of the Territory of Idaho, 1866–1867,* pp. 205–206.
[23] Bird to Browning, April 1, 1867.
[24] Sells to Cox, April 29, 1870.
[25] Cox to Sells, June 16, 1870.
[26] Sells to Cox, August 15, 1870.

NOTES TO CHAPTER XX

[1] Samuel Flagg Bemis, "Captain John Mullan and the Engineers' Frontier," *The Washington Historical Quarterly,* XIV (July, 1923), p. 202.

Bibliography

BIBLIOGRAPHY

MANUSCRIPTS AND UNPUBLISHED SOURCES

Manuscript Collections

The National Archives:

Cartographic Records Division, Maps Prepared by the Corps of Topographical Engineers, 1849–1860.

Cartographic Records Division, Maps Prepared by Wagon Road Superintendents, Office of Pacific Wagon Roads, 1857–1871.

Interior Department Records, El Paso–Fort Yuma Wagon Road Letters Received, 1857–1861.

Interior Department Records, Fort Kearny, South Pass, and Honey Lake Wagon Road, Letters Received, 1857–1861.

Interior Department Records, Fort Ridgely–South Pass Wagon Road, Letters Received, 1857–1861.

Interior Department Records, Office of Indian Affairs, Letter Books, 1855–1856.

Interior Department Records, Records of the Secretary of the Interior Related to Wagon Roads, Letters Sent, 1857–1871, and Register of Letters Received, 1857–1867.

United States Senate and House of Representatives Records, Legislative Reference Division, Memorial and Petitions Received Relative to Road Construction, 1849–1870.

War Department Records, Letter Book of the Chief of Topographical Engineers to the Secretary of War, 1850–1854.

War Department Records, Bureau of Topographical Engineers, Letters Received, and Letters Sent, 1849–1860.

War Department Records, Bureau of Topographical Engineers, Register of Letters Received.

The Library of Congress:

Lander, General Frederick W., Papers.

The University of Chicago Library:

Douglas, Stephen A., Papers.

The Bancroft Library, University of California:

Documents Relative to the Edward Jenner Steptoe March to California by way of Salt Lake, 1854–1855.

The Statement from Gen. E. F. Beale for Mr. Hubert Howe Bancroft. Biographical sketch.

The William Robertson Coe Collection, Yale University Library:

Private Journal kept by Henry K. Nichols during the Construction of Fort Kearny, South Pass and Honey Lake Overland Wagon Road, May 12–Sept. 18, 1857.

The Minnesota Historical Society Library:

Ramsey, Alexander, Papers, 1849–1869.

Sibley, Henry H., Papers, 1815–1891.

Unpublished Dissertations and Theses

Amundson, C. J. "History of the Willamette Valley and Cascade Mountain Road Company." Master's thesis, University of Oregon, 1928.

Bender, Averam B. "Government Explorations, Other Than Railroad Surveys, in the Far Southwest, 1848–1860." Ph.D. dissertation, Washington University, St. Louis, 1932.

Bruce, H. S. "A History of the Oregon Central Military Wagon Road Company, with Reference to the Histories of Four Other Land Grant Companies in the State of Oregon." Master's thesis, University of Oregon, 1936.

Cross, Jack L. "Federal Wagon Road Construction in New Mexico Territory, 1849–1860." Master's thesis, The University of Chicago, 1949.

Larsen, Arthur J. "The Development of the Minnesota Road System." Ph.D. dissertation, University of Minnesota, 1938.

Oviatt, Alton B. "The Movement of a Northern Trail: the Mullan Road, 1859–1869." Ph.D. dissertation, University of California (Berkeley), 1948.

Warrick, W. Sheridan. "James Hervey Simpson, Military Wagon Road Engineer in the Trans-Mississippi West, 1849–1867." Master's thesis, The University of Chicago, 1949.

GOVERNMENT RECORDS

Federal (Chronological arrangement)

U.S. Statutes at Large, Vols. IV–XX, 18–45 Congress, 1824–1879.

Congressional Globe, Vols. XI–XLII, 24–41 Congress, 1835–1869.

Annual Reports of the Chief of Engineers in *Annual Reports of the Secretary of War:*

1833—*Senate Document 1* (serial 238), 74–143, or *House Executive Document 1* (serial 254), 51–118, 23 Congress, 1 Session, 1833–1834.

1834—*Senate Document 1* (serial 266), 100–187, and *House Executive Document 2* (serial 271), 99–179, 23 Congress, 2 Session, 1834–1835.

1835—*Senate Document 1* (serial 279), 100–200, or *House Executive Document 2* (serial 286), 102–200, 24 Congress, 1 Session, 1835–1836.

1836—*House Executive Document 2* (serial 301), 187–297, 24 Congress, 2 Session, 1836–1837.

1837—*Senate Document 1* (serial 314), 284–405, or *House Executive Document 3* (serial 321), 307–467, 25 Congress, 2 Session, 1837–1838.

1838—*Senate Document 1* (serial 338), 180–365, or *House Executive Document 2* (serial 344), 154–338, 25 Congress, 3 Session, 1838–1839.

Annual Reports of the Chief of Topographical Engineers in *Annual Reports of the Secretary of War:*

1839—*Senate Document 58* (serial 355), 10–271, or *House Executive Document 2* (serial 363), 632–893, 26 Congress, 1 Session, 1839–1840.

1840—*Senate Document 1* (serial 375), 171–189, or *House Executive Document 2* (serial 382), 171–189, 26 Congress, 2 Session, 1840–1841.

1845—*Senate Document 1* (serial 470), 290–399, or *House Executive Document 2* (serial 480), 290–399, 29 Congress, 1 Session, 1845–1846.

1847—*Senate Executive Document 1* (serial 503), 656–678, or *House Executive Document 8* (serial 515), 656–678, 30 Congress, 1 Session, 1847–1848.

1849—*Senate Executive Document 1* (serial 549), 294–354, or *House Executive Document 5* (serial 569), 294–354, 31 Congress, 1 Session, 1849–1850.

1850—*Senate Executive Document 1* (serial 587), 385–462, or *House Executive Document 1* (serial 595), 385–462, 31 Congress, 2 Session, 1850–1851.

1851—*Senate Executive Document 1* (serial 611), 386–443, or *House Executive Document 2* (serial 634), 386–443, 32 Congress, 1 Session, 1851–1852.

1852—*Senate Executive Document 1* (serial 659), 217–229, or *House Executive Document 1* (serial 674), 217–229, 32 Congress, 2 Session, 1852–1853.

1853—*Senate Executive Document 1* (serial 692), 3–262, or *House Executive Document 1* (serial 712), 3–262, 33 Congress, 1 Session, 1853–1854.

1854—*Senate Executive Document 1* (serial 747), 168–349, or *House Executive Document 1* (serial 778), 168–349, 33 Congress, 2 Session, 1854–1855.

1855—*Senate Executive Document 1* (serial 811), 272–538, or *House Executive Document 1* (serial 841), 272–538, 34 Congress, 1 and 2 Sessions, 1855–1856.

1856—*Senate Executive Document 5* (serial 876), 357–373, or *House Executive Document 1* (serial 894), 357–373, 34 Congress, 3 Session, 1856–1857.

1857—*Senate Executive Document 11* (serial 920), 283–534, or *House Executive Document 2* (serial 943), 283–534, 35 Congress, 1 Session, 1857–1858.

1858—*Senate Executive Document 1* (serial 976), 1021–1303, or *House Executive Document 2* (serial 999), 1021–1303, 35 Congress, 2 Session, 1858–1859.

1859—*Senate Executive Document 2* (serials 1024–1025), 684–1099, 36 Congress, 1 Session, 1859–1860.

1860—*Senate Executive Document 1* (serial 1079), 292–960, 36 Congress, 2 Session, 1860–1861.

1861—*Senate Executive Document 1* (serial 1118), 118–569, 37 Congress, 2 Session, 1861–1862.

Annual Reports of the Quartermaster General in *Annual Reports of the Secretary of War:*

1850—*Senate Executive Document 1* (serial 587), 120–332, or *House Executive Document 1* (serial 595), 120–332, 31 Congress, 2 Session, 1850–1851.

1851—*Senate Executive Document 1* (serial 611), 216–332, or *House Executive Document 2* (serial 634), 216–332, 32 Congress, 1 Session, 1851–1852.

1853—*Senate Executive Document 1* (serial 691), 129–136, or *House Executive Document 1* (serial 711), 129–136, 33 Congress, 1 Session, 1853–1854.

1855—*Senate Executive Document 1* (serial 811), 145–168, or *House Executive Document 1* (serial 841), 145–168, 34 Congress, 1 and 2 Sessions, 1855–1856.

1856—*Senate Executive Document 5* (serial 876), 251–257, or *House Executive Document 1* (serial 894), 251–257, 34 Congress, 3 Session, 1856–1857.

1865—*House Executive Document 1* (serial 1249), 82–890, 39 Congress, 1 Session, 1865–1866.

Annual Reports of the Secretary of the Interior:

1850—*Senate Executive Document 1* (serial 587), 19–34, or *House Executive Document 1* (serial 595), 19–34, 31 Congress, 2 Session, 1850–1851.

1857—*Senate Executive Document 11* (serial 919), 57–77, or *House Executive Document 2* (serial 942), 57–77, 35 Congress, 1 Session, 1857–1858.

1858—*Senate Executive Document 1* (serial 974), 73–98, or *House Executive Document 2* (serial 997), 73–98, 35 Congress, 2 Session, 1858–1859.

1859—*Senate Executive Document 2* (serial 1023), 91–113, 36 Congress, 1 Session, 1859–1860.

1860—*Senate Executive Document 1* (serial 1078), 29–48, 36 Congress, 2 Session, 1860–1861.

1865—*House Executive Document 1* (serial 1248), i–xxvii, 39 Congress, 1 Session, 1865–1866.

1866—*House Executive Document 1* (serial 1284), 1–24, 39 Congress, 2 Session, 1866–1867.

1867—*House Executive Document 1* (serial 1326), 1–27, 40 Congress, 2 Session, 1867–1868.

"Petition of Samuel Dickens *et al.*, Road, Memphis to Little Rock," *House Executive Document 151* (serial 219), 22 Congress, 1 Session, 1831–1832.

"Road—Chicot County to Little Rock," *House Report 374* (serial 226), 22 Congress, 1 Session, 1831–1832.

"Documents Relating to Bill Making Appropriations for Certain Roads in Arkansas," *Senate Document 300* (serial 282), 24 Congress, 1 Session, 1835–1836.

"Documents Relating to Arkansas Roads," *Senate Document 40* (serial 297), 24 Congress, 2 Session, 1836–1837.

Senate Report on Arkansas Road Bill, February 11, 1839, *Senate Document 211* (serial 340), 25 Congress, 3 Session, 1838–1839.

"Memorial of the Legislative Assembly of the Territory of Iowa, praying an appropriation for the completion of the Road from Dubuque to the Northern Boundary of the State of Missouri," *Senate Document 95* (serial 356), 26 Congress, 1 Session, 1839–1840.

Tilghman, Richard C. "Report on the survey, location, and construction of roads and canals in the Territory of Iowa," *Senate Document 598* (serial 361), 26 Congress, 1 Session, 1839–1840.

"Memorial of the Legislature of Iowa praying for an appropriation to construct a military road from Bloomington to Iowa City," *House Document 53* (serial 383), 26 Congress, 2 Session, 1840–1841.

Barney, Joshua. "Road in Iowa," *House Executive Document 28* (serial 464), 28 Congress, 2 Session, 1844–1845.

Abert, James W. "Journal of Lieutenant J. W. Abert from Bent's Fort to St. Louis, in 1845," *Senate Executive Document 438* (serial 477), 29 Congress, 1 Session, 1845–1846.

Barney, Joshua. "Public Works in Iowa," *House Executive Document 98* (serial 483), 29 Congress, 1 Session, 1845–1846.

Abert, James W. "Report of Lieutenant J. W. Abert of His Examination of New Mexico, in the Years 1846–47," *Senate Executive Document 23* (serial 506), or *House Executive Document 41* (serial 517), 419–548, 30 Congress, 1 Session, 1847–1848.

Cooke, Philip St. George. "Report of Lieut. Col. P. St. George Cooke of His March from Santa Fe, New Mexico, to San Diego, Upper California," *House Executive Document 41* (serial 517), 551–556, 30 Congress, 1 Session, 1847–1848.

Emory, William H. "Notes of a Military Reconnaissance, from Fort Leavenworth, in Missouri, to San Diego, in California, including part of the Arkansas, Del Norte, and Gila Rivers," *House Executive Document 41* (serial 517), 30 Congress, 1 Session, 1847–1848.

Johnston, A. R. "Journal of Captain A. R. Johnston, First Dragoons," *House Executive Document 41* (serial 517), 30 Congress, 1 Session, 1847–1848.

Cooke, Philip St. George. Official Journal of the March from Santa Fe to San Diego, *Senate Document 2* (serial 547), Special Session, March, 1849.

Simpson, James H. "Report of Exploration and Survey of Route from Fort Smith, Arkansas, to Santa Fe, New Mexico," *Senate Executive Document 12* (serial 554), or *House Executive Document 45* (serial 577), 31 Congress, 1 Session, 1849–1850.

Pope, John. "Report of an Exploration of the Territory of Minnesota," *Senate Executive Document 42* (serial 558), 31 Congress, 1 Session, 1849–1850.

"Reports of the Secretary of War, with Reconnaissances of Routes from San Antonio to El Paso . . . and the Report of Lieutenant W. H. C. Whiting's Reconnaissances of the Western Frontier of Texas," *Senate Executive Document 64* (serial 562), 31 Congress, 1 Session, 1849–1850.

Simpson, James H. "Journal of a military reconnaissance from Santa Fe, New Mexico, to the Navajo country," *Senate Executive Document 64* (serial 562), 55–168, 31 Congress, 1 Session, 1849–1850.

Whiting, William H. C. "Report to General J. G. Totten, June 10, 1849," *House Executive Document 5* (serial 569), 31 Congress, 1 Session, 1849–1850.

"Report of the Department of War, Including Military Correspondence Relative to California and New Mexico, 1846–1850," *House Executive Document 17* (serial 572), 31 Congress, 1 Session, 1849–1850.

Marcy, Randolph B. "Report of Captain R. B. Marcy," *House Executive Document 45* (serial 577), 31 Congress, 1 Session, 1849–1850.

Michler, Nathaniel. "Routes from the Western Boundary of Arkansas to Santa Fe and the Valley of the Rio Grande," *House Executive Document 67* (serial 577), 31 Congress, 1 Session, 1849–1850.

Thurston, Samuel R. "Roads in Oregon," *House Report 348* (serial 584), 31 Congress, 1 Session, 1849–1850.

"Memorial of the Legislature of Oregon, July 20, 1849," *Senate Miscellaneous Document 5* (serial 592), 31 Congress, 2 Session, 1850–1851.

Stansbury, Howard. "Exploration and Survey of the Valley of the Great Salt Lake of Utah, Including a Reconnoissance [*sic*] of a New Route through the Rocky Mountains," *Senate Executive Document 3* (serial 608), 32 Congress, Special Session, March, 1851.

Sitgreaves, Lorenzo. "Report of an Expedition down the Zuni and Colorado Rivers," *Senate Executive Document 59* (serial 668), 32 Congress, 2 Session, 1852–1853.

Davis, Jefferson. "Instructions Respecting Military Roads in Oregon," *Senate Executive Document 1,* Part II (serial 691), 33 Congress, 1 Session, 1853–1854.

Reno, J[esse] L. "Survey, etc. [*sic*], of Road from Mendota to Big Sioux River," *House Executive Document 97* (serial 725), 33 Congress, 1 Session, 1853–1854.

Stevens, Isaac I. "Report of Explorations for a Route for the Pacific Railroad near the Forty-Seventh and Forty-Ninth Parallels of North Latitude from St. Paul to Puget Sound," *Senate Executive Document 78,* I (serial 758), or *House Executive Document 91,* I (serial 791), 33 Congress, 2 Session, 1854–1855.

Lander, Frederick W. "Synopsis of a Report of the Reconnaissance of a Railroad Route from Puget Sound via South Pass to the Mississippi River," *Senate Executive Document 78,* II (serial 759), or *House Executive Document 91,* II (serial 792), 33 Congress, 2 Session, 1854–1855.

Parke, John G. "Report of Explorations for that Portion of a Railroad Route, near the Thirty-Second Parallel of North Latitude, Lying between Dona Ana, on the Rio Grande, and Pimas Villages, on the Gila," *Senate Executive Document 78,* II (serial 759), or *House Executive Document 91,* II (serial 792), 33 Congress, 2 Session, 1854–1855.

Pope, John. "Report of Exploration of a Route for the Pacific Railroad near the Thirty-Second Parallel of North Latitude, from the Red River to the Rio Grande," *Senate Executive Document 78,* II (serial 759), or *House Executive Document 91,* II (serial 792), 33 Congress, 2 Session, 1854–1855.

Whipple, Amiel W. "Report of Exploration for a Railway Route near the Thirty-Fifth Parallel of North Latitude from the Mississippi River to the Pacific Ocean," *Senate Executive Document 78,* III (serial 760), or *House Executive Document 91,* III (serial 793), 33 Congress, 2 Session, 1854–1855.

Parke, John G. "Report of Explorations for Railroad Routes from San Francisco Bay to Los Angeles, California, West of the Coast Range, and from the Pimas Villages on the Gila to the Rio Grande, near the 32d Parallel of North Latitude," *Senate Executive Document 78,* VII (serial 764), or *House Executive Document 91,* VII (serial 797), 33 Congress, 2 Session, 1854–1855.

Humphreys, Andrew A., and Jefferson Davis. "Conclusion of the Official Review of the Reports upon the Explorations and Surveys for Railroad Routes from the Mississippi River to the Pacific Ocean," *Senate Executive Document 78,* VII (serial 764), or *House Executive Document 91,* VII (serial 797), 33 Congress, 2 Session, 1854–1855.

Warren, Gouverneur K. "Memoir to Accompany the Map of the Territory of the United States from the Mississippi River to the Pacific Ocean," in "Reports of the Explorations and Surveys, to Ascertain the Most Practicable and Economical Route for a Railroad from the Mississippi River to the Pacific Ocean, 1853–6," *Senate Executive Document 78,* XI (serial 768), or *House Executive Document 91,* XI (serial 801), 33 Congress, 2 Session, 1854–1855.

"Military Roads in Kansas," *House Report 36* (serial 808), 33 Congress, 2 Session, 1854–1855.

Warren, Gouverneur K. "Explorations in the Dacota Country in the Year 1835," *Senate Executive Document 76* (serial 822), 34 Congress, 1 Session, 1855–1856.

"Contracts with the War Department, 1855," *House Executive Document 17* (serial 851), 34 Congress, 1 Session, 1855–1856.

Manypenny, George W. "Report of the Commissioner of Indian Affairs, 1856," *Senate Executive Document 5* (serial 875), 554–575, 34 Congress, 3 Session, 1856–1857.

"Reports from the Department of the Pacific," in *Report of the Secretary of War, 1856, Senate Executive Document 5* (serial 876), 147–203, or *House Executive Document 1* (serial 894), 147–203, 34 Congress, 3 Session, 1856–1857.

Davis, Jefferson. "Military Road in Nebraska," *House Report 180* (serial 914), 34 Congress, 3 Session, 1856–1857.

"Expenditures in the Territories," *House Executive Document 79* (serial 956), 35 Congress, 1 Session, 1857–1858.

Beale, Edward F. "Wagon Road from Fort Defiance to the Colorado River," *House Executive Document 124* (serial 959), 35 Congress, 1 Session, 1857–1858.

"Affairs in Utah," in *Report of the Secretary of War, 1858, Senate Executive Document 1* (serial 975), 28–223, or *House Executive Document 2* (serial 998), 28–223, 35 Congress, 2 Session, 1858–1859.

Simpson, James H. "Wagon Routes in Utah Territory," *Senate Executive Document 40* (serial 984), 35 Congress, 2 Session, 1858–1859.

Warren, Gouverneur K. "Preliminary Report . . . to Captain A. A. Humphreys . . . in charge of Office of Explorations and Surveys, War Department, November 24, 1858," *House Executive Document 2* (serial 998), 620–670, 35 Congress, 2 Session, 1858–1859.

Campbell, Albert H. "Report upon the Pacific Wagon Roads," *House Executive Document 108* (serial 1008), 35 Congress, 2 Session, 1858–1859.

Hartz, Edward L. Diary of the Camel Expedition through the Country between the Pecos and Rio Grande Rivers, 1859, *Senate Executive Document 2* (serial 1024), 424–441, 36 Congress, 1 Session, 1859–1860.

Wallen, Henry D., *et al.* Reports of the Expedition Made in 1859 to Open a Wagon Road from the Dalles of the Columbia River to the Great Salt Lake, *Senate Executive Document 34* (serial 1031), 36 Congress, 1 Session, 1859–1860.

Beale, Edward F. "Wagon Road—Fort Smith to Colorado River," *House Executive Document 42* (serial 1048), 36 Congress, 1 Session, 1859–1860.

"Affairs in Oregon," *House Executive Document 65* (serial 1051), 36 Congress, 1 Session, 1859–1860.

Beale, Edward F. "Estimate—Completion of Wagon Road from Fort Smith to the Colorado," *House Miscellaneous Document 98* (serial 1066), 36 Congress, 1 Session, 1859–1860.

Echols, William H. Report of a Reconnaissance West of Camp Hudson, Texas, 1860, *Senate Executive Document 1* (serial 1079), 36–50, 36 Congress, 2 Session, 1860–1861.

Humphreys, Andrew A. "Report of the Office of Explorations and Surveys, November 12, 1860," in *Report of the Secretary of War, 1860, Senate Executive Document 1* (serial 1079), 146–148, 36 Congress, 2 Session, 1860–1861.

Raynolds, William F. "Preliminary Report on the Yellowstone Expedition," in *Report of the Secretary of War, 1860, Senate Executive Document 1* (serial 1079), 152–155, 36 Congress, 2 Session, 1860–1861.

Mullan, John. "Military Road from Fort Benton to Fort Walla-Walla," *House Executive Document 44* (serial 1099), 36 Congress, 2 Session, 1860–1861.

Lander, Frederick W. "Additional Estimate for Fort Kearney [*sic*], South Pass, and Honey Lake Wagon Road," *House Executive Document 63* (serial 1100), 36 Congress, 2 Session, 1860–1861.

"Maps and Reports of the Fort Kearney [*sic*], South Pass, and Honey Lake Wagon Road," *House Executive Document 64* (serial 1100), 36 Congress, 2 Session, 1860–1861.

Mullan, John. "Report on the Construction of a Military Road from Fort Walla Walla to Fort Benton," *Senate Executive Document 43* (serial 1149), 37 Congress, 3 Session, 1862–1863.

Fisk, James L. "Expedition from Fort Abercrombie to Fort Benton," *House Executive Document 80* (serial 1164), 37 Congress, 3 Session, 1862–1863.

Fisk, James L. "Expedition of Captain Fisk to the Rocky Mountains," *House Executive Document 45* (serial 1189), 38 Congress, 1 Session, 1863–1864.

Simpson, James H. *Report of Lieut. Col. James H. Simpson on the Union Pacific Railroad and branches, Central Pacific Railroad of California, Northern Pacific Railroad, Wagon Roads in the Territories of Idaho, Montana, Dakota, and Nebraska, and the Washington Aqueduct to Honorable James Harlan, Secretary of the Interior, November 23, 1865*. Washington: Government Printing Office, 1865.

Sawyers, James A., *et al.* "Wagon Road from Niobrara to Virginia City," *House Executive Document 58* (serial 1256), 39 Congress, 1 Session, 1865–1866.

Pope, John. "Report . . . of the Condition and Necessities of the Department of the Missouri, February 25, 1866," *House Executive Document 76* (serial 1263), 39 Congress, 1 Session, 1865–1866.

"Wagon Roads in Western Territories," *House Executive Document 105* (serial 1263), 39 Congress, 1 Session, 1865–1866.

"Protection Across the Continent," *House Executive Document 23* (serial 1288), 39 Congress, 2 Session, 1866–1867.

Raynolds, William F. "Report . . . on the Exploration of the Yellowstone and Missouri Rivers in 1859–1860," *Senate Executive Document 77* (serial 1317), 40 Congress, 2 Session, 1867–1868.

"Annual Report of the Commissioner of the General Land Office," in *Report of the Secretary of the Interior, 1875, House Executive Document 1,* Part 5 (serial 1680), 27–428, 44 Congress, 1 Session, 1875–1876.

Macomb, John N. *Report of the Exploring Expedition from Santa Fe, New Mexico, to the Junction of the Grand and Green Rivers of the Great Colorado of the West, 1859.* Washington: Government Printing Office, 1876.

Simpson, James H. *Report of Explorations Across the Great Basin of the Territory of Utah for a Direct Wagon-Route from Camp Floyd to Genoa, in Carson Valley in 1859.* Washington: Government Printing Office, 1876.

Warren, Gouverneur K. "Epitome of Warren's Memoir, 1800–1857. Memoir Giving a Brief Account of Each of the English Expeditions since A.D. 1800." Part II of Vol. I of *United States Geographical Surveys West of 100th Meridian.* 7 vols. Washington: Government Printing Office, 1889.

State and Territorial

Journal of the Council of the Territory of Minnesota, 1849.

Minnesota Territorial Laws, 1854–1857.

"Speech of the Hon. Wm. H. Nobles together with other Documents, Relative to An Emigrant Route to California and Oregon through Minnesota Territory, published by order of the House of Representatives, February 10, 1854."

Laws, Memorials and Resolutions of the Territory of Dakota, 1865–1867.

Laws of the Territory of Montana, 1865–1868.

Laws of the Territory of Idaho, 1865–1869.

Oregon House Journal, Special Session, 1885.

Oregon Senate Journal, Special Session, 1885.

Laws of the Territory of Oregon, 1853–1858.

Statutes of the Territory of Washington, 1854–1867.

Annual Report of the Surveyor General of the State of California, Assembly Document 5, 1856.

California Assembly Journal, 1855–1859.

California Senate Journal, 1855–1859.

California Statutes, 1855–1858.

The People *v.* Johnson, *California Supreme Court Reports,* VI, pp. 499–506.

Laws of the Territory of New Mexico, 1851–1861.

Journal of the House of Representatives of the Legislative Assembly of New Mexico, 1861–1866.

Newspapers

The Deseret News (Salt Lake City). Files from January 27, 1858, to February 27, 1861, Newberry Library, Chicago, Illinois.

Fort Smith Herald. Files from December 22, 1847, to March 7, 1851, examined. Microfilm copies located in The University of Chicago Library.

Kirk Anderson's Valley Tan (Salt Lake City). Title continued as *Valley Tan* from May 24, 1859. Files from November 12, 1858, to November 9, 1859, Chicago Public Library.

Minnesota Democrat. Files from January 1, 1853, to October 31, 1855, Minnesota Historical Society Library.

Sacramento Daily Union. Files from June 1, 1855, to November 1, 1855, Bancroft Library, University of California.

Sioux City Journal (Sioux City, Iowa). Files from January 1, 1865, to January 1, 1867, Historical Department of Iowa, Des Moines.

SECONDARY WORKS

Articles

Baker, James H. "History of Transportation in Minnesota," *Collections of the Minnesota Historical Society*, IX (April, 1901), 1–34.

Beers, Henry P. "A History of the U. S. Topographical Engineers, 1813–1863," *The Military Engineer*, XXXIV (June and July, 1942), 287–291, 348–352.

Bemis, Samuel Flagg. "Captain John Mullan and the Engineers' Frontier," *The Washington Historical Quarterly*, XIV (July, 1923), 201–205.

Bender, Averam B. "Government Explorations in the Territory of New Mexico, 1846–1859," *The New Mexico Historical Review*, IX (January, 1934), 1–32.

Bender, Averam B. "Opening Routes Across West Texas, 1848–1850," *The Southwestern Historical Quarterly*, XXXVII (October, 1933), 116–135.

Bender, Averam B. "The Texas Frontier, 1848–1861," *The Southwestern Historical Quarterly*, XXXVIII (October, 1934), 135–147.

Bieber, Ralph P. "California Gold Mania," *The Mississippi Valley Historical Review*, XXXV (June, 1948), 3–28.

Bieber, Ralph P. "The Southwestern Trails to California in 1849," *The Mississippi Valley Historical Review*, XII (December, 1925), 342–375.

Bradley, Lieutenant James H. "Account of the Building of Mullen's [*sic*] Military Road," *Contributions to the Historical Society of Montana*, VIII (1917), 162–169.

Branch, E. Douglas. "Frederick West Lander, Road-Builder," *The Mississippi Valley Historical Review*, XVI (September, 1929), 172–187.

Burlingame, Merrill G. "The Influence of the Military in the Building of Montana," *The Pacific Northwest Quarterly*, XXIX (April, 1938), 135–150.

Burlingame, Merrill G. "John M. Bozeman, Montana Trailmaker," *The Mississippi Valley Historical Review*, XXVII (March, 1941), 541–568.

Burns, Lee. "The National Road in Indiana," *Indiana Historical Society Publications,* VII (1923), 209–237.

Catlin, George B. "Michigan's Early Military Roads," *Michigan History Magazine,* XIII (Spring, 1929), 196–207.

Cheetham, F. T. "El Camino Militar," *The New Mexico Historical Review,* XV (January, 1940), 1–11.

Clark, Robert Carlton. "Military History of Oregon, 1849–59," *The Oregon Historical Quarterly,* XXXVI (March, 1935), 14–59.

Cole, H. E. "The Old Military Road," *Wisconsin Magazine of History,* IX (September, 1925), 47–62.

Crane, R. C. "Some Aspects of the History of West and Northwest Texas Since 1845," *The Southwestern Historical Quarterly,* XXVI (July, 1922), 30–43.

Creer, Leland H. "The Explorations of Gunnison and Beckwith in Colorado and Utah, 1853," *The Colorado Magazine,* VI (September, 1929), 184–192.

Elliott, T. C. "The Mullan Road: Its Local History and Significance," *The Washington Historical Quarterly,* XIV (July, 1923), 206–209.

Farquhar, Francis P. (ed.) "The Topographical Reports of Lieutenant George H. Derby," *Quarterly of the California Historical Society,* XI (June, September, and December, 1932), 99–123, 247–265, and 365–382.

Gamble, James. "Wiring a Continent," *The Californian, A Western Monthly Magazine,* III (June, 1881), 556–563.

Grinnell, George Bird. "Bent's Old Fort and Its Builders," *Kansas Historical Collections,* XV (1923–1925), 28–91.

Hanford, Cornelius H. "Pioneers of Iowa and of the Pacific Northwest," *The Annals of Iowa,* X, 3rd Series (January–April, 1912), 331–342.

Hartman, Amos W. "The California and Oregon Trail, 1849–1860," *The Quarterly of the Oregon Historical Society,* XXV (March, 1924), 1–35.

Howard, Addison. "Captain John Mullan," *The Washington Historical Quarterly,* XXV (July, 1934), 185–202.

Jackson, W. Turrentine. "The Army Engineers as Road Builders in Territorial Iowa," *Iowa Journal of History,* XLVII (January, 1949), 15–33.

Jackson, W. Turrentine. "The Army Engineers as Road Surveyors and Builders in Kansas and Nebraska, 1854–1858," *The Kansas Historical Quarterly,* XVII (February, 1949), 37–59.

Jackson, W. Turrentine. "Federal Road Building Grants for Early Oregon," *Oregon Historical Quarterly,* L (March, 1949), 3–29.

Jackson, W. Turrentine. "The Fisk Expeditions to the Montana Gold Fields," *The Pacific Northwest Quarterly,* XXXIII (July, 1942), 265–282.

Larsen, Arthur J. "Roads and the Settlement of Minnesota," *Minnesota History,* XXI (September, 1940), 225–244.

Larsen, Arthur J. "Roads and Trails in the Minnesota Triangle, 1849–1860," *Minnesota History,* XI (December, 1930), 387–411.

Ledyard, Edgar M. "American Posts," *Utah Historical Quarterly,* I (April, July, and October, 1928), 56–64, 86–96, 114–127; II (January, April, July, and October, 1929), 25–30, 55–64, 90–96, 127–128; III (January, April, and July, 1930), 27–32, 59–64, 90–96; V (April, July, and October, 1932), 65–80, 113–128, 161–176; VI (January and April, 1933), 29–48, 64–80.

Maloney, Alice B. "The Richard Campbell Party of 1827," *Quarterly of the California Historical Society,* XVIII (December, 1939), 347–354.

Martin, Mabelle Eppard. "California Emigrant Roads Through Texas," *The Southwestern Historical Quarterly,* XXVIII (April, 1925), 287–301.

"The Memphis Convention," *The Commercial Review of the South and West,* edited by J. D. B. DeBow, VIII (March, 1850), 217–232.

Mills, Randall V. "A History of Transportation in the Pacific Northwest," *Oregon Historical Quarterly,* XLVII (September, 1946), 281–312.

"Mullan Road Markers," *The Washington Historical Quarterly,* XVII (January, 1926), 76–77.

Myers, Alice V. "Wagon Roads West: The Sawyers Expeditions of 1865, 1866," *The Annals of Iowa,* XXIII (January, 1942), 213–237.

Neill, Edward D. "James Hervey Simpson," *The Northwest Review,* I (April, 1883), 74–81.

Overmeyer, Philip Henry. "George B. McClellan and the Pacific Northwest," *The Pacific Northwest Quarterly,* XXXII (January, 1941), 3–60.

"Pacific Railway Convention at St. Louis," *The Western Journal,* III (November, 1849), 71–75.

Pray, Carl E. "An Historic Michigan Road," *Michigan History Magazine,* XI (January, 1927), 325–341.

Prosch, Thomas W. "The Military Roads of Washington Territory," *The Washington Historical Quarterly,* II (January, 1908), 118–126.

Prosch, Thomas W. "Notes from a Governmental Document on Oregon Conditions in the Fifties," *Quarterly of the Oregon Historical Society,* VIII (June, 1907), 190–200.

Quattlebaum, Charles B. "Military Highways," *Military Affairs,* VIII (Fall, 1944), 225–238.

"Railroad to the Pacific—Senator Benton's Bill," *The Western Journal,* II (January, 1849), 189–195.

Russel, Robert R. "The Pacific Railway in Politics Prior to the Civil War," *The Mississippi Valley Historical Review,* XII (September, 1925), 187–201.

Shippee, Lester B. "The Federal Relations of Oregon—VII," *Quarterly of the Oregon Historical Society,* XX (December, 1919), 345–370.

Simpson, James H. "Annual Address, delivered before the Minnesota Historical Society, January 19, 1852," *Annals of the Minnesota Historical Society,* III (1852), 5–19.

Simpson, James H. "Coronado's March," *Annual Report of the Board of Regents of the Smithsonian Institution* (1871), 309–340.

Talkington, Henry L. "Mullan Road," *The Washington Historical Quarterly,* VII (October, 1916), 301–306.

Underhill, W. M. "The Northern Overland Route to Montana," *The Washington Historical Quarterly,* XXIII (July, 1932), 177–195.

Van der Zee, Jacob. "The Roads and Highways of Territorial Iowa," *The Iowa Journal of History and Politics,* III (April, 1905), 175–225.

White, Chester Lee. "Surmounting the Sierras: The Campaign for a Wagon Road," *Quarterly of the California Historical Society,* VII (March, 1928), 3–19.

White, Hiram F. "The Career of Samuel R. Thurston in Iowa and Oregon," *The Iowa Journal of History and Politics,* XIV (April, 1916), 239–264.

Winther, Oscar O. "The Development of Transportation in Oregon, 1843–1849," *Oregon Historical Quarterly,* XL (December, 1939), 315–326.

Winther, Oscar O. "Early Commercial Importance of the Mullan Road," *Oregon Historical Quarterly,* XLVI (March, 1945), 22–35.

Winther, Oscar O. "Inland Transportation and Communication in Washington, 1844–1859," *The Pacific Northwest Quarterly,* XXX (October, 1939), 371–386.

Winther, Oscar O. "The Place of Transportation in the Early History of the Pacific Northwest," *The Pacific Historical Review,* XI (December, 1942), 383–396.

Winther, Oscar O. "The Roads and Transportation of Territorial Oregon," *Oregon Historical Quarterly,* XLI (March, 1940), 40–52.

Edited Documents

Clark, Pal (ed.). "Journal from Fort Dalles O. T. to Fort Wallah Wallah W. T. July 1858. Lieut. John Mullan U. S. Army," *The Frontier: A Magazine of the Northwest,* XII (May, 1932), 368–375.

"The Expeditions of Capt. Jas. L. Fisk, 1864–1866," *Collections of the State Historical Society of North Dakota,* II (1908), Part I, 421–461.

"The Expeditions of Jas. L. Fisk to the Gold Fields of Montana and Idaho in 1862 and 1863," *Collections of the State Historical Society of North Dakota,* II (1908), Part II, 35–85.

Foreman, Carolyn T. "Report of Captain John Stuart on the Construction of a Road from Fort Smith to Horse Prairie on Red River," *Chronicles of Oklahoma,* V (September, 1927), 333–347.

Foreman, Grant (ed.). "Survey of a Wagon Road from Fort Smith to the Colorado River," *Chronicles of Oklahoma,* XII (March, 1934), 74–96.

Golder, Frank A., Thomas A. Bailey, and J. Lyman Smith (eds.). *The March of the Mormon Battalion from Council Bluffs to California.* New York: The Century Company, 1928.

Holman, Albert M. (ed.). "Niobrara-Virginia City Wagon Road," *Pioneering in the Northwest.* Sioux City: Deitch & Lamar Co., 1924.

"Official Correspondence Pertaining to the War of the Outbreak, 1862–1865," *South Dakota Historical Collections,* VIII (1916), 100–588.

Sawyers, James A. "Niobrara-Virginia City Wagon Road: Report of Col. James A. Sawyers, Superintendent," *South Dakota Historical Review,* II (1936–1937), 3–48.

Smith, Lewis H. "Diary of the Sawyers' Expedition, 1865," in Harvey Ingham, *Northern Border Brigade* (n.d.), n.p.

Whiting, William H. C. "Diary of a March from San Antonio to El Paso," *Publications of the Southern History Association,* VI (July and September, 1902), 283–294, 389–399.

Books

Albright, George Leslie. *Official Explorations for Pacific Railroads, 1853–1855.* Vol. VII of University of California Publications in History, edited by Herbert E. Bolton. Berkeley: University of California Press, 1921.

Alter, J. Cecil. *James Bridger: Trapper, Frontiersman, Scout and Guide.* Salt Lake City: Shepard Book Company, 1925.

Alter, J. Cecil. *Utah, The Storied Domain: A Documentary History of Utah's Eventful Career.* 3 vols. Chicago and New York: The American Historical Society, 1932.

Bancroft, Hubert Howe. *History of California, 1542–1890.* Vol. V. San Francisco: The History Company, 1886.

Bancroft, Hubert Howe. *History of Oregon, 1848–1888.* 2 vols. San Francisco: The History Company, 1888.

Bancroft, Hubert Howe. *History of Utah, 1540–1886.* San Francisco: The History Company, 1889.

Beers, Henry Putney. *The Western Military Frontier, 1815–1846.* Philadelphia: University of Pennsylvania, 1935.

Bieber, Ralph P., and Averam B. Bender (eds.). *Exploring Southwestern Trails, 1846–1854.* Vol. VII of *The Southwest Historical Series.* Glendale: The Arthur H. Clark Company, 1938.

Bonsal, Stephen. *Edward Fitzgerald Beale: A Pioneer in the Path of Empire, 1822–1903.* New York: G. P. Putnam's Sons, 1912.

Brindley, John E. *History of Road Legislation in Iowa.* Iowa City: The State Historical Society of Iowa, 1912.

Bruce, Robert. *The National Road*. Brooklyn: National Highways Association, 1916.

Burlingame, Merrill G. *The Montana Frontier*. Helena: State Publishing Company, 1942.

Camp, Charles L. (ed.). *Henry R. Wagner's The Plains and the Rockies, A Bibliography of Original Narratives of Travel and Adventure, 1800–1865*. San Francisco: Grabhorn Press, 1937.

Carey, Charles H. *A General History of Oregon Prior to 1861*. 2 vols. Portland: Metropolitan Press, 1935.

Carroll, Charles C. *The Government's Importation of Camels: A Historical Sketch*. U. S. Department of Agriculture, Bureau of Animal Industry, No. 53. Washington: Government Printing Office, 1904.

Chapman, Arthur. *The Pony Express*. New York: G. P. Putnam's Sons, 1932.

Clark, Robert Carlton. *History of the Willamette Valley, Oregon*. Vol. I. Chicago: S. J. Clarke Publishing Company, 1927.

Conkling, Roscoe P., and Margaret B. *The Butterfield Overland Mail, 1857–1869*. Glendale: The Arthur H. Clark Company, 1947.

Creer, Leland H. *The Founding of an Empire: The Exploration and Colonization of Utah, 1776–1856*. Salt Lake City: Bookcraft, 1947.

Creer, Leland H. *Utah and the Nation*. Seattle: University of Washington Press, 1929.

Cullum, George W. *Biographical Register of the Officers and Graduates of the U. S. Military Academy*. 2 vols. New York: D. Van Nostrand, 1868.

Dickson, Arthur Jerome. *Covered Wagon Days*. Cleveland: The Arthur H. Clark Company, 1929.

Dunbar, Seymour. *A History of Travel in America*. Vol. IV. Indianapolis: The Bobbs-Merrill Company, 1915.

Egan, Howard R. *Pioneering the West 1846 to 1878*. Richmond, Utah: Howard R. Egan Estate, 1917.

Folwell, William W. *A History of Minnesota*. Vol. 1. St. Paul: Minnesota Historical Society, 1921.

Foreman, Grant. *Marcy & the Gold Seekers: The Journal of Captain R. B. Marcy, with an Account of the Gold Rush Over the Southern Route*. Norman: University of Oklahoma Press, 1939.

Fowler, Harlan D. *Camels to California*. Stanford: Stanford University Press, 1950.

Gilbert, Edmund W. *The Exploration of Western America, 1800–1850: An Historical Geography*. Cambridge, C. U. P. (England), 1933.

Gue, Benjamin F. *History of Iowa*. Vol. I. New York: The Century History Company, 1903.

Gunnison, John W. *The Mormons or Latter-Day Saints in the Valley of the Great Salt Lake*. Philadelphia: Lippincott, Grambo and Company, 1852.

Hafen, LeRoy R. *The Overland Mail, 1849–1869*. Cleveland: The Arthur H. Clark Company, 1926.

Hasse, Adelaide R. *Reports of Explorations Printed in the Documents of the United States*. Washington: Government Printing Office, 1899.

Heap, Gwinn Harris. *Central Route to the Pacific*. Philadelphia: Lippincott, Grambo, and Company, 1854.

Hebard, Grace R., and E. A. Brininstool. *The Bozeman Trail: Historical Accounts of the Blazing of the Overland Routes into the Northwest, and Fights with Red Cloud's Warriors*. 2 vols. Cleveland: The Arthur H. Clark Company, 1922.

Heitman, Francis B. *Historical Register and Dictionary of the U. S. Army from its Organization, September, 1789, to March 2, 1903*. 2 vols. Washington: Government Printing Office, 1903.

Holcombe, Return I. (ed.). *Minnesota in Three Centuries*, II. Publishing Society of America, 1908.

Holt, W. Stull. *The Office of the Chief of Engineers of the Army: Its Non-Military History, Activities, and Organization*. Baltimore: The Johns Hopkins Press, 1923.

Jordan, Philip D. *The National Road*. Indianapolis: Bobbs-Merrill Company, 1948.

Kearny, Thomas. *General Philip Kearny, Battle Soldier of Five Wars, Including the Conquest of the West by General Stephen Watts Kearny*. New York: G. P. Putnam's Sons, 1937.

Kelly, Charles. *Salt Desert Trails*. Salt Lake City: Western Printing Company, 1930.

Keleher, William A. *The Fabulous Frontier: Twelve New Mexico Items*. Santa Fe: The Rydal Press, 1945.

Lesley, Lewis Burt (ed.). *Uncle Sam's Camels*. Cambridge: Harvard University Press, 1929.

Marcy, Randolph B. *Thirty Years of Army Life on the Border*. New York: Harper and Brothers, 1866.

Meisel, Max. *A Bibliography of American Natural History: The Pioneer Century, 1769–1865*. 3 vols. Brooklyn: The Premier Publishing Company, 1924–1929.

Morton, J. Sterling. *Illustrated History of Nebraska,* Vol. I. Lincoln: Western Publishing and Engraving Company, 1911.

Morton, J. Sterling, and Albert Watkins. *History of Nebraska.* Edited and revised by Augustus O. Thomas, James A. Beattie, and Arthur C. Wakeley. Lincoln: Western Publishing and Engraving Company, 1918.

Moseley, Henry N. *Oregon: Its Resources, Climate, People, and Productions.* London: Edward Stanford, 1878.

Mullan, John. *Miners' and Travelers' Guide to Oregon, Washington, Idaho, Montana, Wyoming, and Colorado via the Missouri and Columbia Rivers.* New York: William Franklin for the author, 1865.

Nash, Wallis. *Two Years in Oregon.* New York: D. Appleton and Company, 1882.

Neff, Andrew Love. *History of Utah, 1847–1869.* Edited by Leland H. Creer. Salt Lake City: Deseret News Press, 1940.

Potter, David M. *Lincoln and His Party in the Secession Crisis.* New Haven: Yale University Press, 1942.

Richman, Irving Berdine. *Ioway to Iowa: The Genesis of a Corn and Bible Commonwealth.* Iowa City: The State Historical Society of Iowa, 1931.

Robinson, Doane. *History of South Dakota.* Vol. I. B. F. Powers, 1904.

Robinson, Edward Van Dyke. *Early Economic Conditions and the Development of Agriculture in Minnesota.* Studies in the Social Sciences, Bulletin No. 3. Minneapolis: University of Minnesota, March, 1915.

Russel, Robert R. *Improvement of Communication with the Pacific Coast as an Issue in American Politics, 1783–1864.* Cedar Rapids: The Torch Press, 1948.

Sabin, Edwin L. *Building the Pacific Railway.* Philadelphia: J. B. Lippincott Company, 1919.

Shambaugh, Benjamin F. *The Old Stone Capital Remembers.* Iowa City: The State Historical Society of Iowa, 1939.

Simpson, James H. *The Shortest Route to California.* Philadelphia: J. B. Lippincott and Company, 1869.

Taylor, Emerson Gifford. *Gouverneur Kemble Warren: The Life and Letters of an American Soldier, 1830–1882.* New York: Houghton Mifflin Company, 1932.

Thompson, Robert Luther. *Wiring a Continent: The History of the Telegraph Industry in the United States, 1832–1866.* Princeton: Princeton University Press, 1947.

Trottman, Nelson. *History of the Union Pacific: A Financial and Economic Survey.* New York: The Ronald Press Company, 1923.

Tullidge, Edward W. *Life of Brigham Young, or Utah and Her Founders*. New York: Tullidge and Crandall, 1877.

Twitchell, Ralph Emerson. *The Leading Facts of New Mexican History*. 2 vols. Cedar Rapids: The Torch Press, 1911–1912.

Waite, C. V. *The Mormon Prophet*. Cambridge: Riverside Press, 1866.

Whitney, Orson F. *History of Utah*. Vol. I. Salt Lake City: George Q. Cannon and Sons Company, 1892.

Winther, Oscar Osburn. *The Old Oregon Country: A History of Frontier Trade, Transportation, and Travel*. Stanford: Stanford University Press, 1950.

Young, Jeremiah Simeon. *A Political and Constitutional Study of the Cumberland Road*. Chicago: The University of Chicago, 1902.

Index

INDEX

Abbot Company of Concord, involved in scandals of El Paso–Fort Yuma road, 230
Abert, Lieutenant James W.: topographical officer with Kearny, 18; return from Bent's Fort to St. Louis, 25; reconnaissance of upper Rio Grande, 107–108
Abert, Colonel John J.: chief topographical engineer, 2; interest in Iowa roads, 9; comments on Minnesota roads, 52; estimates needs for Minnesota roads, 54; threatens to shut down Minnesota operations due to lack of funds, 56; orders Reno to examine the Mendota–Big Sioux military road, 58; refers Minnesota land seizure problem to Attorney General, 62; supports Simpson in Minnesota, 65; ordered to establish superintendency of Pacific roads, 75; prepares Oregon estimates for Davis, 80; makes recommendations concerning Utah constructions, 145–146; advisory capacity to Secretary of the Interior, 177; advises Davis in appointments for Pacific Railway surveys, 243, 321
Adams, Thomas, guide of Mullowney west of South Pass, 195
Albuquerque-Tecolaté road: appropriation, 112; improvement, 113–115
Alexander, Colonel E. B., escorts Magraw party, 198
Ali Hadji, in charge of camels of Beale's expeditions, 254
Allen, Edward J., directs Washington residents in building road across the Cascades, 95
Allen, Captain James, enlists Mormon volunteers, 17
Alta California, proprietors request to Simpson, 153
Alvord, Major Benjamin, surveys route between Rogue and Umpqua rivers, 73
American Fur Company: Cache Valley rendezvous, 32; in Minnesota politics, 56; steamer *St. Mary* on Missouri River, 136
Andrews, Colonel George, travel along Bryan and Stansbury surveys, 146
Annan, James R.: disbursing agent on Central Overland wagon road, 196; in Salt Lake City; 197; asked for testimony about Magraw, 199
Applegate, Jesse: assists Alvord in survey of Rogue and Umpqua river route, 73; road contractor, 73
Arbuckle, General Mathew, 5; supports Arkansas–New Mexico road, 23; ordered to provide emigrant escort to New Mexico, 24
Arkansas: road system, 7; enthusiasm for roads to New Mexico, 22–23; state legislature memorializes Congress, 23; Fort Smith meeting to petition for roads, 23
Armstrong, Moses K., Dakota politician offers to organize party to ride from Missouri Valley to Fort Connor, 303
Army Engineers: history, 2; construction of National Road, 3; early activities in trans-Mississippi West, 5–6. *See also* Corps of Topographical Engineers; United States Army; War Department
Army of Utah: headquarters established, 146; Magraw and men join, 199; Lander authorized to recover wagon road supplies from, 206–207
Arnold, Lieutenant Richard, supervises reconnaissance from Steilacoom to Walla Walla, 95
Astoria-Salem road: approved by Congress, 75; surveyed by Derby, 75–76; problems of improvement, 77
Atchison, Topeka, and Santa Fe Railway: follows part of Cooke's Mormon Battalion route, 22; follows in part Santa Fe–Fort Union road, 120; follows Santa Fe–Doña Ana road, 120; constructs line along Beale's wagon-road route, 256
Atlantic and Pacific Railroad, receives charter to build along Beale's wagon road, 256
Austin, Texas, citizens interested in road to El Paso, 37

Bache, Major Hartman: in charge of Pacific roads, 75, 96; protests Davis' removal of Derby, 79, 101; relieved of duty, 84; correspondence with Derby and Davis over road contracts, 99–100; reports on New Mexico roads as new Chief of Topographical Bureau, 120
Baker, James, guide of Wagner near South Pass, 195
Barney, Joshua, civilian engineer in Iowa, 11
Beale, Edward Fitzgerald: appointed superintendent of Fort Defiance–Colorado River road, 244; biographical sketch, 244–245; appointment makes possible merging of camel experiment and wagon-road build-

Beale, Edward Fitzgerald (*Continued*) ing, 245; march from San Antonio to Albuquerque, 245; arrival and ceremony at Fort Defiance, 245; march from Fort Defiance to Colorado River, 247; evaluation of camels, 246–249; greeted by Mohave Indians, 250; trip to Los Angeles, 250; return march to Fort Defiance, 250, recommends additional appropriation in report to Secretary of War, 250–251; march from the Arkansas frontier to Albuquerque, 252–253; reports on bridge building, 252; enjoys hunting with greyhounds, 253; explores east of Santa Fe and inspects artesian water well, 253; buys sheep to drive to California, 253; second march from Albuquerque to Colorado River, 253–255; return to Albuquerque, 255; urges additional congressional appropriations, 255; strong support of his route for a railroad, 256; appointed surveyor general of California and Nevada, 256

Beale, George: assistant road superintendent, 254; locates emigrant crossing at Little Colorado, 254

Bean, William, guide of Simpson in Utah, 149

Beckwith, Captain Edward G.: supervises Kansas and Nebraska roads, 131, 134; requests additional funds, 134–135

Bee, Colonel Fred A.: president of Placerville and St. Joseph Telegraph Company, 153; confers with Simpson about route for telegraph lines, 153; company uses route of Simpson, 156

Beers, Henry P., 12

Bell, John: Tennessee senator, 169; quoted in opposition to the Fort Ridgely–South Pass Wagon road, 169

Bell, Colonel Peter H., receives report of Hays-Highsmith expedition, 37

Bemis, Samuel Flagg, quoted, 328

Benton, Thomas Hart, recognizes relationship between wagon-road and railroad building, 241–242

Bent's Fort: Peck and Abert leave, 25; Bryan consultation over road routes, 124

Berford and Company, involved in scandals of El Paso–Fort Yuma road, 230

Bernhisel, John M.: Utah delegate, 140; introduces appropriation to improve southern route to California, 140; reports Steptoe to succeed Young, 141; introduces bill to improve Bridger's Pass–Salt Lake route, 144–145

Bestor, Norman, civilian engineer with Kearny, 18

Bieber, Ralph P., quoted, 22

Big Sioux River: appropriation for bridge, 282–283; preliminary work, 305–306; Dakota politicians ask for large appropriation, 306

Biggs, Asa: North Carolina senator, 169; opposes Fort Ridgely–South Pass wagon road, 169

Bigler, John: California governor, 164; message to California legislature urging action on wagon roads, 164–165; accused of using political influence in road legislation, 166

Bird, Wellington: appointed superintendent of Virginia City–Lewiston road, 312; ordered to Washington for conference, 313; considers possible routes, 313; travels along Lolo forks, 315–316; attempts improvements along Lolo forks; 317; return to Washington and an investigation, 317–318; urges Interior Department to continue road work, 318

Bishop, Francis A: appointed chief engineer on western division of Central Overland wagon road, 176; explorations near Carson City, 203; smokes peace-pipe with Indians, 204; divides western section of Central Overland wagon road, 205; commended by Thompson, 206

Black, J. S.: Attorney General, 175; supports appointment of Magraw as road superintendent, 175

Black Beaver: guide of Marcy, 26; Michler fails to get services, 43; refuses to go on Beale's sceond expedition to Colorado River, 251

Blake, Major George: confers with Reynolds, 267; command moves over Mullan road, 268–269

Bonneville, Major Benjamin L. E.: supports Arkansas–New Mexico road, 23; commander of New Mexico Military Department, 119

Bonnycastle, Lieutenant John C., assists Wallen expedition, 86

Bonsal, Stephen, biographer of Beale, 245

Borland, Solon G., Arkansas senator urges military escort for emigrants west of Fort Smith, 24

Bortheaux, Frederick, death, 129

Bowman, Lieutenant Alexander H., supervises Memphis–St. Francis River road, 8

Bozeman, John M.: biographical sketch, 283; explores cutoff from Oregon Trail into Montana, 284; rivalry with Bridger as guide, 284
Bozeman Trail, 284
Bridger, Jim: guide to Stansbury, 31, 33; guide to Raynolds, 267; rivalry with Bozeman as guide, 284
Brodhead, Richard: Pennsylvania senator, 170; inquiry about responsibility for Pacific wagon roads, 170; proposes Pacific railroad surveys, 321
Brookings, W. W.: named superintendent of Minnesota–Big Cheyenne River wagon road, 297; instructions, 298; preparation and exploration, 298–299; survey from Missouri River to eastern Minnesota boundary, 301; reports to Interior Department, 301; ordered to suspend work on Big Cheyenne road, 303; ordered to transfer property to Niobrara road, 303; urges renewal of road survey, 304; urges completion of James River bridge, 310
Brooks, James, New York congressman discusses constitutionality of federal road program, 55
Brown, Captain A. E., provides relief escort for Sawyers' expedition, 292
Browning, Orville H.: Secretary of the Interior, 304; considers resumption of work on Big Cheyenne road, 304; refuses, 305
Brule Sioux Indians: protests Warren's invasion of lands, 138; reservation established, 303
Bryan, Lieutenant Francis T.: surveys upper road across West Texas, 42; examines San Antonio–Fort Merrill road, Austin–Fort Mason road, and reconnaissance between San Antonio and Fort Belknap, 45; in charge of Kansas and Nebraska roads, 123; surveys Fort Riley–Arkansas River road, 123–125; winter quarters in St. Louis, 125; reports completion of road from Fort Riley to Bent's Fort, 126; reprimanded by Abert, 127; survey to Bridger's Pass, 127–129; organizes labor party to improve route to Bridger's Pass, 130–131; relieved of command in Kansas and Nebraska, 131; attacked by Weller in Senate, 163
Buchanan, James: pledged to sponsor Pacific wagon roads, 171; supports Magraw for road superintendency, 175; appoints Medary territorial governor of Minnesota, 183; correspondence from Magraw, 200–201; urged to place Fort Defiance–Colorado River road under Interior Department, 241
Bullock, Isaac, guide to Simpson in Utah, 147
Burleigh, Walter A.: Dakota territorial delegate, 297; biographical sketch, 297; dominates road appointments, 297; nominates Todd for superintendency of Big Cheyenne River road, 304; gets Congress to transfer Big Cheyenne River road funds to James River bridge, 310; continuous correspondence with officials, 310–311
Butler, Andrew P.: South Carolina senator, 170; opposition to Fort Ridgely–South Pass road provokes bitter debate, 170

California: petitions for roads quoted, 161–162; mass meeting in San Francisco to secure better overland communications, 164; Placerville and Sacramento newspapers crusade for wagon road across Sierra, 165; citizens of Calaveras and Placerville aid wagon road financially, 165; citizens urged to buy stock in mail and stage company, 167
  Legislature: authorizes road from Sacramento across Sierra, 165; constitutionality of road law questioned, 166; urged by Sacramento citizens to authorize county appropriations for road construction, 166; passes joint resolution asking Congress for roads across Plains, 167
California Argonauts: use Army trails in West Texas, 43; aided by wagon roads west, 326–327
California Emigrant Road Committee, 175
California emigrants: use Simpson survey from Camp Floyd to California, 155, 156; escorted along Humboldt by Kirk, 204; use Lander's road, 214
California Mail Company: agent at Pleasant Grove, 152; provides wood for raft for Simpson party, 152
California Rangers from Plumas county, greet emigrants along the Humboldt, 204
Camels: used by Beale on Fort Defiance–Colorado River road, 245; Beale evaluates work on march from San Antonio to Albuquerque, 246; Beale's interest in, 247–249; work in snows of Sierra Nevada, 250; used to carry mail along thirty-fifth parallel, 254–255
Campbell, Albert H.: appointed general superintendent of Pacific Wagon Road Office, 178; Minnesota lake named for, 181; quar-

Campbell, Albert H. (*Continued*)
rels with Nobles, 182; threatened by land interests, 183; attempts to settle financial accounts of Fort Ridgely–South Pass wagon road, 188–189; appraisal of relations with Nobles, 190; urges Magraw to depart, 196, receives charges against Magraw from engineers, 197; prepares brief of charges against Magraw and notifies him, 199–200; orders Lander to Washington to report on road work, 209; recommendations for construction of El Paso–Fort Yuma road, 218–219; political considerations of appointment as general superintendent of wagon roads, 220; instructions to Leach, 221–222; fails to testify against Woods and withdraws charges against Leach, 231; instructs Sites in financial matters, 235; report on Nebraska wagon road, 238

Campbell, J. C.: assistant engineer with Lander, 208; explores road through Cache Valley, 208; in charge of expedition property at Salt Lake, 212; surveys west of City Rocks, 212

Campbell, Lewis D.: Ohio congressman, 145; defends Utah appropriation, 145

Campbell, Richard, trader between Santa Fe and San Diego, 28

Cañada-Abiquiu road: appropriation, 112; improvement and importance, 115

Carrington, Albert, examines Salt and Utah lakes, 33

Carrington, Colonel Henry B.: constructs forts along Powder River road, 286; Sawyers' second road party joins at site of new fort, 295

Carrington, W. P. C.: meteorologist, 138; explores Black Hills with Warren, 138

Carson, Kit: meeting with Kearny, 18; associated with Beale, 245, 253

Carter, Thomas J.: road contractor north of Cowlitz Landing, 102, contract disapproved by Davis, 102

Cass, Lewis: Secretary of War, 2; orders survey between Forts Leavenworth and Smith, 6; as Michigan senator speaks for Central Overland route, 171

C. E. Hedges and Company, freighting company with shipments on Sawyers' expedition, 287

Central Overland wagon road: proposed, 163; Senate considers road via Great Salt Lake, 171; route changed to Fort Kearny–Honey Lake, 171; superintendent and engineers appointed, 175–176; preliminary activities, 192–193; first season work reported successful, 195; Kirk's successful survey of western sector, 203–206; Lander's cutoff described, 208; promotional work, 213–214; used by emigrants, 214–215; Honey Lake–Humboldt River trace improved, 216

Chadwick, Stephen W., represents topographical engineers in Oregon courts, 73

Chapin, Lieutenant Gurden, assists Simpson on Utah survey, 149

Chapman, Bird B.: Nebraska territorial delegate, 233; sponsors Platte River–Niobrara River road, 233; recommends route, 233–234; secures appointment of Smyth as engineer, 234

Chapman, W. W., protests to War Department about road locations, 73

Charbonneau, Kearny's guide, 20

Chatfield, Andrew G.: associate justice of Minnesota Supreme Court, 168; presides over mass meeting favoring wagon road, 168

Chicago, Milwaukee, St. Paul, and Pacific Railroad, follows Mullan road in part, 278

Chief Washakie: wagon-road parties seek flour from tribe, 198–199; praised by Lander, 209

Chippewa, arrives at Fort Benton, eastern terminus of the Mullan road, 264

Chisholm, Jesse, guide to Beale's second expedition to Colorado River, 251, 252

Cho-Kup, chief of Shoshones, 151–152

Chorpenning, George: mailcarrier, 150; takes party over Simpson route, 150; improvements utilized by Simpson, 151; failure of men to find satisfactory route from Ruby Valley to Genoa, 151; supports appointment of Leach, 176

Chouteau, Charles P.: St. Louis trader, 260; to push steamboat up Missouri River to head of navigation, 260

Claiborne, Captain Thomas, explores route from Fort Stanton to Pecos River, 119

Clay, Clement C.: Alabama senator, 170; opposes transfer of Fort Ridgely–South Pass road from War to Interior department, 170

Cobb, Howell, Secretary of the Treasury, 177; use of patronage, 177

Cogswell, Lieutenant Milton, commands escort to Macomb expedition in New Mexico, 117

Collamer, Jacob: Vermont senator, 169; demands reading of Minnesota memorial for roads, 169

Colorado miners, backwash from "Pike's Peak" meet Lander's expedition, 212
Commissioner of Indian Affairs: builds Minnesota roads, 61; provides funds for Pacific wagon-road superintendents, 178, 179, funds for Lander, 212
Committee on Military Affairs: Senate committee considers Arkansas memorial, 23–24; House committee kills bill for Minnesota road, 65; requests information on Oregon roads from War Department, 79, 83; recommends continuation of Washington projects, 105; Senate committee approves Salt Lake City–southern California road, 140–141; House committee considers Bridger's Pass–Salt Lake City road, 145; Senate committee reports on Army improvement of Central Overland route, 171; House committee approves appropriation for Fort Defiance–Colorado River road, 251
Committee on Post Office and Post Roads: Senate committee strikes out provision for wagon-road construction in California bill, 163; hearing relative to mail route from Fort Smith to San Diego, 242
Committee on Roads and Canals: House committee reports on Minnesota roads, 49–50; considers Oregon transportation needs, 71; Santa Fe–Taos road not a necessity, 110; considers bill for road from Missouri River to Virginia City, Montana, 282; reports on additional funds for Niobrara River road, 293
Committee on Territories: House committee approves Minnesota road appropriations, 59; accused of favoritism for Oregon and discrimination against Kansas, 81; approves location of Central Overland route, 171; agrees to Dakota bridge appropriation, 310
Congressional appropriations: National Road, 2–3; Arkansas, 7–8; Iowa, 9–11; routes from Mississippi to the Pacific, 24; Minnesota, 49, 51–52, 56–57, 60, 61, 67, 69; Oregon, 72, 75, 80, 82; Northern Pacific railroad survey, 89; Washington, 96, 103, 106; New Mexico, 109, 112, 120; Mormon Trail in Nebraska, 121; Kansas and Nebraska, 123; Kansas and Nebraska denied, 135; Salt Lake City–southern California road, 140–141; Fort Ridgely–South Pass, 170–171; Pacific wagon roads, 173–174; funds for relief of Nobles, 189–190; Nebraska road request, 238; Pacific railroad surveys, 242–243; thirty-fifth parallel road, 255, Missouri River to Walla Walla, 259; second appropriation for Mullan road, 260; Overland Escort Service, 275; publication of Mullan's report, 276; roads to the Montana–Idaho gold region, 282–283; Dakota bridge, 308, 310; Dakota Territory, 311; history of, 320–321, 322, 324, 325, 326
Congressional Globe: letter from Emerson to Rice (Minnesota delegate) printed, 60; Commissioner of Indian Affairs' letter printed, 66; letters from Mendell and Abert printed, 105; Sumner quoted, 110; Quitman objects to being misquoted, 145.
Connell, John, refuses appointment as superintendent of Virginia City–Lewiston road, 312
Conness, John: California senator, 310; gets agreement on Dakota bridge appropriation, 310
Connor, General Patrick E., Sawyers sends out search party to establish contact, 290
Cook, N. P.: Arizona engineer on El Paso–Fort Yuma wagon-road expedition, 221; dispatched to buy stock for expedition, 223; directs laborers from Pima villages to Fort Yuma, 225
Cooke, Captain Philip St. George: instructed to wait for Mormon battalion, 18; march to California, 20–22; order of congratulation to Mormons, 22. *See also* Mormon battalion
Cooper, J. C.: physician with Magraw, 197; in Salt Lake City, 197
Corps [Bureau] of Topographical Engineers, 2; early road-building assignments in trans-Mississippi West, 12; knowledge of road building obtained from Cooke, Marcy-Simpson, and Stansbury surveys, 35; work criticized in Minnesota, 53; achievements in Minnesota during 1853, 59; continues to recommend appropriations for Minnesota roads, 69; reviews complaints of disappointed Washington contractors, 102; reports on necessity for new Washington roads, 105; urges Washington roads, 105–106; assumes responsibility for New Mexico roads, 110; receives complaint about location of Nebraska road, 132; opinion requested on Utah roads, 145; receives Andrews report on Bryan and Stansbury routes, 146; attacked by Weller in Senate, 163; comparison with civilian road building by Interior Department, 232; Pacific Railroad Convention urges explorations in trans-Mississippi West, 242; relationship of

Corps of Topographical Engineers (*Contd.*) railroad and wagon-road surveys, 243; Washington legislature requests improvement of Mullan road, 277–278; contribution to nation's development, 328. *See also* Army Engineers; United States Army; War Department

Cowlitz Convention, memorializes Congress for separation of Washington and Oregon, 89

Cox, James D.: Secretary of the Interior, 311; orders completion of James River bridge, 311; refuses to authorize further expenditures on the Virginia City–Lewiston road, 318

Crawford, Captain Medoram, provides overland escort service along Oregon Trail, 284

Cress, W. Drayton, assistant engineer on El Paso–Fort Yuma road, 220

Crittenden, John: Kentucky senator, 171; urges caution in debate on Central Overland route, 171

Crow Wing–Leech Lake road: appropriation, 61; completed, 67

Cullen, W. J., appointed to board of appraisers, 188

Cumberland Road, 2

Curtis, Major General S. R., negotiates Indian treaty with Dakota Indians, 302

Cushing, Caleb: Attorney General, 62; opinion against land seizures for road construction in Minnesota, 62

Daily Minnesota Pioneer: Simpson pays tribute to Sibley, 64; attacks Rice, 65

Dakota Land Company: interested in sites along Nobles' wagon road, 183; deal with Nobles, 186

Dakota Republican, editor urges renewal of Big Cheyenne survey, 305

Dakota Territory: residents urge additional funds for Big Cheyenne road, 301; legislature requests mail route from Fort Sully to Virginia City, Montana, 301; legislature asks Army to build forts connecting Forts Sully and Connor, 301–302

Davis, Jefferson: issues orders to Withers for Oregon roads, 73; approves work of Withers, 75; notifies Abert to establish superintendency of Pacific roads in San Francisco, 75; rejects Oregon road contracts, 79–80; speaks in Senate for Oregon appropriations, 83; instructions to McClellan, 90; instructions to Derby, 96; correspondence with Bache over road contracts, 99; disapproves Washington contract, 102; orders topographical engineers to direct New Mexico roads, 110; instructions to Scammon, 111; instructions to Dickerson, 132; orders Steptoe to build southern California road, 141; Republican House objects to supervision of Pacific wagon roads, 171; recommends Campbell appointment, 178; interest in southern route, 218; guided by Topographical Bureau in naming officers for Pacific Railway surveys, 243; quoted on advantages of southern route, 244; interest in camel experiment, 244–245; spokesman for Pierce administration on wagon-road and railroad surveys, 321–323

Day, Sherman: California senator, 165; makes survey across the Sierra, 165

DeLacy, W. W.: surveys the Fort Steilacoom–Bellingham Bay road, 103; inspects and approves road from Cowlitz Landing to Ford's Prairie, 104; with Mullan, 261, 263, 265

Democratic Party: 1840 platform on internal improvements, 319–320; changes in program at Cincinnati convention, 1856, 323–324

Dempsey and Hockaday road: west of South Pass, 195; plans for survey, 206

Denver, Frank: disbursing agent from Kirk, 203; explorations near Carson City, 203

Derby, Lieutenant George H.: instructed to survey Astoria-Salem road, 75–76; quoted, 76, 77; estimates funds needed for completion, 77–79; problems in issuing contracts, 79; relieved of Oregon assignment, 79, 101; ordered to Fort Vancouver, 96; problems of building Fort Vancouver–The Dalles road, 97–98; plans for construction of Fort Vancouver–Fort Steilacoom road, 98–99; problems over contracts, 99–191; removal from Washington, 101

Deseret News, reprints story from *New York Times* attacking Thompson, 185

Dick, Delaware hunter and guide, 251; on Beale's second expedition to Colorado River, 251, 253

Dickerson, Lieutenant John H.: appointed superintendent of Omaha–Fort Kearny road, 127; makes survey, 131–134; requests additional appropriations, 134–135

Dillon, Lyman, plows furrow from Dubuque to Iowa City, 9

Dixon, Lieutenant Joseph: topographical officer appointed to accompany Wallen expedition, 85–87; reports, 86; proposes new route to Oregon, 88

Dodd, William B., builds road from St. Peter to St. Paul in Minnesota, 58

Dodge, Augustus C.: Iowa senator, 51; defends Minnesota road appropriations, 51

Douglas, Stephen A.: Illinois senator, 51; defends Minnesota road appropriations, 51; supports final Minnesota appropriation, 67; support sought for Utah roads, 144; supports Magraw for road superintendency, 175; committee reports on Albuquerque–Colorado River road, 251

Drew, George, road builder on Fort Vancouver–Fort Steilacoom, 104

Dufond, Baptiste: guide to Sawyers' expedition, 289; abandoned by military escort, 289; Williford recommends dismissal, 290

Dunbar, Seymour, quoted as transportation historian, 1

Eaton, John H., Secretary of War, 2

Echols, William H., organizes camel corps at Camp Hudson, Texas, 46

Eckles, D. E.: Mormon chief justice, 214; champions Simpson's road, 214

Edmunds, George F.: Vermont senator, 310; challenges appropriations for Dakota bridges, 310

Edmunds, Newton: Dakota territorial governor, 302; urged by Army officers to support Big Cheyenne route, 302; negotiates treaty with Teton Sioux, 302; urges renewal of Big Cheyenne survey, 304

Edwards, N. B., in charge of bridge building along the Canadian River, 252

El Camino Militar, 115, 120

El Paso–Fort Yuma road: proposed, 163; Senate approves Army appropriations for improvements, 171; superintendent and engineers appointed, 176; preparations and departure, 222–223; progress across Arkansas, Indian Territory, and Texas, 223–224; problems of ox train, 225; survey and location of road by Hutton, 226; improvements, 226–227; additional appropriation requested from Congress, 227; work by Stone, 232

Emerson, Charles L.: political ally of Rice, 57; dismissed by Simpson and becomes editor of *Minnesota Democrat,* 59

Emigrant Overland Escort Service: funds administered by Secretary of War, 274; sponsors Fisk expeditions, 274, 275; Fisk disaster leads to curtailment of service, 276; attempts to answer emigrants' needs, 284–285

Emigrant's Guide: published by Lander, 209–210; supplement prepared for western sector of Central Overland route, 213

Emory, Lieutenant William H., topographical officer with Kearny, 18, 19

Engelmann, Henry: geologist with Bryan on survey to Bridger's Pass, 127; assists in preparation of Bryan's report, 130; associated with Simpson on Utah explorations, 148, 149, 150, 156

Engle, P. M.: topographer, 138; explores Black Hills with Warren, 138; with Mullan, 261, 263

Estes, Ben F.: guide to Sawyers' expedition, 290; Williford recommends dismissal, 290

Faribault, Alexander, Minnesota fur trader, 49

Farley, E. Wilder: Maine congressman, 59; urges continuation of Minnesota road appropriations, 59–60

Faulkner, Charles J.: Virginia congressman, 110; questions necessity of Santa Fe–Taos road appropriation, 110

Federal inspector: reports on Fort Ridgely–South Pass wagon road, 184, 186–187. *See also* Gillis, John R.; Sayles, Welcome B.

Ferguson, Lieutenant Samuel W., commands escort to Simpson in Utah, 147

Fetterman Massacre, 286

Ficklin, B. F.: seeks routes in Big Sandy–Green River desert, 195; winters in Rocky Mountains, 198; tries to buy flour from Indians, 198–199

Fillmore, Millard: signs Minnesota road appropriation bill, 56–57; signs Oregon road appropriation bill, 72

Fisk, Captain James L.: leads emigrant party over North Overland route, 1862, 274–275; emigrant party of 1863, 275; disastrous third expedition, 275–276; emigrant party of 1865, 276

Fitch, Graham N.: Indiana senator, 234; urges wagon-road appointment for Indiana Democrats, 234

Fitzhugh, E. C., constructs road from Whatcom to Bellingham Fort, 104

Flagler, Thomas S.: New York congressman, 110; requests New Mexico delegate's views, 110

Floyd, John B.: Secretary of War, 83–84; orders Pacific Road Constructions Office to Fort Vancouver, 83–84; approves Macomb's plans for New Mexico, 112; refers Abert's report on Utah roads to Congress, 146; approves survey from Camp Floyd to California, 150; use of patronage, 177; recommends Campbell appointment, 178; endorses Smyth as engineer, 234; appoints Beale for thirty-fifth parallel road superintendent, 244
Ford, John S.: Austin, Texas, doctor leads road party, 37–39; expedition illustrates military-civilian coöperation, 39
Fort C. F. Smith: on Powder River road, 286; troops withdraw from, 305
Fort Colville: gold discoveries increase Oregon labor prices, 76–77; miners agitate Indians into hostilities, 260
Fort Connor: temporary establishment on the Powder River road, 286; supply wagon for Sawyers' expedition to, 289; report of conflict between Sawyers and Williford, 291
Fort Defiance–Colorado River road: Beale's survey and report on potential use by wagons, 247; return march, 250; promotional work by Beale, 250–251; route and water supply improved, 254
Fort Kearny: Stansbury's arrival, 29; Bryan arrives, 127; Dickerson attends Pawnee council, 133
Fort Laramie: Stansbury's arrival, 29; Nobles to receive instructions there, 181
Fort Leaton, Texas: Hays-Highsmith camp, 37; Whiting party arrives at this trading post, 39
Fort Leavenworth, 6; "The Army of the West" assembled, 17; Stansbury preparation, 29; Stansbury return, 33–34; frontier depot, 121; Lander's expedition, 193
Fort Philip Kearny: on Powder River road, 286; troops withdraw, 305
Fort Reno: on Powder River road, 286; emigrants and road builders assemble here for protection, 294
Fort Ridgely, Minnesota–South Pass wagon road: proposed, 163; passes House, 169; debate in Senate, 169–170; bill passed and signed by President, 170–171; Nobles in charge, 175, 176, 179
Fort Riley: frontier outpost, 121; cholera outbreak, 123
Fort Riley–Arkansas River road: appropriation, 123; survey by Bryan, 123–125; used by Kansas settlers, 126
Fort Riley–Bridger's Pass road: appropriation, 123; reconnaissance by Bryan, 127–129; bridges built, 131; excellent condition for travel, 131
Fort Smith Herald, quoted in favor of roads, 22–23, 24
Fort Steilacoom–Bellingham Bay road: construction urged by Washington legislature, 102–103; surveyed, 103
Fort Tejon, Beale's party rests at, 250
Fort Vancouver–Fort Steilacoom road: appropriation approved, 96; Gibbs survey, 96; Derby's plans for construction, 98–99; additional appropriation urged, 101; work progresses during 1858 and new appropriation asked, 104; appropriation granted, 106
Fort Vancouver–The Dalles road: appropriation approved, 96; Derby examines route, 96–97; road work at Cascades, 97–98; Portage Road considered excellent, 101; new appropriation asked, 104
Fort Washita, Michler begins survey to Pecos River, 43
Fraser River gold discoveries, causes Oregon laborers to abandon road work, 83
Frémont, John C.: active committeeman urging roads to California, 164; Beale as witness during trial of, 245
French, Captain Samuel G., assistant quartermaster in Texas, quoted, 41
Frontier demands for transportation aid, 1, 319

Gallegos, Jose Manuel, New Mexico delegate, 110
General Jesup, Beale's expedition crosses Colorado on, 250
Gibbs, George: civilian engineer, 96; surveys trail along the Columbia and Cowlitz rivers, 96; checks route proposed by settlers, 100
Gilbert's Trading Post, Lander stations soldiers at South Pass to provide emigrant information, 213
Gillis, John R.: federal inspector of Dakota roads, 307; relieved of responsibility for wagon roads, 309
Goddard, George H., engineering survey across the Sierra, 165
Goodale, Tim, fist fight with Magraw, 197
Gore, Sir George, hunting party meets Warren along Yellowstone River, 136

Gorin, Jerome R.: disbursing agent in Minnesota, 182; notified road funds exhausted, 182; administrative difficulties over finances, 184; abandons Fort Ridgely–South Pass expedition, 185; financial embarrassment, 185; liable for embezzlement, 187
Gorman, Willis A.: Minnesota governor, 63; advises Simpson, 63; involved in fist fight with Nobles, 183
Grant, General U. S.: refuses Army escort for Sawyers' second expedition along Niobrara, 293–294; orders escort protection along Big Cheyenne route, 304; signs Dakota appropriation bill as President, 310
Great Northern Railroad, follows in part Mullan road, 278
Green, R. Frank: mail carrier from Missouri to Santa Fe, 252; accompanies Beale's second expedition, 252
Gregg, Josiah, trading along Canadian River in 1839 and 1840, 25
Grow, Galusha A.: Pennsylvania congressman, 171; pushes through amendment to change location of Central Overland wagon road, 171–172; interested in Dakota land promotions, 183; article attacking Buchanan administration, 186
Guadalupe Hidalgo, Treaty of, terms, 23, 36
Guernsey, Orrin, on commission to negotiate Indian treaty, 302
Gunnison, Lieutenant John W.: guides Stansbury's command along Mormon Trail to Salt Lake, 31; examines Salt and Utah lakes, 33; killed by Indians, 141

Harlan, James: Secretary of Interior, 293; reactivates wagon-road office, 293; orders suspension of work on Big Cheyenne road pending Indian treaty negotiations, 303; defends action in closing work on Big Cheyenne road, 304; to name superintendent of Virginia City–Lewiston road, 312
Harney, General William S.: organizes topographical parties for Texas exploration, 40; orders exploration for roads from Great Basin to Oregon, 84–85; plans for exploration in 1860 fail, 88; expedition between Fort Laramie and Fort Pierre, 135; instructs Warren, 136
Hartz, Lieutenant Edward L.: seeks improvement of Fort Davis–El Paso road, 45; escorts camel corps in West Texas, 46
Hastings' route, coincides with parts of Simpson's survey, 151
Haven, Solomen G.: New York congressman, 171; speaks in favor of Central Overland wagon road, 171; debates federal aid to internal improvements, 172–173
Hawley, William J., Utah road builder, 143
Hayden, F. V., explores Black Hills with Warren, 138
Hays, John C.: Texas Ranger, 36; in charge of San Antonio–El Paso expedition, 36–37
Hebard, Alfred, receives bridge-building contract on Fort Riley–Bridger's Pass route, 131
Hedges, Nat D., 290
Hedspeth's road, traveled by engineers west of South Pass, 213
Hendricks, Thomas A.: Commissioner of General Land Office, 234; urges wagon-road appointment for Indiana Democrats, 234
Herbert, Philemon T.: California congressman, 172; quoted in favor of Central Overland wagon road, 172
Highsmith, Captain Samuel, as Texas Ranger commands escort, 36–37
Hitchcock, Phineas W.: Nebraska delegate, 286; interested in Omaha branch of roads to the Montana-Idaho region, 286
Hollinshead, William, quarrels with Simpson over Minnesota roads and Indian treaties, 56
Hooker, Colonel Joseph, in charge of southern Oregon roads, 82–83
Hooper, W. H.: Utah territorial delegate, 157; negotiates with Simpson over purchase of road down Timpanogos Canyon, 157
Houghton, Joad, civilian assistant in New Mexico, 116
Houston, George S., Alabama congressman opposed to Minnesota road appropriations, 54–55
Howard, Conway R., civil engineer with Mullan, 261
Hubbard, Asahel W.: Iowa congressman, 282; introduces bill for road from Missouri to Virginia City, Montana, 282; use of patronage in road appointments, 286; secures escort for Sawyers' expedition, 286; requests additional appropriations for Niobrara River road, 293; requests Army escort for second Sawyers' expedition, 293; protests failure of Moody to build Sioux City–Big Cheyenne road, 306–307
Hugo, Samuel, employed by Mullan to build bridges on eastern section of Mullan road, 272

Humphreys, A. A.: topographical engineer in charge of Office of Explorations and Surveys, 137; urges construction of wagon road along thirty-fifth parallel, 244; appoints Mullan as road builder, 260; convinced by Raynolds' reports that Platte River valley could be connected with Mullan road, 268

Huntington, O. B., refuses to go with Steptoe as guide, 144

Hutton, N. Henry: appointed chief engineer on El Paso–Fort Yuma road, 176–177, 220; takes party to locate water well site, 225; reports on survey and improvements, 226–227

Idaho Territorial legislature: petitions for action on Virginia City–Lewiston road, 312–313; urges additional appropriation for Virginia City–Lewiston road, 318

Independence, Missouri, outfitting headquarters for Central Overland wagon road, 192

Ingalls, Rufus: interest in improving Oregon Trail from Salt Lake to Northwest, 84; secures guide for Wallen expedition, 85; directs supply delivery to Pacific Northwest, 143–144

Ingle, J. H.: sent through McDougall's Pass by Lander, 195; attends to business in Salt Lake, 208

Interior Department: Minnesota delegate seeks aid in road appropriations, 61; work in Minnesota praised, 67–68; total expenditure on Minnesota roads, 69; faces new responsibility with Pacific wagon-road program, 174; instructions for Pacific wagon-road surveys, 177–178; sends inspector to report on Fort Ridgely–South Pass wagon road, 184; suspends Nobles and appoints McAboy, 187; forced to reinstate Nobles, 187; suspends funds for eastern division of Central Overland wagon road, 196; investigates Magraw's conduct, 199; tries to settle Magraw's accounts, 210; obtains funds from Commissioner of Indian Affairs for tribes near Lander's road, 212; issues instructions for building El Paso–Fort Yuma wagon road, 221–222; investigates affairs of El Paso–Fort Yuma road expedition, 229–230; lack of interest in taking responsibility for Fort Defiance–Colorado River road, 241; problem of policing plains, 281; responsibility for wagon roads to the Montana-Idaho gold region, 283; construction of roads to Montana and Idaho, 286–287; champions Niobrara road, 303–304; recommends appropriations for Dakota bridges, 308; failure of federal road program in Dakota Territory, 311; investigation of affairs on Virginia City–Lewiston road, 317–318; difficulty of coöperation with War Department, 326. *See also* Pacific Wagon Road Office

Internal improvements, question of constitutionality, 319–320

Iowa: "agency road," 9; "military road," 9, 11; improvement of Mississippi River crossing at Burlington, 9–11

Izard, Mark W., Nebraska governor complains about location of road, 132

Jackson, Lieutenant William H., surveys between Anton Chico and Santa Fe Trail, 120

Jacobs, John M., Montana trader and associate of Bozeman, 284

James, William, Fort Vancouver–Fort Steilacoom road builder, 104

James River bridge: source of bickering among Dakota politicians, 308–309; Yankton County residents protest delay, 310; Interior Secretary orders completion, 311

Jesup, Quartermaster General Thomas S.: suggests road from Fort Snelling to Fort Towson, 6; accuses Scammon of financial irregularity, 111

Johnson, Andrew: assigns Simpson to Interior Department, 293; proclaims treaty with Dakota Indians, 303

Johnson, J. Neely, California governor, 166

Johnson, Louis, disqualified as road contractor, 102

Johnson, Lieutenant Robert, assists Wallen expedition, 86

Johnson, Robert W.: Arkansas senator, 67; protests Minnesota road appropriation, 67; urges construction of military road from Fort Smith to Albuquerque, 251

Johnson, W. W.: civil engineer with Mullan, 261; sent as special messenger to Washington, D.C., 265; reports renewal of appropriation, 266

Johnston, General Albert S.: necessity of providing supplies from Northwest to Army of Utah in the Great Basin, 84; urges topographical officer to report at Camp Floyd, 146–147; interested in exploration of routes from Salt Lake to California, 149; orders Simpson to explore from Timpanogos Valley to Green River, 155; reports importance of Utah roads, 156; Magraw offers service to Army of Utah, 199. *See also* Army of Utah

Johnston, Lieutenant Colonel Joseph E., chief topographical officer of Texas attached to Van Horn's command, 41–42

Jones, Lieutenant E. C., directs working party on road from Camp Floyd to Fort Bridger, 148

Jones, George W.: Tennessee congressman opposed to internal improvements, 51; quoted in opposition to Oregon road appropriations, 80–81; opposes Pacific wagon-road program, 173

Kallecki, Theodore, topographer with Mullan, 261

Kansas settlers, use Fort Riley–Arkansas River road, 126

Keach, Philip, opens trail from Seattle to Whatcom, 105

Kearny, Colonel Stephen W.: commands "The Army of the West," 17; instructs Philip St. George Cooke to wait for Mormons, 18; march to California, 19; trouble with New Mexico roads, 107

Kelly, Moses: acting Secretary of the Interior, 189; friendly to Nobles, 189

Kirby Smith, Lieutenant J. L.: assistant to Simpson in Utah, 150; returns over last section of route from California to Carp Floyd, 154–155

Kirk, John: appointed superintendent of western division of Central Overland wagon road, 176; instructions and preparations for expedition, 201; expedition from Honey Lake to City Rocks, 203; escorts California emigrants, 204; reservations about Honey Lake–Lassen's Meadows route and proposed changes in route, 204–205; criticism, 205; proposes detailed construction, 205–206; commended by Thompson, 206

Kittson, Norman W., fur trader and Minnesota legislator, 63; advises Simpson, 63

Lambert, John: topographer with Bryan survey to Bridger's Pass, 127–129; assists in preparation of Bryan's report, 130

Lander, Frederick W.: civilian engineer with Stevens in Washington Territory, 92–93; appointed chief engineer on Central Overland wagon road, 176; reports interest in road location west of South Pass, 191–192; gets authorization for separate engineering corps, 192; problems of supply, 192–193; reports to Magraw on route to South Pass, 193; exploration west of South Pass, 193–195; reports to Magraw on first season's operations, 195–196; urged to assume command of the expedition, 197; joins Magraw at South Pass, 197; rebuke to Nichols, 197–198; urges Alexander to use his road, 198; requests delay in investigation of Magraw, 199; offered Magraw's position, 199; accepts superintendency, 200; chief advisor to Thompson, 206; receives instructions from Thompson, 207; road building during second season, 208; disbands road party at close of second season, 209; publishes *Emigrant's Guide,* 209–210; bitter conflict with Magraw, 210–211; requests commission to return to field a third season, 212; difficulties with traders at South Pass, 213; requests appropriation for bridge across Green River, 213; promotional interest in wagon road, 213–214; debate with Simpson over relative merits of roads, 214–215; suggests means of improving western sector of Central Overland route, 216; improves Honey Lake–Humboldt River trace, 216–217; fights Paiutes, 216; resignation and tribute for service, 217

Lane, Joseph: Oregon delegate, requests federal road appropriations, 72, 89; writes Davis urging construction of Astoria-Salem road, 75; confers with Derby about appropriations, 77; defends Oregon appropriations, 81–82

Larpenteur, Auguste L., builds roads from St. Peter to St. Paul, 58

Larsen, Arthur J., quoted, 52, 69–70

Lassen, Peter, explorations near Carson City, 203

Lawrence, John, engineer on James River bridge, 310–311

Lazelle, Lieutenant Henry M., commands escort of Texas–New Mexico boundary commission, 120

Leach, James B.: contracts to improve road from Salt Lake City to southern California, 143; appointed superintendent of El Paso–Fort Yuma road, 176, 220; recommendations for construction of El Paso–Fort Yuma road, 219–220; receives instructions, 221–222; purchases supplies, 222; expedition's progress across Arkansas, 223; separates party from ox train, 223; illness, 223; arrives in El Paso, 225; directs road work

Leach, James B. (*Continued*)
from El Paso to Mesilla, 225–226; difficulties over finances, 227–228; problem of communication with Washington, D.C., 228; charged with misconduct, 228; defends Woods, 230; indicted, 230–231; trial postponed and charges withdrawn, 231

Leco: guide to Beale's expedition, 249; criticized for incompetence, 249

Letcher, John: Virginia congressman, 59; inquires about necessity for Minnesota road appropriations, 59; quoted in opposition to roads recommended by Commissioner of Indian Affairs, 66–67; quoted in opposition to Oregon roads, 81; speaks against New Mexico roads, 110; objects to Utah appropriation, 145

Lincoln, Abraham: signs bill for roads across the northern plains, 282; desires coöperation between War and Interior departments in road building, 283; appoints treaty commission to Indian country, 302

Lombard, Coote: civilian engineer, 125; supervises bridge building on Fort Riley–Arkansas River route, 125–126

Long, M.M., left by Lander to assist Magraw on Central Overland wagon road, 193

Lowell, A. C., road contract rejected, 104

Lowry, Walter: newspaper correspondent, 153; starts return march with Simpson, 153; death, 156

McAboy, William: civilian road superintendent, 67; appointed by Interior Department in Minnesota, 67; temporarily appointed Nobles' successor in Minnesota, 187

McCay, J. R., appointed physician on El Paso–Fort Yuma wagon-road expedition, 220

McClellan, George B.: in charge of western survey party for railroad route in Washington, 90; directs exploration, 91–92; conflict with Stevens, 92–93; ordered to turn over properties of road, 95

McClelland, Robert: Secretary of Interior, 174; administrative responsibility and action for wagon roads, 174

Macomb, Captain John N.: in charge of New Mexico roads, 112; surveys boundaries of Fort Stanton military reservation, 112; difficulty in obtaining specie, 112–113; work on New Mexico roads, 113–116; assigned to survey duty at Fort Craig, 116; asks release from New Mexico assignment, 117; explores San Juan River for wagon-road route, 117; summarizes achievements on New Mexico military roads, 118–119; released from New Mexico assignment, 119

McDonald, Angus, Hudson's Bay factor at Fort Colville, 92

Magraw, William M. F.: candidate for wagon-road superintendency, 175; suggestions about Central Overland wagon road, 191; receives instructions as superintendent, 192; delays departure from Independence, 196; fist fight with Goodale, 197; conflict with engineering corps, 197; requests Lander to investigate, 198; travels with Alexander and goes into winter camp, 198, volunteers for service with Army of Utah, 199; notified of charges by Thompson, 199–200; corresponds with officials in Washington, D.C., 200; charges against Lander and bitter conflict, 210–211

McKinnon, M. A.: appointed disbursing agent on El Paso–Fort Yuma road, 221; accompanies ox train across Texas, 223; delay in executing bonds causes difficulties, 227; inability to administer finances, 228

Manypenny, George W.: Commissioner of Indian Affairs, 66; supports Minnesota roads, 66; relations with Rice, 167; appoints Nobles as superintendent of Fort Ridgely–South Pass wagon road, 179

Marcy, Oliver: scientist on Virginia City–Lewiston expedition, 313; reports on Montana road system from Bitterroot Valley to Virginia City, 316

Marcy, Captain Randolph B.: commands Fort Smith–Santa Fe expedition, 24–27; return march from Santa Fe to Fort Smith, 28–29; route forms junction with Cooke's route, 29; contacts Michler at Fort Washita, 43

Marcy, William L., Secretary of War, 17, 24

Marlette, S. H.: California surveyor general, 165; on wagon-road survey across Sierra, 165; reservations against route desired by governor, 165

Marysville *Herald*, criticizes Kirk, 205

Mason, Major Richard B., 6

Maury, Matthew F.: head of Naval Observatory, 242; quoted in favor of wagon roads, 242

Maynadier, Lieutenant Henry E.: in charge of detachment from Raynolds' commnad, 267; ordered to Sioux City for emigrant escort duty, 285

Medary, Sam: chief engineer on Fort Ridgely–South Pass wagon road, 177, 181; involved in Dakota land promotions, 183; territorial governor of Minnesota, 183; sent to St. Paul for howitzer, 184; appraisal of work, 190

Memphis, Tennessee, railroad promotion meeting urges military wagon-road construction, 242

Memphis and St. Francis River road, 7–8

Mendell, George H.: supervises Oregon roads, 82–83; recommends appropriations, 83; ordered to Great Lakes survey, 84; reports on Washington roads, 101; continues improvements, 102; states inability to survey Fort Steilacoom–Bellingham Bay road, 103; issues contracts for construction near Seattle, 104

Mendota–Big Sioux River military road: appropriation for, 52; additional funds, 57; failure of Simpson to construct, 58; contracts for cutting timber on eastern segment, 62; Thom dissatisfied with state of construction, 68

Mendota-Wabasha road: appropriation for, 52; Potter survey, 53; Simpson delay in survey, 54; Simpson interest following Treaty of Traverse des Sioux, 56; additional funds, 57; resurvey and construction contracts granted, 57; traveled entire length, 62

Mexican War, calls attention to western roads, 17

Michigan roads, 4

Michler, Lieutenant Nathan H.: topographical officer, 43; road survey from Austin, Texas, to Fort Washita, 43; from Red River to Pecos River and return, 43–44; reconnaissance from Corpus Christi to Fort Inge, 44; from San Antonio to Ringgold Barracks, 44

Miller, A. B.: engineer on Sioux City–Big Cheyenne River road, 307; resurveys and improves Dakota road, 308; works on Big Sioux and Vermillion River bridges, 309–310; death, 310

Miller, Charles H.: mountaineer stationed at South Pass by Lander, 209; murdered, 213

Miners' and Travelers' Guide, published by Mullan, 277

Minnesota–Big Cheyenne River wagon road: appropriation approved, 282–283; survey by Brookings' road party, 299; survey from Missouri River to Minnesota boundary, 301; protest at closing down work, 303–304; pressure to continue improvements, 304–305; Treaty of Fort Laramie ends hope of resuming operations, 305

Minnesota Democrat: attacks Simpson, 57; charges Simpson with incompetence, 61; quoted, 65; editor urges legislature to memorialize Congress for road to Fort Laramie, 168

Minnesota Territory: settlement and road pattern in 1849, 48; preterritorial trails, 48–49; agitation for improved roads, 49; St. Croix Valley settlers protest location of road, 52–53; legislature requests additional road engineers, 53; continuous demands by legislature for road appropriations, 65–66; generous apropriations, 69; mass meeting in St. Paul favors road across Plains, 168; promoters support Nobles' plan, 168; legislative memorial for road from Minnesota to Pacific, 168–169; legislature backs Nobles' appointment, 175; request additional funds, 187

Minnesotian: St. Paul paper presents views of Simpson, 57; attacks Thompson, 183

Missouri Overland Mail and Transportation Company, news of organization reaches California, 167

Mix, Charles H., appointed to board of appraisers, 188

Montana gold discoveries, 273, 275

Moody, Gideon C.: named superintendent of Sioux City–Big Cheyenne River wagon road, 297; instructions, 298; preliminary work on Big Sioux River bridge, 305–306; misuse of road funds, 306; travels with federal inspector over route of survey, 307; notified to make no futher expenditures, 308

Moore, A. W., superintendent of Steilacoom–Fort Walla Walla road, 92

Mormon battalion: march to California, 19–21; use of its route, 22

Mormon Trail: proposal to improve between Omaha and New Fort Kearny, 121; Bryan requests civilian engineer to supervise, 126; improved by Dickerson, 132–133; problems of bridging streams, 134

Mormons: desire road from Rocky Mountains to Great Salt Lake and improved connections with Pacific Coast, 140; migrations back to Salt Lake City, 204; provide laborers for Lander's road building, 207–208

Morton, Oliver P.: Indiana senator, 304; aid enlisted in seeking renewing of Big Cheyenne road survey, 304

Mullan, John: explores with Stevens' party, 258; discovers Mullan's Pass, 258; early interest in wagon roads, 259; takes reports and memorials to Washington, 259; preliminary organization of working party, 260; attached to Wright's command for reconnaissance, 260; new appropriation and instructions, 261; tribute to workmen, 264; spends winter preparing reports for War Department, 264; seeks food during winter, 265; urges federal officials to authorize military detachment to march from Fort Benton to Fort Walla Walla, 265; improvements for Blake's command on Mullan road, 268–269; plans for third season's work, 269; road building in 1861, 270–271; inadequate preparations for the winter, 271; work as road builder completed, 272; reports on Montana gold discoveries, 273; aids emigrants using his road, 274, 276; publication of final reports, 276–277; publication of *Miners' and Travelers' Guide*, 277; moves to Pacific Northwest, 277

Mullan road: organization of party, 260; new road party organized, 261; survey from Walla Walla to Coeur D'Alene mission, 261–263; road-building difficulties across the Bitterroots, 263–264; expedition suffers from winter cold in mountains, 264–265; second year's operations begun, 265; improvements into Fort Benton, 266; construction during 1861 season, 270–271; suffering of men during winter of 1861–62, 271; final location of road, 272; use by emigrants and gold seekers, 273–274, 276; inadequacies of road, 277; appraisal of importance, 278; used by Bird expedition, 316

Mullowney, John F.: examines Sublette's cutoff, 193; improvements at Platte River bridge, 197; depositions from men, 198; depositions considered by Interior Secretary, 199

Murry, Lieutenant Alexander: commands escort to Simpson, 151; takes Simpson's wagon train over road to Fort Bridger, 156

Naches Pass road, never popular with emigrants, 95

"National Road," Engineers' noteworthy achievement, 2

Nebraska Territory: settlement along wagon-road routes, 133–134; legislative committee reports favorably on Platte River–Niobrara River wagon road, 238

Neighbors, Major Robert S.: United States Indian Agent for Texas, 37; aids Ford road survey, 37–39

New Mexico: roads neglected by Mexican government, 107; road pattern at time of American occupation, 108; legislative assembly petitions for improved roads, 108–109; artesian well experiment related to road building, 116–117

New Mexico Military Department, exploration in, 117

New York Daily Times: attacks Thompson for unfair administration of wagon roads, 185–186; renews attack, 187

New York Tribune, article attacking Buchanan cabinet, 186

Nichols, Henry K.: assistant engineer, 193; left by Lander to assist Magraw on Central Overland wagon road, 193; protests to Lander about treatment by Magraw, 197; starts from Missouri frontier, 197; relations with Lander, 197–198

Nicholson, George B.: engineering assistant to Bird on Virginia City–Lewiston road, 313; examines Nez Percés trail across Bitterroots, 316; surveys Lolo route, 317

Niobrara–Virginia City wagon road: appropriation approved, 282–283; renewed attempt to connect Central Overland route with Montana gold fields, 283; early Army opposition to Niobrara River route, 285; party assembled, 287; march along Niobrara, 287–289; attacked by Indians, 290, 291; refusal of civilian employees to travel without escort, 291; arrival in Virginia City, 292; improvements by second Sawyers' expedition, 294; arrival in Virginia City, 295; route can not meet competition of Union Pacific Railroad, 296; evaluation of road's importance, 296

Nobles, William H.: discovers Nobles' Pass across Sierra, 168; role of congressional lobbyist, 168; moves to Minnesota, 168; expenses to Washington paid by St. Paul citizens, 168; candidate for road superintendency, 175; appointed superintendent of Fort Ridgely–South Pass wagon road, 179; preliminary survey, 179; reappointed superintendent, 179; initial activities and reports, 181; quarrel with Thompson, 182–183; involved in Dakota land promotions, 183; renews attack on Campbell, 183; rebuked by Thompson, 184; interested in suppressing Indians, 184–185; ordered to deposit road property, 185; removed and reinstated, 187; campaigns for additional appropriations, 187; resigns and made liable to charges, 188; court decision against, 188; appeals to Congress, 189; appraisal, 190

North Californian (Oroville), criticizes Kirk, 205

Northern Pacific Railroad, follows Mullan road in part, 278

Office of Explorations and Surveys: finances exploration of San Juan River for wagon road, 117; resolves to explore west of Missouri River, 137; sponsors Mullan road, 260; sends out Raynolds' expedition, 266; work in trans-Mississippi West, 321, 325–326

Old Spanish Trail: Simpson advised to explore, 27; major New Mexico highway, 108; used by Macomb on San Juan River exploration, 118

Oregon and California Stage, uses part of southern Oregon military road, 75

Oregon Territory: preterritorial road pattern, 71; methods used in building roads, 71–72; legislature memorializes Congress for transportation aid, 72; Oregon emigrants use Lander's road, 214

Oregon Trail: improved by Ingalls, 84; only continuous route to Pacific Northwest in 1850, 257; cutoff becomes concern of private traders, 283–284

Otero, Miguel O.: New Mexico delegate, 116; requests additional financial support from Congress, 116; proposes appropriation for Fort Defiance–Colorado River, 172–173; wants Interior Department to construct Fort Defiance–Colorado River road, 241; introduces new appropriation for Fort Defiance–Colorado River road, 251

Overland Stage, uses part of Stansbury's route, 34

Overmeyer, Philip H., quoted, 93–95

Owen, John, convinces Bird of importance of Deer Lodge–Helena–Fort Benton route, 316

Pacific Emigrant Society: organized to promote wagon roads to California, 164; favors road via Sacramento and Placerville, 165

Pacific Railroad Convention, St. Louis meeting endorses wagon roads for emigrants, 242

Pacific Road Constructions Office: topographical engineers establish, 75; moved from San Francisco to Fort Vancouver, 83–84, 104

Pacific Wagon Road Office: established in Interior Department, 178; appraisal of its accomplishments, 232; purpose of Congress in establishment, 244, 325

Parke, Lieutenant John G., quoted on wagon roads, 243

Peck, Lieutenant William G.: topographical officer with Kearny, 18; return from Bent's Fort to St. Louis, 25; reconnaissance of upper Rio Grande, 107–108

Pete: Ute Indian guide for Simpson in Utah, 150–151, 152; locates important mountain pass, 154

Pfeiffer, Albert H.: subagent for Utah Indians, 117; interprets for Macomb, 117

Phelps, John S.: Missouri congressman, 173; supports Fort Defiance–Colorado River road, 173; supports appointment of Leach, 176; wants Interior Department to construct Fort Defiance–Colorado River road, 241

Phillips, Philip, Alabama congressman and spokesman for Committee on Territories, 110

Pierce, Franklin: names governor for Washington Territory, 89; disappointment of Californians over internal improvement policies, 164; signs Pacific wagon-road bill, 173–174; attitude of administration toward wagon-road program, 174; views on internal improvements, 322

Pioneer and Democrat (Minnesota), attack upon Thompson, 183

Platte River–Niobrara River road: appropriation for, 174; appointment of superintendent and engineer, 234; organization and reconnaissance, 234–235; bridge construction along route, 237; destruction and repair of bridges, 237; emigrants use road, 238

Poinsett, Joel R., Secretary of War, 2

Point Douglas–Fort Gaines [Ripley] road: appropriation for, 52; Simpson survey, 53; survey approved by Minnesotans, 55; additional funds, 57; contracts for improvements, 57; open for wagon and stage travel, 62

Point Douglas–St. Louis River road: appropriation for, 51; Simpson surveys, 53; debate over location, 55; additional funds, 57; improved during 1853, 57; Simpson cuts trail through forest, 1854, 60; attempt to transfer terminus to Wisconsin, 60; im-

Point Douglas–St. Louis, etc. (*Continued*) provements during 1855, 62; Thom dissatisfied with state of construction, 68
Polk, Trusten: Missouri senator, 251; introduces bill to complete Albuquerque [Fort Defiance]–Colorado River road, 251
Pony Express: uses part of Stansbury's route, 34; uses part of Simpson's route to California, 156, 327
Pope, Captain John: recommends El Paso–Fort Yuma road, 218; quoted on relationship of wagon roads to railroads, 243; major general commanding Department of the Northwest, 275; instructions to emigrants, 275; recommendations relative to policing northern plains, 281; orders escort of Sawyers' expedition, 287; withdraws escort of Sawyers' second expedition, 294
Portage Road: along Columbia, 97, 98; excellent improvements, 101
Potter, John S., Minnesota civilian engineer employed by Topographical Bureau, 52–53
Powder River road, Army determined to build from Fort Laramie to Virginia City, 285
Powers, T. P., civilian road superintendent in Oregon, 84
Propper, George N.: surveyor and engineer with Brookings, 298; examination of terrain from Fort Sully to mouth of the Big Cheyenne River, 299
Prudhomme, Gabriel, gives information to Mullan, 258
Purviance, Samuel A.: Pennsylvania congressman, 171; reports bill for Central Overland wagon road, 171; discusses proposal, 172
Putnam, Lieutenant Haldiman L., assistant to Simpson in Utah, 150

Quartermaster Corps: maintenance of Arkansas roads, 8–9; disturbed over status of New Mexico roads, 108
Quitman, J. A.: chairman of House Military Affairs Committee, 79–80; debates Oregon appropriation measure, 80–81; quoted in behalf of Utah appropriation, 145

Ragan, Mathew J., builds bridges on road from Omaha to Fort Kearny, 134
Ramsey, Alexander: as Minnesota governor urges federal road improvements, 49; pressures War Department for action, 53; as United States senator urges roads from Minnesota to Northwest, 282
Raynolds, William F.: expedition along Yellowstone River, 266; instructions to seek new wagon-road routes to join Mullan road, 266; skirts Yellowstone Park area, 267; arrives at Fort Benton and confers with Blake, 267; appraisal of work, 267; recommends routes from Platte Valley to headwaters of Missouri, 268
Red Cloud: as Sioux chief protests military encroachment on reservation, 285–286; warriors harass Sawyers' expeditions, 290, 294–295; forces concessions from United States, 305
Reed, Henry W., on commission to negotiate Indian treaty, 302
Reese, John: guide for Simpson in Utah, 150–151; lost but befriended by Digger Indians, 152; advance guide, 153–154; guide for Kirby Smith, 155
Reno, Jesse L., surveys Mendota–Big Sioux military road, 58–59
Republican Party: committed to federal construction of wagon roads and railroads, 171; control of House of Representatives, 323; favors wagon roads to Pacific Coast, 324
Rice, Henry M.: political leader in Minnesota, 56; writes Abert recommending Emerson, 57; elected congressional delegate, 59; cooperates with Simpson in securing appropriations, 59; involved in railroad land-grant promotion, 60; presents charges against Simpson to War Department, 61; renews efforts for appropriations, 61; works with Department of Interior, 61; complains to War Department about Simpson, 65; urges appropriation for Winona–Fort Ridgely road, 65; quoted in favor of territorial roads, 66; leads movement for road from Minnesota across Plains, 167; interested in Dakota land promotions, 183
Richards, John: mountain trader, 207; promises Lander subsistence, 207
Roberts, Jesse, road contractor in Oregon, 73
Robinson, Doane, South Dakota historian quoted, 306
Robinson, Lewis, Utah road builder, 143
Rockwell, Peter, serves as explorer for Steptoe west of Salt Lake, 144
Roop, Major Isaac: provisional governor of Nevada Territory, 216; urges Lander to fight Paiutes, 216
Rum River–Mille Lacs road: appropriation, 61; completed, 67

Rusk, Thomas J.: Texas senator, 178; recommends Campbell appointment, 178; chief promoter of thirty-second parallel route, 218–219
Russell, Majors & Waddell: agent confers with Simpson about using road to Camp Floyd for freight wagons, 148; reports on route, 148; drive cattle along Simpson's route from Camp Floyd to California, 155; ask Mullan about shipping supplies on Yellowstone and Big Horn rivers, 259; aided by wagon roads west, 327
Russel, Robert R., quoted, 243

Sacramento Daily Union, quoted in favor of wagon roads, 166
St. Anthony Express, criticizes work of Army engineers in Minnesota, 53
St. Anthony–Fort Ridgely road: appropriation to cut timber, 61; work at a standstill, 62; War Department denies appropriation, 65; cutting completed, 68
St. Croix Union, accuses Simpson of misappropriation of Minnesota road funds, 60–61
Salt Lake City–southern California road, appropriation approved, 140–141
San Antonio, citizens raise funds for wagon-road exploration, 36
Santa Fe–Doña Ana road: appropriation for, 109; funds squandered by Scammon, 111; operations resumed, 115–116
Santa Fe–Fort Union road: appropriation for, 112; improvement and importance, 113; additional appropriation, 120
Santa Fe–Taos road: appropriation for, 109; funds squandered by Scammon, 111; survey renewed by Macomb, 115; additional appropriation, 120
Santa Fe Trail, 108, 120
Sawyers, James A.: contract for bridge building between Fort Riley and Big Timbers, 125; requests escort from Davis, 125; construction problems, 125–126; superintendent of the Niobrara–Virginia City wagon road, 286; biographical sketch, 286; negotiations with Army over escort, 287; conflict with military escort, 289–290; seeks escort from Connor, 291; discovers plot to abandon expedition, 291–292; arrives in Virginia City, disbands expedition and returns to Sioux City, 292; road work criticized, 292; ordered to report to Harlan, 293; difficulties in organizing second expedition, 294; road building and Indian attacks on second expedition, 294–295; arrival in Virginia City, 295; an appraisal, 296
Sayles, Welcome B., confidential agent for Post Office Department, investigates affairs on El Paso–Fort Yuma wagon-road expedition, 229–230
Scammon, Captain Eliakim P.: superintends New Mexico road building, 111; failure and dismissal from service, 111
Scholl, Louis, guide to Wallen expedition, 85, 87
Scott, General Winfield, 85
Sedgewick, Theodore: United States Attorney in New York, 229; brings charges against Woods and Leach, 229
Sells, D. M.: Idaho Indian agent, 318; urges expenditure of remaining funds on Virginia City–Lewiston road, 318
Seward, William H., New York senator in favor of Central Overland route, 171
Seymour, David L., New York congressman supports federal road aid to territories, 55
Sherman, General William T.: active committeeman urging roads in California, 164; recommendations concerning routes of migration on northern plains, 285; sees wisdom of coöperating with Interior Department, 285
Shumard, B. F., reports on paleontology along Bryan's survey to Bridger's Pass, 130
Sibley, Henry Hastings: to Congress in 1848 to obtain territorial government for Minnesota and to secure road appropriations, 49; reëlected to Congress, 49; presents Minnesota memorial, 49; quoted, 51; summarizes success in Congress, 52; pressures War Department for action, 53; introduces new road appropriation, 54; quoted, 55; retires as congressional delegate, 59; negotiates Indian treaty with Dakota Indians, 302
Simpson, Lieutenant James H.: ordered to Fort Smith as topographical officer, 24; joins Marcy, 25; road survey along Canadian River, 26–27; explorations in northern New Mexico, 27; accompanies expedition against Navajos, 27–28; reports on possible routes from New Mexico to California, 28; ordered to Minnesota for preliminary surveys, 53; first annual report quoted, 54; trouble with Minnesota residents over location of roads, 55–56; eagerness in pressing for congressional appropriations, 56; transfer contemplated, 56; involved in Minnesota politics,

Simpson, Lieutenant James H. (*Continued*) 56; methods used in improving Minnesota roads during 1853, 57; promoted to captaincy, 57; continued quarrel with Rice and Emerson, 57; accused of misappropriating funds, 60; Minnesota road improvements in 1855, 62–63; problems over timber cutting on Minnesota trails, 63; summarizes experience gained in Minnesota road work, 63–64; demonstrates folly of Army officers in politics, 64–65; transferred from Minnesota, 65; reports for duty with Army of Utah, 147; opens new road from Fort Bridger to Camp Floyd, 147–148; reports to Johnston on Camp Floyd–Fort Bridger road, 148; ordered to explore routes from Salt Lake to California, 149; evaluates comparative advantages of routes from Missouri River to Pacific Coast, 150; requests authority for survey from Camp Floyd to California, 150; outward journey from Camp Floyd to California, 151–153; in California, 153; return to Camp Floyd, 154–155; explores from Timpanogos Valley to Green River, 155; urges extension of survey to Denver, 155; reëxamines road to Fort Bridger, 156; urges use of Utah roads and further appropriations, 156–157; debate with Lander over relative merits of roads, 214–215; Stanton places, in charge of wagon-road and railroad construction, 293; struggles with accounts of Sawyers' expedition, 293; order to abandon Niobrara River route issued and cancelled, 294; orders property transferred from Big Cheyenne River Road to Niobrara River road, 303; orders Sioux City–Big Cheyenne River survey, 305; sends inspector to check on Dakota roads, 307; agrees upon new construction plans for Dakota road, 308; relieved of responsibility for wagon roads, 309; rebukes Bird for abandoning Virginia City–Lewiston road, 317

Sioux City–Big Cheyenne River wagon road: appropriation approved, 282–283; Moody ordered to begin survey, 305; road improved to Yankton, 309

Sioux City Journal: promotes Sawyers' expedition, 287; reports progress of expedition, 289; praises Sawyers' work, 295

Sites, George L.: superintendent of Platte River–Niobrara River road, 176, 234; conducts survey and recommends improvements, 235; reports in Washington, D.C., 237; second season's improvements, 237; endorses Nebraska territorial legislature's request for funds, 238

Sitgreaves, Captain Lorenzo, exploration along the Zuñi, 28

Smith, Lieutenant Colonel Andrew J., commands Mormon battalion, 18

Smith, Byron M.: engineer on Sioux City–Big Cheyenne River road, 307; urges Interior Department to resolve conflicting interests of Dakota residents, 307

Smith, Caleb B.: Secretary of the Interior, 190; authorized to pay Nobles, 190; orders legal action against Magraw, 211–212

Smith, Lewis H.: engineer with Sawyers' expedition, 287; seeks better route along Niobrara River, 289

Smith, Martin L., assists Michler on Texas survey, 45

Smith, General Persifer F., supports appointment of Magraw as superintendent, 175

Smith, William: Virginia congressmen, 81; protests favoritism for Oregon roads, 81; presents views of strict-constructionists, 83; proposes El Paso–Fort Yuma road appropriation, 172

Smith, William F., topographical engineer on Whiting survey in Texas, 39–41

Smyth, Henry B.: chief engineer on Platte River–Niobrara River road, 177, 234; recommends bridges across Missouri River tributaries, 235; reports in Washington, D.C., 237; second season's improvements, 237

Snowden, J. H., civilian topographer with Warren's expedition, 138

Sohon: guide and interpreter to Mullan, 258, 261, 263, 265; transferred to Blake's command, 268

Southern Pacific Railroad: follows part of Cooke's Mormon Battalion route, 22; follows lower road in West Texas, 43

Spink, S. L., Dakota territorial delegate fails in effort for further road appropriations, 311

Stansbury, Captain Howard: prepares escort for emigrants from Fort Leavenworth to Fort Hall, 29; ordered to survey road from Fort Hall to Salt Lake, 29; problem of improving central route, 29–30; locates road from Fort Bridger to Salt Lake City, 31; locates road from Salt Lake City to Fort Hall, 31–32; reconnaissance of Cache Valley, 32; examines shores of Salt Lake, 32; winter among Mormons, 32–33; return to

Fort Leavenworth, 33–34; contribution to road location in the trans-Mississippi West, 34–35; in charge of Minnesota roads, 68; work in Minnesota in closing federal road program, 69
Stanton, E. M.: Secretary of War, 283; urged by Lincoln to coöperate with Interior Department on migration-transportation matters, 283; places Simpson in charge of wagon-road and railroad construction, 293
Steen, Lieutenant Alexander H., surveys road from Fort Garland to Bent's Fort, 119
Steilacoom–Fort Walla Walla road: proposed, 72; appropriation approved, 72, 89; road contractor named, 92; work of local residents, 95; Army aids in building, 257
Stephens, Alexander H., Georgia congressman presents strict-constructionist views, 83
Steptoe, Lieutenant Colonel Edward J.: ordered to construct wagon road from Salt Lake to southern California, 141; concern over arrival in Utah, 141; correspondence with Davis about details of construction, 141; contracts awarded, 143; evaluation of road work, 143; difficulties with Mormon guides, 144; defeated by Indians in Washington Territory, 260
Stevens, Isaac I.: named Washington territorial governor, 89; in charge of Pacific railroad survey, 89–90; instructions to McClellan, 90–91; meets McClellan at Fort Colville, 92; conflict with McClellan, 92–93; addresses legislature on transportation needs, 96; requests appropriations as Washington delegate, 104–105; reminds War Department of road from Missouri River to Columbia River, 259
Stockton Argus, quoted in favor of wagon roads, 166
Stone, Captain Charles P., contracted to repair Southern Overland road between Pima villages and Ojo Excavada, 232
Stuart, Charles E.: Michigan senator, 170; breaks deadlock of departmental responsibility for construction of Fort Ridgely–South Pass wagon road, 170; interested in Dakota land promotions, 183
Stutsman, Enos, political opponent of Moody brings charges, 306
Sublette, Milton, fur trader locates old cutoff west of South Pass, 193
Sully, General Alfred: ordered to increase escort for Sawyers' expedition, 287; interest in Minnesota–Big Cheyenne route, 298–299; urges politicians to agree upon a single route across the plains, 302
Sumner, Colonel Edwin V., New Mexico military commander quoted in *Globe,* 110

Tah-tu-tash: Nez Percé interpreter on Bird expedition, 315; examines Nez Percés trail across Bitterroots, 316
Taylor, Edward B.: superintendent of Indian Affairs, 302; on commission to negotiate Indian treaty, 302
Territorial roads, procedure in construction, 12
Texas military frontier, activity in 1850's, 44
Texas Pacific Railroad, parallels northern road in West Texas, 43
The Olympia's Columbian, advertises local efforts to direct Oregon emigrants to Puget Sound, 95
Thom, Captain George: ordered to Minnesota as topographical engineer, 65; difficulties in starting assignment, 67; recommendations for improvements and appropriations, 68; replaced in Minnesota, 68; assigned to Oregon and Washington, 84; liquidates Washington road program, 104–106
Thompson, Jacob: chosen Secretary of the Interior by Buchanan, 174; receives letter from Nobles, 175; appointments for wagon-road superintendencies, 175–176; use of patronage, 177; instructions for Pacific wagon-road surveys, 178; quarrel with Nobles, 182–183; rebukes Nobles, 184; newspaper attack, 185–186; appoints appraisers for Minnesota property, 188; appraisal of attitude toward Nobles, 190; requests construction plans for candidates for wagon-road appointments, 191; writes Lander for report on Magraw's conduct, 199; appoints Lander in Magraw's place, 200; instruction to Kirk, 201; commends Kirk and Bishop, 206; instructions to Lander, 207; tries to settle Magraw's accounts, 210, 211; instructions to Lander for third season of road work, 212; instructions to Leach, 221–222; withdraws authorization to purchase supplies for El Paso–Fort Yuma road, 227; alarmed by reports concerning El Paso–Fort Yuma road, 228; issues instructions to investigator, 229; forwards statement of financial shortages to Leach, 231; attempts to avoid scandal on southern road, 232; sends

Thompson, Jacob (*Continued*)
orders to Sites, 234; orders Sites and Smyth to return to Nebraska road building in 1858, 237
Thrasher, Ira P., directs laborers on Fort Vancouver–Fort Steilacoom road, 106
Thurston, Samuel R., Oregon delegate, 71
Tilghman, R. C., civilian engineer on Iowa roads, 9
Timpanogos River Turnpike Company, road becomes property of Utah Territory, 157
Tingley, D. W., physician with Sawyers' expedition, 287
Tinkham, Abiel W., crosses Yakima Pass, 93
Todd, John B. S.: protests appointments on Dakota roads, 298; nominated for superintendency of Big Cheyenne road, 304
Toohill, P. E., mailcarrier reports on advantages of Clark's Fork to Mullan, 264
Torbert, Lieutenant Alfred T. A., directs working party on road from Camp Floyd to Fort Bridger, 148
Tower, L. J., Washington road contractor, 102
Traverse des Sioux, Treaty of, opening of lands west of Mississippi encourages road building, 56
Treasury Department: reviews confused accounts of Magraw's expedition, 211; legal action taken, 211–212; establishes special account for Leach, 228; attempts to adjust Leach's accounts, 231–232
Treaty of Fort Laramie, checks road-building activities, 305
Truax, Major Sewell: assistant on Virginia City–Lewiston survey, 313; examines the Nez Percés trail across Bitterroots, 316; left in charge of road work by Bird, 317; notified to cease work, 317
Twiggs, Major General David M., orders reconnaissance of trans-Pecos country for new fort, 45–46

Umpqua River Valley–Rogue River Valley road: proposed, 72; appropriation approved, 72; improved by Withers, 73–75; new appropriation considered, 79–80
Umpqua Weekly Gazette, advertises for construction bids for Rogue River–Umpqua River road, 73
Union Pacific Railroad, uses part of Stansbury's route, 34
United States Army: early road builder, 1; procedures for road building in the trans-Mississippi West, 11–12; launches program of road exploration, 36; achievements in West Texas, 42–43; transfers interest to far West Texas, 45; camel experiment, 45–46; contribution in opening West Texas, 46–47; plans for Kansas and Nebraska roads, 123; authorizes Fisk's third expedition, 275; favors Platte River and Big Cheyenne River routes, 285; resolves to build Powder River road, 285; relations with Brookings' road expedition on Big Cheyenne River, 299; champions Big Cheyenne River road, 303; federal government's road builder, 320, 325. *See also* Army Engineers; Corps of Topographical Engineers; War Department
United States Highway 10, follows Mullan road in part, 278
United States Highway 81, follows Bryan's march, 43
Usher, John P.: Secretary of the Interior, 283; urged by Lincoln to coöperate with War Department on migration-transportation matters, 283; patronage consideration in wagon-road appointments, 286, 297

Van Horn, Major Jefferson, leads infantry command to El Paso, 40–41
Vaugh, William D., road contract rejected, 104
Venable, Abraham W., North Carolina congressman opposed to federal road appropriations, 55
Virginia City–Lewiston wagon road: appropriation approved, 282–283; preliminary preparations, 313; expedition equipment, 315; survey of Lolo forks route, 315–316; plans for improvement of Lolo forks route, 317

Wagner, William H.: reconnaissance near South Pass, 195; works on Lander's cutoff, 208; explores road through Cache Valley, 208; leads detachment west, 212, 213; directs laborers building water tanks between Honey Lake and Humboldt River, 216
Wagon roads west, importance in history of trans-Mississippi West, 326–328
Walbridge, David S.: Michigan congressman, 173; proposes road from Lake Superior to Puget Sound, 173
Wallace, William H.: Idaho delegate, 282; proposes road from Virginia City, Montana, to Lewiston, Idaho, 282

Wallen, Captain Henry D.: wagon-road expedition from The Dalles to Great Salt Lake, 85–88; escorts emigrants to Oregon, 87–88; admits failure, 88

War Department, 1; forms Mormon battalion, 17–18; interest in Fort Smith–Santa Fe road, 24; requested to report on progress of Minnesota roads, 54; receives Minnesota petition, 55; reviews conflict between Rice and Simpson, 61; accepts postponement of timber cutting on Minnesota trails, 63; reports against Winona–Fort Ridgely road, 65; closes Minnesota road office, 69; total expenditures on Minnesota roads, 69; receives protest from Oregon, 73; reviews conflicting claims concerning terminus of Astoria-Salem road, 76; orders work in Oregon closed, 83–84; requests aid for New Mexico, 109; proposes to improve Mormon Trail, 120; refuses claims of Kansas bridge-builder for additional compensation, 126; receives report of Bryan on reconnaissance to Bridger's Pass, 129–130; approves bridge contract in Kansas, 131; urges additional funds for Kansas and Nebraska roads, 134–135; notified by Johnston of importance of connecting Utah and Colorado communities, 156; requested to check equipment taken from Magraw's wagon-road expedition, 210; Ordnance Department furnishes arms and ammunition for Leach's wagon-road party, 222–223; furnishes provisions for work party on South Overland road, 232; continues as wagon-road builder along thirty-fifth parallel, 241; includes appropriations for thirty-fifth parallel road in Army Appropriation Act, 251; supports additional appropriations for thirty-fifth parallel wagon road, 255; receives Beale's report, 256; designates North Overland route as important railroad survey, 257; appropriation for road from Missouri River to Walla Walla considered inadequate, 259; obtains additional appropriation for Mullan road, 260; sponsors Emigrant Overland Escort Service, 274; problem of policing northern plains west of Missouri River, 281; to transfer funds from Dakota military road to Interior Department, 283; urged to connect Forts Sully and Connor by military forts, 301–302; assigned responsibility for military roads, 320; difficulty of coöperation with Interior Department, 326. *See also* Army Engineers; Corps of Topographical Engineers; Office of Explorations and Surveys; United States Army

Warbass, E. D., constructs road from Whatcom to Bellingham Fort, 104

Ward, S. E., agrees to transport Lander supplies, 207

Warner, Lieutenant William H., with Kearny in New Mexico, 18

Washington, Colonel John M., campaign against Navajos, 27

Warren, Gouverneur K.: explorations in northern Nebraska, 135–139; from Fort Pierre to Fort Kearny, 135; from Fort Laramie to Fort Pierre, 136; up Missouri River to Yellowstone, 136; seeks routes of travel west of Missouri River, 137–138; compares advantages of various routes from Missouri Valley to Fort Laramie, 139

Washington *Constitution*, Simpson publishes letter relative to Salt Lake–California wagon route, 215

Washington Star, presents Interior Department's views relative to Nobles, 186

Washington Territory: established, 89; early transportation pattern, 89; legislature memorializes Congress for funds, 96, 102–103; citizens petition on location of road from Cowlitz Landing to Ford's Prairie, 100–101; construction along road, 101–102; legislature reminds Congress of inadequacies of Mullan road, 277–278

Washington Union, publishes letter of Nobles to Thompson, 175

Ways and Means Committee, House of Representatives, includes road appropriation in Army bill, 53

Weller, John B.: as California senator presents memorial from California legislature and citizens' petition for roads, 161; remarks in the Senate, 162; attacks methods of topographical engineers and praises civilian road builders, 163; announces intention to sponsor road construction bills, 163; quoted, 164; argues for Fort Ridgely–South Pass wagon road, 169–170; agrees that Central Overland route be improved by the Army, 171; approves Senate passage of El Paso–Fort Yuma road appropriation, 171; accepts House amendment to road legislation, 173

Wheeler, Lieutenant Junius B., supervises Astoria-Salem road, 84

Whipple, Captain Amiel W., remarks on wagon road along thirty-fifth parallel, 244

Whitcomb, Charles F., confesses dishonesty, 229–230
White, James L., escorts road-building party on Mullan road, 265
Whiting, Robert, assists on road to The Dalles, 96
Whiting, William H. C.: surveys route from San Antonio to El Paso, 39–40; journal quoted, 40
Whitney, Asa, crusader for route from Great Lakes to Oregon, 257
Whittlesey, Joseph H., describes New Mexico roads, 108
Wilkinson, Morton S., as Minnesota senator sponsors road bill for northern states, 282
Williams, B. D.: delegate from Jefferson Territory, 157; supports Simpson's plans to connect Utah and Colorado settlements, 157
Williford, Captain George W.: in charge of escort to Sawyers' expedition, goes to Fort Laramie for supplies, 289; refuses to continue with Sawyers and is relieved of duty, 291
Winnebago Indian Agency [Long Prairie] road: appropriation, 52; Simpson survey, 53; additional funds, 57; construction contract granted, 57; completed, 62
Winnemucca: leads Indians on war path, 216; truce arranged with Lander, 217
Winther, Oscar O., quoted, 72
Wisconsin roads, 4; appropriations, 51
Withers, Lieutenant John, supervises Oregon roads, 73–75
Woods, D. Churchill: personal assistant to Leach on El Paso–Fort Yuma road, 221; accompanies Leach on eastern purchasing tour, 222; directs ox train, 225; arrested, 230; to Texas for trial but escapes prosecution, 231
Wool road, 41
Worth, Major William J.: instructs Neighbors to locate road from Austin to El Paso, 37; orders Whiting survey from San Antonio to El Paso, 39; expresses preference relative to roads, 39; death, 40
Wozencraft, O. M.: proposed for superintendency of western division of Central Overland wagon road, 175, 176; represents interests of northern California, 201; criticizes Kirk's survey, 205
Wrenshall, C. C., supervises Mormon workers on Lander's road, 213
Wright, Daniel, road contractor in Oregon, 79
Wright, Colonel George, expedition against Washington Indian tribes, 260

Yates, Major Edward L., in fist fight between Lander and Magraw, 211
Young, Brigham: conversations with Stansbury, 31; plans chain of settlements through southern Utah, 140; tries to influence expenditure of federal funds in Utah, 142; assures Lander Mormons grateful for employment, 209
Young, Jeremiah S., quoted, 3

Zuñi villages: Beale's expedition arrives, 246–247; sell corn to Beale's second expedition, 254